European Journal of Biochemistry

EJB Reviews 1989

Contributing Authors:

J.-P. Battioni, Paris
P. Battioni, Paris
H. Beinert, Milwaukee
T. Boehm, Cambridge
M. Chabre, Grenoble
P. R. Cook, Oxford
P. Deterre, Grenoble
S. Ghisla, Konstanz
L. A. Grivell, Amsterdam
C.-H. Heldin, Uppsala

W. Hummel, Jülich
M. C. Kennedy, Milwaukee
M.-R. Kula, Jülich
D. Mansuy, Paris
V. Massey, Ann Arbor
T. H. Rabbitts, Cambridge
R. Timpl, Martinsried
B. Westermark, Uppsala
R. J. P. Williams, Oxford

Edited by the
Federation of European Biochemical Societies
Published by Springer-Verlag

Professor Dr. P. Christen
Biochemisches Institut der Universität Zürich
Winterthurerstrasse 190
CH-8057 Zürich

Professor Dr. E. Hofmann
Institut für Biochemie der Karl-Marx-Universität Leipzig
Liebigstraße 16
DDR-7010 Leipzig

This work is subject to copyright. All rights are reserved, whether the whole or part of the material is concerned, specifically the rights of translation, reprinting, re-use of illustrations, recitation, broadcasting, reproduction on microfilms or in other ways, and storage in data banks. Duplication of this publication or parts thereof is only permitted under the provisions of the German Copyright Law of September 9, 1965, in its version of June 24, 1985, and a copyright fee must always be paid. Violations fall under the prosecution act of the German Copyright Law.

© Federation of European Biochemical Societies 1989
Printed in Germany

The use of registered names, trademarks, etc. in this publication does not imply, even in the absence of a specific statement, that such names are exempt from the relevant protective laws and regulations and therefore free for general use.

Typesetting and Printing: WGB, Wiesbaden
Binding: Hollmann, Darmstadt
2131/3145-543210 — Printed on acid-free paper

Preface

In the mid-1980s the European Journal of Biochemistry set out to publish review articles. The enterprise proved successful with many high-level reviews written by well-known scientists appearing in the Journal. The reviews are intended to represent emerging and rapidly growing fields of research in fundamental as well as in applied areas of biochemistry, such as medicine, biotechnology, agriculture and nutrition. Novel methodological and technological approaches which stimulate biochemical research are also included. The authors of the reviews are explicitly asked to be critical, selective, evaluative and interdisciplinarily oriented. The reviews should encourage young scientists toward independent and creative thinking, and inform active investigators about the state of the art in a given field.

The good reception of the reviews by the readers of the European Journal of Biochemistry has induced the Editorial Board and Springer-Verlag to publish them annually in a separate booklet, 'EJB Reviews', in order to further their dissemination among biochemists and scientists in related biological and medical disciplines.

P. Christen
Chairman of the
Editorial Board

E. Hofmann
Reviews Editor

Contents

Molecular mechanism of visual transduction
M. Chabre and P. Deterre, 179/2 (Feb. I) .... 255 – 266

Structure and biological activity of basement
membrane proteins
R. Timpl, 180/3 (April I) .... 487 – 502

Mechanisms of flavoprotein-catalyzed reactions
S. Ghisla and V. Massey, 181/1 (April II)................ 1 – 17

Nucleo-mitochondrial interactions in yeast
mitochondrial biogenesis
L. A. Grivell, 182/3 (July I) .... 477 – 493

NMR studies of mobility within protein structure
R. J. P. Williams, 183/3 (Aug. II)...................... 479 – 497

Dehydrogenases for the synthesis of chiral compounds
W. Hummel and M.-R. Kula, 184/1 (Sept. I) .... 1 – 13

Chemical model systems for drug-metabolizing
cytochrome-P-450-dependent monooxygenases
D. Mansuy, P. Battioni and J.-P. Battioni, 184/2 (Sept. II) .... 267 – 285

Growth factors as transforming proteins
C.-H. Heldin and B. Westermark, 184/3 (Oct. I) .... 487 – 496

A chromosomal basis of lymphoid malignancy in man
T. Boehm and T. H. Rabbitts, 185/1 (Oct. II)............. 1 – 17

The nucleoskeleton and the topology of transcription
P. R. Cook, 185/3 (Oct. II)........................... 487 – 501

Engineering of protein bound iron-sulfur clusters
H. Beinert and M. C. Kennedy, 186/1 – 2 (Dec. I)........... 5 – 15

Subject index i

Review

Molecular mechanism of visual transduction

Marc CHABRE and Philippe DETERRE

Laboratoire de Biophysique Moléculaire et Cellulaire (Unité Associée 520 au CNRS), Département Recherche Fondamentale,
Centre d'Études Nucléaires, Grenoble

(Received September 30, 1988) — EJB 88 1153

INTRODUCTION AND OVERVIEW

Visual transduction covers the sequence of photochemical, biochemical and electrophysiological events through which the absorption of a photon in a pigment molecule in a photoreceptor cell generates an electrical cellular response that will be detectable at the level of the synaptic connection. Our understanding of visual transduction has considerably progressed recently and much of the new developments have been in the field of biochemistry and enzymology (for others recents review, see [1−3]).

The best studied system is the rod cell of the vertebrate retina, which provides unrivalled advantages for biochemical studies, due to the exceptional segregation and concentration of the molecular machinery of visual transduction in its outer segment. The initial molecular event, the photoisomerization of retina in rhodopsin, and the physiological cellular response, a hyperpolarization resulting from the closure of a cationic conductance of the cellular membrane, were both well characterized by the late sixties. When less than a hundred photons are absorbed per rod flash, that is in the non-saturating range, the hyperpolarization rises in about 0.2 s in mammalian rods. After reaching its maximum, it decays nearly as fast so that the response is terminated within 0.5 s. In the rod outer segment the disc membrane which contains rhodopsin is not connected to the cellular membrane which contains the cationic channels whose conductance is modulated by light. This morphology imposed the concept of a soluble transmitter, released or modulated at the level of the disc membrane and acting at the level of the cellular membrane.

Early studies had concentrated on the photochemistry of retinal, the molecule acted upon by light, which induces its isomerization from the 11-*cis* to the all-*trans* conformation. As this transconformation is followed by the release of the all-*trans* retinal from its membrane-bound protein cofactor into the disc membrane lipids, it was often speculated that the liberated retinal might play a direct role in the disc membrane, but its dissociation from bleached rhodopsin is much too slow to be significant for the visual transduction process. Photoexcited rhodopsin itself has to be the trigger. The prevailing hypothesis in the seventies was that the soluble transmitters were calcium ions released from the disc, either directly through photoexcited rhodopsin acting as a channel, or indirectly through another channel controlled by photoexcited rhodopsin [4]. It had also been recognized that illumination interferes in the rods with the metabolism of cyclic nucleotides, and more precisely cGMP; this was generally considered as a secondary process, too slow for a role in transduction and more likely to be involved in adaptation.

Biochemical studies concentrated then on rhodopsin. In the late seventies, reasonable agreements were reached on its molecular mass, around 40 kDa, from the fact that it crosses the disc membrane and the predominance of α helices in the transmembrane region. This was confirmed by the determination of the sequence in which could easily be recognized seven hydrophobic stretches corresponding to seven transmembrane helices. The search for a possible effect of photoexcited rhodopsin on calcium release from the disc, however, remained very disappointing, while by contrast evidence had accumulated on the interaction of photoexcited rhodopsin with peripheral GTP- and ATP-dependent enzymes. Three of these enzymes: a GTP-binding protein, (transducin), the rhodopsin kinase and arrestin (also called S antigen or 48K protein), were originally isolated through their specific interaction with photoexcited rhodopsin [5−7]. Transducin is a member of the large class of G proteins that mediate the action of a membrane receptor on internal secondary effectors; it was demonstrated [8] that, in the rod, transducin is responsible for the amplified activation of the cGMP phosphodiesterase [9]. It then became clear that the main, and probably only, role of photoexcited rhodopsin is to trigger an amplifying cascade which activates with high gain and fast kinetics the hydrolysis of cGMP. But the mechanism through which cGMP could modulate the cationic permeability of the cell membrane remained unknown and was assumed to be complex: according to a classical scheme one expected that cyclic nucleotides activated cGMP-dependent kinases, hence the necessity of ATP, and a possible link with a calcium release at a later stage. Fesenko et al. in 1985 [10] checked an improbably 'simplistic' alternative: they demonstrated that cGMP alone acts directly on the cationic conductance of the cell membrane! A cGMP-binding polypeptide that reconstitutes cGMP-dependent conductance in artificial membranes has been isolated [11], so that the molecular path is now complete,

---

This review is dedicated to the memory of our friend Hermann Kühn, a pioneer of the biochemistry of visual transduction.

Correspondence to M. Chabre, Laboratoire de Biophysique Moléculaire et Cellulaire, CENG Boîte Postale 85X, F-38041 Grenoble, France

Abbreviations. ROS, rod outer segment; R*, photoexcited rhodopsin; T_αGDP, T_αGTP, $T_\alpha$empty, α-subunit of transducin bearing a GDP, a GTP, or with an empty site; T_β, T_γ, β and γ subunit of transducin; PDE, cGMP-specific phosphodiesterase; $PDE_{\alpha\beta}$, catalytic subunits of PDE; I, inhibitory subunit of PDE; GDP[βS], guanosine 5'-[β-thio]diphosphate; GTP[γS], guanosine 5'-[γ-thio]triphosphate; GuoPP[NH]P, guanosine 5'-[β,γ-imino]triphosphate.

Enzymes. cGMP-specific phosphodiesterase (EC 3.1.4.35); guanylate cyclase (EC 4.6.1.2).

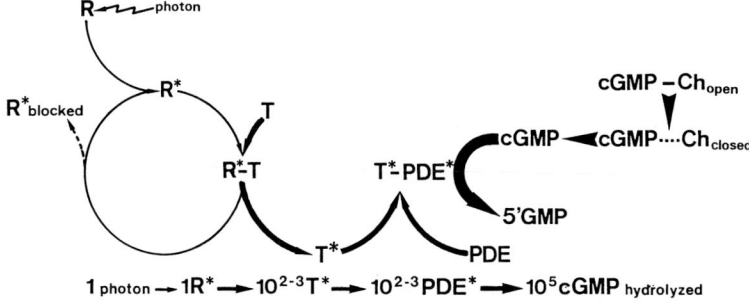

Fig. 1. *The basic scheme of the phototransduction cascade*

from the light-absorbing chromophore to the response-generating channel.

But if the simple scheme of Fig. 1 accounts well for the kinetics of the rising phase of a response to a short flash, it does not account for the fast decay of this response, nor for the slower adaptation processes that lower the sensitivity after prolonged illumination. The light-triggered cascade that activates the degradation pathway of the cGMP pool seems insensitive to calcium. By contrast the synthesis of cGMP, which is required for the regeneration of dark-adapted levels, is ensured by a guanylate cyclase that is not directly acted upon by light but is sensitive to calcium: it is inhibited by micromolecular concentrations of calcium [12, 13], probably through the action of an inhibitory calcium-binding protein [14]. There at last the elusive calcium ion comes into the story, but in a totally different way and with an opposite effect to that originally postulated: intracellular calcium decreases after illumination, as a secondary effect of the closure of the cGMP-sensitive cationic channels [15]. In their open state the channels also have a low permeability for calcium. In the dark, the calcium homeostasis was maintained by the counteracting effect of Na^+/Ca^{2+} exchangers present in the outer membrane of the rod outer segment. After illuminaton the closure of the cGMP-dependent cationic channels blocks the influx of calcium. But the exchangers still extrude calcium, probably at an even faster rate, since the cell hyperpolarizes and the intracellular sodium level bocomes lower. The ensuing decrease in intracellular calcium induces the activation of the guanylate cyclase, hence an increase of cGMP synthesis. This counteracts the effect of the earlier activation of the cGMP phosphodiesterase, tends to restore the dark level of cGMP and to close the light-sensitive channels.

This feedback reaction scheme of the cyclase is very recent and still partially tentative [16]. By contrast, the light-triggered cascade that controls cGMP hydrolysis has been thoroughly analyzed: the proteins involved have been isolated, cloned, most of them are sequenced, and their interactions have been studied in reconstitution experiments. The elucidation of the amplification and regulation mechanisms of this cascade is a major biochemical achievement of the last few years and serves as a model for the very general problem of coupling by G proteins of membrane transducers to intracellular effectors [17, 18]. Indeed the overall visual transduction mechanism has become a model not only for other sensory processes, as in olfactory and gustatory transduction, but even for a large class of *a priori* unrelated types of hormonal or neuronal signal transduction. The analogy is striking at the receptor level: rhodopsin is the prototype of a large class of 'seven transmembrane α helical' transducers, that includes the various adrenergic and muscarinic receptors and the receptor to substance K; many others are coming into the picture. The analogy may even start at the ligand-receptor interaction site: one pictures now the catecholamine ligand binding in the adrenergic receptor in a site analogous to that of retinal in rhodopsin.

The retinal system is however unique in many respects, for example by the very high segregation and concentration of the components of the transduction chain, the large excess number of receptor molecules with respect to the other components, transducin and cGMP phosphodiesterase, and the unusual solubility of these membrane-associated proteins. This may be related to the extreme sensitivity of the rods, implying an enormous amplification, and their short response time (compared to other G-protein-coupled receptors) implying fast kinetics.

INTRACELLULAR COMPARTMENTATION
IN ROD OUTER SEGMENTS:
SOLUBLE, MEMBRANE-ASSOCIATED
AND INTRINSIC PROTEIN COMPONENTS

The compact stack of discs fills most of the volume of the outer segment an divides the cytoplasm into interdiscal layers about 15 nm thick, communicating only around the disc edge with the continuous extradiscal layer underneath the outer cell membrane. The intradiscal compartment probably has an 'extracellular' ionic composition but no well documented exchange with the cytoplasm, at least on the time course of the response to a flash. The rhodopsin molecules are intrinsic components of the disc membrane and the two major proteins of the cGMP cascade, transducin and cGMP phosphodiesterase are only peripherally attached to the disc membrane, from which they can easily be dissociated by lowering the ionic strength of the medium. In its activated form, the GTP-binding subunit of transducin is even physiologically soluble, but it will very soon interact with a disc-membrane-bound phosphodiesterase. The two regulatory components of the cascade, rhodopsin kinase and arrestin, are soluble but have to bind and interact with rhodopsin [19]. The cascade triggered by one photoexcited rhodopsin and controlling the hydrolysis of cGMP develops therefore entirely whithin one interdiscal compartment, on a time course too short to allow any significant exchange of disc-membrane-associated proteins between different interdiscal compartments. The small messenger and metabolite molecules (ATP, GTP, cGMP and calcium) exchange rapidly between the interdiscal compartments and the continuous cytoplasmic layer that surrounds the stack of discs. The guanylate cyclase is not associated with the discs but with the axoneme of the ciliary structure of the outer segment [12]; the cGMP-sensitive cationic channel is an intrinsic component of the outer cell membrane. The pool of cGMP is

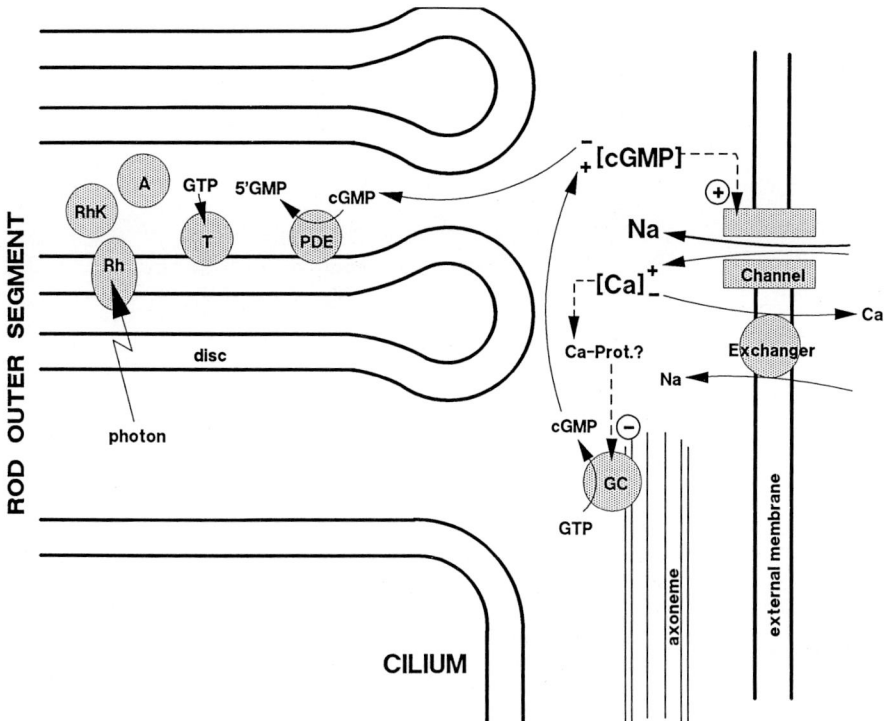

Fig. 2. *Intracellular compartments of the rod outer segment and the main molecular effectors*. Rh, rhodopsin; RhK, rhodopsin kinase; A, arrestin; GC, guanylate cyclase. PDE and GC reciprocally control the cGMP pool, while the Ca^{2+} content is balanced between entry by the channel and extrusion by the Na^+/Ca^{2+} exchanger. The negative control of GC by Ca^{2+} ensures the correlation between Ca^{2+} and cGMP concentration

Table 1. *The main proteins involved in the phototransduction cascade*

Protein	Relation to membrane	Molecular mass	Equivalent concentration in cytoplasm	Number of molecules implicated in one photon event	Reference for sequence
		kDa	µM		
Rhodopsin	intrinsic	39	—	1	[2, 21, 22]
Transducin ($\alpha + \beta + \gamma$)	peripheral or soluble	80 (39 + 37 + 6)	500	$10^2 - 10^3$	[3, 36]
Phosphodiesterase ($\alpha + \beta + I + I$)	peripheral	200 (88 + 84 + 13 + 13)	150	$10^2 - 10^3$	[3, 85, 86]
Rhodopsine kinase	soluble	65	5	1	—
Arrestin	soluble	48	500	1	[75 – 77]
Guanylate cyclase	attached to cytoskeleton	?	?	?	—
cGMP-activated channel	intrinsic	66	?	$10^2 - 10^3$	—
Na^+/Ca^{2+} exchanger	intrinsic	?	?	?	—

therefore controlled negatively by the cascade in the interdiscal space, positively by the cyclase in the outer cytoplasmic layer and acts on the cationic channel. The pool of calcium is controlled at the level of the outer membrane, by the conductance of the cGMP channels and the activity of the Na^+/Ca^{2+} exchanger, and it regulates the cyclase in the extradiscal compartment (Fig. 2 and Table 1).

RHODOPSIN: A SEVEN-HELICES RECEPTOR FOR RETINAL

Rhodopsin is an integral membrane protein. The transmembrane core is made up of a bundle of hydrophobic α-helical segments, adding up to half of the total length of the polypeptide chain (Fig. 3). The predominance of transmembrane helices had long been suggested by a variety of biophysical measurements: X-ray and neutron small-angle diffraction, circular dichroism, diamagnetic anisotropy, near infrared linear dichroism [20]. Seven hydrophobic stretches corresponding to the transmembranes helices, separated by regions rich in hydrophilic and polar amino acids, appeared clearly on the hydropathy pattern of the sequence of bovine rhodopsin [21, 22] and are a constant feature of all sequenced visual pigments, much in invertebrates as well as in vertebrates [2]. Retinal is linked by a protonated-Schiff-base bond to a lysine located in the middle of the last helix and lies therefore in the middle

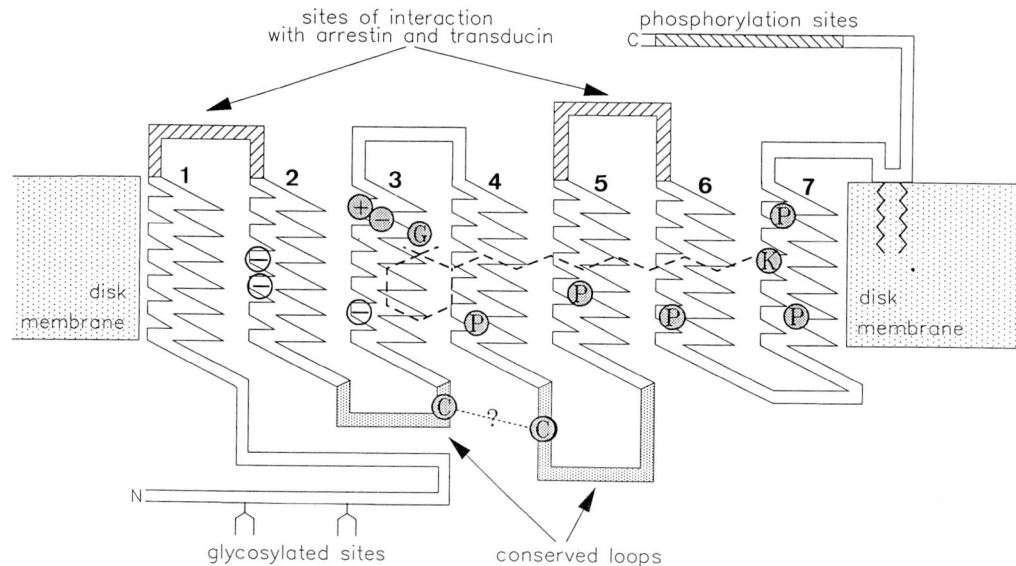

Fig. 3. *Structural model of bovine rod rhodopsin.* Some residues conserved among known G-protein-related receptors are shown (shadowed circles). Empty circles indicate the three negative residues specifically mutated in cone pigments. The C-terminal is anchored in the membrane by palmitoylation of two cysteines [87]

of the membrane, nearly parallel to its surface. Glycines and prolines are found in the central region of five out of the seven helices. As these are assumed to allow or induce 'kinks' in the helical structure, this could help provide a pocket for retinal. This seven-helices model has now been carefully checked by proteolytic attack, by labelling with hydrophobic probes, and is solidly established. But the packing and lateral ordering of the helices remains totally speculative. Rhodopsin is often modelled by analogy with bacteriorhodopsin, the photo-energized proton pump of *Halobacterium halobium*. But although bacteriorhodopsin has seven transmembrane helices, and also a retinal cofactor (hence the name) attached to a lysine in the last transmembrane helice, one must stress that its sequence has no detectable analogy with that of the visual rhodopsin and its function is not that of a signal transducer. Many receptor proteins of the 'seven-transmembrane-helices' superfamily have significant sequence analogy with rhodopsin and have analogous intracellular function; bacteriorhodopsin is the only one that cannot be related structurally as well as functionally to rhodopsin. In the sequence of all known rhodopsins and related seven-helices receptors few features are strictly conserved: for example a sequential pair of positively and negatively charged residues (Arg-Glu or Arg-Asp) located at a fixed position on the cytoplasmic end of helix 2. These are not found in bacteriorhodopsin. A too-long phylogenetic separation does not account for this lack of detectable homology: bovine rhodopsin cDNA probes have detected putative genes of analogous visual pigments in distant species and one is even found, with low stringency, in *Halobacterium* [23], but it is not the gene of bacteriorhodopsin, probably that of a sensory rhodopsin. The structural analogies between visual pigments and bacteriorhodopsin may therefore result from converging evolution and may suggest that the 'magic number' of seven helices be required to cage a retinal oriented perpendicularly to the helices.

The polypeptide loops connecting the helices constitute two hydrophilic poles of about equal size. The hydrophilic carboxyl-terminal polypeptide is on the cytoplamic side of the membrane. In all visual pigments, it is always rich in serine and threonine residues, 6–10 of them being concentrated in the final block of 10–20 amino acids. In photoexcited rhodopsin these hydroxylated residues will become phosphorylation sites for the rhodopsin kinase. Their sequence is, however, not conserved within the different visual pigments. The site of interaction with transducin is mainly on the polypeptide loops connecting helices 5 to 6 and helices 2 to 3 [7].

The amino-terminal polypeptide is on the intradiscal side (or extracellular side in cones). It varies in length and sequence between visual pigments, but always carries glycosylation sites for one or two oligosaccharides. Two of the polypeptide loops (2–3 and 4–5) are highly conserved in all visual pigments and two conserved cysteine residues probably establish a disulfide link between these two loops. This conserved structure must correspond to an important function yet to be determined. It has been suggested that the loop 4–5 could represent a calcium-binding site.

INTERACTION OF RETINAL WITH RHODOPSIN IN THE DARK AND AFTER ILLUMINATION

Rhodopsin is a membrane receptor for the photoactivable ligand retinal. The protein environment of retinal in the transmembrane region fulfills multiple roles.

a) In the dark, the site stabilizes the 11-*cis* conformation of the bound chromophore and hinders its thermal isomerization, hence protecting the system against spurious triggering. While the energy difference is only about 2 kJ/mol between the 11-*cis* and the all-*trans* conformations of free retinal, it requires more than 125 kJ/mol to isomerize a retinal bound in its rhodopsin site. This estimate is based on a microcalorimetric measurement of the energy uptake upon illumination at liquid nitrogen temperature [24]. At this low temperature the chromophore can still be forced to change conformation upon absorbing the large energy (209 kJ/mol) of a visible photon but the protein movements are blocked, and the photoinduced process stops at the bathorhodopsin stage after the isomerization of retinal: most of the energy uptake must be due to the movement of the protonated Schiff base end of the chromophore apart from a fixed negative counterion in the protein. A candidate counterion has been proposed, Asp-83 on helix 2 but it is not present in all cone pigments [25]. A sequential pair of charged amino acids (Arg-Asp or Arg-Glu) is found at a constant position close to the cytoplasmic end of helix 3 in all visual pigments, as well as in the analogous adrenergic and muscarinic receptors. Evidence indicates that in these receptors the agonist ligand binds in the membrane-spanning region at a location analogous to that of retinal in visual pigment [26]. Both epinephrine and acetylcholine have a protonated quarternary ammonium and that seems essential for their interaction with the receptor and could play the same role as the protonated Schiff base in rhodopsin. It is therefore tempting to speculate that the critical interaction of all these analogous ligands is with these charged residues found on helix 3 in all the 'seven-helices' receptors [2].

b) The distribution of charged groups around the chromophore modulates the wavelength of light absorption. Few charged residues are found in the predominantly nonpolar transmembrane core of all rhodopsins. According to the 'point charge model' of Honig et al. [27], an anionic group located close to the middle of the polyene chain could strongly modulate the red shift of the pigment. Comparison of the sequences of all the human cone and rod pigments confirms that supplementary anionic groups are found in the transmembrane core of the more red-shifted pigment, mostly in helices 2 and 3. Charged groups do not however explain everything: the human red and blue cone pigments, which have diverged very recently from a common ancestor and have very close absorption wavelengths, have kept exactly the same charged group distribution in the transmembrane region. But they have different aromatic residues. It has been proposed that aromatic residues lying with suitable orientation close to the end of the polyene chain of retinal could also shift its absorption [28]. In frog and cattle rhodopsin the orientation of one tryptophan in an hydrophobic environment is perturbed upon the isomerization of retinal [29].

c) The local perturbation generated in a few picoseconds by the photoisomerization of retinal in the membrane-spanning region propagates on a slower time scale until it reaches the distant cytoplasmic surface of the protein where it induces the formation of sites of specific interaction with peripheral or soluble proteins. This is the critical transduction step in which the information of the capture of a photon in the membrane is memorized into a metastable protein structure to be recognized by the internal enzymatic machinery of the cell. This process lasts in the order of 0.1 ms at physiological temperature. It is correlated with spectral changes of the chromophore and decrease of the free energy of the protein. Thus the charge separation energy loaded in the fast initial displacement of the protonated Schiff-bonded lysine relaxes by inducing movements of charged residues close by in the protein. Finally a 'meta II rhodopsin' state is reached in which all-*trans* retinal is still bound to the lysine, with an orientation close to its original one in the site, but with an absorption wavelength close to that of free retinal. In this state, stable for seconds before the hydrolysis of the Schiff base bond releases the chromophore, the interaction between the chromophore and its site are nearly totally relaxed, and new conformations have been generated on the cytoplasmic surface of the protein: the occurrence of a structural change is detectable by X-ray or neutron diffraction [30, 31] and by light scattering [46], however, these physical techniques do not give precise information on the localization and extent of the changes. Long-established biochemical evidence for a functional change of the cytoplasmic surface was the early observation that, after illumination, the serine and threonine residues near the C-terminal become susceptible to phosphorylation by rhodopsin kinase which has no action on the dark-adapted pigment [32, 33]. This correlates with an increased sensitivity of this terminal polypeptide to proteolytic cleavage after illumination [34]. The most important change however, which functionally defines the active state R*, is the appearance on the cytoplasmic surface of a binding site for transducin: if GTP has been suppressed from the medium, this binding remains permanent, it is released under the later addition of GTP. These observations of Kühn [6] provided a very elegant purification procedure for transducin and gave a critical clue about the catalysis by photoexcited rhodopsin (R*) of GDP/GTP exchange in transducin. The cytoplasmic loop connecting helices 1 and 2 and that connecting 5 and 6 are part of the binding site, which does not include the C-terminal polypeptide [7].

Are there any effects of retinal isomerization on the conformation of the intradiscal pole of rhodopsin? From the prolific and somewhat deceiving literature on calcium release from rod-outer-segment (ROS) membrane preparations, one may extract the observation [35] that rod disc vesicles can store calcium in the dark and release some of it upon illumination, provided that a calcium ionophore has been inserted into the membrane. The stoichiometry for release upon illumination is of the order of one molecule calcium/bleached rhodopsin molecule. This calcium cannot take part directly in the phototransduction mechanism especially since *in situ*, without ionophore, it would remain trapped into the discs. But this artificial observation may reveal a real effect of the chromophore isomerization on the conformation of the intradiscal pole of rhodopsin to which this calcium is probably bound.

TRANSDUCIN ON DARK-ADAPTED MEMBRANES: THE 'INACTIVE' GDP-BOUND HOLOENZYME

When no rhodopsin is photoexcited, transducin is bound to the disc membrane, but the membrane attachment is weak and very sensitive to the ionic strength and composition of the solution: transducin can be eluted quantitatively from broken ROS suspension by low-ionic-strength washing and is solubilized as an undissociated heterotrimeric complex $T_\alpha GDP\text{-}T_{\beta\gamma}$. This contrasts with all other G proteins, which, in their 'inactive' GDP-bound state remain tightly membrane-attached, probably through their γ subunit. The β subunit of transducin is strictly identical to the β subunit of other G proteins like G_s or G_i, but the γ subunits, which cannot be dissociated from their β partners without denaturing the complexes, are different [36]: T_γ is very acidic but not particularly hydrophobic, and no transmembrane stretch is detectable on its sequence [37]. The other G_γ are more hydrophobic, but none have been sequenced yet. Some G proteins also have a post-translational myristoyl residue on their α subunit which contributes to their membrane attachment [38]; it is not found on T_α. Even in a physiological medium the binding of transducin is not very tight, and the membrane-bound pool can be significantly depleted, if the outer cell membrane has been made leaky and the ROS suspension is diluted to below 0.5 mg/ml [39]. *In situ* however, the disc membrane concentration is so high inside a native ROS that 99.7% of transducin is membrane-bound. To what membrane site? Probably not specifically to rhodopsin, the only intrinsic membrane protein to be present in sufficient amount. The surface charge of rhodopsin may however contribute unspecifically to the binding, as do the charged lipids. Based on the observation that purified transducin in solution tends to aggregate, there have been suggestions that transducin might be polymeric on the membrane [40]; however, this aggregation could as well as be due to interaction between the uncovered hydrophobic surfaces that normally bind to the membrane.

The lateral mobility of transducin on the disc membrane has never been directly measured and it is often implicitly assumed that rhodopsin, whose lateral diffusion is unusually high for an intrinsic membrane protein, would be the most mobile component in the transduction cascade. This is unlikely: transducin diffuses on the membrane surface in a cytoplasmic environment that is by two orders of magnitude more fluid than the lipid bilayer in which rhodopsin is embedded. As discussed later, the lateral diffusion of transducin is probably predominant because of its collision rate with photoexcited rhodopsin. It is noteworthy that in the invertebrate visual cells rhodopsin appears fixed in quasicrystalline arrays and therefore the G protein must be the only mobile component controlling the collision rate [41].

Magnesium ions are important for the binding of the nucleotide to T_α, for the association of T_α-GDP to $T_{\beta\gamma}$ and the binding of the heterotrimeric complex to the membrane. This multiplicity of effects, which *in vitro* can perturb the kinetics of the light-activated process involving transducin, sometimes led to speculation of a putative regulatory role for magnesium on transducin or other G-protein-mediated processes. One must keep in mind that *in vivo* the magnesium concentration is buffered in the millimolar range. Effects that are only observable *in vitro* far above or below this concentration range cannot have a physiological significant regulatory function. A magnesium ion participates in the binding of GDP to T_α and interacts with the β-phosphate of the bound nucleotide [42]. Its affinity is in the 10–100-μM range. *In vitro*, in the absence of divalent ion chelators, sufficient residual magnesium is usually present to ensure its persistence. Its removal by EDTA facilitates the release of GDP from T_α, increases the release of the inactive T_αGDP-$T_{\beta\gamma}$ holoenzyme from the membrane and impairs the light-triggered nucleotide-exchange reaction. On the other hand, high concentrations of magnesium, above 10 mM (i.e. above the physiological range), induce the dissociation of the holoenzyme and the release of the isolated T_αGDP subunit from the membrane [43]. This effect can be reproduced with similar concentrations of calcium. It looks therefore like an unspecific divalent-ion effect on protein–protein and/or protein–lipid interactions, as is often observed for membrane-bound peripheral proteins, rather than a specific action of magnesium on a regulatory site of the protein. At physiological concentrations of magnesium, the dissociation rate constant of the GDP bound to T_α is extremely slow, in the range of hours [44]. The concept of affinity of the nucleotide for T_α is therefore not very meaningful as an equilibrium is never reached and the state devoid of nucleotide is never obtained under physiological conditions, except when transducin is bound to R*.

THE FIRST AMPLIFYING STEP: COLLISION COUPLING OF PHOTOEXCITED RHODOPSIN WITH TRANSDUCIN AND CATALYSIS OF NUCLEOTIDE EXCHANGE

The successive catalytic actions of a photoexcited rhodopsin, R*, on many transducins transmit and amplify the information of the capture of a photon by rhodopsin. R* binds to T_αGDP-$T_{\beta\gamma}$ and induces the opening of the nucleotide site on T_α. The formerly locked-in GDP becomes rapidly exchangeable. It can be released and eventually replaced in

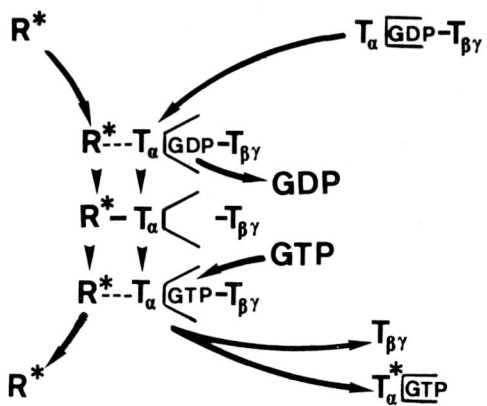

Fig. 4. *Steps of the catalysis by R* of GDP:GTP exchange in transducin*

the site by a new GTP molecule. The binding of GTP induces a transconformation of T_α that makes the process quasi-irreversible: T_αGTP loses its affinity for R* as well as for $T_{\beta\gamma}$ and it is instantly released into the cytoplasm. The nucleotide site locks again onto the bound GTP in the isolated T_αGTP subunit. R* is then free again to bind to a new transducin, until it is inhibited by the combined action of rhodopsin kinase and arrestin [19, 45]. Let us discuss in more detail the successive steps of this R*-catalyzed exchange reaction (Fig. 4).

R*-T binding and opening of the nucleotide site

The binding step is best studied where the exchange reaction is prevented by suppressing GTP from the medium. Kühn first observed that, under such conditions, full illumination of ROS membrane induced the quantitative binding of transducin that becomes unextractable even in low-ionic-strength media and in the presence of EDTA [6]. The transducins bind tightly to photoexcited rhodopsins. *In vitro* this reaction was shown [46] to correlate with rapid changes of the turbidity of ROS fragment suspension, observable by near-infrared light scattering [47]. This techniques helped demonstrate that the binding is stoichiometric (one molecule R* binds one molecule transducin) and rapid, in the millisecond time range [46]. The signal observed does not originates directly from the molecular binding event, but from a change in electrostatic interaction between the disc membranes after the formation of many R*-T complexes on their surface [48]. On morphologically intact ROS this results in a slight change in disc spacing which can also be monitored by neutron diffraction [31]. The charge modification does not simply result from the release of GDP from T_α: it is observed even in the presence of concentrations of GDP of GDP[βS] high enough to prevent this release. Although R* acts only on the conformation of T_α, the presence of the complete holoenzyme T_αGDP-$T_{\beta\gamma}$ is necessary for transducin to recognize R* properly and bind to it [7, 49]. It is not clear whether this requirement for $T_{\beta\gamma}$ reflects the need of a direct interaction of the $\beta\gamma$ subunit with R* for binding the complex, or is due to a necessary modification of the α subunit upon its association to $\beta\gamma$, before it can bind to R*, or even simply reflects a kinetic enhancement of the reaction resulting from the binding and proper orientation of T_αGDP on the membrane through its interaction with the membrane-bound $T_{\beta\gamma}$.

The interaction between R* and transducin modifies the conformation of T_α so that the nucleotide site 'opens' and the bound GDP becomes rapidly exchangeable (Fig. 4). The K_{off} for GDP release is reduced from hours to milliseconds or less. It must be stressed that the replacement of the released GDP by GTP is not particularly favored over the simple GDP/GDP exchange which can be observed by illumination in the presence of radioactive GDP and absence of GTP [44]: very rapidly, all the previously bound unlabelled GDP molecules are replaced by radioactive ones. The affinity of R*-bound transducin for GDP is in the range of 10 µM. GDP[βS] can also bind into the 'open' site and if added in large excess will kinetically compete with GTP. What will eventually drive the reaction toward the GTP-bound state is not a much higher K_{on} rate for GTP than for GDP in the open site, but the fact that the binding of GTP induces a transconformation of T_α that causes the dissociation of the rhodopsin–transducin complex into three independent components: R*, $T_{\beta\gamma}$ and T_αGTP, the latter one being released from the membrane into the cytoplasm.

The tight intermediate complex $R^* - T_\alpha empty - T_{\beta\gamma}$

As long as a GDP molecule remains present in the site of T_α, if, in the absence of GTP, a sufficient concentration of GDP is maintained in the medium, the R*-T complex is short-lived, transducin exchanges rapidly on a given R*. But if GDP as well as GTP have been suppressed from the medium, the 'open' site remains empty, the binding of $T_\alpha$empty to R* becomes very tight and the $R^* - T_\alpha empty - T_{\beta\gamma}$ complex remains stable, extensive washing in low-ionic-strength medium only extracts a little of $T_{\beta\gamma}$ dissociated from T_α that remains attached to R*. Under the unphysiological condition of absence of GTP, the system is 'frozen' in the normally transient intermediate state of catalytic activation of transducin by the receptor. Interestingly in this tight $R^* - T_\alpha empty - T_{\beta\gamma}$ complex not only is the conformation of T modified by the coupling to R* but conversely T reacts on the conformation of R* at the level of the chromophore site. It was first observed that at low temperature the interaction with transducin enhances the meta II state over the earlier state meta I, this being the basis for the identification of meta II as the active R* state [50, 51]. At physiological temperature, the meta II state is transient for free rhodopsin and decays in a few seconds towards another spectral state, meta III, followed by the release of the chromophore from the protein; this decay was found to be much slower for transducin-bound rhodopsin [52]. We have recently confirmed [84] that when rhodopsin is blocked in an $R^* - T_\alpha empty - T_{\beta\gamma}$ complex totally devoid of nucleotide, the R* state is stabilized, both in its spectral and its functional characteristics: after a 30-min incubation at physiological temperature the chromophore has not decayed and if one liberates R* from its bound transducin (by addition of GTP) it is still able (after washing the GTP) to bind to and activate another transducin molecule. The retinal is in the transmembrane core of rhodopsin and transducin binds on its cytoplasmic surface. These two distant sites therefore remain allosterically coupled in the R* state: it is the isomerization of the chromophore in the transmembrane region that triggered the formation of the transducin site on the surface, but conversely the binding of transducin on the cytoplasmic surface blocks the chromophore in its internal site. These observations have an important correlate in other receptor–G-protein systems: the binding of transducin (a G protein) to R* (an activated receptor) hinders the release of the bound chromophore, the equivalent of a liganded agonist. The phenomenological observation is that the release of the agonist from the receptor is slower in the absence of GTP than in its presence. For a hormone receptor, in which the ligand is not bound prior to activation, a slower ligand release means a higher binding affinity. The observed GTP dependence of chromophore decay in rhodopsin is therefore equivalent to the GTP dependence of ligand affinity of hormone receptors. The case of rhodopsin tells us that this GTP dependence results from the 'post-coupling' of the G protein (transducin) to the agonist-activated receptor (R*) rather than from a putative 'pre-coupling' to inactivated receptors.

Dissociation of the complex upon GTP binding and release of 'active' $T_\alpha GTP$

The addition of GTP or a non-hydrolyzable analog to illuminated ROS membranes (or flash illumination in the presence of GTP) induces instantly the release of transducin from R* and its dissociation into two independent subunits T_αGTP and $T_{\beta\gamma}$ [8, 53]. In low-ionic-strength media, both subunits of transducin go into solution. If the ionic composition is close to that of cytoplasm, only T_αGTP goes into solution while $T_{\beta\gamma}$ remains membrane-bound. Light-scattering studies have demonstrated that, in well preserved ROS in the presence of the physiological concentration of GDP and GTP, the release of T_αGTP is instantaneous and the total duration of an exchange reaction is of the order of a millisecond [48]. The kinetic limit might be that of the rate of successive encounter of R* with the membrane-bound transducins in the T_αGDP-$T_{\beta\gamma}$ state. It is probable that in this process the mobility of transducin predominates over that of rhodopsin. No direct measurement of transducin lateral diffusion has yet been published, but Phillips and Cone [54] reported that the kinetics of the photoresponse in the intact isolated rods varied linearly with the cytoplasm viscosity, suggesting that diffusion in the cytoplasm rather than in the membrane is a rate-limiting process. Our recent kinetic measurements (Bruckert F. and Chabre M., unpublished) using the light-scattering technique and photoexcitation by fringe interference laser flash also suggest that the lateral mobility of rhodopsin is not rate-limiting for the formation of T_αGTP. For a G-protein-mediated process, when discussing collision coupling models, the emphasis is generally on the receptor mobility. It is however likely that, as in the visual cascade, the most mobile protein component is the G protein.

THE ACTIVE MESSENGER T_α GTP AND ITS ANALOGS: $T_\alpha GTP[\gamma S]$ OR $T_\alpha GDP-AlF_3$

The dissociation of T_αGTP from $T_{\beta\gamma}$ and its instantaneous solubilization may be physiologically important as they allow the activated T_α to diffuse freely through the cytoplasm and, in the native structure, also to reach the cGMP phosphodiesterase on the facing disc surface. The solubility of T_αGTP, or of hydrolysis-resistant analogs such as $T_\alpha GTP[\gamma S]$ suggests that the conformation of the protein has been significantly modified upon the binding of GTP. Another indication of a structural change is the decreased proteolytic sensitivity of T_αGTP versus that of T_αGDP [55], suggestive of a more compact conformation for the GTP-bound protein. One usually thinks of T_αGTP as an 'activated' state of the G protein, since it can couple to the effector enzyme and activate it. But this does not imply that the γ-phosphate of the bound GTP has yet transferred any energy to the protein: non-hydrolyzable analogs such as Guo*PP*[NH]*P* or GTP[γS], which cannot transfer energy from γ-phosphate phosphoryl-bond, confer to T_α a permanently 'active' T_α conformation.

Another way of 'activating' T_α, by fluoro-metallic complexes, provides further insight into the role of GTP. Sternweis and Gilman [56] demonstrated that the long known activating effect of fluorides on G proteins depends on the presence of traces of aluminum that is etched from the glassware by NaF or KF solution, and suggested that the real activator was the ionic complex AlF_4^-. We demonstrated that a prerequisite for the activation of one T_α molecule by one molecule of fluoroaluminate complex was the presence of GDP in the site of T_α: the activation observed in dark-adapted ROS membrane was suppressed when the GDP was allowed to be released from transducin by illumination and subsequent R*-T complex formation [57, 58]. Analogs of GDP bound in the site are also efficient for fluoroaluminate activation provided their β-phosphate remains unsubstituted. Noticing structural analogies between AlF_4^- and a phosphate group, we proposed that the fluoroaluminate complex binds next to the β-phosphate and simulates the presence of the γ-phosphate of GTP, hence conferring to T_α its 'active' conformation. The value of this model has now been ascertained and extended to other types of proteins interacting with nucleoside polyphosphates or simply with phosphates [59, 60].

The aluminofluoride activation bypasses the requirement for catalysis by a receptor, since the guanosine part of the nucleotide has not to be exchanged: the GDP is directly 'complemented' in the non-exchangeable site. As a consequence, and by contrast with receptor-catalyzed permanent activation by GTP[γS], activation by aluminofluoride is reversible. The presence of the $T_{\beta\gamma}$ subunit is not needed for activation but rather hinders or reverses activation by pushing to the right the reversible equilibrium [58]:

$$T_\alpha GDP\text{-}AlF_3 + T_{\beta\gamma} \rightleftharpoons T_\alpha GDP\text{-}T_{\beta\gamma} + AlF_4.$$

The aluminofluoride complex must lose one fluorine upon entering the site, to bind to the β-phosphate oxygen. This view is reinforced by the similar efficiency of beryllo-fluoride complexes, which are also isomorphous to phosphates. In contrast to aluminum, which like phosphorus can be penta- or hexa-coordinated, beryllium is strictly tetra-coordinated and must bind as such in the site. That the binding of a tetrahedral beryllium confers to $T_\alpha GDP$ the conformation observed with bound GTP confirms that in the 'activated' state the γ-phosphate of GTP is not constrained in a penta- or hexa-coordinated conformation as usually found for an intermediate state of interaction between a nucleoside triphosphate and a protein: the 'energy-rich' γ-phosphate has not yet conferred any of its potential energy to the protein.

SECOND STEP OF THE CASCADE:
ACTIVATION OF cGMP PHOSPHODIESTERASE BY T_αGTP

As is common for a fast hydrolytic enzyme, activation of cGMP phosphodiesterase is indeed the release of an inhibition. Miki et al. [61] first observed that PDE activity could be elicited by tryptic treatment of ROS membrane, suggesting that the PDE is maintained in its basal state by an easily proteolyzable inhibitor. From the purified enzyme Hurley and Stryer [62] isolated and purified a heat-resistant 13-kDa subunit (hereafter called I) which could inhibit with high affinity the 85 + 88-kDa catalytic complex $PDE_{\alpha\beta}$ of the trypsin-activated enzyme. The natural activation of the PDE was therefore expected to result from an interaction of T_αGTP with the inhibitor. In the low-ionic-strength eluate of ROS membranes illuminated in the presence of GTP[γS], Deterre et al. [63] isolated a stable complex of I and T_α, dissociated from the PDE catalytic complex. Gel filtration studies and polypeptide analysis suggested a molecular mass of 50 kDa and a 1:1 stoichiometry for this T_α-I complex. But one does not know whether under physiological conditions the T_αGTP-I complex separates physically from $PDE_{\alpha\beta}$ on the membrane. The fact, however, that these two components appear dissociated upon low-ionic-strength extraction suggests that they are not tightly bound together on the membrane.

In the absence of precise data on the inhibitor abundance, a 1:1:1 stoichiometry had usually been assumed for the α, β and I subunits in the inactive PDE holoenzyme, but further analysis of the T-PDE interaction demonstrated the existence of two inhibitory subunits per native molecule: the chromatography of an activated ROS extract resolved not one, but two peaks of activated PDE, beside the excess native inactive peak [64]. Only the last peak is devoid of inhibitor, the I/$PDE_{\alpha\beta}$ ratio in the first active peak being close to 50% of that found for the native enzyme. The simplest interpretation is that the three PDE peaks, with 2/1/0 relative amount of inhibitor for a normalized amount of catalytic units, correspond respectively to inactive I_2-$PDE_{\alpha\beta}$ and to two different active states I-$PDE_{\alpha\beta}$ and $PDE_{\alpha\beta}$. The same two states of activated PDE can be isolated upon progressive mild trypsin proteolysis of the native enzyme, which degrades and successively eliminates the two inhibitors. The specific activity for the I-$PDE_{\alpha\beta}$ peak fraction is about half of that measured on totally stripped $PDE_{\alpha\beta}$. But by mixing purified $PDE_{\alpha\beta}$ with native $I_2PDE_{\alpha\beta}$ exchange of inhibitor is observed: $PDE_{\alpha\beta} + I_2PDE_{\alpha\beta} \rightleftharpoons 2$ I-$PDE_{\alpha\beta}$. The possibility of this exchange, the kinetics of which is not yet known, precludes an unambiguous estimate of the enzymatic properties (V and K_m) of the intermediate complex I-$PDE_{\alpha\beta}$. As the existence of two inhibitory subunits were not suspected, their relative binding affinities for the catalytic complex have not yet been properly analyzed.

When tested in solution, the specific activities are the same for the trypsin-activated species as for the corresponding transducin-activated PDE: of order of 1000 cGMP molecular hydrolyzed/s for the fully stripped $PDE_{\alpha\beta}$ complex. However in the presence of membrane and at physiological ionic strength, the trypsin-activated species no longer binds to membrane as the transducin-activated one does: before digesting the inhibitor, trypsin cleaves a very short terminal peptide that seems necessary for membrane attachment of the catalytic unit [65] (and Catty and Deterre, unpublished results). The loss of membrane attachment may influence the efficiency of the catalytic units. The tryptic treatment of the PDE can therefore no longer be used for quantitative analysis of the 'natural' activation process by T_α.

OUTPUT SIGNAL:
CLOSURE OF cGMP-DEPENDENT CHANNELS ON THE CELL MEMBRANE

We have seen that one R* molecule could catalyze the formation of T_αGTP at the rate of about 1 ms^{-1} and that two T_αGTP molecules are required to activate fully one $PDE_{\alpha\beta}$ molecule which then hydrolyzes cGMP at a rate of more than 10^3 s^{-1}; within 200 ms, the photoexcitation of one rhodopsin molecule can induce the hydrolysis of at least 10^5 cGMP molecules. The diffusion time of cGMP over distances of the order of micrometres is in the range of milliseconds. The free cGMP concentration in a dark-adapted ROS is not accurately known: a substantial fraction of the total cGMP content in ROS, around 60 μM, seems to be bound, leaving only between 5–30 μM free cGMP [1]. To obtain an order of magnitude, one can calculate that in a 2-μm-diameter ROS, the removal of 10^5 molecules of cGMP could decrease the cGMP concentration by 10 μM over a length of about 5 μm. The kinetics and gain of the cascade can therefore account for the closure within 200 ms of all the channels upon a sizeable fraction of the ROS length, assuming these channels have an affinity in the 10 μM range for their controlling cGMP.

The direct control by cGMP of the light-sensitive conductance, discovered by Fesenko et al. [10], has a mid-point sensitivity of 30 μM cGMP and a Hill coefficient of 1.8. In normal Ringer, that is in the presence of millimolar Ca^{2+}, the unit conductance of the putative channels was too low to allow observation of single-channel activity from an excised patch. A value of 100 fS was inferred from noise analysis [66]: this might then have resulted from the activation of a transporter rather than the opening of a pore. This conductance is not very selective for Na^+, divalent cations are also transported with not much lower efficiency. However, in the absence of divalent cations in the bathing medium, the cGMP-dependent flux of monovalent ions is highly increased and characteristic single channel openings can be recorded, with a mean conductance of 25 pS and a Hill coefficient around 3 [67, 68]. The low conductance to Na^+ in the physiological medium is therefore probably limited by the slow passage of Ca^{2+} ions which hinders the faster entry of Na^+. A cGMP-dependent Ca^{2+} release from ROS membrane vesicles had been detected as early as 1979 by Cavaggioni and coworkers [69, 70] and later by Koch and Kaupp [71], in broken ROS preparations which included both disc and outer cell membranes. In the context of the then prevailing calcium hypothesis, these interesting observations had been generally misunderstood as related to a cGMP-dependent calcium release from the discs. It is not clear whether some light-sensitive channels are also present in the disc membrane, but from the same preparation Kaupp and coworkers [11, 72] have now isolated and purified a 66-kDa polypeptide with which they could reconstitute, in artificial lipid vesicles, a cGMP-dependent conductance with a Hill coefficient around 3 and properties close to those of the plasma membrane channels. After insertion of the purified polypeptide into a planar lipid bilayer, they observed cGMP-stimulated single-channel activity with conductance of 26 pS in the absence of divalent cation. It is therefore likely that the plasma membrane channel is constituted by at least three units of the 66-kDa polypeptide. The drug l-cis-diltiazem, that blocks the cGMP conductance in excised patches, was however ineffective on the reconstituted channels, suggesting that an additional component might be missing. Matesic and Liebman [73] had also obtained cGMP-dependent conductance in vesicles reconstitution with a preparation enriched in a protein of 39 kDa. But the purification and characterization of this 39-kDa polypeptide are still insufficient to ascertain its participation in the light-sensitive channel.

TERMINATION OF THE CASCADE AND REGULATION OF THE PHOSPHODIESTERASE ACTIVITY

The response to a single photon decays nearly as fast as it raised (see [1]): within 0.5 s the cGMP-dependent channels reopen, the cell recovers its initial depolarized state and it is almost immediately able to respond to a new photon. This requires that all the 'active' intermediates of the cascade have decayed or have been blocked. Upon strong flashes or continuous illumination adaptation phenomena are observed, which modify the gain and kinetics of responses to subsequent flashes. These are too complex to be analyzed yet at the molecular level, but the termination of the response to weak flashes is now quite well understood.

Inactivation of R* by R*-specific kinase and arrestin

The spontaneous decay of the metastable R* state is much too slow to account for the termination of the response to a weak flash: at physiological temperature, in the absence of ATP and soluble enzymes, meta II rhodopsin remains able to activate new transducins for tens of seconds. It loses this capacity only when decaying to meta III and other spectral intermediates that precede the release of retinal from the protein. Liebman and Pugh [74] first observed that in vitro, the cGMP response was shortened in the presence of ATP and suggested the implication of rhodopsin-kinase. This enzyme is active only on photoexcited rhodopsin: Kühn and Hargrave [34] had observed that the C-terminal polypeptide becomes more susceptible to proteolysis by trypsin after illumination. This demonstrated an 'uncovering' of this C-terminal polypeptide which contains the seven phosphorylatable serine and threonine residues among its 15 last amino acids: concomtant with the formation of the binding site for transducin, photoexcitation induces the release of phosphorylation sites for the kinase. Both sites are on the cytoplasmic surface of rhodopsin and a competition starts between the two enzymes for access to their respective sites as steric hindrance prevents them from binding simultaneously (Fig. 5, steps 3–5). Upon weak illumination in the presence of ATP but without GTP, when transducin in excess remains permanently bound on all the R* formed, the ATP-dependent kinase remains inactive. It becomes active only if micromolar GTP is added [52]. By contrast, upon strong illumination, when R* is in excess over transducin, no GTP is required for kinase activation. Hence, the enzyme is not activatable by GTP, which is required only to prevent the permanent binding of transducin on R*, but only by ATP and the presence of its substrate, R*. In vivo, both GTP and ATP are present at high concentration and phosphorylation could be very fast.

The phosphorylation of R* is but the first step of its inactivation, it reduces but does not fully block the rate of catalytic coupling to transducin. Full blocking requires the intervention of a very soluble and very abundant protein 'arrestin' which had been isolated 10 years ago under diverse names: '48K protein' or 'S antigen' [19]. Its new name relates to the demonstration [45] that its bind specifically to phosphorylated R* and definitely blocks the access to transducin. In the sequence of arrestin, near the C-terminal, limited but significant analogy with transducin has been detected [75–77]. Both proteins may recognize a common site on R*. This is not the phosphorylated region, whose elimination by trypsin prevents the binding of arrestin but not the coupling of transducin to R* [7].

This process of receptor inactivation is of broad interest as, once more, it seems to be general to the regulation of the activity of other G-protein-coupled receptors [78]. Rhodopsin kinase is the prototype of a new class of 'activated receptor-specific kinases', not dependent on any external activator, and arrestin is the prototype of 'phosphorylated receptor-specific blocker'. The β-adrenergic and muscarinic receptors also have a serine + threonine-rich C-terminal which can be phosphorylated by rhodopsin kinase when these receptors are agonist-

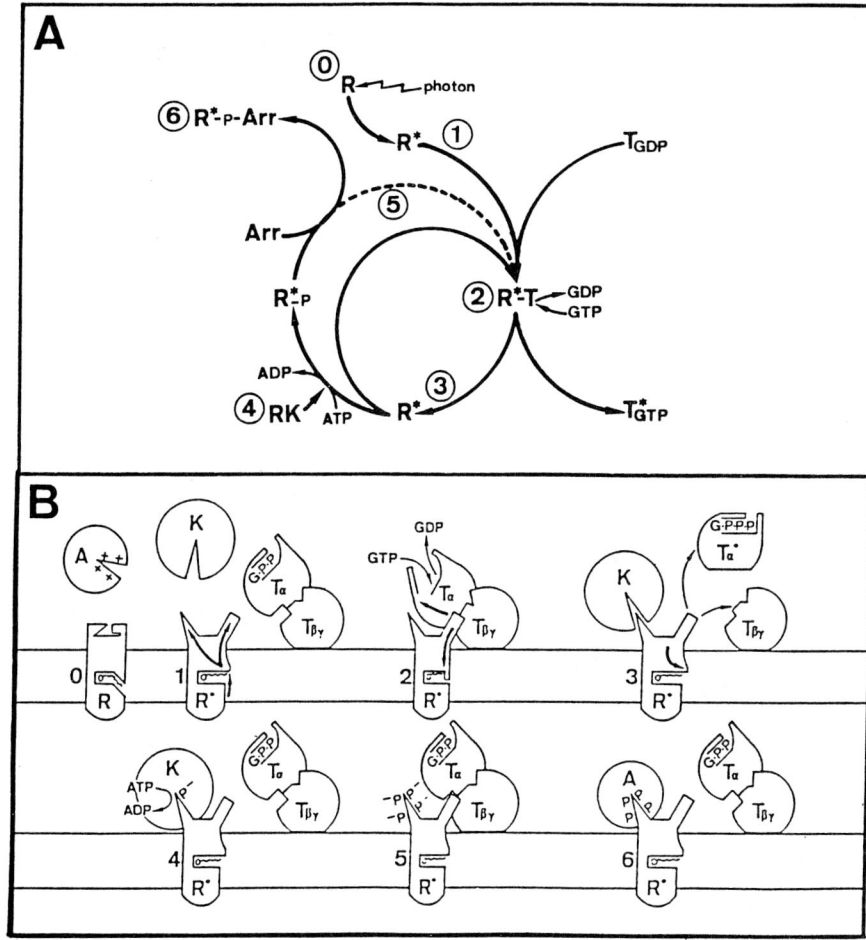

Fig. 5. *Regulation by rhodopsin kinase and arrestin of the catalytic action of R* on transducin.* The numbers in A and B correspond to the same stages of reaction

liganded and further blocked by arresting after phosphorylation [79].

Inactivation of $T_\alpha GTP$: the problem of the GTPase rate

Blocking the activated receptor is not sufficient to arrest the cascade: the $T_\alpha GTP$ already formed will maintain the PDE activity by remaining bound to inhibitors until they lose this capacity after the hydrolysis of GTP. The hydrolysis of the γ-phosphate is coupled to a change of conformation that regenerates the $T_\alpha GDP$ form, with low affinity for the PDE inhibitor and high affinity for $T_{\beta\gamma}$: $T_\alpha GDP$ dissociates from the PDE inhibitor which immediately rebinds to the catalytic units of PDE and blocks their activity. An apparent problem is that even upon strong illumination, when the number of R* molecules exceeds that of transducin, the turn-over rate of GTP hydrolysis, measured *in vitro*, remains surprisingly low: of the order of 3 molecules GTP hydrolyzed/molecule transducin $\times$ min^{-1} [8, 42, 53]. At first sight this seems to imply much too long a lifetime, about 20 s, for the 'active' $T_\alpha GTP$ state, to be compared to the shut-off time of less than a second for the cascade. But the GTPase rate measures the duration of a full cycle of transducin shuttling from R* to PDE and back to R* to load a new GTP. The association to $T_{\beta\gamma}$ is required for $T_\alpha GDP$ to be able to bind to R* and undertake a new GDP/GTP exchange. Fung [49] observed, however, that the GTPase in reconstituted systems was not rate-limited by $T_{\beta\gamma}$: the saturation rate of 3 mol/mol $T_\alpha \times$ min^{-1} was already reached with 0.1 mol $T_{\beta\gamma}$/mol T_α. A possible explanation could then be that the limiting step would be the reassociation of $T_\alpha GDP$ to the disc membrane. *In vitro*, when the cell membrane is disrupted the excess soluble $T_\alpha GTP$ dilutes in the extraneous solution. After the hydrolysis of GTP, it must get back onto the disc membrane to which $T_{\beta\gamma}$ has remained bound. Indeed, significant differences between the GTPase rate measured by various groups, 1 – 5 mol GTP/mol transducin $\times$ min^{-1}, seem to be related to differences of dilutions and/or ionic compositions of the suspensions. Indirect measurements by light-scattering technique suggest that in very concentrated broken rods suspensions [80] or in resealed ROS from which $T_\alpha GTP$ does not leak out [8], the GTPase rate is reduced to a few seconds. Very recently, direct measurements at high GTP concentration and 37°C suggest also a fast GTPase rate [88].

Another suggestion that could make a slow GTPase rate compatible with a fast termination of the cascade would be that in a first step $T_\alpha GTP$ takes on a conformation $T_\alpha GDP$-P with low affinity for the PDE, and only later does it regain its high affinity for $T_{\beta\gamma}$ upon slowly releasing the hydrolyzed γ-phosphate. Such sequences of fast internal hydrolysis followed by slow phosphate release are observed, for example, in the polymerization/depolymerization cycle of tubulin [59] and in the ATPase cycle of actin [82]. For transducin the total GTPase cycle time would then be divised in a

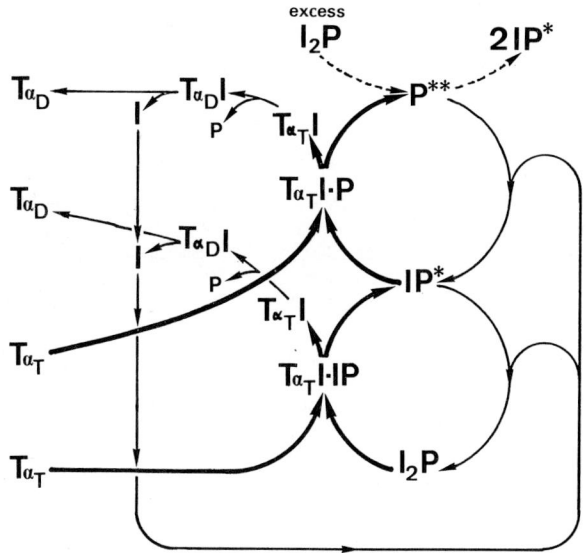

Fig. 6. *Tentative scheme of cGMP-specific phosphodiesterase activity regulations.* P represent the catalytic units of phosphodiesterase ($PDE_{\alpha\beta}$). The stars symbolize the two possible states of activation, the thick lines the activation steps, the thin ones the inactivation reactions. In dotted lines, the possible desactivation mechanism by back exchange

short active time followed by a longer 'dead time' during which $T_\alpha GDP\text{-}P$ would not be reactivable. Recent data from Arshavski et al. [83] do not favor this hypothesis, which however deserves further investigation. A way to look for possible intermediate stages during the GTP hydrolysis is to analyze, by time-resolved microcalorimetry, the time course of the enthalpy release within the cycle, in the absence of cGMP. One molecule GTP is hydrolyzed to $GDP + P_i$, and therefore 33 kJ are released for each complete cycle of GDP/GTP exchange followed by GTP hydrolysis and regeneration of $T_\alpha GDP\text{-}T_{\beta\gamma}$. With a specially designed microcalorimeter, reaching a time resolution of 0.5 s, we have detected the heat pulse following the flash illumination of a ROS suspension in the presence of GTP and of high concentration of hydroxylamine that helps the fast quenching of R*: the length of the heat pulse resulting from the hydrolysis of GTP is then reduced to about 3 s at 23 °C [84]. Preliminary measurements of the GTPase turn-over rate, as measured classically through the rate of GDP production upon saturating illumination in samples of high membrane concentration, also suggest a much faster rate than those published before on more dilute suspensions.

A diffusion-controlled rapid inactivation of the phosphodiesterase?

The existence of two inhibitory subunits per PDE complex might also allow for further modulation of the PDE activity. The exchangeability of inhibitors between inactive I_2-PDE and fully activated $PDE^{**}_{\alpha\beta}$ (Fig. 6) raises the possibility of a rapid diffusion-controlled switch-off mechanism of the PDE activity after a flash, which would by-pass the requirement for GTP hydrolysis [64]: *in situ*, within a few hundred milliseconds, all the transducins surrounding a photoexcited rhodopsin are activated and the local concentration of solubilized $T_\alpha GTP$ in the cytoplasmic cleft between two discs may reach 500 µM (Table 1). This high concentration may be required to overcome the high affinity of the last inhibitor for the PDE and fully stripped $PDE^{**}_{\alpha\beta}$. But as soon as R* is blocked by phosphorylation and arrestin binding, the pool of soluble T_α dilutes in the cytoplasm and its concentration may become rapidly too low to produce more $PDE^{**}_{\alpha\beta}$. The $PDE^{**}_{\alpha\beta}$ formed initially also diffuses away into the membrane area where it encounters excess native $I_2\text{-}PDE_{\alpha\beta}$ from which it can regain one inhibitor by back exchange: $PDE^{**}_{\alpha\beta} + I_2\text{-}PDE_{\alpha\beta} \rightleftharpoons 2\ I\text{-}PDE^{*}_{\alpha\beta}$. If $I\text{-}PDE^{*}_{\alpha\beta}$ has a lower specific activity or a higher K_m than $PDE^{**}_{\alpha\beta}$, this process could provide a rapid quenching of the PDE activity of cGMP hydrolysis that would precede the inactivation resulting from the GTP hydrolysis and release of inhibitor by T_α.

CONCLUSION

In visual transduction the cycle of transducin shuttling between rhodopsin and cGMP-specific phosphodiesterase has been elucidated in more detail than for any other transduction cascade process. One is still, however, far from understanding the interactions between the three major proteins, rhodopsin, transducin and phosphodiesterase, at the molecular level. This will require the determination of their three-dimensional structures. The analysis of the two key protein—protein interactions would also demand crystallographic studies of the two complexes $R^*-T_\alpha empty-T_{\beta\gamma}$ and $T_\alpha GTP\text{-}I$. The mechanism of action of the guanine nucleotide on transducin also needs to be analysed further, perhaps by drawing on analogies not only with other classes of GTP-binding proteins, such as elongation factors and tubulin, but even with ATP-binding proteins like actin. Notwithstanding peculiar characteristics related to the nature of the visual signal, the transducin cascade has proved and will further prove to be a very instructive model for G-protein-mediated transduction processes.

REFERENCES

1a. Stryer, L. (1986) *Annu. Rev. Neurosci. 9*, 87–119.
1b. Nathans, J. (1987) *Annu. Rev. Neurosci. 10*, 163–194.
2. Applebury, M. L. & Hargrave, P. A. (1986) *Vision Res. 26*, 1881–1895.
3a. Pugh, E. N. (1987) *Annu. Rev. Physiol. 49*, 715–742.
3b. Liebman, P. A., Parker, K. R. & Dratz, E. A. (1987) *Annu. Rev. Physiol. 49*, 765–792.
3c. Hurley, J. B. (1987) *Annu. Rev. Physiol. 49*, 793–812.
4. Hagins, W. A. (1972) *Annu. Rev. Biophys. Bioeng. 1*, 131–158.
5. Kühn, H. (1978) *Biochemistry 17*, 4389–4395.
6. Kühn, H. (1980) *Nature (Lond.) 283*, 587–589.
7. Kühn, H. (1984) in *Progress in retinal research* (Osborne, N. & Chader, J., eds) pp. 123–153, Pergamon, New York.
8. Fung, B. K. K., Hurley, J. B. & Stryer, L. (1981) *Proc. Natl Acad. Sci. USA 78*, 152–156.
9. Yee, R. & Liebman, P. A. (1978) *J. Biol. Chem. 253*, 8902–8909.
10. Fesenko, E. E., Kolesnikov, S. S. & Lyubarsky, A. L. (1985) *Nature (Lond.) 313*, 310–313.
11. Cook, J. N., Hanke, W. & Kaupp, U. B. (1987) *Proc. Natl Acad. Sci. USA 84*, 585–589.
12. Fleischman, D. & Denisevich, M. (1979) *Biochemistry 18*, 5060–5066.
13. Lolley, R. N. & Racz, E. (1982) *Vision Res. 22*, 1481–1486.
14. Koch, K. W. & Stryer, L. (1988) *Nature (Lond.) 334*, 64–66.
15. Yau, K. W. & Nakatani, K. (1985) *Nature (Lond.) 313*, 579–581.
16. Pugh, E. & Altman, J. (1988) *Nature (Lond.) 313*, 16–17.
17. Stryer, L. & Bourne, H. R. (1986) *Annu. Rev. Cell Biol. 2*, 391–419.
18. Dohlman, H. G., Caron, M. G. & Lefkowitz, R. J. (1987) *Biochemistry 26*, 2657–2664.

19. Pfister, C., Chabre, M., Plouet, J., Tuyen, V. V., De Kozak, Y., Faure, J. P. & Kühn, H. (1985) *Science (Wash. DC)* 228, 891–893.
20. Chabre, M. (1985) *Annu. Rev. Biophys. Chem.* 14, 331–360.
21. Ovchinnikov, Y. A. (1982) *FEBS Lett.* 148, 179–181.
22. Hargrave, P. A., McDowell, J. H., Curtis, D. R., Wang, J., Juszezak, E., Fong, S.-L., Rao, J. K. M. & Argos, P. (1983) *Biophys. Struct. Mechanism* 9, 235–244.
23. Martin, R. L., Wood, C., Baehr, W. & Applebury, M. L. (1986) *Science (Wash. DC)* 232, 1266–1269.
24. Cooper, A. (1979) *Nature (Lond.)* 282, 531–533.
25. Nathans, J., Thomas, D. & Hogness, D. S. (1986) *Science (Wash. DC)* 232, 193–202.
26. Dixon, R. A. F., Sigal, I. S., Rands, E., Register, R. B., Candelare, M. R., Blake, A. D. & Strader, C. D. (1987) *Nature (Lond.)* 326, 73–77.
27. Honig, B., Dinur, U., Nakanishi, K., Balogh-Nair, V., Gavinowicz, M. A., Arnaboldi, M. & Motto, M. G. (1979) *J. Am. Chem. Soc.* 101, 7084–7086.
28. Katikani, H., Katikani, T., Rodman, H. & Honig, B. (1985) *Photochem. Photobiol.* 41, 471–479.
29. Chabre, M. & Breton, J. (1979) *Photochem. Photobiol.* 30, 295–299.
30. Chabre, M. & Cavaggioni, A. (1975) *Biochim. Biophys. Acta* 382, 336–343.
31. Vuong, T. M., Pfister, C., Worcester, D. L. & Chabre, M. (1987) *Biophys. J.* 52, 587–594.
32. Kühn, H. & Dreyer, W. J. (1972) *FEBS Lett.* 20, 1–6.
33. Bownds, D., Dawes, J., Miller, J. & Stahlman, M. (1972) *Nature (Lond.)* 237, 125–127.
34. Kühn, H. & Hargrave, P. A. (1981) *Biochemistry* 20, 2410–2417.
35. Kaupp, U. B., Schnetkamp, P. P. M. & Junge, W. (1981) *Biochemistry* 20, 5511–5516.
36. Hildebrandt, J. D., Codina, J., Rosenthal, W., Birnbaumer, L., Neer, E. J., Yamazaki, A. & Bitensky, M. W. (1985) *J. Biol. Chem.* 260, 14867–14872.
37. Hurley, J. B., Fong, H. K. W., Teplon, D. B., Dreyer, W. J. & Simon, M. I. (1984) *Proc. Natl Acad. Sci. USA* 81, 6948–6952.
38. Buss, J. E., Mumby, S. M., Casey, P. J., Gilman, A. G. & Sefton, B. M. (1987) *Proc. Natl Acad. Sci. USA* 84, 7493–7497.
39. Bruckert, F., Vuong, T. M. & Chabre, M. (1988) *Eur. J. Biophys.* 16, 207–218.
40. Baehr, W., Morita, E. A., Swanson, R. J. & Applebury, M. L. (1982) *J. Biol. Chem.* 257, 6452–6460.
41. Saibil, H. (1982) *J. Mol. Biol.* 158, 435–456.
42. Yamanaka, G., Eckstein, F. & Stryer, L. (1985) *Biochemistry* 24, 8094–8101.
43. Deterre, P., Bigay, J., Pfister, C. & Chabre, M. (1984) *FEBS Lett.* 178, 228–232.
44. Bennett, N. & Dupont, Y. (1985) *J. Biol. Chem.* 260, 4156–4168.
45. Wilden, U., Hall, S. W. & Kühn, H. (1986) *Proc. Natl Acad. Sci. USA* 83, 1174–1178.
46. Kühn, H., Bennett, N., Michel-Villaz, M. & Chabre, M. (1981) *Proc. Natl Acad. Sci. USA* 18, 6873–6877.
47. Hofmann, K. P., Uhl, R., Hoffmann, W. & Krentz, W. (1976) *Biophys. Struct. Mechanism* 2, 61–77.
48. Vuong, T. M., Chabre, M. & Stryer, L. (1984) *Nature (Lond.)* 311, 659–661.
49. Fung, B. K. K. (1983) *J. Biol. Chem.* 258, 10495–10502.
50. Emeis, D., Kühn, H., Reichert, J. & Hofmann, K. P. (1982) *FEBS Lett.* 143, 29–34.
51. Bennett, N., Michel-Villaz, M. & Kühn, H. (1982) *Eur. J. Biochem.* 127, 97–103.
52. Pfister, C., Kühn, H. & Chabre, M. (1983) *Eur. J. Biochem.* 136, 489–499.
53. Kühn, H. (1981) *Curr. Top. Membr. Transp.* 15, 171–201.
54. Phillips, E. S. & Cone, R. A. (1986) *Biophys. J.* 49, 277a.
55. Fung, B. K. K. & Nash, C. R. (1983) *J. Biol. Chem.* 258, 10503–10510.
56. Sternweis, P. C. & Gilman, A. G. (1982) *Proc. Natl Acad. Sci. USA* 79, 4888–4891.
57. Bigay, J., Deterre, P., Pfister, C. & Chabre, M. (1985) *FEBS Lett.* 191, 181–185.
58. Bigay, J., Deterre, P., Pfister, C. & Chabre, M. (1987) *EMBO J.* 6, 2907–2913.
59. Carlier, M. F., Didry, D., Melki, R., Chabre, M. & Pantaloni, D. (1988) *Biochemistry* 27, 3555–3559.
60. Lunardi, J., Dupuis, A., Garin, J., Issartel, J. P., Michel, L., Chabre, M. & Vignais, P. V. (1988) *Proc. Natl Acad. Sci. USA* 85, 9858–9862.
61. Miki, N., Baraban, J. M., Keirns, J. J., Boyce, J. J. & Bitensky, M. W. (1975) *J. Biol. Chem.* 250, 6320–6327.
62. Hurley, J. B. & Stryer, L. (1982) *J. Biol. Chem.* 257, 11094–11099.
63. Deterre, P., Bigay, J., Robert, M., Kühn, H. & Chabre, M. (1986) *Proteins Struct. Funct. Genet.* 1, 188–193.
64. Deterre, P., Bigay, J., Forquet, F., Robert, M. & Chabre, M. (1988) *Proc. Natl Acad. Sci. USA* 85, 2424–2428.
65. Wensel, T. G. & Stryer, L. (1986) *Proteins Struct. Funct. Genet.* 1, 90–99.
66. Detwiller, P. B., Conner, J. D., Bodoia, R. D. (1982) *Nature (Lond.)* 300, 59–61.
67. Haynes, L. W., Kay, A. R. & Yau, K.-W. (1986) *Nature (Lond.)* 321, 66–70.
68. Zimmerman, A. L. & Baylor, D. A. (1986) *Nature (Lond.)* 321, 70–72.
69. Caretta, A., Cavaggioni, A. & Sorbi, R. T. (1979) *J. Physiol. (Lond.)* 295, 171–178.
70. Cavaggioni, A. & Sorbi, R. T. (1981) *Proc. Natl Acad. Sci. USA* 78, 3964–3968.
71. Koch, K.-W. & Kaupp, U. B. (1985) *J. Biol. Chem.* 260, 6788–6800.
72. Hanke, W., Cook, J. N., Kaupp, U. B. (1988) *Proc. Natl Acad. Sci. USA* 85, 94–98.
73. Matesic, D. & Liebman, P. A. (1987) *Nature (Lond.)* 326, 600–603.
74. Liebman, P. A. & Pugh, E. N. (1980) *Nature (Lond.)* 287, 734–736.
75. Wistow, G., Katial, A., Craft, C. & Shinohara, T. (1986) *FEBS Lett.* 196, 23–28.
76. Yamaki, K., Takahashi, Y., Sakuragi, S. & Matsubara, K. (1987) *Biochem. Biophys. Res. Commun.* 142, 904–910.
77. Shinohara, T., Dietzschold, B., Craft, C. M., Wistow, G., Early, J. J., Donoso, L. A., Horwitz, J. & Tao, R. (1987) *Proc. Natl Acad. Sci. USA* 84, 6975–6979.
78. Sibley, D. R., Benovic, J. L., Caron, M. G. & Lefkowitz, R. J. (1987) *Cell* 48, 913–922.
79. Benovic, J. L., Kühn, H., Weyand, I., Codina, J., Caron, M. G. & Lefkowitz, R. J. (1987) *Proc. Natl Acad. Sci. USA* 84, 8879–8882.
80. Dratz, E. A., Lewis, J. W., Schaechter, L. E., Parker, K. R. & Kliger, D. S. (1987) *Biochem. Biophys. Res. Commun.* 146, 379–386.
81. Wagner, R., Ryba, N. & Uhl, R. (1988) *FEBS Lett.* 234, 44–48.
82. Korn, E. D., Carlier, M. F. & Pantaloni, D. (1987) *Science (Wash. DC)* 238, 638–644.
83. Arshavsky, V. Y., Antoch, M. P. & Philippov, P. P. (1987) *FEBS Lett.* 224, 19–22.
84. Chabre, M., Bigay, J., Bruckert, F., Bornancin, F., Deterre, P., Pfister, C. & Vuong, T. M. (1988) *Cold Spring Harbor Symp. Quant. Biol.* 53.
85. Ovchinnikov, Y. U., Lipkin, V. M., Kumarev, V. P., Gubanov, V. V., Khramtsov, N. V., Akhmedov, N. B., Zagranichny, V. E. & Muradov, K. G. (1986) *FEBS Lett.* 204, 288–292.
86. Ovchinnikov, Y. U., Gubanov, V. V., Khramtsov, N. V., Ischenko, K. A., Zagranichny, V. E., Muradov, K. G., Shuvaeva, T. M. & Lipkin, V. M. (1987) *FEBS Lett.* 223, 169–173.
87. Ovchinnikov, Y. A., Abdulaev, N. G. & Bogachuk, A. S. (1988) *FEBS Lett.* 230, 1–5.
88. Sitaramayya, A., Casadevall, C., Bennett, N. & Hakki, S. I. (1988) *Biochemistry* 27, 4880–4887.

Review

Structure and biological activity of basement membrane proteins

Rupert TIMPL

Max-Planck-Institut für Biochemie, Martinsried

(Received November 7, 1988) — EJB 88 1288

Collagen type IV, laminin, heparan sulfate proteoglycans, nidogen (entactin) and BM-40 (osteonectin, SPARC) represent major structural proteins of basement membranes. They are well-characterized in their domain structures, amino acid sequences and potentials for molecular interactions. Such interactions include self-assembly processes and heterotypic binding between individual constituents, as well as binding of calcium (laminin, BM-40) and are likely to be used for basement membrane assembly. Laminin, collagen IV and nidogen also possess several cell-binding sites which interact with distinct cellular receptors. Some evidence exists that those interactions are involved in the control of cell behaviour. These observations have provided a more defined understanding of basement membrane function and the definition of new research goals in the future.

All multicellular animals possess various extracellular matrices. They determine body shape and stability, compartmentalization of organs and several cellular activities. These matrices include ubiquitously occurring basement membranes which are 20–200-nm-broad deposits of some specific proteins in close proximity to epithelial, muscle, fat and nerve cells. Basement membranes are found in vertebrates and invertebrates except sponges and are produced as the first matrix during embryonic development. They are considered to control cell phenotype, tissue invasion of cells and filtration of macromolecules through glomeruli [1–4]. This implies a defined supramolecular architecture for basement membranes and a distinct repertoire of biological activities expressed as the sum of their individual constituents. Yet, electron microscopy has so far revealed few details of basement membrane structure which, after staining, presents as two amorphous zones (the lamina lucida and the lamina densa) distinguished by different electron densities [5].

Basement membranes are normally but not exclusively produced and deposited by cells which then remain in close contact with these structures. These contacts are mediated by specific cellular receptors which bind to some defined extracellular ligands [1–3]. Other noncovalent molecular interactions maintain the matrix structure and covalent cross-links make basement membranes rather insoluble even in denaturing solvents. The increasing use of tumor models, such as the Engelbreth-Holm-Swarm (EHS) mouse tumor [6], which produce large amounts of soluble basement-membrane material has over the past decade allowed a comprehensive characterization of the major components. The progress in our understanding of the structure, biology and pathology of basement membranes [1–5, 7–10] as well as some methodological aspects [10, 11] have been recently reviewed. The present article will be restricted to a review of the structural properties of the major basement-membrane proteins (Table 1) and some accessory components and discuss their possible structure/function relationships.

A COLLAGEN-TYPE-IV NETWORK AS SCAFFOLD

Monomer structure and sequence

Many previous studies have shown that basement membranes are highly insoluble and possess a distinct stability against mechanical forces. This was correlated with the presence of large amounts of a collagenous protein which, as originally shown by Kefalides [43], differs from the fiber-forming collagens I–III and, thus, was referred to as collagen IV. The unique nature of collagen IV was confirmed by various methods, including cloning and complete sequence analysis of its constituent α1(IV) and α2(IV) chains (Table 1). Studies with large collagen IV fragments [44, 45] indicated a heterotrimer composition [α1(IV)]$_2$[α2(IV)] for most, if not all, individual collagen IV molecules. These chains are arranged as shown by rotary shadowing into a 390-nm-long triple-helical rod which is terminated at its C-terminus by a globular domain, NC1 (Fig. 1). A 30-nm-long segment at the N-terminus of the triple helix is separately referred to as the 7S domain [46] because of its unique role in oligomer formation (see below).

The domain model is in excellent agreement with amino acid sequence data. The sequence of the 7S domain and the remaining triple-helical segment shows the α1(IV) and α2(IV) chain to have many Gly-Xaa-Yaa repeats but quite a few nontriplet deviations. These include Gly to Ala substitutions and interruptions of the repeats by 2–24 amino acid residues. In human collagen IV [15, 19, 28, 29] there are 21 interruptions in the α1(IV) and 23 in the α2(IV) chain and these match each other in location in most cases, giving rise to a total number of 25 triple-helical imperfections rather evenly distributed along the helix (Fig. 1). A comparison of mouse and human

---

Correspondence to R. Timpl, Max-Planck-Institut für Biochemie, D-8033 Martinsried, Federal Republic of Germany

Abbreviations. EHS, Engelbreth-Holm-Swarm; Gla, 4-carboxyglutamic acid; EGF, epidermal growth factor.

Table 1. *Amino acid sequence analyses of basement membrane proteins*
Amino acid numbers are given for processed chains with those for signal peptides indicated in brackets. An asterisk denotes sequences with no or a tentative identification of the signal-peptide-cleavage site

Polypeptide chain	Species	Amino acid number (signal peptide)	Reference
Collagen IV			
α1(IV) chain	mouse	1669*	[12–14]
	human	1642 (27)*	[15–20]
	Drosophila	1752 (23)*	[21]
α2(IV) chain	mouse	partial	[22–25]
	human	1712*	[26–29]
Laminin			
B1 chain	mouse	1765 (21)*	[30]
	human	1765 (21)*	[31]
	Drosophila	1784*	[32]
B2 chain	mouse	1574 (33)*,a	[33, 34]
	human	1576 (33)*	[35]
A chain	mouse	3060 (24)*	[36]
	human	partial	—b
Nidogen (entactin)	mouse	1217 (28)	[37]
	rat	partial	[38]
BM-40 (SPARC, osteonectin)	mouse	285 (17)	[39, 40]
	human	286 (17)	[40]
	bovine	287 (17)	[41]
Heparan sulfate proteoglycan (low-density form)	mouse	partial	[42]

a Two residues are missing in a second analysis [34] but present in the human polypeptide.
b D. R. Olsen et al., unpublished results.

collagen IV (Table 1) shows invariance in their positions but not necessarily in the sequence of the interruptions, suggesting that they are of functional importance. Their presence implies a higher flexibility of the collagen-IV triple helix compared to fiber-forming collagens, as indicated by electron microscopy [47], and provides sites sensitive to proteolytic digestion (i.e. for pepsin; see Fig. 1). Another unique feature of the triple-helical domain is the presence of seven or eight Cys residues, mainly located in interruptions of the α1(IV) and α2(IV) chains and in the N-terminal telopeptide, which are involved in intra- and intermolecular cross-links. Intramolecular bonds are mainly formed in the more central region [19, 29] and are close to a site cleaved by a collagenase which is specific for collagen IV [48]. A particular pair of Cys residues in the α2(IV) chain forms, in addition, an intrachain loop (22 residues) which is not present in the α1(IV) chain. The overall similarity of the triple-helical sequences of the α1(IV) and α2(IV) chain is, rather low except for that of the regular Gly residues, indicating an early divergence during evolution.

A striking sequence similarity was found for the NC1 segments (227–229 residues) which constitute the C-terminal globular domain [13, 17, 23, 25]. The α1(IV) and α2(IV) segments of NC1 both show a distinct internal repeat, with each having six Cys residues in invariant positions. Each set of Cys residues forms three disulfide bonds, thus generating a symmetrical folding pattern into two subdomains within the NC1 segments [49]. There is, as yet, no evidence of interchain disulfide bonding between individual NC1 segments within collagen IV monomers although that would provide an internal stabilization of the globular domain [50, 51]. The data also demonstrated considerable sequence similarity between NC1 segments from man, mouse and *Drosophila* (Table 1).

Oligomer structures

The terminal domains, 7S and NC1, provide the crucial sites in the formation of defined collagen IV oligomers [46]. Dimers can be generated by interactions between globular NC1 domains while tetramers are formed by the lateral association of 7S segments which are aligned alternatively in parallel and antiparallel fashion (Fig. 1). These two forms of oligomeric structures undergo, *in situ*, a rapid stabilization by covalent cross-linking and can be conveniently isolated in the form of defined fragments (7S collagen, NC1) from tissues digested with bacterial collagenase [50, 52]. Reduction of disulfide bonds in the 7S structure produces a series of oligomeric peptide segments indicating the presence of additional, nonreducible cross-links [52, 53]. These are likely to be derived from oxidized lysine and/or hydroxylysine residues, since inhibition of lysyl oxidase by lathyrogens increases the solubility of collagen IV [54, 55]. The NC1 structure was shown to consist only of monomeric and dimeric chain segments in noncovalent association. The dimers apparently arise by linkage through disulfide bonds of monomeric segments originating from two collagen IV molecules. Dimers, but not monomer segments, retain the potential to reform intact globular structures suggesting that formation of collagen IV dimers is accompanied by distinct conformational changes within the NC1 domain [50]. A further function of the NC1 domain may include the selection and alignment of newly synthesized α1(IV) and α2(IV) chains since some evidence indicates that folding of the triple-helical domain starts from its C-terminus [56].

Sequence analysis of the 7S segment [16, 26] provided clear predictions on molecular details of the assembly process. The N-terminus of both collagen IV chains are short nonhelical sequences possessing several potential cross-linking sites (Cys, Lys residues). If aligned to a second chain in an antiparallel direction they oppose another potential cross-linking structure located about 80 amino acid residues away within the triple-helical region (Fig. 1). This arrangement of antiparallel molecules allows, in addition, maximal hydrophobic interactions between opposing strands and is also in good agreement with electron microscope data. A three-dimensional evaluation of the amino acid sequence also indicated that the azimuthal orientation of two 7S segments is determined by a hydrophobic reaction edge of about 90° width. These interaction predictions are compatible with the orientation of two pairs of segments in parallel/antiparallel orientation and explains why the assembly process is limited to the tetramer stage [26].

The localization of disulfide bonds in the dimeric NC1 segments [49] revealed the surprising observation that they were identical to those intrachain bonds present in the monomer. It strongly indicates that dimer stabilization occurs by reshuffling of one or two disulfide bonds within one folding subdomain to exactly the same acceptor residues which are, however, present in the opposing monomer segment. It was also found that monomeric NC1 segments contain free sulfhydryl groups which increase in number upon dimerization [57]. Presumably, they represent the intrinsic catalysts for reformation of disulfide bonds between molecules. Completely reduced monomers can also be reoxidized to form globular NC1 domains indistinguishable from genuine NC1 [57]. All data together suggest that the formation of collagen IV dimers

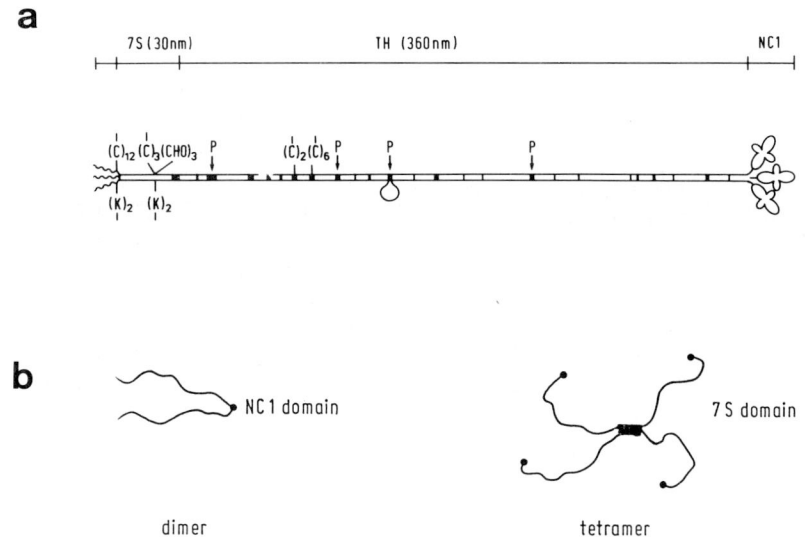

Fig. 1. *Schematic model of collagen-IV monomer (a) and its modes of assembly into oligomers (b)*. The monomer is composed of two α1(IV) and one α2(IV) chains which are aligned in parallel through the major triple helix (TH), the N-terminal 7S domain and the C-terminal globular domain NC1. Black bars along the triple helix indicate the positions of Gly-Xaa-Yaa imperfections. Potential cross-linking sites include cysteines (C) and lysine/hydroxylysine (K) residues. CHO: N-linked carbohydrate, P: pepsin-sensitive peptide bonds. The disulfide-bonded loop in the center is unique for the α2(IV) chain. Based on [29, 46]; (a) is reproduced with permission from [29]

and tetramers depends initially on noncovalent contacts followed by a concerted disulfide-exchange reaction which stabilizes the final products. The order of reactions of NC1 or 7S segments in the self-assembly of collagen IV and whether these interactions are the first molecular recognition events is unknown. Studies with collagen IV monomers from cell cultures showed a concentration-dependent, reversible assembly to tetramers connected by 7S domains which, upon extended incubation, became disulfide-linked [58]. No indications were found for covalent dimerization of NC1 under the same conditions or in tissue culture [51, 58]. However, preformed collagen-IV dimers obtained from tissue extracts [55], still possess the potential for lateral aggregation and the formation of polygonal networks, compatible with the interpretation that dimers could be intermediates in the assembly process.

Network models

Collagen IV is essentially insoluble *in situ*, implying that all intermediates in assembly are eventually converted into huge supramolecular aggregates. One of the earlier structural proposals was based on electron microscopy of oligomers and suggested a rather open but unlimited network which can be formed just by the alternating use of the 7S and NC1 cross-linking domains [46]. This model did not include lateral interactions along the major portion of the triple helix. That such lateral interactions indeed exist was subsequently shown in thermal reconstitution experiments with collagen IV dimers which form irregular, polygonal networks [55]. Individual strands within these networks were composed of two to four triple-helical segments and were intersected at a distance of 100 – 200 nm by the globular domain. Other studies with the isolated NC1 globule showed its binding to the triple-helical domain at regular distances, a process which also disturbed network formation by the collagen molecules [59]. This indicates another recognition site in NC1 probably not identical to those used in dimer formation. Evidence for the existence of such irregular networks *in situ* was obtained recently for

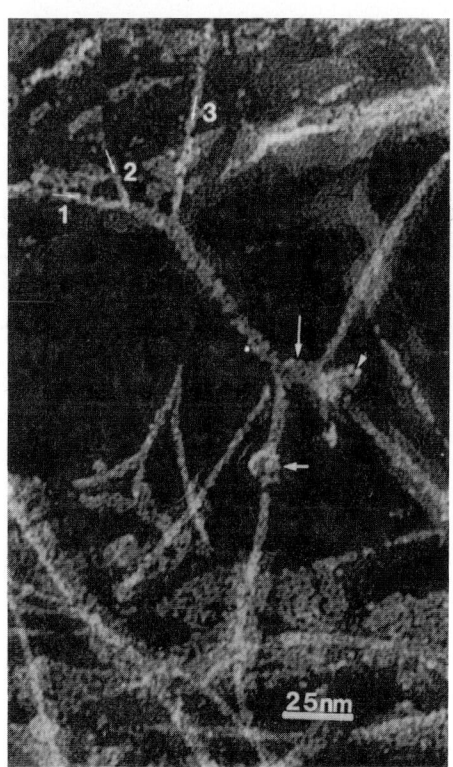

Fig. 2. *Section of the polygonal collagen-IV network in amniotic basement membranes*. Noncollagenous glycoproteins were removed by extraction and the remaining structure was visualized in the electron microscope after unidirectional metal shadow casting. Three narrow filaments (labeled 1 – 3) represent single triple helices and show extensive branching and supercoil formation. The horizontal arrow marks a globular domain, NC1. Reproduced with permission from [60]

partially extracted basement membrane after unidirectional metal shadow casting [60]. Electron microscopy (Fig. 2) revealed a complex, three-dimensional meshwork of highly branched filaments and the typical appearance of NC1 globules at an average distance of 220 nm. In addition it showed that laterally associated collagen-IV molecules form a supertwisted helix. This extensive twisting and branching is certainly dependent on the rather flexible triple-helical domains for which the many helical imperfections (Fig. 1) may be responsible. Supertwisting would also imply that lateral associates are formed first during assembly and are then locked in place by the interactions occurring at the ends of the molecule.

The formation of large supportive structures is certainly one of the major functions of collagen IV. However, collagen IV also possesses binding activities for laminin, proteoglycan, nidogen and cells which will be discussed in the following sections. Such interactions are obviously required for the completion of the supramolecular architecture of basement membranes.

Collagen IV may be more complex than has been assumed. Studies of NC1 monomers demonstrated minor amounts of two new components (M2 and M3) which appear to correspond to analogous segments of two other polypeptide chains, $\alpha3(IV)$ and $\alpha4(IV)$ [61, 62]. In addition, M2 was found to possess the Goodpasture epitope [63, 64], a cryptic epitope recognized by some human autoantibodies [64]. The structural consequences of these observations remain unclear at present.

Genes and evolution

Collagen IV differs from other collagens not only at the structural but also the genomic level. Exon analyses, still incomplete [65–69], demonstrated a variable size for those exons encoding triple-helical sequences while a uniform size of 54 bp or multiples of it are found for exons encoding fiber-forming collagens. The genes for human $\alpha1(IV)$ and $\alpha2(IV)$ chain were found to be syntenic on chromosome 13 [70]. They are located on opposite strands and share a 120-bp promoter region [71]. This implicates interesting regulatory mechanisms in the transcription of both genes. The promoter itself is, however, not sufficiently activating and is dependent on enhancer (and possibly silencer) elements present in the first large intron (> 36 kb) of the $\alpha1(IV)$-chain gene [72].

The recent comprehensive analysis of the collagen IV analogue of *Drosophila* basement membranes [21, 73] has provided interesting insights into the evolution of collagen IV. The *Drosophila* protein contains only $\alpha1(IV)$ chains and this chain is somewhat larger than the mammalian $\alpha1(IV)$ chain. However, it shows significant similarity in the 7S and NC1 domains but less correspondence in the positions of the imperfections and the Cys residues along the triple-helical domain. The latter are arranged in three pseudorepeats of about 489 amino acid residues which could force the molecules in several lateral microfibrillar arrangements stabilized by disulfide bonds. It was also suggested that the assembly of collagen IV from *Drosophila* corresponds to a primordial supramolecular structure of collagens which later evolved into networks and fibrils. The gene of *Drosophila* collagen IV is also quite different to the mammalian genes and consists of nine large exons separated by relatively small introns [21].

THE CELL-BINDING PROTEIN, LAMININ

Laminin was originally identified as a major noncollagenous protein present in neutral buffer extracts of the mouse EHS tumor [74] and soon after shown to have a unique cross-shaped structure [75] and to possess distinct cell-binding properties [76]. It is an ubiquitous basement membrane component [2], appears as the first extracellular matrix protein to be produced during embryogenesis (two-cell stage; [77, 78]) and has been detected in several invertebrate tissues [79–81]. This, together with the variety of other biological activities ascribed to laminin [1–4], indicates a central structural and functional role within basement membranes.

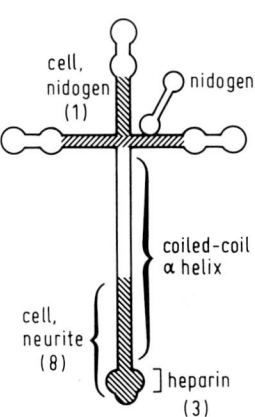

Fig. 3. *Schema of the laminin-nidogen complex as revealed by electron microscopy [90] and localization of some binding activities.* Shadowed areas represent the proteolytic fragments 1 and 8 involved in cell-binding. The heparin-binding domain 3 represents a portion of the globular domain present at the distal end of the long arm

Domain structure and sequence

As determined by ultracentrifugation [75], mouse laminin has a molecular mass of 900 kDa which is in good agreement with sequence data (Table 1) indicating that the protein consists of three polypeptide chains which are linked to each other by disulfide bonds [74, 75, 82]. The chains were previously identified [82, 83] as B1 and B2 (each about 220 kDa) and A chains (400 kDa). The assembly product of these chains has a cruciform structure which, after visualization by rotary shadowing or negative staining [75], appears as three similar short arms and a distinct long arm (Fig. 3). The 37-nm short arms consist of two globular domains connected by rod-like elements while the long arm appears as a 77-nm-long rod terminating at its distal end into a complex globular domain. A similar structure has been reported for laminin from human placenta [84] although this protein contains an additional component, the M chain. Laminin molecules isolated from *Drosophila* [81], sea urchin [79] and leech [80] have a similar shape except for a slight extension (15–30 nm) in the length of the rod-like domain of the long arm. The multidomain structure of laminin is also reflected by diverse conformations including 20–30% α helix, some β structure and aperiodic elements [85, 86].

The domain model of laminin was in part confirmed in studies with large proteolytic fragments [75, 85–93] which were useful for mapping functional activities of laminin and for the assignment of the constituent chains within the cross (Figs 3, 4). Based on these data and complete sequence analysis (Table 1) it became clear that the three short arms are formed from individual N-terminal segments of the B1, B2 and A chain, respectively. The three chains are then aligned in parallel through the rod of the long arm where the B1 and B2 chains terminate with an internal disulfide bridge. The globule of the long arm is exclusively formed from the C-terminal segment of the A chain (Fig. 4).

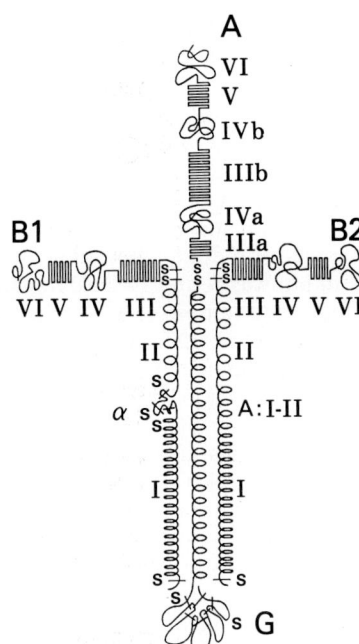

Fig. 4. *Arrangement of the B1, B2 and A chains and of individual domains in the cross-shaped laminin structure.* The N-termini of each chain are at the distal ends of individual short arms. The C-termini of the B1 and B2 chains are at the end of domain I and of the A chain at the end of domain G. Domains I and II are α-helical and are interrupted in B1 chain by a short Cys-rich segment (α, S). Domains III and V are composed of several consecutive EGF-like repeats. Globular structures are denoted as IV, VI and G. S indicates some of the disulfide bonds. Reproduced with permission from [36]

The crucial finding in the sequence work was the high sequence similarity in domains predicted for the N-terminal 1100–1300 residues of all three constituent chains of mouse laminin [30, 33, 36, 91]. These domains included globular structures VI and IV (200–250 residues) at and close to the N-terminus which possess no, or only a few, Cys residues (Fig. 4). These correspond in size and location to the globular domains in the short arms of laminin (Fig. 3). The rod-like segments correspond to domains V and III and are each composed of several consecutive Cys-rich repeats of about 50 residues. They show similarity to epidermal growth factor (EGF) and many EGF-like repeats in other proteins [94] but contain eight instead of six Cys residues/repeat. A further predicted globular domain IVa which interrupts domain III in the A chain (Fig. 4) has so far not been confirmed by electron microscope or fragmentation studies. Sequences of C-terminal domains II and I in B1 and B2 chain and in a central position of A chain (each about 600 residues) predict α-helical conformation, and, because of heptad repeats in domains I of B1 and B2, a coiled-coil structure. The C-terminal segment of the A chain (1000 residues) forms the large globule G with five internal repeats with about 25% identity. A 90% identical sequence with similar domain organizations was predicted for the B1 and B2 chain of human laminin [31, 35]. A similar degree of similarity was also found for *Drosophila* B1 chain [32] with greater variability between different domains (25–50% sequence identity).

Among the various proteolytic fragments of laminin characterized so far [85–93], fragments 1 and 8 (Fig. 3) of about 200–250 kDa, were of particular importance in structural and biological studies. Fragment 8 from the long arm has been isolated from limited elastase digests and was shown to consist of domain I from all three chains and large portions of domain G from the A chain [86, 90, 92]. The disulfide-linked B1–B2-chain segment showed characteristic properties of a coiled-coil α helix with a reversible thermal transition around 58 °C [86] and thus confirmed predictions from cDNA sequences [95]. It is, however, not yet clear how the corresponding α-helical A chain segment participates in this structure. The globular domain could be separated into subfragments 3, T1 and T2 [92] in agreement with three globules visualized at the end of laminin's long arm by electron microscopy [86].

Fragment 1 is usually obtained from extensive pepsin digests of laminin and consists of three disulfide-linked rod-like segments from the proximal portions (domain III) of the short arms [74, 75, 85]. Several larger variants of this structure, including those with full-length short arms have been described after digestions with neutral proteases [85, 87, 89, 91, 93]. Similar fragments were also interpreted to consist of two short-arm segments and one long-arm (domain II) segment [96] based on antibody localizations. This interpretation would implicate a distinct α-helical content which has so far not been found for such fragments [85]. It is also not compatible with the high protease sensitivity of the long arm of laminin [85, 87] or with sequence analysis of fragment 1 [97] (R. Deutzmann, personal communication).

Mouse laminin contains about 13% carbohydrate, very probably all of it being in N-linked conjugates. Structural analyses demonstrated about 20 different bi-, tri- and tetra-antenary complex-type components, with large variations at the nonreducing ends due to extensions by lactosamine, α-galactose and sialic acid [97, 98]. They occupy most of the NXT/NXS-acceptor sites (in the one-letter amino acid code) in laminin with 13 in the B1 chain, 14 in the B2 chain and 43 in the A chain [30, 33, 36]. A particularly high abundance of acceptor sites is found in domains I and II of the A chain. Their glycosylation may be important for stabilizing the long arm against proteolytic attack. Terminal α-galactose in laminin was also found to react with normal human antibodies and some antibodies arising in parasitic infections, such as Chaga's disease and leishmaniasis [99].

Biosynthesis and regulation

Laminin is produced by a large variety of cultured cells [2] and mRNA from these cells was used to clone and sequence the three polypeptide chains (Table 1). mRNAs of about 10 kb, 6 kb and 8 kb encode the A, B1 and B2 chains, respectively [100–102]. The corresponding genes have been localized to human chromosomes 7 (B1; [31]), 1 (B2; [103]) and 18 (A; M. L. Chu, personal communication). Studies with genomic clones demonstrated a large size (> 60 kb) for the mouse B1 gene and the presence of more than 35 exons of variable size [104]. Production of all three laminin chains is coordinately regulated in F9 teratocarcinoma cells after stimulation with retinoic acid [100, 101] indicating control at the transcriptional level. This tight regulation seems not to exist in a variety of other cells [100, 102] which show a low abundance of A-chain mRNA, in particular. This could indicate production of laminin isoforms (see below) or that A-chain translation is the rate-limiting step in laminin biosynthesis. Promoter regions such as those for the mouse B2-chain gene have been recently characterized [105] and will allow a more precise analysis of these regulatory mechanisms.

Studies with cultured cells demonstrated disulfide-linked B1-B2 chain dimers as intermediates in laminin biosynthesis [106, 107]. The addition of the A chain is apparently the crucial event controlling secretion of laminin. Immunoprecipitation of laminin from culture media demonstrated the frequent association of laminin with a 150-kDa polypeptide, referred to variously as C chain [82], entactin [108] or nidogen [78]. The production of laminin and nidogen are not tightly coordinated in a variety of cultured cells [78]. Structural properties of these complexes will be discussed later.

Laminin self-assembly and binding to other matrix components

Like collagen IV, laminin undergoes a thermal polymerization reaction which is, however, dependent on a critical minimum protein concentration (0.1 mg/ml) and on the presence of calcium [109]. Chelating agents interfere with this process in a reversible manner by terminating the reaction at the level of small oligomers. The same heat gelation was also observed with the laminin-nidogen complex [89, 90]. Electron microscopy of laminin oligomers demonstrated associations between globular domains either from the end of the long arm or from outer segments of the short arms [109]. This was supported by showing that antibodies to fragment 3 (see Fig. 3) inhibit the reaction [110]. Whether laminin polymerization plays a significant role *in situ* is unknown, although the observation [90] showing that chelating agents greatly enhance laminin solubility would be consistent with this.

The binding of laminin to collagen IV was indicated in cell-binding studies [76, 87] and shown in other studies to be dependent on an intact triple helix in collagen IV [111]. Complexes formed between both components in solution were visualized by electron microscopy [112, 113] and indicated the location of two or three binding sites along the triple helix, being about 80 nm, 140 nm and 220 nm away from the C-terminal NC1 domain (see Fig. 1). A similar diversity of binding sites was also indicated on laminin, including structures within its short arms [113] and the globular domain (fragment 3) at the end of the long arm [110, 112]. Similar studies with the laminin-nidogen complex (M. Aumailley et al., unpublished results) demonstrated mainly specific associations between short-arm structures (including nidogen) and 80-nm and 180-nm sites (see above) on the collagen-IV triple helix. This suggests that nidogen mediates the binding of laminin to collagen IV. So far, none of the isolated laminin domains which are implicated in collagen-IV binding have been found to interact with collagen IV. The molecular nature of the complexes formed between collagen IV and laminin, therefore, still remains unclear.

A further binding potential of laminin includes the recognition of heparin and heparan sulfate chains [114]. Binding studies indicated at least two different binding sites on laminin with dissociation constants of the order of 100 nM [115]. A major binding site was mapped to fragment 3 at the end of the long arm (Fig. 3) with some weaker binding activities also observed for short-arm structures [85]. Fragment 3 corresponds to the C-terminal 400-residue sequence of the A chain [92] which contains several clusters of basic amino acid residues, suggesting that they may be involved in heparin binding. Two more heparin-binding sites were mapped close to the center of the cross and to a more distal short-arm structure of laminin in inhibition studies with monoclonal antibodies [115]. A synthetic 20-residue peptide designed according to a cationic, hydrophilic sequence of domain IV of the laminin B1 chain was found to completely block the binding of laminin to heparin [116]. This result is at present difficult to reconcile with other data since this sequence is not present in fragment 3 [92] and domain IV of the B1 chain, when used as an elastase fragment, does not bind to heparin [93].

Isoforms of laminin

The possible existence of structural variants of laminin was originally indicated in biosynthetic studies with muscle [117], Schwann cells [118] and early embryos [77], which demonstrated the absence or reduced levels of a 400-kDa A chain. Schwann-cell laminin was found to lack fragment-3 but not fragment-1 epitopes and to possess a Y-shaped rather than a cross-shaped structure [119, 120]. Such laminins still possess considerable ability to stimulate neurite outgrowth, although this may require an associated proteoglycan. During early development, kidney mesenchyme expresses laminin B1 and B2 chains but no A chains. The production of A chain is induced at the onset of tubule formation and accompanied by the appearance of fragment-3 epitopes [121]. These epitopes partially disappear upon further maturation. Since low expression of a 10-kb A-chain mRNA has been observed in various cultured cells and tissues (see above), isoforms which lack a 400-kDa chain may have a broad occurrence. Whether they consist of B1 – B2-chain dimers or are associated with a variant A chain is unknown. The size of Schwann-cell laminin and the presence of a small globular domain at the end of its long arm [120] would favour the existence of a small A chain (≤ 200 kDa) although this chain has not been demonstrated directly.

NIDOGEN/ENTACTIN FORMS A STABLE COMPLEX WITH LAMININ

Nidogen was first identified as an 80-kDa fragment obtained from the EHS tumor [122] which was shown to originate from a dumbbell-shaped 148-kDa protein [123]. Sequence analysis of nidogen and its fragments identified it as a single, unique polypeptide chain [93, 123]. Entactin was characterized as a 158-kDa polypeptide band in biosynthetic studies, shown to contain sulfate [124] and, similar to nidogen, to bind to laminin [108]. A comparison of both components [125] demonstrated co-chromatography in a variety of systems and the presence of tyrosine O-sulfate which accounted for the entire sulfate label. Both proteins have been cloned and found to be virtually identical (Table 1). Entactin has so far been purified only by electrophoresis and not, as yet, analyzed by peptide sequencing [124]. Several indications exist from studies with EHS tumor [125] (and unpublished results) and with culture medium of *Drosophila* [81] that this electrophoretic region may not only contain nidogen/entactin but some other proteins. The nature of these extra proteins and their potential binding to laminin remains to be established.

Domain structure and sequence

Nidogen consists of 1217 amino acid residues (Table 1) and about 5% carbohydrate (both in *N*- and *O*-linked form), which is in good agreement with its molecular mass (148-kDa) determined by ultracentrifugation [123]. Electron microscopy of nidogen after rotary shadowing or negative staining demonstrated a molecule containing two globular domains of unequal size and a connecting, 17-nm-long flexible rod [90, 123]. Fragmentation studies located the large and the small globule to the N-terminus and the C-terminus, respectively, with the rod occupying a central position [93]. This domain model was confirmed and refined by sequence analysis

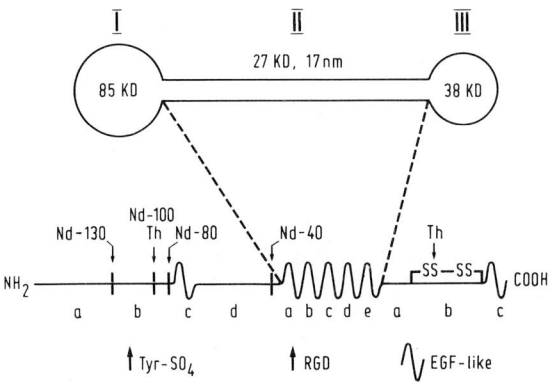

Fig. 5. *Domain model of nidogen (top) and correlation with its amino acid sequence schematically outlined at the bottom*. Subdomains of the morphologically obvious domains (I−III) are denoted by small letters and include seven EGF-like repeats. Positions of cleavage by endogenous proteases (Nd) and thrombin (Th), of tyrosine sulfation (Tyr-SO$_4$) and of a cell-binding sequence (RGD) are indicated. SS denotes extra disulfide bonds not present in the EGF-like repeats. Reproduced with permission from [37]

(Fig. 5). Based on these data the rod domain is composed of five consecutive EGF-like repeats with six Cys residues/repeat. Two more EGF-like repeats and several other structures constitute the globular domains, which can be divided into several subdomains. The rod-like domain contains an Arg-Gly-Asp sequence and several permutations of this structure and two Asn residues within consensus sequences for β-hydroxylation, suggesting binding sites for cells and calcium. Consensus sequences for Tyr sulfation were exclusively located in the N-terminal globular domain [37].

The conformation of nidogen is mainly aperiodic and lacks distinct α-helical and β structures [90, 123]. The sequence predicts many β turns in the EGF-like repeats, as is the case for similar regions in laminin. The globular domains apparently contain exposed peptide segments which are readily cleaved by endogenous and exogenous proteases yielding characteristic fragments of about 130 kDa, 100 kDa and 80 kDa [123, 126]. Most of the cleavage sites have been localized to the N-terminal globular domain (Fig. 5). The high protease sensitivity of nidogen when compared to laminin was considered to be essential for basement membrane remodelling [126, 127].

Laminin-nidogen complex and other interactions

Evidence for the existence of such complexes came from immunoprecipitation studies (see above) and tissue analyses [78] indicating an equimolar ratio of nidogen and laminin and high-affinity binding between the components ($K_d = 1-10$ nM). Yet, it proved initially difficult to isolate the complex in intact form due to proteolytic degradation of nidogen [126]. In fact, such degradation very probably occurred during the initial purification of laminin from the EHS tumor [74] and other sources, yielding materials essentially free of nidogen. Rapid extraction with physiological buffers containing a chelating agent (1−10 mM EDTA) has recently overcome these difficulties and allowed the purification of the laminin-nidogen complex from EHS mouse tumor [90] and leech ganglions [80] and indicates that the anchorage of the complex in the matrix is dependent on calcium or other divalent cations.

Studies with mouse laminin-nidogen complex demonstrated an equimolar ratio of both components and the need for denaturing agents (i.e. 2 M guanidine · HCl) for their separation. Electron microscopy also revealed the binding of nidogen by one of its globular domains to a proximal rod-like segment of one of the short arms of laminin [90]. These morphological observations were confirmed in radioligand-binding studies [93], demonstrating binding sites on fragment 1 of laminin and on the C-terminal globular domain of nidogen, respectively. Further preliminary studies also indicated that the B1-chain segment (domain III) of fragment 1 is most active in binding, with weaker activities found for the B2- and A-chain segments [128]. Binding is also accompanied by increased protease resistance of nidogen [127].

Other binding assays demonstrated a specific interaction of nidogen with collagen IV while laminin, when separated from the complex, was inactive [128] (M. Aumailley et al., unpublished results). The binding of nidogen to collagen IV was weaker than to laminin and could also be mapped to the C-terminal globular domain of nidogen. Nidogen also bound, apparently, to the same two triple-helical sites in collagen IV as the whole laminin-nidogen complex. Thus, the data indicate that nidogen mediates the binding of laminin to collagen IV and could be a major function of the protein. So far, no binding of nidogen to heparin or proteoglycans has been shown. Some data exist that nidogen can aggregate into nest-like structures [122] by the interaction of the N-terminal globular domains.

HEPARAN SULFATE PROTEOGLYCANS PROVIDE POLYANIONIC SITES IN BASEMENT MEMBRANES

Polyanionic sites were discovered in basement membranes stained with cationic dyes. They were destroyed by heparitinase or nitrous acid, indicating that they contained heparan sulfate chains. These polyanionic sites were also implicated in the ionic control of filtration through basement membranes [129]. These heparan sulfate chains are bound to protein cores [130]. The characterization of heparan sulfate proteoglycans from several basement membranes [130−138] has initially provided conflicting results on their size and composition. Studies with the EHS tumor indicated the existence of at least two forms, which differ in size and heparan sulfate content and thus in their buoyant density [131, 133, 137]. Biosynthetic studies demonstrated, for the low-density form, a uniform protein-core precursor of about 400 kDa [139]. Immunological studies with antibodies against this protein core indicated a widespread occurrence in cultured cells and tissues, but limited to basement membranes [132, 138]. Thus, a large proteoglycan of low buoyant density seems to be a genuine product present in basement membranes while the nature and origin of smaller forms is still controversial.

Low-density heparan sulfate proteoglycan

The proteoglycan is apparently tightly bound to the EHS tumor matrix and requires the use of denaturing agents (7 M urea, 6 M guanidine · HCl) for its extraction and further purification [133, 140]. This proteoglycan was comprehensively characterized by ultracentrifugation, electron microscopy using three imaging techniques and fragmentation [140]. The molecular mass of the proteoglycan was variable (620−720 kDa) presumably due to posttranslational modifications but more uniform (about 500 kDa) for the protein core. A model based on these data proposes a 80-nm-long, multidomain protein core which terminates at one end into three heparan sulfate chains (Fig. 6). The images of the protein

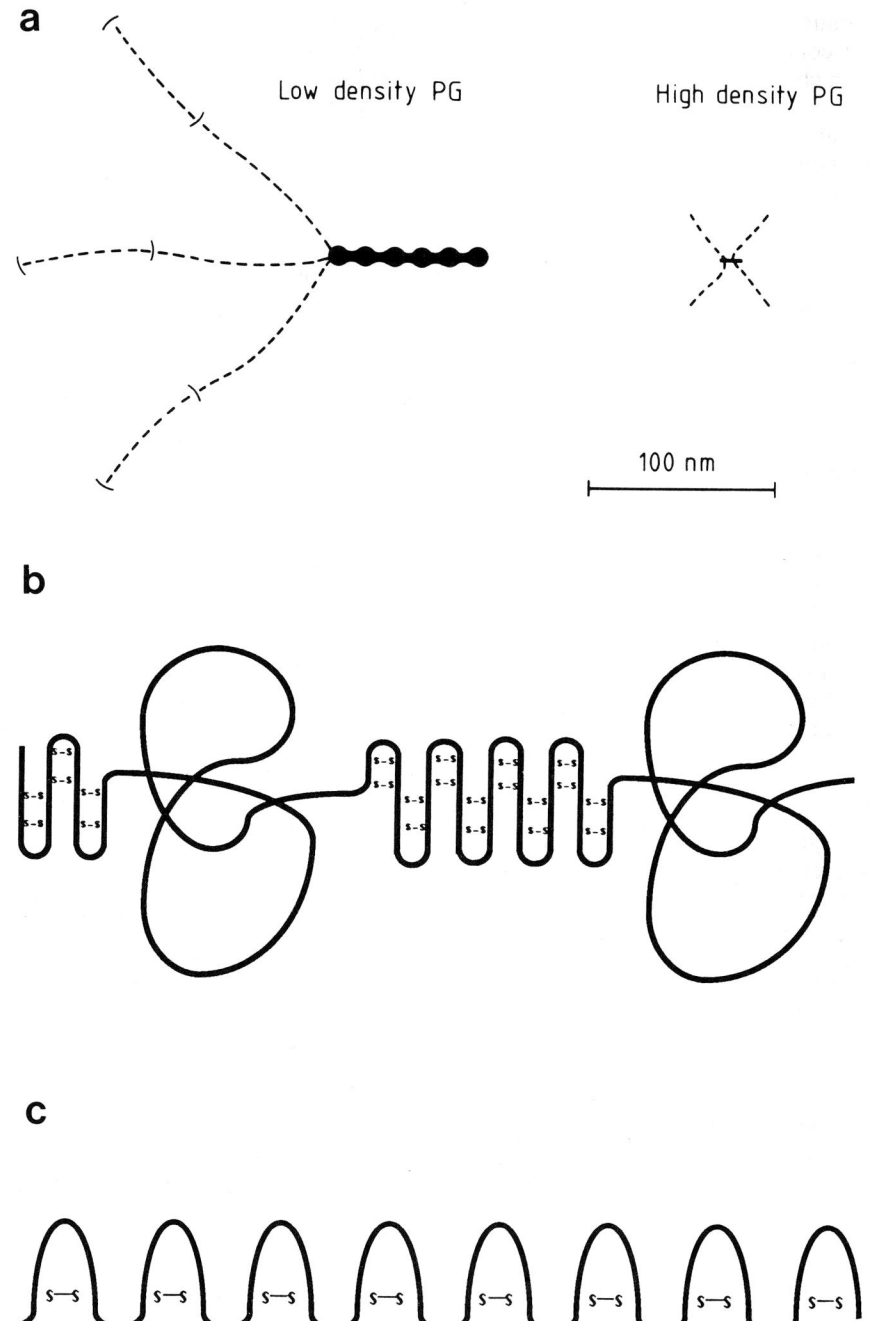

Fig. 6. *Shapes of basement membrane heparan sulfate proteoglycans (a) of either low [140] or high [131] buoyant density.* Protein cores are outlined by thick, black lines and heparan sulfate chains by dashed lines. Sequence-based predictions [42] of protein domains present in the low-density proteoglycan (PG) include EGF-like repeats alternating with globular domains (b) and repeats of disulfide-bonded loops similar to N-CAM (c). Reproduced with permission from [42, 140]

core show some variability, but are best summarized by a row of six globular domains occasionally connected by neck-like segments. The terminal clustering of heparan sulfate chains was confirmed by isolating a tryptic fragment containing all the heparan sulfate chains, but only 5% of the protein. The multidomain nature was also shown by protease cleavage, which produced a 200-kDa fragment devoid of heparan sulfate chains and some smaller fragments [141]. The data also indicated that the protein core is a single polypeptide chain and thus one of the largest gene products identified. Conformational analysis demonstrated mainly aperiodic elements with some β structure and a collapse to random-coil structures in 6 M guanidine · HCl [140]. It remains questionable, therefore, whether the isolated proteoglycan still possesses its native conformation.

Considerable progress in the understanding of the proteoglycan structure can be expected from cDNA clones which hybridize to a 12-kb mRNA [42]. Sequence analysis of two partial clones comprising about 40% of the total mass revealed two different structures (Fig. 6). One was composed of globular domains connected by cysteine-rich repeats similar to those found in laminin. The other was an eightfold repeat of structures similar to the cell-adhesion protein, N-CAM, comprising of peptide loops with two Cys residues, which are typical for the immunoglobulin-gene superfamily. This proteoglycan structure contained also a fourfold repeat of a DSGEY sequence which could represent heparan sulfate acceptor sites [42].

Heparan sulfate proteoglycans with a large protein core were also isolated from glomerular basement membranes [136, 142]. Biosynthetic and immunological studies demonstrated that they comprise various fragments of the molecule identified in the EHS tumor [142]. Whether these fragments are specific processing products or artifacts of isolation remains uncertain. Similar fragments have been described for the EHS tumor [132, 133] indicating that the structure is sensitive to endogenous proteolysis.

Other proteoglycans

A small, high-density heparan sulfate proteoglycan (130 kDa) was isolated from NaCl extracts of the EHS tumor and characterized [131]. The data indicated the connection of four heparan sulfate chains to a small protein core (Fig. 6). A similar proteoglycan was also described for glomerular basement membranes [134, 135]. It was argued that this structure arises as a degradation product of the major low-density proteoglycan [139], compatible with the demonstration of shared epitopes [132, 139]. Yet, the assembly pattern of heparan sulfate chains on the protein core (Fig. 6), as well as their structural properties [131], are not in favor of this interpretation. Another study [137] demonstrated a distinct heterogeneity of high-density forms, both in protein-core size (21–34 kDa) and in the presence of chondroitin sulfate. The peptide maps of these protein cores are distinct from that of the low-density proteoglycan. It therefore remains an open question whether the high- and low-density forms represent different gene products.

The possible existence of proteoglycans with chondroitin sulfate/dermatan sulfate chains in basement membranes was indicated from chemical analyses [2] and from immunological localizations [143]. Biosynthetic studies showed that they are of small size [125]. Whether their protein cores represent unique gene products or are related to those of heparan sulfate proteoglycans [137] remains to be determined.

Interactions promoted by the protein core and the heparan sulfate chains

The protein core probably plays a crucial role in the integration of low-density heparan sulfate proteoglycan into basement membranes, since the proteoglycan cannot be dissociated by high-salt concentrations which would interfere with ionic interactions. The strong binding may include self-assembly of the molecule to oligomers, as observed with purified material heated to 37°C [144]. Association to dimers and to more complex stellate clusters occurs by binding between terminal domains of the core which are most distal to the heparan-sulfate-attachment region (Fig. 6). Proteolytic removal of these domains abolishes self-assembly. Binding of the proteoglycan to the end of the long arm of laminin and to two regions in the collagen-IV triple helix (210 nm and 80 nm away from the NC1 domain) has been detected by electron microscopy [113] but does not show whether binding occurs through the protein core or the heparan sulfate chains.

Heparan sulfate chains exhibit a versatile repertoire of interactions which are weak and sensitive to moderate salt concentrations. Such weak binding ($K_d = 1$ μM) to laminin and the NC1 domain of collagen IV was in fact observed for heparan sulfate chains obtained from the high-density proteoglycan of EHS tumor [131]. The binding was also not very specific since dextran sulfate and, to some extent, chondroitin sulfate displaced it from its ligands. Binding properties can vary depending on the sulfate content as shown for heparan sulfate obtained from EHS tumor and Reichert's membrane, an extra-embryonic basement membrane found in rodents [145]. Only the latter material showed a strong binding to anti-thrombin which was comparable in its salt sensitivity to the binding of heparin. Strong binding was correlated with a high content of 3-O-sulfated glucosamine known to be essential for this interaction. This also suggested that some heparan sulfate proteoglycans may be involved in the control of serine protease activity which could be crucial for the maintenance and remodelling of basement membranes [145]. Additional functions may include the storage of angiogenic stimuli, such as basic fibroblast growth factor [146], which may be released during physiological or pathological neovascularization.

CALCIUM-BINDING STRUCTURES OF BASEMENT MEMBRANES

Recent evidence showed that some basement membrane proteins (laminin, BM-40) can bind several calcium ions with dissociation constants in the order of 1–100 μM. As discussed elsewhere [89], this concentration range is distinctly above the normal cytoplasmic level but below the extracellular calcium concentration suggesting that these proteins change from a calcium-free to a fully saturated form upon secretion. Yet, the actual concentration of free calcium in basement membranes is unknown and may be due to the binding of polyanions being lower than that in the extracellular fluids (≈ 1 mM). Thus, it may be close to the K_d values and subtle changes in calcium concentrations in the matrix could, as known for many intracellular calcium-binding proteins, be a major regulatory principle. It was also argued that such proteins are required to prevent calcification of basement membranes [147].

The calcium-dependent polymerization of laminin [109] and the solubilization of laminin-nidogen complex by chelating agents [90] were the first indications that laminin has an affinity for divalent cations. This was then shown in direct binding studies [89], indicating 1 or 2 high-affinity calcium-binding sites ($K_d \approx 7$ μM) and a larger number of low-affinity sites ($K_d \approx 0.3$ mM). Only the high-affinity sites are involved in laminin polymerization. The calcium-binding sites of laminin have not been localized but some with K_d values in the range 50–150 μM are present on fragments 1 and 8. Whether binding of calcium changes conformation is unknown. It stabilizes, however, a structure in the short arms of laminin against proteolytic degradation [148].

BM-40/osteonectin/SPARC

A novel 40-kDa basement membrane protein, BM-40, was identified in 6-M-guanidine·HCl extracts of the EHS tumor, found to be present in a variety of tissues and cells but not accessible in quite a few tissues when probed by antibody staining [149]. It could be also extracted together with the laminin-nidogen complex by chelating agents and was shown to have a unique sequence [40, 150]. This sequence was identical to that of the endoderm protein SPARC [39] and, more surprisingly, to that of bone osteonectin [41]. The latter was

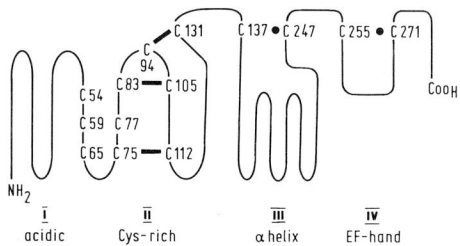

Fig. 7. *Domain model of BM-40/osteonectin/SPARC as deduced from the amino acid sequence [40, 41, 147].* The domains include a Glu-rich region (I) and an EF hand (IV) suggested to be involved in the binding of calcium. A Cys-rich domain (III) shows similarity in the disulfide pattern (black bars) to ovomucoid and other protease inhibitors. Domain III with predicted α helix is stabilized like the EF hand by a single disulfide bond (black circles). Position numbers of Cys residues (C) follows the mouse sequence. Reproduced with permission from [40]

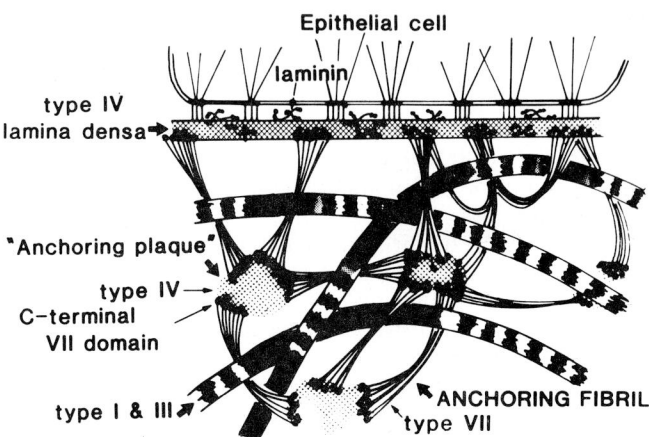

Fig. 8. *Schematic representation of the anchoring fibril network which connects the lamina densa present underneath squamous epithelium with the underlying stroma [162].* Anchoring plaques refer to basement membrane-like structures present in the stroma. They are connected to each other and to the lamina densa by anchoring fibrils consisting of collagen-VII dimers which bind to these structures by their C-terminal globular domains: I, III, fiber-forming collagens; IV, basement membrane collagen. Reproduced with permission from [162]

before considered to be a bone-specific protein with binding activities for calcium, hydroxyapatite and collagen I [151].

Sequence data [40, 41, 147] indicates that the protein consists of four domains I–IV (Fig. 7) and that the two terminal domains (I, IV) represent potential calcium-binding sites. Domain I contains about 16 Glu residues which, however, unlike the Gla domains of other calcium-binding proteins, are not modified by γ-carboxylation. Domain IV has a typical EF-hand structure and is stabilized by a disulfide bond. A second EF-hand motif is postulated to be present in the α-helical domain III [41]. The central region (domain II) is rich in disulfide bonds and shows similarity to ovomucoid and other protease inhibitors. This domain contains phosphoserine [147].

Studies with BM-40 demonstrated 25–30% α-helical conformation, which was distinctly reduced after removal of calcium [147]. The data also indicated the cooperative binding of up to seven calcium ions which is close to the nine binding sites predicted from the sequence. BM-40, obtained after extraction with 6 M guanidine · HCl [147], and osteonectin which was extracted with 0.5 M EDTA [152] showed little or no α helix and no modulation of conformation by calcium, indicating that both preparations were denatured. A single calcium-binding site ($K_d = 0.3$ μM) still found in osteonectin [152] was attributed to the EF-hand domain [147].

Complementary-DNA clones, encoding portions of the protein, demonstrated a major 2-kb mRNA species by Northern hybridization [39–41]. These probes and antibodies against the protein showed expression of BM-40/osteonectin/SPARC [149, 153–155] in bone, basement membranes and a wide variety of other extracellular matrices during embryonic development and in mature tissues. The protein was also found to be stored in platelets [156] and is encoded by a single, large gene which in mouse (26.5 kb) and cattle (11 kb) contains nine or ten exons [157, 158].

The widespread occurrence of BM-40/osteonectin/SPARC suggests an important function in matrix structure. It exists in the EHS tumor in stoichiometric amounts with laminin and its binding to the matrix is apparently mediated by calcium [150]. The components involved in binding have so far not been identified but may not include the laminin-nidogen complex.

ANCHORING PROTEINS AND OTHER COMPONENTS

Most basement membranes face, at one site of their lamina densa, loose connective tissue to which they are firmly attached [5]. The molecular nature of these connections is not known in most cases, except for basement membranes underlying squamous epithelium in skin, esophagus, cornea and the amnion. These basement membranes are linked by numerous, short anchoring fibrils which are bound at one site to the laminina densa and at the other site to the stroma (Fig. 8). A major structural component of these anchoring fibrils was recently identified as collagen VII [159].

Collagen type VII

Collagen VII was first discovered as a pepsin fragment comprising its triple-helical domain. The material obtained was actually a dimer formed from 450-nm-long monomers with a 60-nm-long disulfide-linked overlap of their N-termini [160]. The genuine protein was subsequently shown to be a procollagen consisting of three identical α1(VII) chains. Each chain contains a N-terminal globular segment NC2 (30 kDa), a triple-helical segment (170 kDa) and at the C-terminus another globular segment, NC1 (150 kDa), indicating that the precursor has a molecular mass of about 1000 kDa. The NC1 domain has a complex structure and is composed of three rods and several globules. The NC2 domain is required for aligning the dimers but is removed prior to the deposition of the molecules and thus resembles a typical procollagen peptide [161].

The presence of collagen-VII dimers in anchoring fibrils was strongly indicated by their close correspondence in length and cross-striation patterns and was confirmed by immunostaining [159]. This implies that several collagen-VII dimers have to associate in a lateral fashion to account for the whole width of individual fibrils which should contain identical NC1 epitopes at both ends [162]. Whether lateral association is an intrinsic property of collagen VII or requires additional components is still unknown. The insertion site in the stroma was identified as an anchoring plaque of basement-membrane-like appearance which also contained collagen IV (Fig. 8). The ligand to which the NC1 domain of collagen VII is bound is not yet known. Bridges of anchoring fibrils were not only found between the lamina densa and anchoring plaques but also between individual plaques (Fig. 8). This highly interwoven network of fibrils is obviously well able to stabilize basement membrane connections in tissues which are exposed to considerable shearing forces [162]. Such connections are severly disturbed in patients with epidermolysis bullosa acquisita, a severe and chronic blistering disease, apparently due to autoantibodies against the NC1 domain of collagen VII [163].

How many more components?

Several more proteins including fibronectin, amyloid P, acetylcholinesterase and some complement components (reviewed in [2]) have been identified in basement membranes but usually show a restricted distribution. These components undoubtedly have specific functions where present. Other components with restricted localization include a developmentally regulated 80-kDa protein, merosin [164], and agrin which is thought to promote acetylcholinesterase aggregation in neuromuscular junctions [165]. Von Willebrand factor (reviewed in [166]) may also belong to these components and is exclusively produced by endothelial cells and megakaryocytes. The processed factor (250 kDa) has a complex multidomain structure and forms large, thread-like polymers by disulfide-bonding at both terminal domains. Most of von Willebrand factor is stored in specific intracellular granules but some is found in plasma or deposited in extracellular matrices, including basement membranes. The major function of von Willebrand factor is, however, to provide a linkage between the matrix and platelets after vascular injury [166].

The major components reviewed here, with the exception of proteoglycans, are found in comparable molar proportions in the EHS tumor matrix [90, 131, 150] and may, as suggested from immunogold labeling, exist in similar proportions in other basement membranes [167]. It is unlikely that these are the only constitutive basement membrane proteins since electrophoretic analyses of various tumor extracts [90, 123, 168] show in fact numerous, as yet unidentified, polypeptides. Their characterization will be important for a comprehensive analysis of basement membrane structure.

MOLECULAR BASIS OF INTERACTIONS WITH CELLS

Basement membranes and some of their individual constituents have been shown in previous studies [1−4] to influence various cellular activities such as adhesion and spreading, differentiation, polarization, proliferation, locomotion, tissue invasion, chemotactic responses and the expression of distinct gene products. It was also observed that complex substrates containing several components in defined or arbitrary arrays [1, 168, 169] could provoke qualitatively different responses compared to those of single proteins alone. The mechanisms underlying these diverse reactions are insufficiently known. This review will be restricted to biochemical data related to the initial event, the binding of extracellular ligands to cellular receptors.

Cell-binding sites of laminin

More than 80% of the cells examined in the past [2] and more recently have been found to attach and spread on laminin substrates. Initial mapping studies showed a high-affinity site on fragment 1 (see Fig. 3) [87, 88, 170, 171], hepatocyte attachment to some small fragments not yet mapped [88] and neurone stimulation by fragment 8 [172].

Subsequent investigations demonstrated that most cells attach and spread equally well on laminin and its fragment 1 and 8 [173, 174], with only a few cells not binding to fragment 1. The affinity of binding in radioligand studies was comparable for laminin and the two fragments ($K_d \approx 1-10$ nM), with receptor numbers in the range $10^4 - 10^5$/cell [173]. However, competition assays indicated that each of the two laminin fragments possesses a distinct recognition sequence implying the existence of at least two different high-affinity cell receptors. Yet, both fragments prevented laminin binding to cells which was shown to be either of a competitive (fragment 8) or noncompetitive (fragment 1) nature [175]. The data were interpreted as indicating that fragment 8 possesses the same high-affinity binding site(s), also recognized on laminin, while the fragment-1 site is latent, requires proteolytic activation (i.e. by pepsin) and interferes with laminin binding, either by steric hindrance or an allosteric effect.

Some evidence exists that the fragment-8 sites are important for cellular recognition of laminin both *in vitro* and *in vivo*. Antibodies against fragment 8 and the related fragment 3 (Fig. 3) block most of the cell adhesion to laminin [173, 174]. Fragment-3 epitopes but not those of fragment 1 were found, in some basement membranes, to be close to the cell-contact region [176]. Antibodies to fragments 8 or 3, but not against other laminin domains, were able to block cell polarization during kidney development [121], which coincides with the expression of laminin A chain. The need for a latent high-affinity cell-binding site on fragment 1 is not immediately apparent. Yet, cells respond to this structure by spreading, which implies a distinct rearrangement of the cytoskeleton [177] and thus resembles a typical phenomenon observed in cell-matrix interactions. It was also found that fragment 1 is more active in stimulating collagenase production than laminin itself [178].

The high-affinity-binding sequences on fragments 1 and 8 have so far not been identified. Antibody inhibition studies [173], as well as studies with a globular structure obtained from human laminin [179], would locate the fragment-8 sites to the large globular domain G of the A chain (Fig. 4). The binding structure is very sensitive to moderate heating [174] and is destroyed by 2 M guanidine · HCl, indicating that conformation plays an important role. Activation of the latent fragment-1 site is dependent on the protease used [128, 175].

There are several indications for additional cell-binding sites in laminin, not related to the two high-affinity-binding structures. One is apparently located in the short-arm structures but not recognized ubiquitously [174]. Domain IIIb (Fig. 4) of mouse laminin [36] contains an Arg-Gly-Asp (RGD) sequence and another RGD structure has been identified in domain G of human laminin (D. Olsen et al., unpublished results). Since RGD sequences were identified in fibronectin and some more proteins as cell-binding structures [180, 181], their role in cellular recognition of laminin remains an interesting possibility. The possible involvement of carbohydrate structures of laminin in cell-binding was indicated in other studies [182].

Several studies have also shown that laminin possesses growth-promoting activity implicating cell-binding as the initial step. It was, however, argued [1] that spreading of cells on laminin substrates may expose receptors for growth factors and thus, indirectly, mediate the observed effects. Studies with soluble substances localized the growth factor effects to the fragment-1 structure of laminin [183, 184]. The effect observed was comparable, in several parameters, to that produced by EGF but fragment 1 did not compete with EGF for binding to the EGF receptor [184]. There was also no correlation between cell-adhesion and growth-promoting activity indicating that the latter activity is not due to the high-affinity cell-binding sites of laminin.

Quite a few studies (reviewed in [2]) have also shown that laminin is one of the most potent agents in promoting survival of cultured neurones and in stimulating neurite outgrowth. A major activity was located to fragment 8 and a minor one to a short-arm structure of laminin [172]. The response could be inhibited by antibodies to fragment 3, which was confirmed by monoclonal antibodies of similar specificity [185]. The same sites were also implicated in the stimulation of tyrosine hydroxylase activity in neuron-like, chromaffin cells [186]. It is possible that the laminin structures responsible for stimulation are those which promote cell attachment and spreading.

Other cell-binding basement membrane proteins

Despite the overwhelming evidence for cellular recognition of laminin in basement membranes, other proteins may be of equivalent importance. This is, in particular, indicated for collagen IV [2]. Binding of cells was localized to the triple-helical but not the 7S and NC1 domains of collagen IV and dependent on an intact triple-helical conformation [187]. Studies with fragments indicated two cell-binding sites present in the N- and C-terminal portions of the triple helix (M. Aumailley et al., unpublished results). The triple helix contains many RGD structures [19, 29] but as shown for a neuron-like cell line this sequence may not be involved in cell-binding [188].

A single RGD sequence has been detected in nidogen [37]. Binding of cells to this structure was shown in adhesion studies and found to be specific for RGD by inhibition assays with synthetic peptides. It may, however, not represent one of the major cell-binding sites present in basement membranes [37]. So far no evidence exists for binding of cells to proteoglycans and BM-40.

Cellular receptors

Evidence that the binding of cells to laminin involves integrin receptors came initially from antibody inhibition studies [188–190]. At the same time it was recognized that integrins represent a gene superfamily, encoding membrane-spanning α/β heterodimers involved in cell-cell and cell-matrix interactions [180, 181, 191]. These cellular receptors often show specificity for the RGD sequence and are able to distinguish various extracellular ligands, such as fibronectin, vitronectin and collagen. A human integrin, apparently specific for laminin, was isolated by affinity chromatography from a glioma hybrid cell line [192] and shown to consist of a novel 150-kDa α subunit and a 120-kDa β_1 subunit, the latter being very probably shared with the fibronectin receptor. Since this cell line possesses at least two different laminin receptors [174] the site on laminin recognized by integrin is not obvious, but could be the fragment-8 site. A mixture of integrins consisting of 180/140-kDa α subunits and the β_1 subunit was purified from a neuron-like rat cell line and, after insertion into liposomes, was shown to bind to laminin and/or collagen IV but not to fibronectin [193]. Nerve growth factor stimulated the expression of the 180-kDa α subunit which is apparently laminin-specific [194]. This integrin bound to lamin fragment 1 but not fragment 8 and could thus represent the high-affinity laminin receptor predicted from ligand-binding studies (see above). Integrins seems also to be involved in neurite growth promotion since anti-integrin antibodies inhibited the response to laminin but were themselves stimulatory when used as a substrate [189].

The very first affinity chromatography studies on laminin using membrane extracts from various tissues and cells have consistently identified a laminin-binding 67-69-kDa polypeptide (reviewed in [195]). This laminin-binding protein can also be inserted into liposomes and monoclonal antibodies raised against it stained cell surfaces but also intracellular structures and inhibited the interaction with laminin weakly [196, 197]. Yet, the chemical nature of this component is insufficiently known and it could be still a heterogenous mixture of proteins including serum albumin, as indicated by amino acid sequence data. Partial cloning of the laminin-binding protein from human carcinoma, however, demonstrated a unique sequence [198]. The sequence was recently completed [199] (Y. Yamada, personal communication) and predicted a 295-residue polypeptide, which lacks a signal peptide and a typical hydrophobic transmembrane sequence. Antibodies raised against fusion proteins were able to detect a 32-kDa, together with a 67-kDa, peptide (Y. Yamada, personal communication). The relationship between both components, as well as their involvement as cellular receptors, is still unclear.

Other laminin-binding proteins restricted to muscle and heart tissue were isolated as 56-kDa or 66-kDa components, shown to be distinct from the proteins discussed above and to insert into liposomes [201]. The 406-residue-long sequence contains a signal peptide and an unusual poly(Asp) sequence at the C-terminus but lacks a hydrophobic domain [202]. Immunogold staining localized the protein to the extracellular matrix, in close vicinity to muscle cells. Yet, it was not possible to decide whether it is a peripheral plasma membrane protein (and thus possibly a receptor) or a new matrix component with distinct laminin-binding properties [201]. Several more laminin-binding proteins of 120 kDa and 180 kDa were detected in neurone-like cell lines and fibroblasts by overlay techniques and/or affinity chromatography [203, 204]. They are still insufficiently characterized and their possible relationship to integrin proteins has not yet been clarified.

Binding of laminin to cell surfaces may also include interactions with phospholipids and membrane-bound heparan sulfate (reviewed in [2, 3]). The binding of sulfatides is apparently very specific and of high affinity [205]. Such interactions could be involved in stabilizing receptor-mediated binding of laminin.

Studies with synthetic peptides

Another approach to characterizing the binding of cells to laminin was based on synthetic peptides designed according to available sequence data (Table 1) or on synthetic sequences shown for other proteins to bind to cells [180, 181]. The sequence CDPGYIGSR which is present in domain III of the laminin B1 chain [30] was shown to be active in cell attachment and inhibition assays at high concentrations (10^4-10^5-fold molar excess over laminin), to promote cell migration and to inhibit experimental metastasis formation [206–208]. The active sequence is represented by the pentapeptide YIGSR-amide and is converted to inactive substances when modified in the Y or R positions [209]. The pentapeptide or related structures failed, however, to inhibit radioligand binding of the two high-affinity laminin fragments, 1 and 8, (D. Edgar, unpublished results) and to interfere with the binding of a laminin-specific integrin [192] and the growth-factor-like activity of laminin [184]. It appears, therefore, that this sequence corresponds to a cell-binding site of laminin not yet properly identified in other studies. It could be related to those sites recognized by the 67-kDa laminin-binding protein [206, 209].

Other studies showed that RGD-containing peptides were able to block spreading on laminin with the inverted sequence being even more active [210] and to inhibit partially endothelial cell adhesion on laminin [197]. No effects were observed in some further investigations with laminin-binding integrin receptors [188, 191]. Even though laminin possesses one or two RGD sequences (see above) and also a DGR sequence (domain III of B2 chain) [33] their participation in cellular recognition still remains to be clarified.

CONCLUSIONS AND PERSPECTIVES

One of the major research goals in the comprehensive characterization of basement membrane proteins is the understanding of the principles which determine supramolecular organization. Assembly patterns seem to have some plasticity, as shown for the polygonal collagen IV network [60] and by the diverse interaction potentials observed for most of the major structural elements. These data and the known diversity of basement membranes have led to the proposal of a polymorphic polymerization model [211] where the concentrations of individual components during neosynthesis and remodelling determine the extent of homopolymeric and heteropolymeric interactions and, thus, the final structure. This implies that regulatory events at the pre- and post-translational level are of crucial importance. Most basement membranes also appear as anisotropic structures indicating that at least some components show preferred topological orientations. High-resolution immunoelectron microscopy based on domain-specific antibodies [167, 176] has supported this possibility and will be instrumental in further studies.

Over the past ten years there was considerable progress in the structural and functional characterization of six major constituents but several more are likely to exist. They may include variants of collagen IV and laminin, small proteoglycans, other glycoproteins and some components being unique to only a few basement membranes. Other large composites of proteins (i.e. viral capsids, intracellular cytoskeleton) are known to depend, in their assembly and dynamics, on regulatory components occurring in substoichiometric amounts compared to the major structural elements. It is likely that such components also exist in basement membranes, including proteases, growth factors and other noncatalytic regulatory proteins. The only example of such a component known so far is the collagenase which specifically cleaves collagen IV [212].

The fast completion of even large protein sequences, such as in laminin [30, 33, 36], has provided us with a wealth of information which is still far from completely exploited. In the future, this information will assist in the precise identification of those sequences involved in various biological activities. Their roles *in situ* could then be studied by the use of synthetic peptides, authentic fragments or antibodies directed against these structures. The application of such tools could overcome one of the current problems in basement membrane research, which is the lack of appropriate functional assays.

The use of recombinant-DNA technology in structural studies (Table 1) has also set the stage for studying gene organization, regulation of basement membrane synthesis and the application of designed mutation products in functional assays. Developmental models such as those described in [77, 121] indicate the value of such approaches. They may not only help to understand the role of basement membranes during embryogenesis and in mature animals but also to approach the molecular pathology of inherited and acquired basement membrane diseases.

I thank Drs Y. Yamada, J. R. Hassell, J. H. Fessler, D. Edgar, M. L. Chu and R. Deutzmann for access to unpublished data, Drs G. R. Martin, K. Kühn and K. von der Mark for helpful comments and critique and Mrs. C. Wesse for preparing the manuscript. Own work cited in the article was supported by the *Deutsche Forschungsgemeinschaft* and *Fritz Thyssen Stiftung*.

REFERENCES

1. Kleinman, H. K., Cannon, F. B., Laurie, G. W., Hassell, J. R., Aumailley, M., Terranova, V. P., Martin, G. R. & Dubois-Dalcq, M. (1985) *J. Cell. Biochem. 27*, 317–325.
2. Timpl, R. & Dziadek, M. (1986) *Int. Rev. Exp. Pathol. 29*, 1–112.
3. Martin, G. R. & Timpl, R. (1987) *Annu. Rev. Cell Biol. 3*, 57–85.
4. Liotta, L. A., Rao, C. N. & Wewer, U. (1986) *Annu. Rev. Biochem. 55*, 1037–1057.
5. Martinez-Hernandez, A. & Amenta, P. S. (1983) *Lab. Invest. 48*, 656–677.
6. Orkin, R. W., Gehron, P., McGoodwin, E. B., Martin, G. R., Valentine, T. & Swarm, R. (1977) *J. Exp. Med. 145*, 204–220.
7. Shibata, S. (1985) *Basement membranes. Proceedings of the International symposium on basement membranes* (Shibata, S., ed.) Elsevier Science Publ. (Mishima Japan), Amsterdam.
8. Martin, G. R., Timpl, R. & Kühn, K. (1988) *Adv. Protein Chem. 39*, 1–50.
9. Paulsson, M. (1987) *Collagen Relat. Res. 7*, 443–461.
10. Engel, J. & Furthmayr, H. (1987) *Methods Enzymol. 145*, 3–78.
11. Timpl, R., Paulsson, M., Dziadek, M. & Fujiwara, S. (1987) *Methods Enzymol. 145*, 363–391.
12. Schuppan, D., Glanville, R. W. & Timpl, R. (1982) *Eur. J. Biochem. 123*, 505–512.
13. Oberbäumer, I., Laurent, M., Schwarz, U. Sakurai, Y., Yamada, Y., Vogeli, G., Voss, T., Siebold, B., Glanville, R. W. & Kühn, K. (1985) *Eur. J. Biochem. 147*, 217–224.
14. Wood, L., Theriault, N. & Vogeli, G. (1988) *FEBS Lett. 227*, 5–8.
15. Babel, W. & Glanville, R. W. (1984) *Eur. J. Biochem. 143*, 545–556.
16. Glanville, R. W., Qian, R. Q., Siebold, B., Risteli, J. & Kühn, K. (1985) *Eur. J. Biochem. 152*, 213–219.
17. Pihlajaniemi, T., Tryggvason, K., Myers, J. C., Kurkinen, M., Lebo, R., Cheung, M. C., Prockop, D. J. & Boyd, C. D. (1985) *J. Biol. Chem. 260*, 7681–7687.
18. Brinker, J. M., Gudas, L. J., Loidl, H. R., Wang, S. Y., Rosenbloom, J., Kefalides, N. A. & Myers, J. C. (1985) *Proc. Natl Acad. Sci. USA 82*, 3649–3653.
19. Brazel, D., Oberbäumer, I., Dieringer, H., Babel, W., Glanville, R. W., Deutzmann, R. & Kühn, K. (1987) *Eur. J. Biochem. 168*, 529–536.
20. Soininen, R., Haka-Risku, T., Prockop, D. J. & Tryggvason, K. (1987) *FEBS Lett. 225*, 188–194.
21. Blumberg, B., MacKrell, A. J. & Fessler, J. H. (1988) *J. Biol. Chem. 263*, 18328–18337.
22. Schwarz, U., Schuppan, D., Oberbäumer, I., Glanville, R. W., Deutzmann, R., Timpl, R. & Kühn, K. (1986) *Eur. J. Biochem. 157*, 49–56.
23. Schwarz-Magdolen, U., Oberbäumer, I. & Kühn, K. (1986) *FEBS Lett. 208*, 203–207.
24. Vogeli, G., Horn, E., Carter, J. & Kaytes, P. S. (1986) *FEBS Lett. 206*, 29–32.
25. Kurkinen, M., Condon, M. R., Blumberg, B., Barlow, D. P., Quinones, S., Saus, J. & Pihlajaniemi, T. (1987) *J. Biol. Chem. 262*, 8496–8499.

26. Siebold, B., Qian, R. G., Glanville, R. W., Hofman, H., Deutzmann, R. & Kühn, K. (1987) *Eur. J. Biochem.* 168, 569−575.
27. Hostikka, S. L., Kurkinen, M. & Tryggvason, K. (1987) *FEBS Lett.* 216, 281−286.
28. Killen, P. D., Francomano, C. A., Yamada, Y., Modi, W. S. & O'Brien, S. J. (1987) *Hum. Genet.* 77, 318−324.
29. Brazel, D., Pollner, R., Oberbäumer, I. & Kühn, K. (1988) *Eur. J. Biochem.* 172, 35−42.
30. Sasaki, M., Kato, S., Kohno, K., Martin, G. R. & Yamada, Y. (1987) *Proc. Natl Acad. Sci. USA* 84, 935−939.
31. Pikkarainen, T., Eddy, R., Fukushima, Y., Byers, M., Shows, T., Pihlajaniemi, T., Saraste, M. & Tryggvason, K. (1987) *J. Biol. Chem.* 262, 10454−10462.
32. Montell, D. J. & Goodman, C. S. (1988) *Cell* 53, 463−473.
33. Sasaki, M. & Yamada, Y. (1987) *J. Biol. Chem.* 262, 17111−17117.
34. Durkin, M. E., Bartos, B. B., Liu, S.-H., Phillips, S. L. & Chung, A. E. (1988) *Biochemistry* 27, 5198−5204.
35. Pikkarainen, T., Kallunki, T. & Tryggvason, K. (1988) *J. Biol. Chem.* 263, 6751−6758.
36. Sasaki, M., Kleinman, H. K., Huber, H., Deutzmann, R. & Yamada, Y. (1988) *J. Biol. Chem.* 263, 16536−16544.
37. Mann, K., Deutzmann, R., Aumailley, M., Timpl, R., Raimondi, L., Yamada, Y., Pan, T., Conway, D. & Chu, M.-L. (1989) *EMBO J.* 8, 65−72.
38. Durkin, M. E., Carlin, B. E., Vergnes, J., Bartos, B., Merlie, J. & Chung, A. E. (1987) *Proc. Natl Acad. Sci. USA* 84, 1570−1574.
39. Mason, I. J., Taylor, A., Williams, J. G., Sage, H. & Hogan, B. L. M. (1986) *EMBO J.* 5, 1465−1472.
40. Lankat-Butgereit, B., Mann, K., Deutzmann, R., Timpl, R. & Krieg, T. (1988) *FEBS Lett.* 236, 352−356.
41. Bolander, M. E., Young, M. F., Fisher, L. W., Yamada, Y. & Termine, J. D. (1988) *Proc. Natl Acad. Sci. USA* 85, 2919−2923.
42. Noonan, D. M., Horigan, E., Ledbetter, S., Vogeli, G., Sasaki, M., Yamada, Y. & Hassell, J. R. (1988) *J. Biol. Chem.* 263, 16379−16387.
43. Kefalides, N. A. (1973) *Int. Rev. Connect. Tissue Res.* 6, 63−104.
44. Trüeb, B., Gröbli, B., Spiess, M., Odermatt, B. F. & Winterhalter, K. H. (1982) *J. Biol. Chem.* 257, 5239−5245.
45. Mayne, R. & Zettergren, J. G. (1980) *Biochemistry* 19, 4065−4072.
46. Timpl, R., Wiedemann, H., van Delden, V., Furthmayr, H. & Kühn, K. (1981) *Eur. J. Biochem.* 120, 203−211.
47. Hofmann, H., Voss, T., Kühn, K. & Engel, J. (1984) *J. Mol. Biol.* 172, 325−343.
48. Fessler, L. I., Duncan, K. G., Fessler, J. H., Salo, T. & Tryggvason, K. (1984) *J. Biol. Chem.* 259, 9783−9789.
49. Siebold, B., Deutzmann, R. & Kühn, K. (1988) *Eur. J. Biochem.* 176, 617−624.
50. Weber, S., Engel, J., Wiedemann, H., Glanville, R. W. & Timpl, R. (1984) *Eur. J. Biochem.* 139, 401−410.
51. Blumberg, B., Fessler, L. I., Kurkinen, M. & Fessler, J. H. (1986) *J. Cell Biol.* 103, 1711−1719.
52. Risteli, J., Bächinger, H. P., Engel, J., Furthmayr, H. & Timpl, R. (1980) *Eur. J. Biochem.* 108, 239−250.
53. Qian, R. G. & Glanville, R. W. (1984) *Biochem. J.* 222, 447−452.
54. Kleinman, H. K., McGarvey, M. L., Liotta, L. A., Gehron Robey, P., Tryggvason, K. & Martin, G. R. (1982) *Biochemistry* 21, 6188−6193.
55. Yurchenco, P. D. & Furthmayr, H. (1984) *Biochemistry* 23, 1839−1850.
56. Dölz, R., Engel, J. & Kühn, K. (1988) *Eur. J. Biochem.* 178, 357−366.
57. Weber, S., Dölz, R., Timpl, R., Fessler, J. H. & Engel, J. (1988) *Eur. J. Biochem.* 175, 229−236.
58. Duncan, K. G., Fessler, L. I., Bächinger, H. P. & Fessler, J. H. (1983) *J. Biol. Chem.* 258, 5869−5877.
59. Tsilibary, E. C. & Charonis, A. S. (1986) *J. Cell Biol.* 103, 2467−2473.
60. Yurchenco, P. D. & Ruben, G. C. (1987) *J. Cell Biol.* 105, 2559−2568.
61. Butkowski, R., Langeveld, J. P. M., Wieslander, J., Hamilton, J. & Hudson, B. G. (1987) *J. Biol. Chem.* 262, 7874−7877.
62. Saus, J., Wieslander, J., Langeveld, J. P. M., Quinones, S. & Hudson, B. G. (1988) *J. Biol. Chem.* 263, 13374−13380.
63. Butkowski, R. J., Wieslander, J., Wisdom, B. J., Barr, J. F., Noelken, M. E. & Hudson, B. G. (1985) *J. Biol. Chem.* 260, 3739−3747.
64. Wieslander, J., Langeveld, J., Butkowski, R., Jodlowski, M., Noelken, M. & Hudson, B. G. (1985) *J. Biol. Chem.* 260, 8564−8570.
65. Kurkinen, M., Bernard, M. P., Barlow, D. P. & Chow, L. T. (1985) *Nature (Lond.)* 317, 177−179.
66. Soininen, R., Tikka, L., Chow, L., Pihlajaniemi, T., Kurkinen, M., Prockop, D. J., Boyd, C. D. & Tryggvason, K. (1986) *Proc. Natl Acad. Sci. USA* 83, 1568−1572.
67. Hostikka, S. L. & Tryggvason, K. (1987) *FEBS Lett.* 224, 297−305.
68. Sakurai, Y., Sullivan, M. & Yamada, Y. (1986) *J. Biol. Chem.* 261, 6654−6657.
69. Soininen, R., Chow, L., Kurkinen, M., Tryggvason, K. & Prockop, D. J. (1986) *EMBO J.* 5, 2821−2823.
70. Griffin, C. A., Emanuel, B. S., Hansen, J. R., Cavenee, W. K. & Myers, J. C. (1987) *Proc. Natl Acad. Sci. USA* 84, 512−516.
71. Pöschl, E., Pollner, R. & Kühn, K. (1988) *EMBO J.* 7, 2687−2695.
72. Killen, P. D., Burbelo, P. D., Martin, G. R. & Yamada, Y. (1988) *J. Biol. Chem.* 263, 12310−12314.
73. Lunstrum, G. P., Bächinger, H. P., Fessler, L. I., Duncan, K. G., Nelson, R. E. & Fessler, J. H. (1988) *J. Biol. Chem.* 263, 18318−18327.
74. Timpl, R., Rohde, H., Gehron Robey, P., Rennard, S. I., Foidart, J. M. & Martin, G. R. (1979) *J. Biol. Chem.* 254, 9933−9937.
75. Engel, J., Odermatt, E., Engel, A., Madri, J. A., Furthmayr, H., Rohde, H. & Timpl, R. (1981) *J. Mol. Biol.* 150, 97−120.
76. Terranova, V. P., Rohrbach, D. H. & Martin, G. R. (1980) *Cell* 22, 719−726.
77. Cooper, A. R. & MacQueen, H. A. (1983) *Dev. Biol.* 96, 467−471.
78. Dziadek, M. & Timpl, R. (1985) *Dev. Biol.* 111, 372−382.
79. McCarthy, R. A., Beck, K. & Burger, M. M. (1987) *EMBO J.* 6, 1587−1593.
80. Chiquet, M., Masuda-Nakagawa, L. & Beck, K. (1988) *J. Cell Biol.* 107, 1189−1198.
81. Fessler, L. I., Campbell, A. G., Duncan, K. G. & Fessler, J. H. (1987) *J. Cell Biol.* 105, 2383−2391.
82. Cooper, A. R., Kurkinen, M., Taylor, A. & Hogan, B. L. M. (1981) *Eur. J. Biochem.* 119, 189−197.
83. Howe, C. C. & Dietzschold, B. (1983) *Dev. Biol.* 98, 385−391.
84. Ohno, M., Martinez-Hernandez, A., Ohno, N. & Kefalides, N. A. (1985) in *Basement membranes* (Shibata, S., ed.) pp. 3−11, Elsevier Science Publishers, Amsterdam.
85. Ott, U., Odermatt, E., Engel, J., Furthmayr, H. & Timpl, R. (1982) *Eur. J. Biochem.* 123, 63−72.
86. Paulsson, M., Deutzmann, R., Timpl, R., Dalzoppo, D., Odermatt, E. & Engel, J. (1985) *EMBO J.* 4, 309−316.
87. Rao, C. N., Margulies, I. M. K., Tralka, T. S., Terranova, V. P., Madri, J. A. & Liotta, L. A. (1982) *J. Biol. Chem.* 257, 9740−9744.
88. Timpl, R., Johansson, S., van Delden, V., Oberbäumer, I. & Höök, M. (1983) *J. Biol. Chem.* 258, 8922−8927.
89. Paulsson, M. (1988) *J. Biol. Chem.* 263, 5425−5430.
90. Paulsson, M., Aumailley, M., Deutzmann, R., Timpl, R., Beck, K. & Engel, J. (1987) *Eur. J. Biochem.* 166, 11−19.
91. Hartl, L., Oberbäumer, I. & Deutzmann, R. (1988) *Eur. J. Biochem.* 173, 629−635.
92. Deutzmann, R., Huber, J., Schmetz, K. A., Oberbäumer, I. & Hartl, L. (1988) *Eur. J. Biochem.* 177, 35−45.

93. Mann, K., Deutzmann, R. & Timpl, R. (1988) *Eur. J. Biochem. 178*, 71–80.
94. Apella, E., Weber, I. T. & Blasi, F. (1988) *FEBS Lett. 231*, 1–4.
95. Barlow, D. P., Green, N. M., Kurkinen, M. & Hogan, B. L. M. (1984) *EMBO J. 3*, 2355–2362.
96. Palm, S. L., McCarthy, J. B. & Furcht, L. T. (1985) *Biochemistry 24*, 7753–7760.
97. Fujiwara, S., Shinkai, H., Deutzmann, R., Paulsson, M. & Timpl, R. (1988) *Biochem. J. 252*, 453–461.
98. Arumugham, R. G., Hsieh, T. C.-Y., Tanzer, M. L. & Laine, R. A. (1986) *Biochim. Biophys. Acta 883*, 112–126.
99. Towbin, H., Rosenfelder, G., Wieslander, J., Avila, J. L., Rojas, M., Szarfman, A., Esser, K., Nowack, H. & Timpl, R. (1987) *J. Exp. Med. 166*, 419–432.
100. Kleinman, H. K., Ebihara, I., Killen, P. D., Sasaki, M., Cannon, F. B., Yamada, Y. & Martin, G. R. (1987) *Dev. Biol. 122*, 373–378.
101. Durkin, M. E., Phillips, S. L. & Chung, A. E. (1986) *Differentiation 32*, 260–266.
102. Boot-Handford, R. P., Kurkinen, M. & Prockop, D. J. (1987) *J. Biol. Chem. 262*, 12475–12478.
103. Mattei, M. G., Weil, D., Pribula-Conway, D., Bernard, M. P., Passage, E., Van Cong, N., Timpl, R. & Chu, M.-L. (1988) *Hum. Genet. 79*, 235–241.
104. Yamada, Y., Sasaki, N., Kohno, K., Kleinmann, H. K., Kato, S. & Martin, G. R. (1985) in *Basement membranes* (Shibata, S., ed.) pp. 139–146, Elsevier Science Publishers, Amsterdam.
105. Ogawa, K., Burbelo, P. D., Sasaki, M. & Yamada, Y. (1988) *J. Biol. Chem. 263*, 8384–8389.
106. Peters, B. P., Hartle, R. J., Krzesick, R. F., Kroll, T. G., Perini, F., Balun, J. E., Goldstein, I. J. & Ruddon, R. W. (1985) *J. Biol. Chem. 260*, 14732–14742.
107. Morita, A., Sugimoto, E. & Kitagawa, Y. (1985) *Biochem. J. 229*, 259–264.
108. Carlin, B. E., Durkin, M. E., Bender, B., Jaffe, R. & Chung, A. E. (1983) *J. Biol. Chem. 258*, 7729–7737.
109. Yurchenco, P. D., Tsilibary, E. C., Charonis, A. S. & Furthmayr, H. (1985) *J. Biol. Chem. 260*, 7636–7644.
110. Charonis, A. S., Tsilibary, E. C., Saku, T. & Furthmayr, H. (1986) *J. Cell Biol. 103*, 1689–1697.
111. Woodley, D. T., Rao, C. N., Hassel, J. R., Liotta, L. A., Martin, G. R. & Kleinman, H. K. (1983) *Biochim. Biophys. Acta 761*, 278–283.
112. Charonis, A. S., Tsilibary, E. C., Yurchenco, P. D. & Furthmayr, H. (1985) *J. Cell Biol. 100*, 1848–1853.
113. Laurie, G. W., Bing, J. T., Kleinman, H. K., Hassel, J. R., Aumailley, M., Martin, G. R. & Feldmann, J. R. (1986) *J. Mol. Biol. 189*, 205–216.
114. Sakashita, S., Engvall, E. & Ruoslahti, E. (1980) *FEBS Lett. 116*, 243–246.
115. Skubitz, A. P. N., McCarthy, J. B., Charonis, A. S. & Furcht, L. T. (1988) *J. Biol. Chem. 263*, 4861–4868.
116. Charonis, A. S., Skubitz, A. P. N., Koliakos, G. G., Reger, L., Dege, J., Vogel, A. M., Wohlhueter, R. & Furcht, L. T. (1988) *J. Cell Biol. 107*, 1253–1260.
117. Kühl, U., Timpl, R. & von der Mark, K. (1982) *Dev. Biol. 93*, 344–354.
118. Cornbrooks, C. J., Carey, D. J., McDonald, J. A., Timpl, R. & Bunge, R. P. (1983) *Proc. Natl Acad. Sci. USA 80*, 3850–3854.
119. Davies, G. E., Manthorpe, M., Engvall, E. & Varon, S. (1985) *J. Neurosci. 5*, 2662–2671.
120. Edgar, D., Timpl, R. & Thoenen, H. (1988) *J. Cell Biol. 106*, 1299–1306.
121. Klein, G., Langegger, M., Timpl, R. & Ekblom, P. (1988) *Cell 55*, 331–341.
122. Timpl, R., Dziadek, M., Fujiwara, S., Nowack, H. & Wick, G. (1983) *Eur. J. Biochem. 137*, 455–465.
123. Paulsson, M., Deutzmann, R., Dziadek, M., Nowack, H., Timpl, R., Weber, S. & Engel, J. (1986) *Eur. J. Biochem. 156*, 467–478.
124. Carlin, B., Jaffe, R., Bender, B. & Chung, A. E. (1981) *J. Biol. Chem. 256*, 5209–5214.
125. Paulsson, M., Dziadek, M., Suchanek, C., Huttner, W. B. & Timpl, R. (1985) *Biochem. J. 231*, 571–579.
126. Dziadek, M., Paulsson, M. & Timpl, R. (1985) *EMBO J. 4*, 2513–2518.
127. Dziadek, M., Clements, R., Mitrangas, K., Reiter, H. & Fowler, K. (1988) *Eur. J. Biochem. 172*, 219–225.
128. Timpl, R., Mann, K., Aumailley, M., Gerl, M., Deutzmann, R., Nurcombe, V., Edgar, D., Chu, M.-L. & Yamada, Y. (1989) in *Structure, interactions and assembly of cytoskeletal and extracellular proteins* (Aebi, U. & Engel, J., eds) Springer series in biophysics 3, in the press.
129. Farquhar, M. G. (1981) in *Cell biology of extracellular matrix* (Hay, E. D., ed.) pp. 335–378, Plenum Press, New York.
130. Hassell, J. R., Gehron Robey, P., Barrach, H. J., Wilczek, J., Rennard, S. I. & Martin, G. R. (1980) *Proc. Natl Acad. Sci. USA 77*, 4494–4498.
131. Fujiwara, S., Wiedemann, H., Timpl, R., Lustig, A. & Engel, J. (1984) *Eur. J. Biochem. 143*, 145–157.
132. Dziadek, M., Fujiwara, S., Paulsson, M. & Timpl, R. (1985) *EMBO J. 4*, 905–912.
133. Hassell, J. R., Leyshon, W. C., Ledbetter, S. R., Tyree, B., Suzuki, S., Kato, M., Kimata, K. & Kleinman, H. K. (1985) *J. Biol. Chem. 260*, 8098–8105.
134. Kanwar, Y. S., Hascall, V. C. & Farquhar, M. G. (1981) *J. Cell Biol. 90*, 527–532.
135. Kanwar, Y. S., Veis, A., Kimura, J. H. & Jakubowski, M. L. (1984) *Proc. Natl Acad. Sci. USA 81*, 762–766.
136. Parthasarathy, N. & Spiro, R. G. (1984) *J. Biol. Chem. 259*, 12749–12755.
137. Kato, M. Y., Oike, Y., Suzuki, S. & Kimata, K. (1987) *J. Biol. Chem. 262*, 7180–7188.
138. Fenger, M., Wewer, U. & Albrechtsen, R. (1984) *FEBS Lett. 173*, 75–79.
139. Ledbetter, S. R., Tyree, B., Hassell, J. R. & Horigan, E. A. (1985) *J. Biol. Chem. 260*, 8106–8113.
140. Paulsson, M., Yurchenco, P. D., Ruben, G. C., Engel, J. & Timpl, R. (1987) *J. Mol. Biol. 197*, 297–313.
141. Ledbetter, S. R., Fisher, L. W. & Hassell, J. R. (1987) *Biochemistry 26*, 988–995.
142. Klein, D. J., Brown, D. M., Oegema, T. R., Brenchley, P. E., Anderson, J. C., Dickinson, M. A. J., Horigon, E. A. & Hassell, J. R. (1988) *J. Cell Biol. 106*, 963–970.
143. Couchman, J. R., Caterson, B., Christner, J. E. & Baker, J. R. (1984) *Nature (Lond.) 307*, 650–652.
144. Yurchenco, P. D., Cheng, Y.-S. & Ruben, G. C. (1987) *J. Biol. Chem. 262*, 17668–17676.
145. Pejler, G., Bäckström, G., Lindahl, U., Paulsson, M., Dziadek, M., Fujiwara, S. & Timpl, R. (1987) *J. Biol. Chem. 260*, 5036–5043.
146. Folkman, J., Klagsbrun, M., Sasse, J., Wadzinski, M., Ingber, D. & Vlodavsky, I. (1988) *Am. J. Pathol. 130*, 393–400.
147. Engel, J., Taylor, M., Paulsson, M., Sage, H. & Hogan, B. L. M. (1987) *Biochemistry 26*, 6958–6965.
148. Paulsson, M., Saladin, K. & Landwehr, R. (1988) *Eur. J. Biochem. 177*, 477–481.
149. Dziadek, M., Paulsson, M., Aumailley, M. & Timpl, R. (1986) *Eur. J. Biochem. 161*, 455–464.
150. Mann, K., Deutzmann, R., Paulsson, M. & Timpl, R. (1987) *FEBS Lett. 218*, 167–172.
151. Termine, J. D., Kleinman, H. K., Whitson, S. W., Conn, K. M., MacGarvey, M. L. & Martin, G. R. (1981) *Cell 26*, 99–105.
152. Romberg, R. W., Werness, P. G., Lollar, P., Riggs, B. L. & Mann, K. G. (1985) *J. Biol. Chem. 260*, 2728–2736.
153. Mason, I. J., Murphy, D., Münke, M., Francke, U., Elliot, R. W. & Hogan, B. L. M. (1986) *EMBO J. 5*, 1831–1837.
154. Holland, P. W. H., Harper, S. J., McVey, J. H. & Hogan, B. L. M. (1987) *J. Cell Biol. 105*, 473–482.
155. Young, M. F., Bolander, M. E., Day, A. A., Ramis, C. I., Gehron Robey, P., Yamada, Y. & Termine, J. D. (1986) *Nucleic Acids Res. 14*, 4483–4497.

156. Stenner, D. D., Tracy, R. P., Riggs, B. L. & Mann, K. G. (1986) *Proc. Natl Acad. Sci. USA 83*, 6892–6896.
157. McVey, J. H., Nomuras, S., Kelly, P., Mason, I. J. & Hogan, B. L. M. (1988) *J. Biol. Chem. 263*, 11111–11116.
158. Findlay, D. M., Fisher, L. W., McQuillan, C. I., Termine, J. D. & Young, M. F. (1988) *Biochemistry 27*, 1483–1489.
159. Sakai, L. Y., Keene, D. R., Morris, N. P. & Burgeson, R. E. (1986) *J. Cell Biol. 103*, 1577–1586.
160. Burgeson, R. E. (1987) in *Structure and function of collagen types* (Mayne, R. & Burgeson, R. E., eds), pp 145–172, Academic Press, Orlando, Florida.
161. Lunstrum, G. P., Kuo, H.-J., Rosenbaum, L. M., Keene, D. R., Glanville, R. W., Sakai, L. Y. & Burgeson, R. E. (1987) *J. Biol. Chem. 262*, 13706–13712.
162. Keene, D. R., Sakai, L. Y., Lunstrum, G. P., Morris, N. P. & Burgeson, R. E. (1987) *J. Cell Biol. 104*, 611–621.
163. Woodley, D. T., Burgeson, R. E., Lunstrum, G., Bruckner-Tuderman, L., Reese, M. J. & Briggaman, R. A. (1988) *J. Clin. Invest. 81*, 683–687.
164. Leivo, I. & Engvall, E. (1988) *Proc. Natl Acad. Sci. USA 85*, 1544–1548.
165. Godfrey, E. W., Nitkin, R. M., Wallace, P. G., Rubin, L. L. & McMahan, U. J. (1984) *J. Cell Biol. 99*, 615–627.
166. Titani, K. & Walsh, K. A. (1988) *Trends Biochem. Sci. 13*, 94–97.
167. Grant, D. S. & Leblond, C. P. (1988) *J. Histochem. Cytochem. 36*, 271–283.
168. Kleinman, H. K., McGarvey, M. L., Hassell, J. R., Star, V. L., Cannon, F. B., Laurie, G. W. & Martin, G. R. (1986) *Biochemistry 25*, 312–318.
169. Kubota, Y., Kleinman, H. K., Martin, G. R. & Lawley, T. J. (1988) *J. Cell Biol. 107*, 1589–1598.
170. Terranova, V. P., Rao, C. N., Kalebic, T., Margulies, I. M. & Liotta, L. A. (1983) *Proc. Natl Acad. Sci. USA 80*, 444–448.
171. Barsky, S. H., Rao, C. N., Williams, J. E. & Liotta, L. A. (1984) *J. Clin. Invest. 74*, 843–848.
172. Edgar, D., Timpl, R. & Thoenen, H. (1984) *EMBO J. 3*, 1463–1468.
173. Aumailley, M., Nurcombe, V., Edgar, D., Paulsson, M. & Timpl, R. (1987) *J. Biol. Chem. 262*, 11532–11538.
174. Goodman, S. L., Deutzmann, R. & von der Mark, K. (1987) *J. Cell Biol. 105*, 589–598.
175. Nurcombe, V., Aumailley, M., Timpl, R. & Edgar, D. (1989) *Eur. J. Biochem. 180*, 9–14.
176. Schittny, J. C., Timpl, R. & Engel, J. (1988) *J. Cell Biol. 107*, 1599–1610.
177. Woods, A. & Couchman, J. R. (1988) *Collagen Relat. Res. 8*, 155–182.
178. Turpeenniemi-Hujanen, T., Thorgeirsson, U. P., Rao, C. N. & Liotta, L. A. (1986) *J. Biol. Chem. 261*, 1883–1889.
179. Dillner, L., Dickerson, K., Manthorpe, M., Rouslahti, E. & Engvall, E. (1988) *Exp. Cell Res. 177*, 186–198.
180. Ruoslahti, E. & Pierschbacher, M. D. (1987) *Science (Wash. DC) 238*, 491–497.
181. Rouslahti, E. (1988) *Annu. Rev. Biochem. 57*, 375–413.
182. Dennis, J. W., Waller, C. A. & Schirrmacher, V. (1984) *J. Cell Biol. 99*, 1416–1423.
183. Terranova, V. P., Aumailley, M., Sultan, L. H., Martin, G. R. & Kleinman, H. K. (1986) *J. Cell Physiol. 127*, 473–479.
184. Panayotou, G., End, P., Aumailley, M., Timpl, R. & Engel, J. (1989) *Cell 56*, 93–101.
185. Engvall, E., Davis, G. E., Dickerson, K., Ruoslahti, E., Varon, S. & Manthorpe, M. (1986) *J. Cell Biol. 103*, 2457–2465.
186. Acheson, A., Edgar, D., Timpl, R. & Thoenen, H. (1986) *J. Cell Biol. 102*, 151–159.
187. Aumailley, M. & Timpl, R. (1986) *J. Cell Biol. 103*, 1569–1575.
188. Tomaselli, K. J., Damsky, C. H. & Reichardt, L. F. (1987) *J. Cell Biol. 105*, 2347–2358.
189. Hall, D. E., Neugebauer, K. M. & Reichardt, L. F. (1987) *J. Cell Biol. 104*, 623–634.
190. Bozyczko, D. & Horwitz, A. (1986) *J. Neurosci. 6*, 1241–1251.
191. Hynes, R. O. (1987) *Cell 48*, 549–554.
192. Gehlsen, K. R., Dillner, L., Engvall, E. & Ruoslahti, E. (1988) *Science (Wash. DC) 241*, 1228–1229.
193. Tomaselli, K. J., Damsky, C. H. & Reichardt, L. F. (1988) *J. Cell Biol. 107*, 1241–1252.
194. Marchisio, P. C., Gavazzi, I., Abbadini, M., Timpl, R., Tarone, G. & Rossino, P. (1988) *J. Cell Biol. 107*, 799a.
195. von der Mark, K. & Kühl, U. (1985) *Biochim. Biophys. Acta 823*, 147–160.
196. Liotta, L. A., Hand, P. H., Rao, C. N., Bryant, G., Barsky, S. H. & Schlom, J. (1985) *Exp. Cell Res. 156*, 117–126.
197. Yannariello-Brown, J., Wewer, U., Liotta, L. & Madri, J. A. (1988) *J. Cell Biol. 106*, 1773–1786.
198. Wewer, U. M., Liotta, L. A., Jaye, M., Ricca, G. A., Drohan, W. N., Claysmith, A. P., Rao, C. N., Wirth, P., Coligan, J. E., Albrechtsen, R., Mudryj, M. & Sobel, M. E. (1986) *Proc. Natl. Acad. Sci. USA 83*, 7137–7141.
199. Yow, H., Wong, J. M., Chen, H. S., Lee, C., Steele, G. D. & Chen, L. B. (1988) *Proc. Natl Acad. Sci. USA 85*, 6394–6398.
200. Reference deleted.
201. Hall, D. E., Frazer, K. A., Hann, B. C. & Reichardt, L. F. (1988) *J. Cell Biol. 107*, 687–697.
202. Clegg, D. O., Helder, J. C., Hann, B. C., Hall, D. E. & Reichardt, L. F. (1988) *J. Cell Biol. 107*, 699–705.
203. Smalheiser, N. R. & Schwartz, N. B. (1987) *Proc. Natl Acad. Sci. USA 84*, 6457–6461.
204. Kleinman, H. K., Ogle, R. C., Cannon, F. B., Little, C. D., Sweeney, T. M. & Luckenbill-Edds, L. (1988) *Proc. Natl Acad. Sci. USA 85*, 1282–1286.
205. Robert, D. D., Rao, C. N., Magnani, J. L., Spitalnik, S. L., Liotta, L. A. & Ginsburg, V. (1985) *Proc. Natl Acad. Sci. USA 82*, 1306–1310.
206. Graf, J., Iwamoto, Y., Sasaki, M., Martin, G. R., Kleinman, H. K., Robey, F. A. & Yamada, Y. (1987) *Cell 48*, 989–996.
207. Iwamoto, Y., Graf, J., Sasaki, M., Kleinman, H. K., Greatorex, D. R., Martin, G. R., Robey, F. A. & Yamada, Y. (1988) *J. Cell. Physiol. 134*, 287–291.
208. Iwamoto, Y., Robey, F. A., Graf, J., Sasaki, M., Kleinman, H. K., Yamada, Y. & Martin, G. R. (1987) *Science (Wash. DC) 238*, 1132–1134.
209. Graf, J., Ogle, R. C., Robey, F. A., Sasaki, M., Martin, G. R., Yamada, Y. & Kleinman, H. K. (1987) *Biochemistry 26*, 6896–6900.
210. Yamada, Y. & Kennedy, D. W. (1987) *J. Cell. Physiol. 130*, 21–28.
211. Furthmayr, H., Yurchenco, P. D., Charonis, A. S. & Tsilibary, E. C. (1985) in *Basement membranes* (Shibata, S., ed), pp. 169–180, Elsevier, Amsterdam.
212. Salo, T., Liotta, L. A. & Tryggvason, K. (1983) *J. Biol. Chem. 258*, 3058–3063.

Review

Mechanisms of flavoprotein-catalyzed reactions

Sandro GHISLA[1] and Vincent MASSEY[2]

[1] Fakultät für Biologie der Universität Konstanz
[2] Department of Biological Chemistry, University of Michigan, Ann Arbor

(Received September 30/December 14, 1988) — EJB 88 1152

Flavoproteins are a class of enzymes catalyzing a very broad spectrum of redox processes by different chemical mechanisms. This review describes the best studied of these mechanisms and discusses factors possibly governing reactivity and specificity.

A large number of flavin-containing enzymes, several hundreds, has been uncovered to date. An unusual feature of flavoproteins is the variety of the catalytic reactions performed, which range from typical redox catalysis such as the dehydrogenation of an amino acid, or the activation of dioxygen, to photochemistry; from 'DNA damage repair' to light emission. These few examples illustrate the fact that the same coenzyme is able to catalyze, or take part in catalytic events which must vary widely from a mechanistic point of view. The chemistry underlying the conversion itself will be quite different from case to case. This versatility sets flavoproteins apart from most other cofactor-dependent enzymes, which, in general, each catalyze a single type of chemical reaction. The activation or 'steering' of a particular activity of the flavin results from the interaction with the protein at the active center. On the other hand, for the vast majority of these enzymes a common feature exists, that at some stage during the catalytic event a transfer of electrons takes place between the substrate and the flavin itself.

The purpose of the present review is not to enumerate the different functions of flavoproteins, this having been done elsewhere [1–4], but to focus on the mechanisms of catalysis, and on the possible ways of interaction of the flavin nucleus with the protein, i.e. on how this brings about chemistry appropriate to the particular enzyme reaction. In order to understand how this is done we think it useful to reiterate briefly the chemical properties of the flavin itself.

CHEMICAL PROPERTIES OF THE FLAVIN

The following scheme will remind the reader of the structure of the redox-active part of the flavin molecule, the 7,8-dimethylisoalloxazine, in its oxidized and fully reduced states, and the numbering of the most important functional groups.

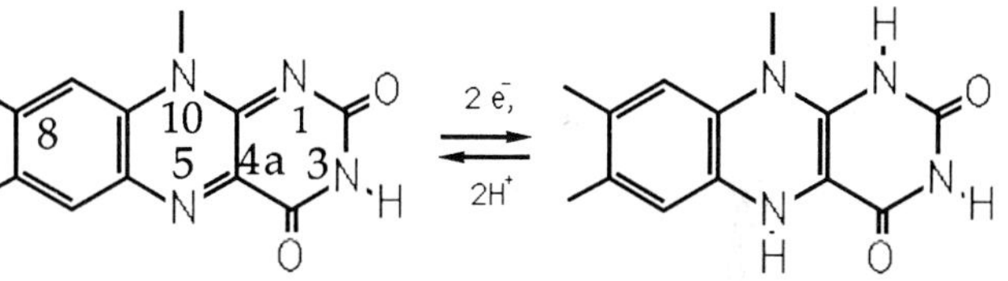

Oxidized flavin (Fl_{ox})   Reduced flavin ($Fl_{red}H_2$)

The modes of modulation of flavin reactivity can be subdivided roughly into effects on the thermodynamics and those on the kinetics of a possible reaction. With respect to the first it should be pointed out that the redox chemistry of the flavin is restricted to the isoalloxazine nucleus, side-chain functions such as the adenine moiety not being involved in catalysis, but serving in anchoring the coenzyme at the active site. Reduction of the flavin occurs reversibly by two one-electron steps, or one two-electron step, involving changes which affect particularly the chemistry of the 'enediamine' subfunction and thus of positions N(1), C(4a) and N(5). The potential of this conversion (E_m, pH 7) is around -200 mV and can be lowered or increased in an approximately 600-mV range through interaction with the protein (from -495 mV for the $FlH^{\cdot}/Fl_{red}H^{-}$ couple in *Azotobacter vinelandii* flavodoxin [5] to $+80$ mV for the $Fl_{ox}/Fl^{\cdot -}$ couple in thiamin dehydrogenase [6]). The pyrimidine nucleus of the three-membered ring system is electron-deficient and can be viewed as an 'electron sink'. In thermodynamic terms, any interaction which tends to lower its electron or negative charge density will increase the redox potential. The reduced flavin consists roughly of an electron-rich phenylenediamine moiety fused with a (4,5-diamino)-uracil. The latter is the moiety in which the negative charge (i.e. two electrons taken up by the oxidized flavin) is localized and, most importantly, stabilized. The de-

---

Correspondence to S. Ghisla, Fakultät für Biologie der Universität Konstanz, D-7750 Konstanz, Federal Republic of Germany

Abbreviations. Fl, flavin; Fl_{ox}, $FlH^{\cdot}$, Fl_{red}, flavin in its oxidized, half-reduced (radical) and fully reduced (dihydro) forms; GSH, glutathione; GSSG, oxidized glutathione; Nbs_2, bis(4-nitrophenyl)-disulfide 3,3'-dicarboxylic acid, formerly named 5,5'-dithiobis(2-nitrobenzoic acid) and abbreviated to DTNB.

Enzymes. Glutathione reductase (EC 1.6.4.2); lipoyl dehydrogenase, dihydrolipoamide dehydrogenase (EC 1.8.1.4); mercuric reductase (EC 1.16.1.1); thioredoxin reductase (EC 1.6.4.5); *p*-hydroxybenzoate hydroxylase (EC 1.14.13.2); melilotate hydroxylase (EC 1.14.13.4); anthranilate hydroxylase (EC 1.14.12.2); 2-methyl-3-hydroxypyridine-5-carboxylic acid oxygenase (EC 1.14.12.4); cyclohexanone monooxygenase (EC 1.14.13.22); glucose oxidase (EC 1.1.3.4); acyl-CoA dehydrogenase (EC 1.3.99.3); D-amino acid oxidase (EC 1.4.3.3); L-lactate dehydrogenase, flavocytochrome b_2 (EC 1.1.2.3); D-lactate dehydrogenase (EC 1.1.2.4); trypanothione reductase (EC 1.6.4.–).

gree to which this negative charge is stabilized or destabilized is an important factor governing the redox potential. Thus while a positive charge in the protein around the pyrimidine ring will contribute to increase the redox potential, the presence of negative charge or of a hydrophobic environment will lower it. These concepts have been put forward earlier [3, 7] and have, in their essence, been verified by recent X-ray crystallographic studies [8 – 11], as will be detailed below. The thermodynamic stabilization or destabilization of intermediate radical forms can be viewed along the same lines. An additional important factor affecting the redox potential is the modification of the flavin nucleus through introduction of functional groups. Thus introduction of an imidazolyl residue at position 8α, such as in 8α-histidyl-flavins, and the corresponding ionisation states, modulate the redox potential [12]. Similar effects are observed when the substitution is introduced directly into the aromatic system such as in 6-cysteinyl, 6-hydroxy or 8-hydroxy flavins [13 – 16]; in the latter two cases the redox potential is additionally affected by the ionization of the substituent.

With respect to the second, broad generalization, it is again important to point out that reactions of the flavin nucleus with a given substrate will occur preferentially at specific loci, which vary depending on the type of reaction. For example, activation of dioxygen to form hydroperoxides occurs through formation of a covalent adduct at position 4a of the flavin [17], oxidation of lactate by lactate oxidase via position N(5) [18]. Thus steric restriction of access to the flavin itself, or to a specific position, can lower the rate of a reaction to essentially zero, while other reactions might remain unaffected. Conversely, facilitation of access or of encounter with specific reactants can make other reactions predominant or exclusive.

MECHANISMS OF FLAVIN-CATALYZED DEHYDROGENATION REACTIONS

Most biological oxidation reactions involve the rupture of (at least) one organic-substrate – H bond, with concomitant transfer of two electrons to a suitable acceptor molecule. In the case of oxidation reactions catalyzed by simple pyridine-nucleotide-linked enzymes, the acceptor is NAD^+ or $NADP^+$ and the resultant NAD(P)H needs to be reoxidized by coupling with a second pyridine-nucleotide-linked enzyme. In the case of flavoproteins, the flavin being tightly bound to the protein, a second substrate serves to reoxidize the reduced flavin to complete the catalytic cycle. Thus catalysis by flavoprotein enzymes always involves a reductive half-reaction, where the enzyme-bound flavin is reduced, and an oxidative half-reaction, where the reduced flavin is reoxidized. In most cases it is possible to study these two half-reactions separately, a feature which has permitted the detailed analysis of catalytic events not possible with most other enzymes.

As detailed previously [3], we can conveniently distinguish four different types of substrates undergoing dehydrogenation reactions with flavoproteins; in each of these classes a different mechanism appears to exist.

Pyridine nucleotides

In these enzymes a direct transfer of a hydride equivalent appears to take place between the C(4) of the pyridine nucleotide and the N(5) position of the flavin, in accord with chemical model studies [19]. This concept is nicely supported by the crystallographic data on glutathione reductase, which shows the nicotinamide ring stacked over the flavin pyrazine ring with the *S*-hydrogen at its C(4) position close to the flavin N(5) position [20]. Such a juxtaposition is essential for hydride transfer, in distinction to a radical mechanism, where orbital overlap between pyridine nucleotide and flavin would not have to be so restrictive.

Substrates activated for carbanion transfer

Many flavoprotein enzymes involve oxidation of substrates with electron-withdrawing activating groups next to the position of dehydrogenation. Such reactions appear to be initiated by abstraction of the relatively acidic α-hydrogen atom as a proton and therefore involve formally a carbanion of the substrate, either as an intermediate or, at least, as a transition state [21, 22]. These mechanisms have been described in detail in previous reviews [1, 3] and the evidence in their favor appears to be compelling. Therefore, here we will reiterate only the main points plus some relevant new information.

The best-characterized flavoprotein reaction appearing to proceed via a carbanion mechanism is the dehydrogenation of α-hydroxy acids as exemplified by the enzymes L-lactate oxidase and flavocytochrome b_2 (yeast L-lactate dehydrogenase). In spite of basic agreement about the overall mechanism, some uncertainties remain which revolve around the mode of transfer of the redox equivalents from the carbanionic species to the flavin and the formation of the primary products. Before discussing this, the topology of the active center of flavocytochrome b_2, which has recently been solved by X-ray crystallography [23, 24] will be discussed in connection with structural evidence derived from work with L-lactate oxidase. As shown in Fig. 1, substrate (or product in the specific case) binds to the *Si*-side of the flavin. Therefore this side specificity seems to be a common theme in α-hydroxyacid dehydrogenation, since it is found also with the recently reported crystal structure of spinach glycollate oxidase [25] and with D-lactate dehydrogenase and L-lactate oxidase [26] (cf. Table 1).

The carboxylate function of substrate is bound to Arg-376 and placed near the benzene ring, while the α-position is located near N(5) (374 pm), and the further α-substituent ($-CH_3$ in the specific case) in a position where it can interact with the flavin C(4a) as required for inactivation with suicide substrates [27 – 29]. This is in nice agreement with the mode of binding and the stereospecificity deduced from studies with L-lactate oxidase [26]. Of equal importance, the positive charge proposed to facilitate flavin reduction thermodynamically and to stabilize the binding of many anionic flavins [3, 7] is found as lysine 349, which interacts with N(1)-C(2) = O in flavocytochrome b_2 (Fig. 1).

Of particular interest is His-373, which is in close contact (261 pm) with the carbonyl of pyruvate. Assuming homology between the two enzymes this would explain perfectly the tight and pH-dependent binding of oxalate to lactate oxidase [30]. His-373 is thus in an appropriate position for α-proton abstraction, this being possibly facilitated also by the interaction of the α-hydroxy group with Tyr-254 [24].

For the transfer of redox equivalents from the assumed carbanion the following alternatives (Scheme 1) are possible:

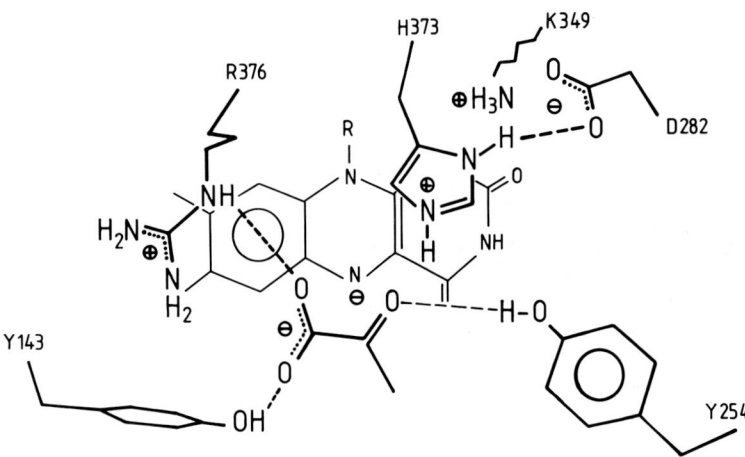

Fig. 1. *Orientation of (dianionic) reduced flavin, pyruvate (product) and of relevant amino acid residues at the active center of yeast L-lactate dehydrogenase (flavocytochrome b₂) as proposed by Lederer and Mathews [24]*. Arg-376, His-373 aspartate-282, and tyrosine-254 are located above the flavin *Si*-plane, while Lys-349 stabilizes the negative charge at the flavin position $N(1)$-$C(2) = O$ (located behind the plane defined by the isoalloxazine ring and below His-373). Adapted with the permission of the authors from [24]

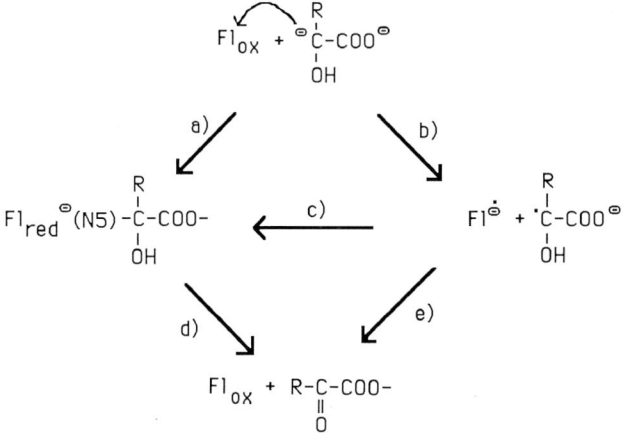

Scheme 1. *Alternative modes of transfer of the redox equivalents from carbanionic substrate (a transition state) to the flavin.* See text for further details

(a) direct nucleophilic attack at the flavin N(5) position to yield a covalent adduct; (b) one-electron transfer, followed by (c) collapse of the radical pair to the flavin N(5)-substrate adduct or (e) by a second one-electron transfer to give oxidized substrate and reduced flavin directly; (d) fragmentation of the covalent adduct yields the same products.

In this context the following experimental data are relevant. (a) The reaction of both the above-mentioned enzymes with the normal substrate, L-lactate, is fast and proceeds to the product, pyruvate, without observable intermediates [31, 32]. (b) In the case of glycollate as a substrate, two covalent intermediates are formed with lactate oxidase [18], which are derived respectively from abstraction of either the *Re*- or the *Si*-proton of the substrate and subsequent addition of the glycollyl moiety to the flavin N(5).

Thus, a covalent adduct lies on a feasible reaction coordinate. Its observation might result simply from the lifetime, i.e. the capacity of the active site to accommodate it and from the thermodynamic difference between adducts of pyruvate and glyoxylate. A differentiation between pathways (a) and (b + c) boils down to semantics if (c) is very fast, as it might be expected to be. Sequence (b + e) might be more feasible for lactate, the difference from glycollate residing in the stability of the two radicals. While the picture presented by Lederer and Matthews is probably an excellent description of the general setup of the active center and of the involvement of a carbanion mechanism for both enzymes, there are some important aspects which we think are incorrect. From simple chemical considerations, the reduced flavin is most unlikely to exist as a dianion with negative charges both at N(1) and N(5) as proposed in [24]. Placing a strong nucleophile such as the proposed anionic N(5) of reduced flavin at about 370 pm distance to a carbonyl without formation of a σ bond appears difficult, especially since with lactate oxidase and glycollate such an adduct is stable. Furthermore His-373 and Tyr-254 could drive the reaction towards adduct formation. The key for the solution might lie in the following telling sentence quoted directly from Lederer and Matthews [24]: '... what strikes immediately is the closeness of His-373 N(3) to the ligand oxygen. It looks as if His-373 were ready to give a proton to an incipient oxyanion arising as the result of hydride donation to substrate C(2) by the reduced flavin'. This description is undoubtedly correct, as well as the deduction. However, as the authors themselves have pointed out, hydride expulsion from N(5)-H is unlikely in the present case. A close look at the original figure in question [24] also reveals that, in the context of the above quotation, addition of the lone pair of N(5) to the substrate C(2) carbonyl would do exactly the same as a hydride transfer, i.e. produce the expected substrate C(2)oxyanion via formation of a covalent adduct (Fig. 2). A mechanism such as that proposed in [24] would require that the covalent adducts formed with lactate oxidase and glycollate [18] would be formed in a non-productive side reaction starting from reduced enzyme flavin and glyoxylate. However, the glycollyl adduct derived from abstraction of the *Si*-hydrogen of glycollate clearly has been shown not to be formed directly from reduced enzyme and glyoxylate. Thus either different mechanisms are operative in the two enzymes, or a general one has to be devised, which can also explain the cases of covalent bond formation. We think that the mechanism proposed in Fig. 2 achieves that goal.

D-Amino acid oxidase is another enzyme for which much similar mechanistic evidence exists in favor of a carbanion mechanism [3]. With D-amino acid oxidase the occurrence of a covalent intermediate was first deduced and the corresponding mechanism proposed by Bright and coworkers [22]. A

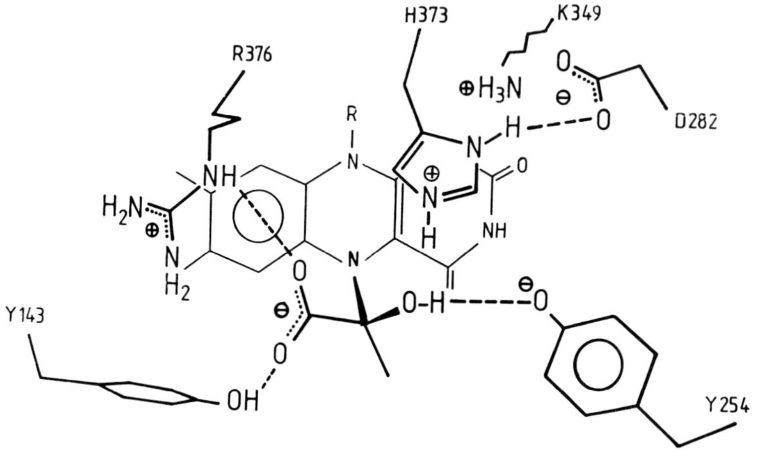

Fig. 2. *Active site of yeast L-lactate dehydrogenase (flavocytochrome b_2), modified from the description of Lederer and Mathews [24] in order to accomodate a covalent adduct between a lactyl residue and the flavin position N(5).* This adduct might occur as a transition state during the oxidation of lactate, and an analogous one has been observed upon reaction of lactate oxidase with glycollate. Note that the difference between Figs 1 and 2 is essentially the presence of a covalent bond in the latter, all other interactions being retained, with exception of the charge being on Tyr-254. (Adapted with the permission of the authors [24])

difference from the α-hydroxy-acid dehydrogenating enzymes is the *Re*-side specificity of the flavin [26] (Table 1). In addition to the other protein residues believed to play a role in the catalytic reaction (see [3, 33] for reviews), a methionine residue located at position 110 in the amino acid sequence has been identified as the residue modified in the inactivation of the enzyme by O-(2,4-dinitrophenyl)hydroxylamine. A possible catalytic role for this residue has been proposed in which it participates in the breakdown of the proposed flavin N(5)-substrate adduct [34].

Substrates undergoing α,β-dehydrogenation

The oxidation of acyl-CoA can be viewed as a special case of the dehydrogenation of an 'activated' substrate, the α position of acyl-CoA substrates clearly being activated, the β one not. Realizing this, Cornforth suggested in the late fifties a carbanion-initiated mechanism [35], which has been substantiated by several pieces of evidence, as reviewed in detail elsewhere [3, 36]. The mechanism can thus be formulated as shown in Scheme 2.

Here abstraction of the α-hydrogen as a proton is initiated by a protein base, possibly a glutamic acid carboxylate [37, 38]. The substrate has been shown to be positioned on the *Re*-face of the flavin from the stereospecificity studies of Manstein et al. [39]. As pointed out previously, several questions can be asked about the mode of transfer of the redox equivalents from the incipient carbanion to the flavin [36]: do covalent adducts or radical intermediates occur, is the carbanion a true intermediate, as opposed to a transition state, how is the second C-H bond broken, e.g. does it occur in a mechanism concerted with rupture of the first C-H bond? There are clear-cut answers to most of these questions: the reaction proceeds without formation of observable flavin radical or covalent flavin intermediates [38] and the deuterium isotope effects on rupture of the α and of the β bond behave multiplicatively as opposed to additively [40, 41]. This clearly supports a concerted breaking of the two bonds. This mechanism is best reconciled with expulsion of hydride from the β position of substrate and its direct addition to the flavin N(5)-position. Indeed studies involving the use of 5-deaza analogs of FAD and general acyl-CoA dehydrogenase confirm this [39, 42].

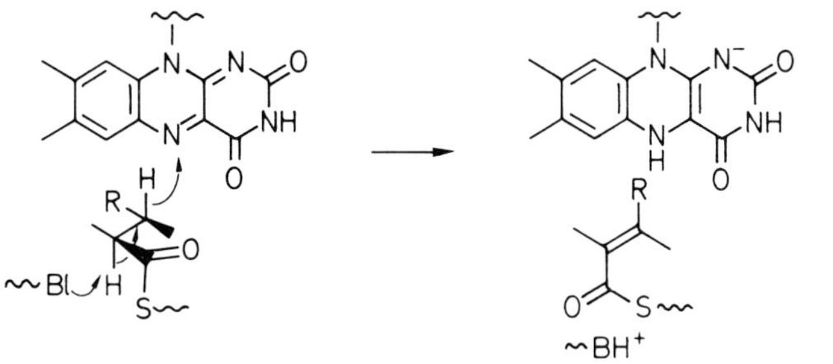

Scheme 2. *Mechanism for the concerted rupture of the C(α)-H and C(β)-H bonds and concomitant transfer of a hydride to the flavin N(5) position as deduced for general acyl-CoA dehydrogenase.* Note that the *Re*-face of the flavin has been found to accept hydride. Adapted from [42]

On the other hand, several acyl-CoA dehydrogenases have been shown to catalyze exchange at the substrate α position without any reduction occurring or being possible [43], a clear-cut proof that a carbanion can be formed. With some substrates exchange might be concurrent with α,β-dehydrogenation. This is considered a typical example of an 'uncoupled-concerted' reaction according to the definition of Jencks [44].

General acyl-CoA dehydrogenase from pig kidney is a typical representative of this class of enzymes and it exhibits several noteworthy properties. These include charge transfer absorption bands due to complexation of oxidized enzyme with substrate modifications rendering them good donors, or of reduced enzyme with acceptors such as enoyl-CoA derivatives [45–47]. The enzyme also shows only partial reduction with some substrates and typically exhibits complex kinetics. These phenomena have recently been clarified by a rapid-reaction kinetics study [38], as summarized in Scheme 3.

The main point of this scheme is the rationalization of the occurrence of substantial amounts of oxidized enzyme at the end of the reaction even in the presence of excess substrate. This results from the sum of the rates of forward and reverse reactions as well as from the ratios of the K_d for binding of substrate and product to oxidized as well as reduced enzyme. The biphasic reaction course is due to reequilibration of the

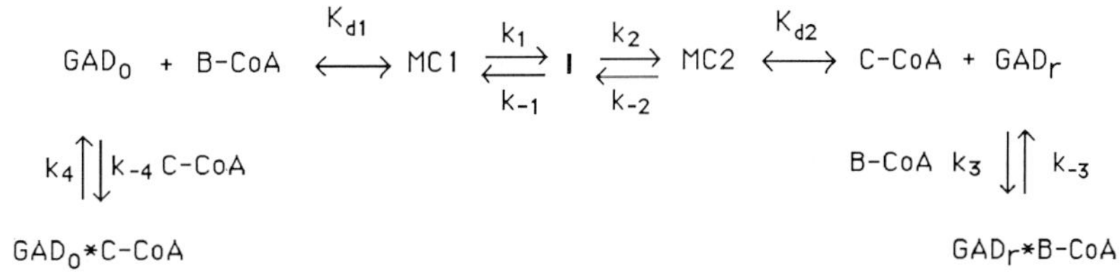

$$GAD_o + B\text{-}CoA \xrightleftharpoons{K_{d1}} MC1 \xrightleftharpoons[k_{-1}]{k_1} I \xrightleftharpoons[k_{-2}]{k_2} MC2 \xrightleftharpoons{K_{d2}} C\text{-}CoA + GAD_r$$

$$k_4 \updownarrow k_{-4}\ C\text{-}CoA \qquad\qquad\qquad\qquad\qquad B\text{-}CoA\ k_3 \updownarrow k_{-3}$$

$$GAD_o * C\text{-}CoA \qquad\qquad\qquad\qquad\qquad GAD_r * B\text{-}CoA$$

Scheme 3. *Kinetic steps involved in the reaction of general acyl-CoA dehydrogenase (GAD_o, GAD_r oxidized, reduced forms of the enzyme) with butyryl-CoA (B-CoA) as a substrate.* C-CoA = crotonyl-CoA, MC1 and MC2 = Michaelis complexes formed between oxidized and reduced enzyme and substrate or products. I = intermediate characterized by a pronounced transient charge absorption spectrum above 500 nm. K_{d1} and K_{d2} are the dissociation constants determined for formation of the corresponding complexes and $k_{(1\ to\ -4)}$ the rates estimated for interconversion of the single species. See text and [38] for details. Adapted from [38]

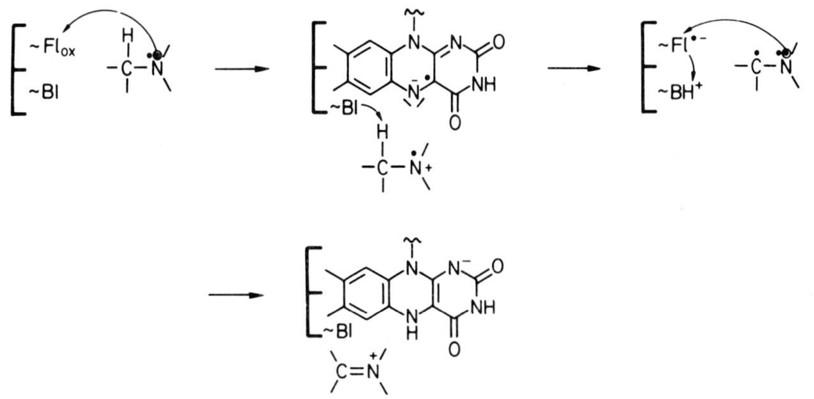

Scheme 4. *Mechanism for the oxidation of 'non-activated' amines as proposed by Silverman.* (Adapted from [49])

system subsequent to a primary reduction event followed by dissociation of product from the reduced-enzyme – product complex.

The enzyme also exhibits an intrinsic (de)hydratase activity which is due to the capacity of the enzyme to abstract a proton reversibly from the α position, thus facilitating expulsion of HO^- from the β position [48].

Non-activated substrates

Flavoproteins such as methanol oxidase and monoamine oxidase carry out oxidation of 'non-activated' substrates, this definition meaning the absence of an inductive or mesomeric effect capable of stabilizing formation of a (transient) negative charge. The removal of the α-hydrogen as a proton being unlikely to initiate the reaction, several alternatives have been put forward, a plausible one being the radical mechanism formulated by Silverman [49] for monoamine oxidase (Scheme 4).

Mechanistic details such as the sequence of abstraction of e^- and of hydrogen have been questioned by Walker and Edmondson, based on the results of a survey of a large number of substrates with catalytic velocities spanning several orders of magnitude, all of which exhibited large primary (H/D) isotope effects and based on the lack of spectral evidence for flavin semiquinone intermediates [50]. On the other hand, the substituent effects observed on the rates are consistent with a hydrogen atom abstraction in the rate-limiting step [50].

It should be emphasized, however, that there is also no compelling evidence against a hydride transfer mechanism, in particular in the case of alcohol oxidases [51, 52]. With alcohol oxidases, the enzyme is set up in such a way as to strongly stabilize flavin radicals, a hint, that this could reflect stabilization of transition states. Therefore, while we think that a radical mechanism is the most likely, the possibility remains open that oxidation of alcohols and amines might proceed via a hydride transfer mechanism.

These considerations stress the versatility of flavin reactivity, and at the same time make it difficult to reach absolute mechanistic conclusions. Flavins can accept and donate redox equivalents via hydride transfer, radical steps, or formation of covalent adducts. Since in the case of pyridine-nucleotide-linked alcohol dehydrogenases hydride transfer is the accepted mechanism, one might wonder why the same does not apply when the flavin is the acceptor. It might be that the different reaction coordinates are energetically not too dissimilar, and that the better capability of the flavin to stabilize radicals might make the difference.

STEREOSPECIFICITY OF FLAVIN-SUBSTRATE INTERACTIONS

It has long been recognized that in redox reactions involving pyridine nucleotides, the H atom involved is transferred stereospecifically, to either the *Re*- or the *Si*-side of the pyridine ring [53]. With the flavin the same basic situation exists, reactions occurring via the p orbitals, i.e. on the flavin plane, with the difference that in the general case with $Fl_{red}H^-$ the side-specific orientation of the hydrogen is easily lost due to exchange of N(5)-H or inversion processes at this center. At the active center of proteins, however, exchange and/or inversion can be restricted by the surrounding protein, and access of substrate to the flavin is likely to occur from only one side.

It is thus obvious that similar stereospecificity of substrate flavin interaction involving either the *Re*- or the *Si*-face of the flavin should exist. This was first shown to be correct for an FMN reductase from luminescent bacteria [54]. From the crystallographic data with glutathione reductase it was clear that pyridine nucleotide was positioned on the *Re*-face of the flavin [9]. This finding has been used as a reference point for

Table 1. *Stereospecificity of flavin − substrate interactions*

Enzyme	Source	Substrate	Flavin face	Reference
Glutathione reductase	human red blood cells	NADP$^+$	*Re*	[9]
Mercuric reductase	*Pseudomonas aeruginosa*	NADP$^+$	*Re*	[39]
Thioredoxin reductase	*Escherichia coli*	NADP$^+$	*Re*	[39]
p-Hydroxybenzoate hydroxylase	*Pseudomonas fluorescens*	NADP$^+$	*Re*	[39]
Melilotate hydroxylase	*Pseudomonas* sp.	NADP$^+$	*Re*	[39]
Anthranilate hydroxylase	*Trichosporum cutaneum*	NADP$^+$	*Re*	[39]
2-Methyl-3-hydroxy-pyridine-5-carboxylic acid oxygenase	*Pseudomonas* sp.	NADP$^+$	*Re*	[26]
Cyclohexanone monooxygenase	*Acinetobacter*	NADP$^+$	*Re*	[26]
Glucose oxidase	*Aspergillus niger*	glucose	*Re*	[39]
General acyl-CoA dehydrogenase	pig kidney	enoyl-CoA BH$_4$	*Re*	[39]
D-Amino acid oxidase	pig kidney	pyruvate + NH$_3^+$	*Re*	[26]
L-Lactate oxidase	*Mycobacterium smegmatis*	pyruvate	*Si*	[26]
D-Lactate dehydrogenase	*Megasphera elsdenii*	pyruvate	*Si*	[26]
Glycollate oxidase	spinach	thioglycollate (inhibitor)	*Si*	[25]
Flavocytochrome b_2	yeast	pyruvate	*Si*	[24]

determining flavin-substrate stereospecificity in other flavoproteins, using 5-deaza-8-hydroxyflavins as replacements of the native flavin, and determining the stereospecificity of ^{3}H transfer between the reduced deazaflavin and substrate relative to that with glutathione reductase [26, 39]. Knowledge of flavin-substrate stereospecificity available to date is summarized in Table 1. Of the eight pyridine-nucleotide-linked flavoproteins so far tested, all use the flavin *Re*-face in their interaction with pyridine nucleotides. In other replacement studies involving flavins with chemically reactive substituents, the same enzymes all showed that the flavin benzene ring was exposed to solvent [7]. With the two pyridine-nucleotide-linked flavoproteins whose crystal structures have been determined (glutathione reductase [9] and p-hydroxybenzoate hydroxylase [10]) there is a clear channel leading from the exposed benzene ring of the flavin along its *Re*-face; it is through this channel that the pyridine nucleotide gains access to the flavin. Thus it would appear that all these enzymes have common structural features, suggesting a common evolutionary pathway. It should be noted, however, that approximately half the flavoenzymes tested remove the *S*-proton from NAD(P)H and half the *R*-proton [55].

FLAVOENZYMES CONTAINING A REDOX-ACTIVE DISULFIDE

An interesting group of flavoenzymes about which much mechanistic and structural information is available is one where a redox-active cystine residue of the protein acts in concert with the flavin to carry out the catalytic reaction. This group of enzymes has a common theme of structural elements, modulated from one enzyme to the other in order to accommodate the particular catalytic task. The group includes lipoyl dehydrogenase, glutathione reductase, thioredoxin reductase, trypanothione reductase and mercuric reductase. Early work on the first three enzymes has been described in detail in a review by Williams [56].

With the exception of thioredoxin reductase, which appears to have a quite different protein structure, there are remarkable similarities between the other known enzymes of the group, one of the most notable characteristics of which is the absorption spectrum of the 2e$^-$-reduced enzyme, which lead to the discovery of the redox-active disulfide in lipoyl dehydrogenase [57, 58] and glutathione reductase [59]. In this EH$_2$ form the flavin is oxidized, the disulfide reduced, and the long-wavelength absorption is due to charge transfer between the thiolate anion of one of the nascent cysteine residues and the oxidized flavin [60−62]. The other remarkable similarity between the enzymes is that of similar amino acid sequence [63−66] and chain folding, resulting in overall similar three-dimensional structures. In the case of glutathione reductase from human erythrocytes, the three-dimensional structure has been solved by X-ray diffraction to 0.2-nm resolution [67, 20] and recently refined to 0.154-nm resolution [68]. Thus this structure has become the prototype for ideas about most of the other enzymes of the group. As will be described later, this approach has been very successful, particularly for mercuric reductase, but clearly has to be done with caution, because the enzymes are after all distinctly different species as shown by the very different behavior of the 6-SCN derivative of FAD introduced into glutathione reductase [69] and mercuric reductase [70]. Fortunately the crystal structures of a lipoyl dehydrogenase from *Azotobacter vinelandii* [71] and the plasmid-encoded *Escherichia coli* mercuric reductase (E. F. Pai, personal communication) are under active investigation, so that conjecture may soon be replaced by fact. Similarly the recent cloning of the genes for *E. coli* and *A. vinelandii* lipoyl dehydrogenase [73, 74] and for *Pseudomonas aeruginosa* glutathione reductase [75] and the site-directed mutagenesis work, which is underway, will do much to clarify similarities and differences within the group. (Extensive gene manipulation has already been performed with mercuric reductase, see later section.)

Glutathione reductase

The steps involved in the reaction catalyzed by glutathione reductase:

$$GSSG + NADPH + H^+ \rightleftharpoons NADP^+ + 2 GSH$$

had been dissected in some detail with the yeast enzyme by a combination of classical spectroscopic and rapid-reaction techniques [59, 76−78] and by specific chemical modifications [79, 80]. The reaction mechanism proposed for the yeast enzyme was confirmed in exquisite detail by the structural studies on the human erythrocyte enzyme, where the locations of the NADPH and glutathione substrates were easily determined, as well as the location and probable role of specific active-site residues [20, 67].

Fig. 3 is a cartoon representation of the active site, showing the location of the FAD prosthetic group, the active-site disulfide (Cys-58−Cys-63), the positions for binding of NADPH and GSSG, and the location of several residues

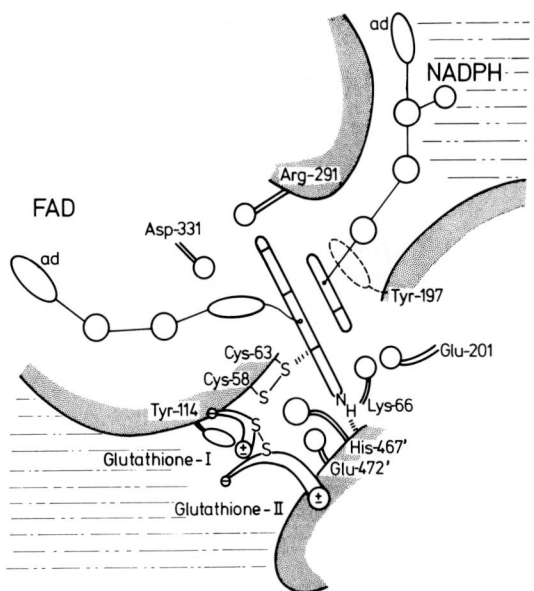

Fig. 3. *A cartoon representation of the catalytic center of glutathione reductase as described in [20] by Schulz and Pai.* Obtained from the authors and reproduced with their permission

d) Following breakage of the labile flavin C(4)-cysteinyl bond the flavin is returned to the normal oxidized state, the disulfide being reduced, with Cys-58 protonated and Cys-63 as the thiolate anion. The Cys-63 thiolate was shown to be the one responsible for the charge-transfer absorption as alkylation of Cys-58 does not affect the charge-transfer interaction [80]. Titration of the thiolate results in loss of the charge-transfer absorption with a pK of 4.8 [80]. The formation of this species is accompanied by the rapid release of $NADP^+$ to form free EH_2. The proposed sequence of reactions 3 and 4 is shown in Scheme 5. The structure of EH_2 (Scheme 5, structure III) has been determined in crystal-soaking experiments, where the sulfur of Cys-58 was found to move closer to the GSSG binding site by approximately 0.1 nm, while there was also a slight movement (≈ 0.01 nm) of Cys-63 toward the flavin [20].

e) The series of reactions described above constitute the reductive half of the catalytic reaction and are very rapid [59, 78, 79]. The catalytic cycle is completed by the reaction of EH_2 with GSSG, via thiol-disulfide interchange reactions, as envisaged in Scheme 6.

In this sequence the tautomeric form of EH_2 shown in structure IV is envisaged as the nucleophile attacking the disulfide of bound GSSG to liberate GSH and forming the mixed Cys-58—S-S-G disulfide of form V [80]. The structure of the latter species has again been determined by X-ray diffraction in crystal-soaking experiments and confirmed that Cys-58 was indeed the residue involved in the thiol-disulfide interchange reaction [20]. This had been predicted from the chemical modification studies, where it was shown that iodoacetamide alkylated Cys-58 preferentially by a factor of approximately 10, and that the resulting EHR was stable to oxidation by O_2, ferricyanide or GSSG [80]. Similarly, in crystal-soaking experiments, no evidence was found for binding of GSSG to EHR [20]. The remaining step in the catalytic cycle shown in Scheme 6 involves the nucleophilic attack of the thiolate of Cys-63 on the mixed disulfide of state V. In a reaction which is probably concerted with deprotonation of His-467′, the second molecule of protonated GSH is liberated, returning the enzyme to its fully oxidized state, ready for the next catalytic cycle.

believed to be important in catalysis [20]. A noteworthy feature of the active site is that it is made up of components from the two separate but identical polypeptide chains of the dimeric protein molecule, i.e. there are two identical active sites, in each case a GSSG is bound by elements of the two separate polypeptide chains, and the His-467′ and Glu-472′ residues shown are located in the polypeptide chain which contributes the Cys-58—Cys-63 pair to the second active site.

Considering catalysis in the predominant physiological direction:

$$GSSG + NADPH + H^+ \rightarrow NADP^+ + 2 GSH$$

the following steps are envisaged or, in many cases, found to occur [20, 80, 81].

a) Binding of NADPH involves a conformational change in which Tyr-197, normally in the position shown by the dotted line in Fig. 3, is displaced to allow entrance of the pyridine nucleotide so that the dihydropyridine ring is positioned roughly parallel to the flavin and on its *Re*-face [67].

b) The C(4) atom of the pyridine nucleotide lies directly above the flavin N(5) at a distance of 0.35 nm. Thus the 4*S*-hydrogen of NADPH should be positioned correctly for hydride transfer to flavin-N(5). Independent evidence for such a specific transfer comes from experiments where the native FAD was replaced with the 8-hydroxy-5-deaza derivative of FAD, and where stereospecific transfer of ^{3}H label between reduced flavin and pyridine nucleotide was shown [39].

c) The next step is envisaged to be reduction of the active-site disulfide (which is located on the *Si*-side of the flavin ring) by the anionic reduced flavin, via a labile flavin C(4a)-cysteinyl linkage. Precedence for such an intermediate comes from observations on the effect of NAD^+ on the EHR form of lipoyl dehydrogenase (where the active-site thiol closest to the N-terminus was alkylated by reaction with iodoacetamide) [82]. This step is accompanied by or rapidly followed by uptake of H^+ from a bound H_2O molecule [68] by a protein base, presumably His-467.

Lipoyl dehydrogenase

As discussed previously by Thorpe et al. [80] and Matthews et al. [83, 84], a set of reactions analogous to those described in Schemes 5 and 6 for glutathione reductase can be formulated for lipoyl dehydrogenase; in fact many of the mechanistic concepts found to apply for glutathione reductase were first formulated in studies with lipoyl dehydrogenase. These included the nature of EH_2 [57, 58], the recognition of an active-site base forming an ion pair to stabilize the thiolate anion of EH_2 and to play its catalytic role in the thiol–disulfide interchange reaction with lipoic acid and lipoyl derivatives [62] and the recognition that the two cysteines of the active site play distinct roles in the catalytic cycle [81]. One apparently different property is that while alkylation of Cys-58 in glutathione reductase has no effect on the charge-transfer absorption of EH_2 [24], the reaction of iodoacetamide with the analogous thiol of lipoyl dehydrogenase (Cys-45) to form EHR, abolishes the charge-transfer interaction between the thiolate of Cys-50 and oxidized FAD [81]. This is probably due to a distortion of the active-site geometry brought about by the alkyl substituent and could be a reflection of the relative sizes of the enzyme pockets binding the disulfide substrate in

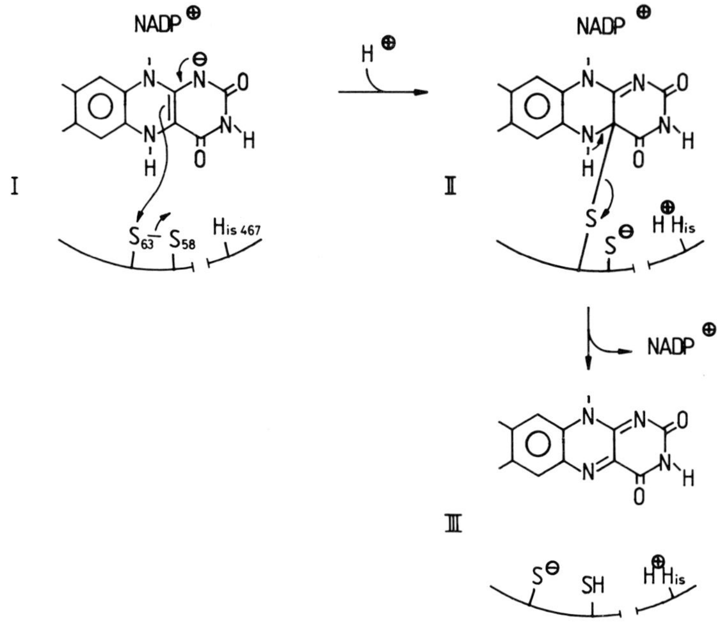

Scheme 5. *Interaction between reduced flavin and the redox-active disulfide at the active site of glutathione reductase*

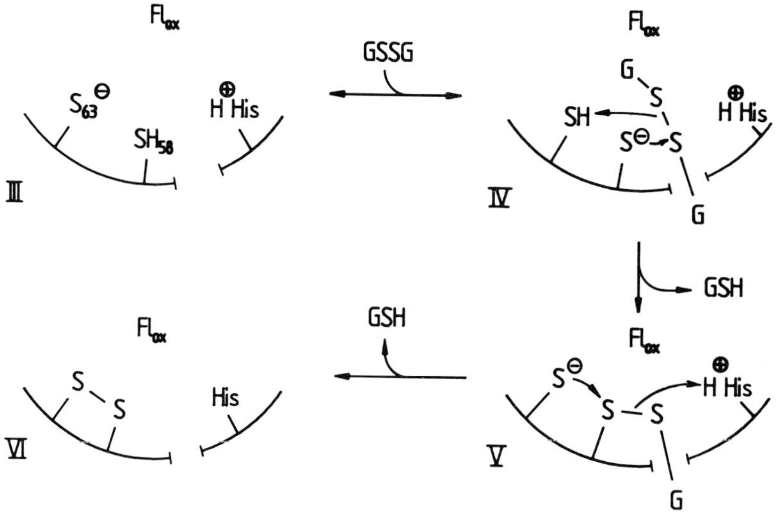

Scheme 6. *Mode of reduction of oxidized glutathione at the active center of glutathione reductase showing the thiol−disulfide interchange reactions involved in catalysis.* See text for details

the two enzymes, there being sufficient room to accomodate the alkyl group in the rather large GSSG-binding pocket of glutathione reductase without disturbing the orientation of Cys-63 with respect to the flavin, but not in the analogous case of lipoyl dehydrogenase [80].

Trypanothione reductase

A recent finding of potential world health significance is that trypanosomes and other protozoal parasites do not possess glutathione reductase, but instead a closely related enzyme with no activity toward glutathione, but with high activity toward the glutathione conjugate trypanothione disulfide [N^1,N^8-bis(glutathionyl)-spermidine disulfide] [85]. The enzyme has been isolated from *Crithidia fasciculata* [86] and *Trypanosoma cruzi* [87] and shown to have very similar properties to those of glutathione reductase, including molec-

ular mass, non-covalently bound FAD, an active-site disulfide with 14 residues identical with glutathione reductase, the formation of a typical EH_2 charge-transfer spectrum and specific alkylation of EH_2 with iodoacetamide at the active-site cysteine closest to the N-terminus [86, 87]. While little structural or mechanistic work has yet been reported, the reaction mechanism is undoubtedly similar to that of glutathione reductase.

Mercuric reductase

Mercuric reductase catalyzes the reduction of $Hg(SR)_2$ to Hg^0 using NADPH as reductant; added thiols are required for efficient catalysis, so that the overall reaction may be written:

$$Hg(SR)_2 + NADPH + H^+ \rightarrow Hg^0 + 2\,RSH + NADP^+.$$

The enzyme was found to contain a redox-active cystine residue, in analogy to lipoyl dehydrogenase and glutathione reductase [88]. It also exhibited a spectrum on $2e^-$ reduction which was remarkably similar to the EH_2 spectra of lipoyl dehydrogenase and glutathione reductase and, like the latter, could be converted to an EHR form with iodoacetamide which still retained the charge-transfer absorbance [89]. It was thus evident that remarkable similarities in structure must exist between these enzymes; this was borne out by the finding that the sequences of the active-site tryptic peptide from EHR were identical for 11 and 12 of the 16 residues of the active-site peptides of pig heart lipoyl dehydrogenase and human erythrocyte glutathione reductase respectively [89]. Further rapid progress with this enzyme was made possible by the cloning and sequencing of the gene encoding the enzyme and the demonstration of extensive sequence similarities with the other enzymes of the group [66]. Thus it was clear by analogy that the active-site disulfide should comprise residues Cys-135 and Cys-140, and that the latter should be the group responsible for the charge-transfer absorption of the $2e^-$-reduced enzyme. This was shown directly in elegant studies of Schultz et al. [90], where two oligonucleotide-directed mutant enzymes were constructed, Ser-135/Cys-140 and Cys-135/Ser-140. As expected, both of these enzymes lacked mercuric reductase activity, but retained NADPH transhydrogenase activity. The Cys-135/Ser-140 enzyme showed a normal flavin absorption spectrum and was reduced fully by uptake of two-electron equivalents to the $FADH_2$ form with no intermediate charge-transfer absorption. On the other hand, the Ser-135/Cys-140 enzyme exhibited charge-transfer absorption in the oxidized state; on lowering the pH, the long-wavelength absorbance decreased and the spectrum became like that of normal flavin [90]. These results demonstrated very beautifully that Cys-140 was indeed the thiolate residue in charge-transfer interaction with the flavin, with a pK of 5.2. A large number of other mutant forms of the enzyme have been constructed to answer specific questions of structure and mechanism [91–93]. Lack of space prevents us describing this work in detail, but it should be mentioned that the changes not only in catalytic properties but also in other physicochemical characteristics of the enzyme flavin, such as absorption spectrum, fluorescence and redox potential which are induced by single amino acid replacements, provide a wealth of information concerning the importance of specific flavin–protein interactions.

Like lipoyl dehydrogenase [84] and glutathione reductase [77], the absorption spectrum of the EH_2 form of mercuric reductase is also influenced by complex formation with $NADP^+$ and NADPH. The spectra of $EH_2 \cdot NADP^+$ and $EH_2 \cdot NADPH$ are differentiated by the latter having appreciably greater absorbance in the 510–560-nm region, while $EH_2 \cdot NADP^+$ has greater absorbance than $EH_2 \cdot NADPH$ at wavelengths beyond 580 nm. These characteristics were initially discovered in rapid-scanning stopped-flow experiments studying the reductive half of the reaction with NADPH as the substrate, where the sequence shown below can be inferred [94]:

$$E + NADPH \underset{}{\overset{k_1}{\rightleftharpoons}} E\text{-}NADPH \overset{k_2}{\rightarrow} EH_2\text{-}NADP^+$$

$$EH_2\text{-}NADP^+ \underset{}{\overset{k_3}{\rightleftharpoons}} EH_2 + NADP^+$$

$$EH_2 + NADPH \underset{}{\overset{k_4}{\rightleftharpoons}} EH_2\text{-}NADPH.$$

In this study, at pH 7.3, 5°C, E-NADPH (presumably a charge-transfer complex with NADPH as donor and FAD as acceptor, and preceding reduction) was formed in the dead time of the stopped-flow apparatus, with EH_2-$NADP^+$ and EH_2-NADPH following sequentially at rates of $43\ s^{-1}$ and $8\ s^{-1}$, independent of the NADPH concentration. The latter rate is probably determined by the 'off' constant, k_3. Further rapid-reaction experiments monitoring the enzyme spectrum during catalytic turnover with Hg(cysteine)$_2$ and NADPH as the limiting substrate indicated that the enzyme was predominantly in the EH_2-NADPH form in the steady state [95] in agreement with the conclusions of Miller et al. [96].

The mutant enzyme studies have shown that the N-terminal cysteine residues, Cys-10 and Cys-13, do not appear to play any catalytic role [91]. On the other hand, when the C-terminal cysteine pair, Cys-558 – Cys-559 is replaced by alanine residues, mercuric reductase activity is lowered drastically, by a factor of approximately 1000 [92]. The C-terminal cysteines have been shown by fluorescence studies to be located close to the flavin and to be in slow redox communication with the active-site Cys-135 – Cys-140 pair. Thus, on storage at 0°C for several days, it is found that the flavin fluorescence of wild-type enzyme is approximately double that when the enzyme is freshly isolated, and that this is due to the oxidation of the Cys-558 – Cys-559 pair to the disulfide [97]. In this form the enzyme exhibits a lag in mercuric reductase activity (see also [98]), due to the necessity of reducing the C-terminal disulfide via the active-site dithiol. The oxidized enzyme, with Cys-558 – Cys-559 as disulfide, on anaerobic titration with NADPH or dithionite gives rapid spectral changes indicative of EH_2 formation on each addition of reducing agent, followed by slower disappearance of the long-wavelength absorption. Spectra taken at equilibrium after each addition show that full formation of EH_2 is not achieved until 1.8 mol dithionite is added/mol enzyme flavin; formation of EH_4 requires 3 mol dithionite/mol. In contrast, enzyme which has been pre-treated with dithiothreitol, and then filtered through Sephadex G-25 to remove excess reagent (the active-site disulfide is reoxidized during this step), requires only one equivalent of dithionite to be reduced to the EH_2 level (or one equivalent of NADPH to reach the EH_2-$NADP^+$ level) and shows no lag in catalysis [97]. Rapid-reaction studies showed that the thiol–disulfide interchange between Cys-135 – Cys-140 and Cys-558 – Cys-559 is too slow to be involved as a catalytic step, rather the reduction of the C-terminal disulfide is a necessary priming step for catalysis.

The catalytic role of the C-terminal cysteine pair was suggested by rapid-reaction studies of the mutant enzyme Ala-558 – Ala-559 [99]. Reaction of the enzyme with an equimolar concentration of HgCl$_2$-EDTA and an excess of NADPH lead to the rapid production of EH_2-NADPH, as with wild-type enzyme, followed by slower conversion ($k \approx 2\ s^{-1}$ at pH 7.3, 4°C) to an inhibited enzyme with a strong charge-transfer interaction between NADPH and oxidized flavin, which is quite distinct from the EH_2 forms of the enzyme. As the complex requires only one equivalent of Hg^{2+}, it is clear that no reduction of Hg^{2+} had occurred and that probably the inhibited form involves complexing of Hg^{2+} with both the active-site thiols, Cys-135 and Cys-140. The role of Cys-558 – Cys-559 in catalysis can therefore be envisaged as essential for efficient binding of Hg^{2+} without formation of an inhibited complex like that described above, involving a bidentate complex with Cys-135 – Cys-140. Such bidentate complexes typically have association constants in the range of $K_a = 10^{37} – 10^{45}$ [100, 101] and would be expected to lower the redox potential of the bound Hg^{2+} to such a level (as low as -475 mV) as to be thermodynamically difficult to reduce by NADPH. Thus it has been proposed that the C-terminal cysteine residues participate with the active-site thiol pair to provide a tri- or tetra-coordinate Hg^{2+} complex at the active site [97]. With the additional ligands to Hg^{2+}, the resulting complex should be energetically destabilized relative to the bidentate complex, primarily due to unfavorable interactions between ligand electrons and those on Hg^{2+}. Thus, each of the Hg-S bonds in the complex should be more ionic in character, with more positive charge being localized on the Hg^{2+}, than in the bidentate complex and hence have a higher and more favorable reduction potential. It should be noted that Brown and coworkers, comparing the sequence homology between mercuric reductase and glutathione reductase, suggested that the C-terminal cysteine pair might be involved in Hg^{2+} binding [66]. Their proposal derived from the assumption that the two proteins are likely to fold in a similar fashion. Since glutathione reductase has the C-terminal of one monomer folded into the active site of the other monomer in order to provide the catalytically essential histidine residue (see previous section), it seemed reasonable to propose that in mercuric reductase the C-terminal cysteines of one subunit may be similarly folded into the active site of the other subunit.

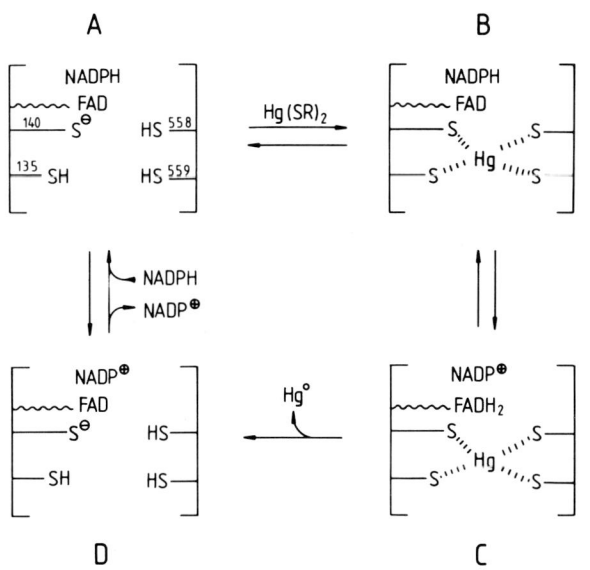

Scheme 7. *Role of the C-terminal cysteine pair, Cys-558 and Cys-559, in conjunction with the active-site cysteine pair, Cys-135 and Cys-140, to form a multidentate complex with Hg^{2+}, and the subsequent reduction of Hg^{2+} to Hg. From [97]*

This prediction was thus borne out very nicely by the later findings.

The proposed role of the C-terminal thiols in the catalytic mechanism is illustrated in Scheme 7. The EH_2-NADPH complex of activated enzyme (form A) is produced, as described earlier, by reaction of 3 mol NADPH with 1 mol unactivated enzyme or 2 mol NADPH with 1 mol pre-activated enzyme. Binding of Hg^{2+} results in a tri- or tetra-coordinate complex (form B). As discussed above, this would be expected to raise the redox potential of the bound Hg^{2+} so that it might be readily reducible by the reduced flavin of form C. The dissociation of Hg^0 would thus result in the activated EH_2-$NADP^+$ complex (form D), which by dissociation of NADP and reaction with NADPH would complete a catalytic cycle [97]. This scheme is a modification of that proposed by Miller et al. [96], to take into account not only the role of the C-terminal cysteines, but also the observations of Distefano et al. [93] for the reduction of Hg^{2+} by reduced flavins. Although the mechanism of reduction of Hg^{2+} is not yet clear, the most likely route, in analogy with the mechanisms of glutathione reductase and lipoyl dehydrogenase, would be via a transitory C(4a) adduct of the flavin with Cys-140, concomitant with an outer-sphere reduction of Hg^{2+}, as shown in Scheme 8. Experimental support for this mechanism comes from rapid-reaction studies of wild-type enzyme with NADPH at low pH values (but in the absence of Hg^{2+}) where spectral changes consistent with the intermediate formation of a flavin C(4a) derivative were observed [102].

In summary, the presence of the C-terminal cysteine pair in mercuric reductase appears to be the main factor differentiating this enzyme from lipoyl dehydrogenase and glutathione reductase, and permitting the efficient reduction of Hg^{2+} as opposed to reduction of disulfide substrates with the other enzymes. It should be noted that the EH_2 forms of both lipoyl dehydrogenase [103] and glutathione reductase [59] react with Hg^{2+}, but form inhibited complexes, probably due to formation of tight bidentate complexes with the active-site thiols. In the absence of the C-terminal cysteine pair, mercuric reductase is restricted in catalytic capacity in ways similar to those of the disulfide reductases, i.e. to transhydrogenation reactions, which involve only the flavin, and to thiol–disulfide interchange reactions, such as NADPH–Nbs_2 reductase activity [93].

ELECTRON TRANSFER MECHANISMS

Reduced flavoproteins, produced by the pathways described in the previous sections, can complete their catalytic cycle in the oxidative half-reaction, either by one- or two-electron transfers. We have already discussed in some detail

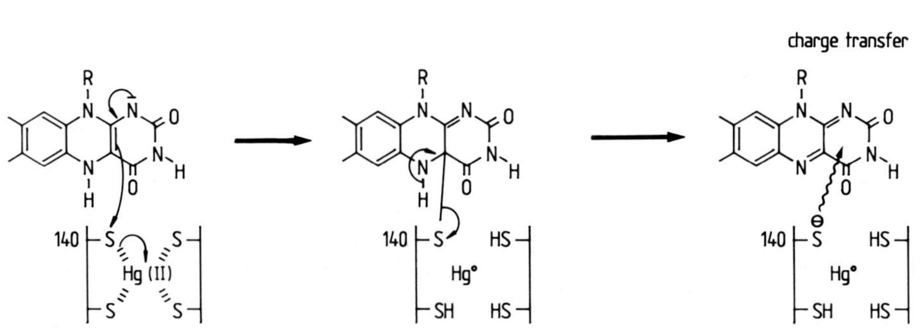

Scheme 8. *Detailed catalytic mechanism proposed for the reduction of Hg^{2+} at the active center of mercuric reductase. From [97]*

the oxidative half-reaction in the case of flavoproteins also having a redox-active disulfide, because in these enzymes the definition of what constitutes the reductive and what the oxidative half-reaction is rather arbitrary, since the reactions are in general quite reversible, and involving two-electron transfer steps. In the final section we will deal with the reaction of reduced flavins and flavoproteins with molecular oxygen, because of the complexities of these reactions and the large number of enzymes involved. In this section we will discuss briefly one-electron transfers involving flavins. The crossover between two-electron transfers, catalyzed by simple pyridine-nucleotide-linked enzymes, and the one-electron transfers, occurring with iron-sulfur proteins and heme proteins, is a phenomenon largely restricted to flavoproteins and quinones and is due to their favorable redox potentials and thermodynamic stabilization of the semiquinone forms.

At first sight it might appear to be relatively simple to reoxidize a reduced flavoprotein in two sequential steps, a process which must exist for example with enzymes which use ferricyanide as a good acceptor:

$$E\text{-Fl}_{red}H_2 + Fe(CN)_6^{3-} \rightarrow E\text{-FlH}^{\cdot} + Fe(CN)_6^{4-}$$

$$E\text{-FlH}^{\cdot} + Fe(CN)_6^{3-} \rightarrow E\text{-Fl}_{ox} + Fe(CN)_6^{4-}.$$

Such a simple two-step mechanism was first documented by Strittmatter in the case of NADH−cytochrome b_5 reductase, who showed that the fully reduced enzyme and its semiquinone form were able to reduce cytochrome b_5 at rates sufficiently high to be consistent with steady-state turnover [104]. A similar two-step mechanism has been demonstrated for reaction between reduced general acyl-CoA dehydrogenase and the electron-transferring flavoprotein of pig kidney [105]. In this case the electron transfer was shown to be rapid only when the enoyl-CoA product was bound to the reduced dehydrogenase, thus identifying the important role of the product in modulating the thermodynamic and kinetic behavior of the dehydrogenase.

The factors governing such electron transfers between two proteins have been explored in a series of rapid reaction studies by Tollin and coworkers, employing the photogenerated flavodoxin semiquinone as reductant and a number of heme proteins and ferric salts as oxidants [106−108]. The rates of electron transfer were concluded to be determined by contributions from three main factors, the thermodynamic driving force between the two partners, protein surface topography and electrostatic potential at the reaction site. Computer modelling showed a very nice fit between the negatively charged region of the protein around the flavin semiquinone and a positively charged region around the heme edge of cytochrome c [106−108], suggesting that this interaction was important for the electron transfer process, and consistent with the observed effects of ionic strength on the reaction rate. This was supported by the opposite effect of increasing ionic strength causing an increased rate of electron transfer to ferricyanide, consistent with a kinetic barrier between two negatively charged partners [106].

A clear limitation on the use of two sequential one-electron transfers in enzyme catalysis would be if the reaction of the flavoprotein semiquinone with the one-electron oxidant was much slower that the first one-electron transfer. A good example of this is given by flavodoxin: while the fully reduced protein at pH 7, 25 °C reacts with molecular oxygen relatively fast to yield the flavodoxin neutral semiquinone and O_2^- ($k \approx 10^4 \text{ M}^{-1}\text{s}^{-1}$), the reactivity of the semiquinone with O_2 is many orders of magnitude lower [109]. This problem seems to have been addressed in nature by the development of multicenter redox proteins with rapid internal electron transfers between the redox centers, permitting specific functions to be carried out by the individual centers. This concept arose from studies of milk xanthine oxidase, a particularly complex example containing, in addition to FAD, a molybdenum center and two discrete iron-sulfur centers of the spinach ferredoxin type (see [110, 111] for reviews). With this enzyme the multiphasic kinetics of reduction by substrate and reoxidation by O_2 were explained on the assumption of xanthine donating two electron equivalents at the same time through the molybdenum centers, followed by the rate-limiting dissociation of urate and the immediate redistribution of reducing equivalents among the other redox centers, the distribution being governed solely by the relative redox potentials of the centers [112]. The validity of this concept was demonstrated by determination of the redox potentials, whose absolute values were in accord with the predictions from the rapid-equilibrium concept [113, 114]. With the Mo^{VI}/Mo^V couple having the lowest potential and one of the iron-sulfur centers the highest, it is thus possible to load the enzyme with a total of six electron equivalents from xanthine, although in the steady state of catalytic turnover it is doubtful that this stage would ever be reached. The reaction of O_2 with the enzyme was shown to be with the flavin, since a deflavoenzyme could be prepared which was still reduced rapidly by xanthine, but reacted extremely slowly with O_2 [115]. The enzyme containing FAD in the fully reduced state was shown to react more rapidly with O_2 than that containing FADH$^{\cdot}$, the former gave rise directly to H_2O_2 as product, while the latter, as expected, gave O_2^- [116−118]. Thus the 'O_2^- flux', the percentage of the total xanthine oxidation giving rise to O_2^-, depends on the extent of electron loading of the enzyme by substrate in the steady state, the greater the degree of reduction the smaller the O_2^- flux. In this enzyme the function of the iron-sulfur centers appears to be to act as electron sinks, allowing the Mo centers to achieve the Mo^{VI} state for reaction with reducing substrate and the flavin to be in the $FADH_2$ or FADH form for reaction with O_2.

The concept of rapid internal equilibration of electrons within a multi-redox-center enzyme seems to apply fairly generally, with the different centers possessing discrete functions. For example, P-450 reductase contains both FAD and FMN, and the two flavins have quite different potentials [119]. The FAD has been shown to be the flavin reduced by NADPH in the catalytic reaction and FMN the flavin reacting with P-450. In catalysis the enzyme appears to function between the $3e^-$- and the $1e^-$-reduced levels, with $FMNH_2$ being the most reactive species with P-450, and rapid internal electron transfer between FADH$^{\cdot}$ and FMNH$^{\cdot}$ at the $2e^-$ level regenerating $FMNH_2$ for reaction with a second molecule of P-450 [120].

Similar concepts apply in the actual interactions between discrete proteins in an intermolecular complex. This has been discussed in detail by Lambeth and Kamin [121] particularly with reference to the adrenal mitochondrial P-450 system (composed of a complex between the flavoprotein, NADPH−adrenodoxin reductase, the [2Fe-2S] protein, adrenodoxin, and a steroid-specific P-450), E. coli sulfite reductase and spinach nitrite reductase. These all have the common theme of reducing equivalents from NADPH entering through the flavin component and finally leaving to the specific acceptor through a second or third redox center of the complex, with internal 'electron shuttling' being an important part of the overall electron transport process.

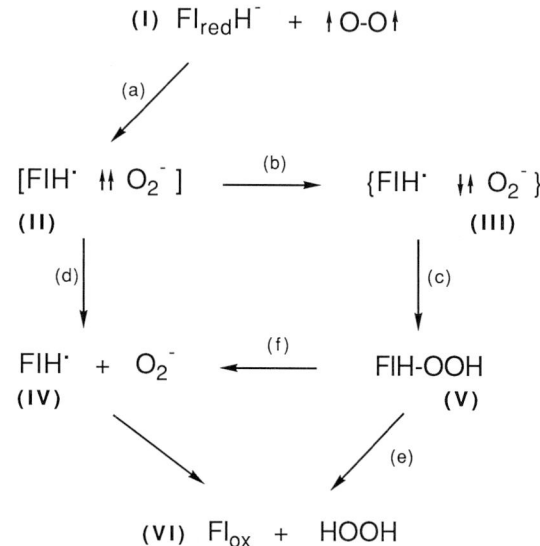

Scheme 9. *Mode of activation of dioxygen by free flavin*

REACTIONS WITH OXYGEN

Reactions of free reduced flavin with oxygen

Reduced flavin is one of the few biocatalysts which can efficiently reduce dioxygen. The mechanism by which this is brought about by free reduced flavin has recently been solved. The critical step in any chemical reaction of paramagnetic dioxygen is the inversion of the spin of one electron, a relatively slow process. Due to extensive delocalisation in a flavin radical, such an inversion is assumed to be comparatively facile. As shown in Scheme 9, which results from the work from the groups of Bruice [122, 123] and Anderson [124, 125], a first encounter results in the transfer of an electron from reduced flavin (I) to O_2 to form a paramagnetic complex of O_2^- and the flavin radical (II, step a). It should be pointed out that transfer of electrons from the flavin to O_2 might occur at a distance of several nanometres and that in (II) the two species do not have to be in close contact.

Spin inversion of the unpaired flavin electron in complex (II), or physical rotation of one partner lead (step b) to the biradical complex (III), in which the two electrons have paired spins and can combine to give the covalent flavin 4a-hydroperoxide (V, step c). This will occur provided distance and orientation of flavin and superoxide is such as to allow formation of a thermodynamically stable bond. It is noteworthy that the point of bond formation corresponds to the locus of highest spin density in FlH˙ [126]. The hydroperoxide (V) itself can dissociate heterolytically via step (e) to yield hydrogen peroxide and Fl_{ox} (VI). As will be detailed below, dissociation of the complex (II) via (d) to yield free FlH˙ (IV) and superoxide is a reaction competing with (b). The latter two species could also arise from homolytic cleavage of the hydroperoxide (V) via step (f), a conversion which, however, is chemically less likely.

Important experimental evidence supporting the sequences of Scheme 9 is obtained from experiments of Anderson [124]. An essentially quantitative formation of free neutral flavin radical can be observed in a pulse radiolysis experiment within 20 μs after the pulse. This species is converted to an intermediate in a very fast reaction, which is directly dependent on the concentration of O_2^- ($k = 7 \times 10^8 \, M^{-1} \, s^{-1}$). The absorption spectrum of the intermediate is consistent with the presence of approximately 80% flavin hydroperoxide (V) and 20% Fl_{ox} (VI). At pH 6.5 the subsequent decay of (V) to (VI) is independent of O_2^- concentration, $k = 260 \, s^{-1}$. It follows that some 20% of (VI) must be formed from (II) bypassing (V). This probably involves direct e^- transfer from (III) as will be discussed in the next section.

Activation of oxygen by flavoproteins

While many reduced flavoproteins react only slowly with molecular oxygen, there are also many which can react efficiently, some of them having the additional capacity of catalyzing the insertion of one atom of O from O_2 into substrate. Accordingly they have been classified as either 'oxidases' or 'monooxygenases'. A characteristic of the 'monooxygenases' is the occurrence of the flavin 4a-hydroperoxide (V, Scheme 9) on the catalytic path. Its formation can be viewed as proceeding via steps (a), (b), and (c) (Scheme 9). This is consistent with recent pulse radiolysis experiments [128], in which immediately upon pulsing glucose oxidase at pH 6.0 and *p*-hydroxybenzoate hydroxylase at pH 6.5 under conditions generating enzyme flavin radicals and O_2^-, a species having the spectral characteristics of the neutral radical (cf. IV) were observed. This then converted to the hydroperoxide (V) in a second-order reaction with O_2^- (rate $\approx 10^9 \, M^{-1} \, s^{-1}$ for glucose oxidase). The hydroperoxides subsequently decayed to the oxidized enzyme (VI) at a rate of $350 \, s^{-1}$ for glucose oxidase, and $70-80 \, s^{-1}$ for *p*-hydroxybenzoate hydroxylase. In sharp contrast to this, pulsing at pH 8.5 lead to exclusive formation of the enzyme-bound flavin radical anions, Fl^{-}, no hydroperoxide (V) being formed subsequently.

The initial product of reaction of aqueous electrons from the radiation pulse is the enzyme anionic flavin radical Fl^{-} followed by rapid protonation to give the neutral FlH˙ (IV). The pK_a of the glucose oxidase radical is 7.3 [129] and that of *p*-hydroxybenzoate hydroxylase ≈ 7.1 (R. F. Anderson and V. Massey, unpublished results). At low pH, neutral FlH˙ will bind O_2^- in an orientation compatible with radical pairing, thus yielding (III). In contrast to this, at high pH, repulsion of the two negatively charged species Fl^{-} and O_2^- appears efficiently to prevent any encounter.

These rather straightforward interpretations have to be compared with the oxygen reaction of reduced glucose oxidase and of reduced hydroxylases (e.g. *p*-hydroxybenzoate hydroxylase) in the absence of substrate or effectors, the so-called 'oxidase' reaction. With these two enzymes a rapid second-order reaction with O_2 ensues which leads to formation of oxidized enzyme and H_2O_2, no radical species, superoxide or covalent hydroperoxides such as (V) being found [128]. Most importantly, rates appreciably faster than the decay of enzyme-bound flavin 4a-hydroperoxide to Fl_{ox} (cf. conversion of V to VI via step e, Scheme 9) can be observed, precluding a pathway over steps (c) and (e), i.e. over (V). On the other hand, a spin inversion must occur at some stage during the formation of these products. Thus, the sequences of Scheme 9 are clearly not sufficient to explain the oxidase reaction and the question arises as to whether a second, parallel, mechanism has to be formulated. The 'direct' reaction of reduced flavin with O_2 to yield a covalent adduct over a 'low lying flavin triplet state' has been proposed [131]. For the discussion of this seeming contradiction the reaction of free reduced flavin can be taken as the standard case, in which no particular effects due to protein interactions will stabilize or destabilize any intermediates. Rapid-reaction

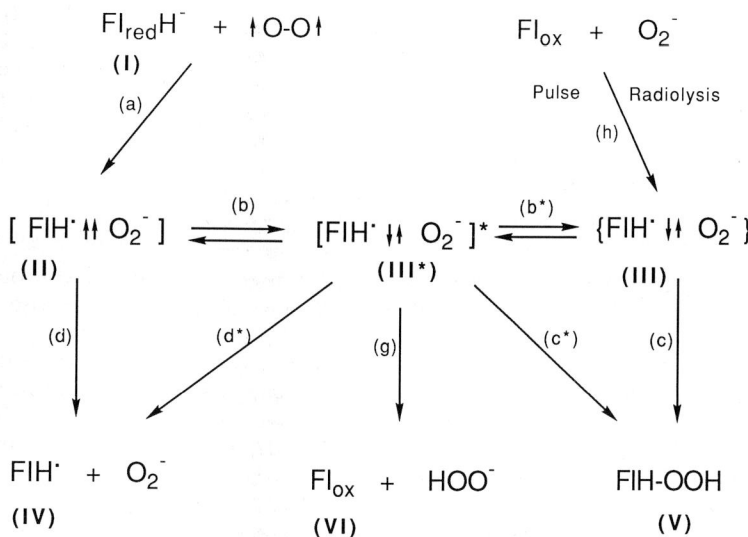

Scheme 10. *Possible routes for the reduction of oxygen by reduced enzymes such as glucose oxidase or p-hydroxybenzoate hydroxylase (in the absence of substrate), and mode of reaction of intermediates by pulse radiolysis.* See text for details

studies on the reaction of free $Fl_{red}H^-$ with O_2 require the formation of an intermediate to account for the observed kinetics [132, 133]. With reduced flavoprotein a hydroperoxide may be stabilized depending on the starting conditions (e.g. presence of substrate or effectors with p-hydroxybenzoate hydroxylase). Thus the simplest interpretation would be: when a covalent hydroperoxide is on the reaction pathway it is likely to be observable; if none can be observed, most probably it is not formed, or, at best, it is a transition state.

There is evidence supporting this interpretation. Reduced flavodoxin reacts in a second-order reaction with O_2 to give the neutral radical $FlH^{\cdot}$ and O_2^-. The $FlH^{\cdot}$ formed does not react further efficiently. From the X-ray structure of flavodoxin it is apparent that the space around the flavin positions which are candidates for formation of a covalent bond with O_2^- is insufficient to accommodate the hydroperoxide product, and that access of O_2 is severely limited [8]. Only part of the benzene moiety of the flavin and the 8-CH_3 function are exposed to solvent. Thus, in this case the reaction will be likely to proceed via steps (a) and (d) of Scheme 9, formation of (III) being possible, but abortive in the sense that the collapse of the radical pair cannot yield (V).

The two different sets of dioxygen reactivities exemplified by p-hydroxybenzoate hydroxylase, and the reaction of reduced glucose oxidase with O_2 can thus be rationalized according to the steps of Scheme 10, an extension of Scheme 9. The right-hand side represents the (artificial) reaction, in which $FlH^{\cdot}$ and O_2^- are formed by pulse radiolysis and then combine to yield a 'productive' complex (III, step h), which collapses to the covalent hydroperoxide (V, step c). This requires that from the beginning of the reaction the active center is set up in order to stabilize (V), i.e. to prevent release of HOO^-.

On the left-hand side the 'oxidase' reaction is shown in which the paramagnetic complex (II) is formed (step a) similarly as in Scheme 9. For formation of the final products $Fl_{ox} + HOO^-$ spin inversion is envisaged as occurring during conversion of (II) to (III*). The crucial point in this scheme would be the difference between the two diamagnetic species (III) and (III*). In our opinion this might lie simply in the different properties of (III) and (III*). Thus (III) would appear to be able to collapse to a thermodynamically metastable hydroperoxide such as (V); (III*), in contrast, would not and thus reacts further by transfer of $1e^- + H^+$ to O_2^- (step g), or dissociates (steps d).

These differences might result from steric restrictions as, for example, in the case of flavodoxin. There the inability to form (V) might force dissociation of either complex (II) or (III*) to yield $FlH^{\cdot} + O_2^-$. In the case of p-hydroxybenzoate hydroxylase, it has been shown [134] that the presence of substrate or product results in substantial changes of the geometry of the flavin, implying modification of the protein structure around it. It is thus easily conceivable that such effects contribute to the stabilization of the hydroperoxide (V), also since this consists largely in preventing the dissociation of HOO^-, i.e. abstraction of the N(5) hydrogen as H^+. Binding of substrate introduces an additional net negative charge at the active center and this might be at the origin of the different reactivities. In this context the course of the reaction might be affected by the different state of charge of the species from which it originates.

ACTIVATION OF FLAVIN-HYDROPEROXIDE OXYGEN

Following the primary step of oxygen activation leading to formation of a flavin hydroperoxide as discussed above, the hydroperoxide moiety requires activation and/or modulation of its chemical activity for reaction with a second substrate. The key molecule is, of course, the flavin 4a-hydroperoxide (cf. also V, Schemes 9 and 10) itself, a molecule of unusual properties:

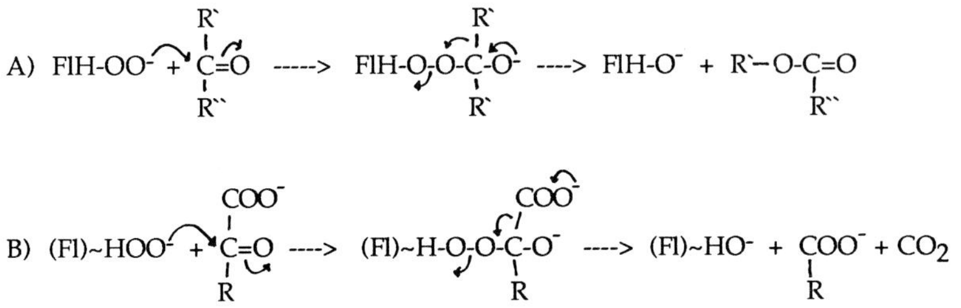

$$A) \quad FlH-OO^- + \underset{R''}{\overset{R'}{C}}=O \longrightarrow FlH-O-O-C-O^- \longrightarrow FlH-O^- + R'-O-C=O$$

$$B) \quad (Fl)\sim HOO^- + \underset{R}{\overset{COO^-}{C}}=O \longrightarrow (Fl)\sim H-O-O-C-O^- \longrightarrow (Fl)\sim HO^- + COO^- + CO_2$$

Scheme 11. *(A) Baeyer-Villiger mechanism formulated by Walsh et al. [138] for the oxidative ring expansion of cyclohexanone monooxygenase. (B) Mechanism for the oxidative decarboxylation of L-lactate oxidase involving enzyme-bound H_2O_2 and oxo acid product*

As found mainly by Bruice and colleagues (see [2, 134–136] for reviews) its chemical reactivity resembles more that of peracids than that of alkylperoxides, and indeed it might be viewed as a derivative of barbituric acid. The first striking property is the unusually low pK_a of 7–8 estimated by Favaudon for the distal oxygen [137], which correlates with the similarly low value of 9.1–9.5 for the 4a-hydroxyl function of the related flavin 4a-hydroxide [134]. This has been attributed to the inductive effects of the substituents around position 4a. Basically the three following mechanisms can be formulated for the reactions of this hydroperoxide: (a) nucleophilic activation, (b) radical activation, (c) electrophilic activation.

The nucleophilic activation does not pose mechanistic difficulties. As has been formulated for cyclohexanone monooxygenase [138], it implies nucleophilic attack of the distal oxygen of the hydroperoxide to a carbonyl residue of the substrate and a subsequent Baeyer-Villiger rearrangement as shown on Scheme 11 (A). With enzymes such as L-lactate oxidase essentially the same scheme could be applied, with the difference that hydroperoxide formed at the active center reacts with the enzyme-bound α-oxoacid before dissociation of the products occurs and induces oxidative decarboxylation by the similar mechanism shown in Scheme 11 (B).

However, as discussed in a previous section, no evidence for intermediate formation of a flavin 4a-hydroperoxide could be obtained with L-lactate oxidase, Fl_{ox} and H_2O_2 being the products formed directly from $E \sim Fl_{red}H^-$ and oxygen. From the mechanistic point of view, the oxygen reactivity of L-lactate oxidase is that of an 'oxidase', although from its overall function it might better be called a 'monooxygenase'.

A special case is most probably represented by the bioluminescent reaction of bacterial luciferase. Also with this FMN-dependent enzyme a metastable flavin 4a-hydroperoxide is formed upon reaction of reduced luciferase-bound $FMNH^-$ with oxygen in a primary step (b) of Scheme 12 [139]. The unique behavior of this enzyme should be noted in this context since here FMN behaves as a true coenzyme, and not as a tightly bound prosthetic group as in the majority of flavin enzymes. Free $Fl_{red}H^-$ is first bound by luciferase (step a), on which it carries out catalysis, then being released as FMN upon completion of the catalytic cycle.

The actual bioluminescent reaction starts upon addition of a long-chain aldehyde $R-CH=O$ (chain length = 8–14 for the *in vitro* reaction, 14 for the natural substrate *in vivo*) and leads to formation of the flavin 4a-hydroxide, carboxylic acid (R-COOH), and light (Scheme 12, reaction c). Subsequently water is split off (step d), FMN dissociates (step e), and can be rereduced by an FMN reductase (see [140] for a review).

Of particular interest here are the processes which lead to the production of light, i.e. the chemical events producing the

a) $L + FMN_{red}H^- \rightleftharpoons L \sim FMN_{red}H^-$

b) $L \sim FMN_{red}H^- + O_2 + H^+ \longrightarrow L \sim FMNH-OOH$

c) $L \sim FMNH-OOH + R-CH=O \longrightarrow L \sim FMNH-OH + R-COOH + h\nu$

d) $L \sim FMNH-OH \longrightarrow L \sim FMN + H_2O$

e) $L \sim FMN \rightleftharpoons L + FMN$

Scheme 12. *Sequence of reactions occurring during the catalytic cycle of bacterial luciferase (L)*

energy which is needed to populate the excited state of the emitter. These events take place during step (c) of Scheme 12 and have not yet been clarified satisfactorily. An early proposal [141], by which a Baeyer-Villiger mechanism is operative as discussed above for the reaction of cyclohexanone monooxygenase, is probably not correct as suggested by recent results [143]. Present indications point towards a so-called CIEEL mechanism (chemically induced electron exchange luminescence), for which an adaptation of a general scheme [142] for the bacterial luciferase reaction has been formulated [143, 144]. Upon binding of luciferase the formation of a peroxyhemiacetal by nucleophilic attack of the hydroperoxide moiety on the aldehyde carbonyl is envisaged to occur [141]. The key step involves an intermediate intramolecular transfer of an electron from the flavin N(5) region to the peroxide, which 'weakens' the peroxide bond, inducing fragmentation. Subsequently the electron is transferred back to an unoccupied orbital of the 4a,5-dihydroflavin radical cation yielding an excited state of the flavin 4a-hydroxide. Relaxation of the latter by emission of light ($\lambda_{max} \approx 490$ nm) concludes the events of step (c) of Scheme 12 [144]. It should be pointed out that fluorescent proteins can take part in the reaction and lead to emission of light with either blue-shifted [145] or red-shifted spectra [146]. Mechanistically the involvement of these proteins requires some adjustments of the basic mechanism of Scheme 12, for the discussion of which we refer to the original literature [144, 145].

In conclusion, the mechanism of bacterial luciferase can be viewed as being initiated by a nucleophilic reaction of the peroxide, most probably followed by a radical-type oxidation of the aldehyde, and concomitant generation of an excited state, thus by a combination of nucleophilic activation and electron transfer.

As to electrophilic activation, there are two typical subclasses of flavin enzymes which are reasonably assumed to function by this mechanism. The first is that exemplified by the microsomal 'mixed-function monooxygenase', an enzyme involved in detoxification processes in the liver [147], which

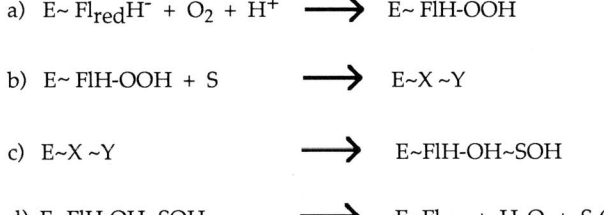

Scheme 13. *Sequence of reactions catalyzed by p-hydroxybenzoate hydroxylase.* X and Y denote chromophores occurring as intermediates during the insertion of oxygen into phenolates, proposed to be derived from the flavin and/or the substrate. The chemical structure of the intermediates has not yet been elucidated

metabolizes a variety of nucleophilic substrates (Nu) having the typical functions R_3N or R-SH (R = alkyl or H), but also others such as I^- [148]. This mechanism is best depicted by a nucleophilic displacement at the distal atom of the hydroperoxide [127]:

$$FlH\text{-}O\text{-}O\text{-}H + Nu \rightarrow FlH\text{-}O^- + HO\text{-}Nu^+.$$

This rather straightforward mechanism contrasts with that of the so-called 'hydroxylases' which, in spite of having been the first to receive extensive attention [149, 150], still resist elucidation in their key step (c) of Scheme 13. The cycle is initiated, as in the previous cases, by formation of the hydroperoxide (V) (step a).

In the presence of substrate, which must be activated by the presence of a phenolic function, the hydroperoxide complex is converted (step b) to a species which stands out by its characteristic absorption spectrum, having a maximum in the region 350–420 nm with high molar absorption [150–152]. Subsequently this species is transformed to the complex of enzyme-bound flavin 4a-hydroxide and product (step c), which can dissociate into the components as shown by step d. The fact that only (nucleophilic) phenolates will react suggests that the reaction is initiated by nucleophilic attack of the latter onto the electrophilic hydroperoxide to form an intermediate complex denoted as $X \sim Y$ in Scheme 13 (where X stands for some product resulting from reaction of the flavin hydroperoxide and Y for some form originating from the substrate which interacts with X). This intermediate has become known as 'intermediate II' in the pertinent literature [150]. The structure of this intermediate is still elusive, but is logically expected to enclose the key for the elucidation of the detailed mechanism. This has fostered several structural and mechanistic proposals, none of which, however, has withstood experimental verification.

An interesting proposal has been put forward recently based on the absorption of the radicals obtained upon 1e^- oxidation of typical substrates of *p*-hydroxybenzoate hydroxylase which, superimposed on that of the flavin 4a-hydroperoxide, yields a calculated spectrum very similar to that of the intermediate $E \sim X \sim Y$ [153]. If true, this assumption would predict that $E \sim X \sim Y$ should have an EPR signal or exhibit magnetic susceptibility. Up to now, however, experiments aimed at testing this have been negative. In view of this it is not reasonable to discuss further such detailed mechanisms in the context of the present review; further ideas and experimental evidence clearly are needed for elucidation of the full catalytic mechanisms of the aromatic hydroxylases.

REFERENCES

1. Walsh, C. T. (1980) *Acc. Chem. Res. 13*, 148–155.
2. Bruice, T. C. (1980) *Acc. Chem. Res. 13*, 256–262.
3. Massey, V. & Ghisla, S. (1983) in *Biological oxidations* (Sund, H. & Ullrich, V., eds) pp. 114–139, Springer, Berlin.
4. Müller, F. (1983) *Top. Curr. Chem. 108*, 71–107.
5. Barman, B. G. & Tollin, G. (1972) *Biochemistry 11*, 4755–4759.
6. Gomez-Moreno, C., Choy, M. & Edmondson, D. E. (1979) *J. Biol. Chem. 254*, 7630–7635.
7. Ghisla, S. & Massey, V. (1986) *Biochem. J. 239*, 1–12.
8. Burnett, R. M., Darling, G. D., Kendall, D. W., LeQuesne, M., Mayhew, S. G., Smith, W. W. & Ludwig, M. L. (1974) *J. Biol. Chem. 249*, 4383–4392.
9. Schulz, G. E., Schirmer, R. H. & Pai, E. F. (1982) *J. Mol. Biol. 160*, 287–308.
10. Weijer, W. J., Hofsteenge, J., Beintema, J. J., Wierenga, R. K. & Drenth, J. (1983) *Eur. J. Biochem. 133*, 109–118.
11. Mathews, F. S. & Zia, Z. x. (1987) in *Flavins and flavoproteins* (Edmondson, D. E. & McCormick, D. B., eds) pp. 123–132, W. deGruyter, Berlin.
12. Williamson, G. & Edmondson, D. E. (1985) *Biochemistry 24*, 7790–7797.
13. Ghisla, S., Kenney, W. C., Knappe, W. R., McIntire, W. & Singer, T. P. (1980) *Biochemistry 19*, 2537–2544.
14. Ghisla, S. & Mayhew, S. G. (1980) *Methods Enzymol. 66*, 241–253.
15. Schöllnhammer, G. & Hemmerich, P. (1974) *Eur. J. Biochem. 44*, 561–577.
16. Mayhew, S. G., Whitfield, C. D., Ghisla, S. & Schuman-Jorns, M. (1970) *Eur. J. Biochem. 44*, 579–591.
17. Vervoort, J., Müller, F., Lee, J., van den Berg, W. A. M. & Moonen, C. T. W. (1986) *Biochemistry 25*, 8062–8067.
18. Ghisla, S. & Massey, V. (1980) *J. Biol. Chem. 255*, 5688–5696.
19. Brüstlein, M. & Bruice, T. C. (1972) *J. Am. Chem. Soc. 94*, 6548–6549.
20. Pai, E. G. & Schulz, G. E. (1983) *J. Biol. Chem. 258*, 1752–1757.
21. Walsh, C. T., Schonbrunn, A. & Abeles, R. H. (1971) *J. Biol. Chem. 246*, 6855–6866.
22. Porter, D. J. T., Voet, J. B. & Bright, H. J. (1973) *J. Biol. Chem. 248*, 4400–4416.
23. Mathews, F. S. & Xia, Z. x. (1987) *Flavins and flavoproteins* (Edmondson, D. E. & McCormick, D. B., eds) pp. 123–132, W. deGruyter, Berlin, New York.
24. Lederer, F. & Mathews, F. S. (1987) *Flavins and flavoproteins* (Edmondson, D. E. & McCormick, D. B., eds) pp. 133–142, W. deGruyter, Berlin, New York.
25. Lindqvist, Y. & Bränden, C. I. (1985) *Proc. Natl Acad. Sci. USA 82*, 6855–6859.
26. Manstein, D. J., Massey, V., Ghisla, S. & Pai, E. F. (1988) *Biochemistry 27*, 2300–2305.
27. Schonbrunn, A., Abeles, R. H., Walsh, C. T., Ghisla, S., Ogata, H. & Massey, V. (1976) *Biochemistry 15*, 1798–1807.
28. Ghisla, S., Olson, S. T., Massey, V. & Lhoste, J.-M. (1979) *Biochemistry 18*, 4733–4742.
29. Pompon, F. & Lederer, F. (1985) *Eur. J. Biochem. 148*, 145–154.
30. Ghisla, S. & Massey, V. (1977) *J. Biol. Chem. 252*, 6729–6735.
31. Lockridge, O., Massey, V. & Sullivan, P. A. (1972) *J. Biol. Chem. 247*, 8097–8106.
32. Lederer, F. (1974) *Eur. J. Biochem. 46*, 393–399.
33. Williams, C. H. Jr, Arscott, L. D. & Swenson, R. P. (1984) in *Flavins and flavoproteins* (Bray, R. C., Engel, P. C. & Mayhew, S. G., eds) pp. 95–109, W. deGruyter, Berlin, New York.
34. D'Silva, C., Williams, C. H. Jr & Massey, V. (1987) *Biochemistry 26*, 1717–1722.
35. Cornforth, J. W. (1959) *J. Lipid Res. 1*, 3–28.
36. Ghisla, S. (1984) in *Flavins and flavoproteins* (Bray, R. C., Engel, P. C. & Mayhew, S. G., eds) pp. 385–401, W. deGruyter, Berlin, New York.
37. Fendrich, G. & Abeles, R. H. (1982) *Biochemistry 21*, 6685–6695.

38. Schopfer, L. M., Massey, V., Ghisla, S. & Thorpe, C. (1988) *Biochemistry 27*, 6599−6611.
39. Manstein, D. J., Pai, E. F., Schopfer, L. M. & Massey, V. (1986) *Biochemistry 25*, 6807−6816.
40. Reinsch, J., Katz, A., Wean, J., Aprahamian, G. & McFarland, J. R. (1980) *J. Biol. Chem. 255*, 9093−9097.
41. Pohl, B., Raichle, Th. & Ghisla, S. (1986) *Eur. J. Biochem. 160*, 109−115.
42. Ghisla, S., Thorpe, C. & Massey, V. (1984) *Biochemistry 23*, 3154−3161.
43. Ikeda, Y., Hine, D. G., Okamura, I. K. & Tanaka, K. (1985) *J. Biol. Chem. 260*, 1326−1337.
44. Jencks, W. P. (1981) *Chem. Soc. Rev. 10*, 345−375.
45. Beinert, H. & Page, E. (1956) *J. Biol. Chem. 225*, 479−497.
46. Thorpe, C. & Massey, V. (1983) *Biochemistry 22*, 2972−2978.
47. Thorpe, C. (1987) in *Flavins and flavoproteins* (Edmondson, D. E. & McCormick, D. B., eds) pp. 149−157, W. deGruyter, Berlin, New York.
48. Lau, S. M., Powell, P., Büttner, H., Ghisla, S. & Thorpe, C. (1986) *Biochemistry 25*, 4187−4189.
49. Silverman, R. B. (1984) *Biochemistry 23*, 5206−5213.
50. Walker, M. C. & Edmondson, D. E. (1987) in *Flavins and flavoproteins* (Edmondson, D. E. & McCormick, D. B., eds) pp. 699−703, W. deGruyter, Berlin, New York.
51. Mincey, T., Tayrien, G., Mildvan, A. S. & Abeles, R. H. (1980) *Proc. Natl Acad. Sci. USA 77*, 7099−7101.
52. Geissler, J., Ghisla, S. & Kroneck, P. M. H. (1986) *Eur. J. Biochem. 160*, 93−100.
53. Fisher, H. F., Ofner, P., Conn, E. E., Vennesland, B. & Westheimer, F. H. (1953) *J. Biol. Chem. 202*, 687−697.
54. Yamazaki, S., Tsai, L., Stadtman, T. C., Jacobsen, F. S. & Walsh, C. (1980) *J. Biol. Chem. 255*, 9025−9027.
55. You, K., Arnold, L. J., Allison, W. S. & Kaplan, N. O. (1978) *Trends Biochem. Sci. 3*, 265−268.
56. Williams, C. H. Jr (1976) *The enzymes*, vol. 13 (Boyer, P. D., ed.) pp. 89−173, Academic Press, New York.
57. Massey, V. & Veeger, C. (1961) *Biochim. Biophys. Acta 48*, 33−47.
58. Searls, R. L., Peters, J. B. & Sanadi, D. R. (1961) *J. Biol. Chem. 236*, 2317−2322.
59. Massey, V. & Williams, C. H. Jr (1965) *J. Biol. Chem. 240*, 4470−4480.
60. Kosower, E. B. (1966) in *Flavins and flavoproteins* (Slater, E. C., ed.) pp. 1−14, Elsevier, Amsterdam.
61. Massey, V. & Ghisla, S. (1974) *Ann. NY Acad. Sci. 227*, 446−465.
62. Matthews, R. G. & Williams, C. H. Jr (1976) *J. Biol. Chem. 251*, 3956−3964.
63. Williams, C. H. Jr, Arscott, L. D. & Schulz, G. E. (1982) *Proc. Natl Acad. Sci. USA 79*, 2199−2201.
64. Rice, D. W., Schulz, G. E. & Guest, J. R. (1984) *J. Mol. Biol. 174*, 483−496.
65. Krohne-Erich, G., Schirmer, R. H. & Untucht-Grau, R. (1977) *Eur. J. Biochem. 80*, 65−71.
66. Brown, N. L., Ford, S. T., Pridmore, R. D. & Fritzinger, D. (1983) *Biochemistry 22*, 4089−4095.
67. Thieme, R., Pai, E. F., Schirmer, R. H. & Schulz, G. E. (1981) *J. Mol. Biol. 152*, 763−782.
68. Karplus, P. A. & Schulz, G. E. (1987) in *Flavins and flavoproteins* (Edmondson, D. E. & McCormick, D. B., eds) pp. 45−54, W. deGruyter, Berlin, New York.
69. Massey, V., Ghisla, S., Ermler, V. & Schulz (1987) in *Flavins and flavoproteins* (Edmondson, D. E. & McCormick, D. B., eds) pp. 79−84, W. deGruyter, Berlin, New York.
70. Massey, V., Miller, S. M., Ballou, D. P., Williams, C. H. Jr, Moore, M., Distefano, M. D. & Walsh, C. T. (1987) in *Flavins and flavoproteins* (Edmondson, D. E. & McCormick, D. B., eds) pp. 41−44, W. deGruyter, Berlin, New York.
71. Schierbeek, A. J., Drenth, J. & Hol, W. G. J. (1984) in *Flavins and flavoproteins* (Bray, R. C., Engel, P. C. & Mayhew, S. G., eds) p. 147, W. deGruyter, Berlin, New York.
72. Reference deleted.
73. Stephens, P. E., Lewis, H. M., Darlison, M. G. & Guest, J. R. (1983) *Eur. J. Biochem. 135*, 519−527.
74. Westphal, A. H. & deKok, A. (1988) *Eur. J. Biochem. 172*, 299−305.
75. Greer, S. & Perham, R. N. (1986) *Biochemistry 25*, 2736−2742.
76. Massey, V., Matthews, R. G., Foust, G. P., Howell, L. G., Williams, C. H. Jr, Zanetti, G. & Ronchi, S. (1970) in *Pyridine nucleotide-dependent dehydrogenases* (Sund, H., ed.) pp. 393−411, Springer, Heidelberg.
77. Bulger, J. E. & Brandt, K. G. (1971) *J. Biol. Chem. 246*, 5570−5577 & 5578−5587.
78. Huber, P. W. & Brandt, K. B. (1980) *Biochemistry 19*, 4568−4575.
79. Mannervik, B., Boggaram, V., Carlberg, I. & Larson, K. (1979) in *Flavins and flavoproteins* (Yagi, K. & Yamano, T., eds) pp. 173−187, University Park Press, Baltimore.
80. Arscott, L. D., Thorpe, C. & Williams, C. H. Jr (1981) *Biochemistry 20*, 1513−1520.
81. Thorpe, C. & Williams, C. H. Jr (1976) *J. Biol. Chem. 251*, 3553−3557.
82. Thorpe, C. & Williams, C. H. Jr (1981) *Biochemistry 20*, 1507−1513.
83. Matthews, R. G., Ballou, D. P. & Williams, C. H. Jr (1977) *J. Biol. Chem. 252*, 3199−3207.
84. Matthews, R. G., Ballou, D. P. & Williams, C. H. Jr (1979) *J. Biol. Chem. 254*, 4974−4981.
85. Fairlamb, A. H. & Cerami, A. (1985) *Mol. Biochem. Parasitol. 14*, 187−196.
86. Shames, S. L., Fairlamb, A. H., Cerami, A. & Walsh, C. T. (1986) *Biochemistry 25*, 3519−3526.
87. Krauth-Siegel, L., Enders, B., Henderson, G. B., Fairlamb, A. H. & Schirmer, R. H. (1987) *Eur. J. Biochem. 164*, 123−128.
88. Fox, B. S. & Walsh, C. T. (1982) *J. Biol. Chem. 257*, 2498−2503.
89. Fox, B. S. & Walsh, C. T. (1983) *Biochemistry 22*, 4082−4088.
90. Schultz, P. G., Au, K. G. & Walsh, C. T. (1985) *Biochemistry 24*, 6840−6848.
91. Walsh, C. T., Moore, M. J. & Distefano, M. D. (1987) in *Flavins and flavoproteins* (Edmondson, D. E. & McCormick, D. B., eds) pp. 13−28, W. deGruyter, Berlin, New York.
92. Moore, M. J. & Walsh, C. T. (1989) *Biochemistry*, in the press.
93. Distefano, M. D., Au, K. G. & Walsh, C. T. (1989) *Biochemistry*, in the press.
94. Sahlman, L., Lambeir, A. M., Lindskog, S. & Dunford, H. B. (1984) *J. Biol. Chem. 259*, 12403−12408.
95. Sandström, A. & Lindskog, S. (1987) in *Flavins and flavoproteins* (Edmondson, D. E. & McCormick, D. B., eds) pp. 33−36, W. deGruyter, Berlin, New York.
96. Miller, S., Ballou, D. P., Massey, V., Williams, C. H. Jr & Walsh, C. T. (1986) *J. Biol. Chem. 261*, 8081−8084.
97. Miller, S., Moore, J. J., Massey, V., Williams, C. H. Jr, Distefano, M. D., Ballou, D. P. & Walsh, C. T. (1989) *Biochemistry*, in the press.
98. Sandström, A. & Lindskog, S. (1987) *Eur. J. Biochem. 164*, 243−249.
99. Moore, M. J., Distefano, M. D., Walsh, C. T., Miller, S., Massey, V., Williams, C. H. Jr & Ballou, D. P. (1987) in *Flavins and flavoproteins* (Edmondson, D. E. & McCormick, D. B., eds) pp. 37−40, W. deGruyter, Berlin, New York.
100. Stankovich, M. T. & Bard, A. J. (1977) *J. Electroanal. Chem. 75*, 487−505.
101. Casa, J. S. & Jones, M. M. (1980) *J. Inorg. Nucl. Chem. 42*, 99−102.
102. Sahlman, L., Lambier, A. M. & Lindskog, S. (1986) *Eur. J. Biochem. 156*, 479−488.
103. Casola, L. & Massey, V. (1966) *J. Biol. Chem. 241*, 4985−4993.
104. Strittmatter, P. (1965) *J. Biol. Chem. 240*, 4481−4487.
105. Gorelick, R. J., Schopfer, L. M., Ballou, D. P., Massey, V. & Thorpe, C. (1985) *Biochemistry 24*, 6830−6839.
106. Simondson, R. P., Weger, P. C., Salemme, F. R. & Tollin, G. (1982) *Biochemistry 24*, 6366−6375.

107. Tollin, G., Cheddar, G., Watkins, J. A., Meyer, T. E. & Cusanovich, M. A. (1984) *Biochemistry 23*, 6345−6349.
108. Weber, P. C. & Tollin, G. (1985) *J. Biol. Chem. 260*, 5568−5573.
109. Mayhew, S. G. & Ludwig, M. L. (1975) *The enzymes* (Boyer, P. D., ed.) pp. 57−118, Academic Press, New York.
110. Bray, R. C. (1975) *The enzymes* (Boyer, P. D., ed.) pp. 300−419, Academic Press, New York.
111. Hille, R. & Massey, V. (1985) in *Molybdenum enzymes* (Spiro, T. P., ed.) pp. 443−518, Academic Press, New York.
112. Olson, J. S., Ballou, D. P., Palmer, G. & Massey, V. (1974) *J. Biol. Chem. 249*, 4363−4382.
113. Barber, M. J., Bray, R. C., Cammack, R. & Coughlan, M. P. (1977) *Biochem. J. 163*, 279−289.
114. Porras, A. G. & Palmer, G. (1982) *J. Biol. Chem. 257*, 11617−11626.
115. Komai, H., Massey, V. & Palmer, G. (1969) *J. Biol. Chem. 244*, 1692−1700.
116. Olson, J. S., Ballou, D. P., Palmer, G. & Massey, V. (1974) *J. Biol. Chem. 249*, 4350−4362.
117. Hille, R. & Massey, V. (1981) *J. Biol. Chem. 256*, 9090−9095.
118. Porras, A. G., Olson, J. S. & Palmer, G. (1981) *J. Biol. Chem. 256*, 9096−9103.
119. Iyanagi, T., Makino, N. & Mason, H. S. (1974) *Biochemistry 13*, 1701−1710.
120. Vermilion, J. L., Ballou, D. P., Massey, V. & Coon, M. J. (1981) *J. Biol. Chem. 256*, 266−277.
121. Kamin, H. & Lambeth, J. C. (1982) in *Flavins and flavoproteins* (Massey, V. & Williams, C. H. Jr, eds) pp. 665−666, Elsevier, New York.
122. Kemal, C., Chan, T. W. & Bruice, T. C. (1977) *J. Am. Chem. Soc. 99*, 7272−7286.
123. Eberlein, G. & Bruice, T. C. (1982) *J. Am. Chem. Soc. 104*, 1449−1452.
124. Anderson, R. F. (1982) in *Flavins and flavoproteins* (Massey, V. & Williams, C. H. Jr, eds) pp. 278−283, Elsevier, New York.
125. Anderson, R. F. (1984) in *Flavins and flavoproteins* (Bray, R. C., Engel, P. C. & Mayhew, S. G., eds) pp. 57−60, W. deGruyter, Berlin, New York.
126. Müller, F., Ghisla, S. & Bacher, A. (1988) in *Vitamine II* (Isler, O., Brubacher, G., Ghisla, S. & Kräutler, B., eds) pp. 94−95, Thieme Verlag, Stuttgart, New York.
127. Ballou, D. P. (1984) in *Flavins and flavoproteins* (Bray, R. C., Engel, P. C. & Mayhew, S. G., eds) pp. 605−618, W. deGruyter, Berlin, New York.
128. Massey, V., Schopfer, L. M. & Anderson, R. F. (1988) in *Oxidases and related enzyme systems* (King, T. E., Mason, H. S. & Morrison, M. M., eds) pp. 147−166, Alan Liss, New York.
129. Stankovich, M. T., Schopfer, L. M. & Massey, V. (1978) *J. Biol. Chem. 253*, 4971−4979.
130. Reference deleted.
131. Hemmerich, P. & Wessiak, A. (1976) in *Flavins and flavoproteins* (Singer, T. P., ed.) pp. 9−22, Elsevier, Amsterdam, New York.
132. Gibson, Q. H. & Hastings, J. W. (1962) *Biochem. J. 83*, 368−377.
133. Massey, V., Palmer, G. & Ballou, D. P. (1973) in *Oxidases and related systems* (King, T. E., Mason, H. S. & Morrison, M., eds) pp. 25−43, University Park Press, Baltimore.
134. Bruice, T. C. (1982) in *Flavins and flavoproteins* (Massey, V. & Williams, C. H. Jr, eds) pp. 265−277, Elsevier, New York.
135. Bruice, T. C. (1984) *Israel J. Chem. 24*, 54−61.
136. Bruice, T. C. (1984) in *Flavins and flavoproteins* (Bray, R. C., Engel, P. C. & Mayhew, S. G., eds) pp. 45−55, W. deGruyter, Berlin, New York.
137. Favaudon, V. (1977) *Eur. J. Biochem. 78*, 293−307.
138. Ryerson, C. C., Ballou, D. P. & Walsh, C. T. (1982) *Biochemistry 21*, 2644−2655.
139. Hastings, J. W., Balny, C., LePeuch, C. & Douzou, P. (1973) *Proc. Natl Acad. Sci. USA 70*, 3468−3472.
140. Hastings, J. W. & Nealson, K. H. (1977) *Annu. Rev. Microbiol. 31*, 549−595.
141. Eberhard, A. & Hastings, J. W. (1972) *Biochem. Biophys. Res. Commun. 47*, 348−353.
142. Schuster, G. B. (1979) *Acc. Chem. Res. 12*, 366−373.
143. Macheroux, P., Eckstein, J. & Ghisla, S. (1987) in *Flavins and flavoproteins* (Edmondson, D. E. & McCormick, D. B., eds) pp. 613−619, W. deGruyter, Berlin, New York.
144. Ghisla, S., Eckstein, J. & Macheroux, P. (1987) in *Flavins and flavoproteins* (Edmondson, D. E. & McCormick, D. B., eds) pp. 601−612, W. deGruyter, Berlin, New York.
145. Lee, J., O'Kane, D. J. & Visser, J. W. G. (1985) *Biochemistry 24*, 1476−1483.
146. Macheroux, P., Schmidt, K. U., Steinerstauch, P., Buntic, G., Hastings, J. W. & Ghisla, S. (1987) *Biochem. Biophys. Res. Commun. 146*, 101−106.
147. Poulsen, L. L. & Ziegler, D. M. (1979) *J. Biol. Chem. 254*, 6449−6455.
148. Jones, K. & Ballou, D. P. (1984) in *Flavins and flavoproteins* (Bray, R. C., Engel, P. C. & Mayhew, S. G., eds) pp. 619−622, W. deGruyter, Berlin, New York.
149. Spector, T. & Massey, V. (1972) *J. Biol. Chem. 247*, 5632−5636.
150. Entsch, B., Ballou, D. P. & Massey, V. (1976) *J. Biol. Chem. 251*, 1550−1563.
151. Wessiak, A., Schopfer, L. M. & Massey, V. (1984) *J. Biol. Chem. 259*, 12547−12556.
152. Detmer, K. & Massey, V. (1985) *J. Biol. Chem. 160*, 5998−6005.
153. Anderson, R. F., Patel, K. B. & Stratford, M. R. L. (1987) *J. Biol. Chem. 262*, 17475−17479.

Review

Nucleo-mitochondrial interactions in yeast mitochondrial biogenesis

Leslie A. GRIVELL

Section for Molecular Biology, Department of Molecular Cell Biology, University of Amsterdam

(Received December 5, 1988/March 21, 1989) — EJB 88 1400

For what once were probably free-living aerobic bacteria, mitochondria show a surprising degree of domestication in terms of their dependence on the genetic system of the cells in which they reside. In most organisms, only a handful of the hundred or so mitochondrial proteins are encoded by mitochondrial DNA (mtDNA). The remainder, including most of the components of the mitochondrion's own genetic system, are encoded by nuclear genes, which also specify all other enzymes necessary for the synthesis, import, processing and modification of these proteins and associated lipid components. The assembly of a functional mitochondrion is possibly one of the most complex logistic exercises faced by the eukaryotic cell. In a facultative anaerobic organism like the yeast *Saccharomyces cerevisiae*, which is capable of tailoring the level of mitochondrial biosynthesis to its specific needs in response to its environment, several hundred nuclear genes are likely to be involved in one way or another with mitochondrial biogenesis. The way in which the expression of these genes is regulated, how they in turn regulate the expression of genes in mitochondrial DNA (mtDNA) and to what extent cross-talk occurs between the two genetic systems are subjects of this review. The main focus will be on the yeast *S. cerevisiae*, with occasional excursions to other organisms. Earlier reviews, which also cover historic and genetic aspects of this topic can be found elsewhere [1, 2].

A. YEAST mtDNA: FEATURES OF GENE ORGANIZATION AND EXPRESSION

In *S. cerevisiae* mtDNA is a circular molecule, which, dependent on strain, varies in size from about 73 kb to 82 kb (reviewed in [3-5]). In common with the much smaller mtDNAs from metazoa, it contains genes coding for the two ribosomal RNAs, a set of tRNAs and some components of the respiratory enzymes of the inner membrane (Table 1). In addition, there are genes without counterparts in metazoan mtDNAs. These code for at least one endonuclease involved in intron transposition, proteins required for RNA splicing (RNA maturases), a protein associated with the small subunit of the mitochondrial ribosome and an RNA that forms part of an RNase-P-like tRNA-processing enzyme. Three reading frames identified by DNA sequence analysis (URFs 1–3) have not yet been assigned to known protein products, but

Correspondence to L. A. Grivell, Section for Molecular Biology, Department of Molecular Cell Biology, University of Amsterdam, Kruislaan 318, NL-1098 SM Amsterdam, The Netherlands

Table 1. *Mitochondrial genes in yeast and mammals*

Mitochondrial component	Mitochondrial gene product in	
	yeast	mammals
Cytochrome *c* oxidase subunits I, II, III	+	+
QH_2–cytochrome *c* reductase cytochrome *b*	+	+
ATP synthase		
subunits 6, 8	+	+[a]
subunit 9	+	−
NADH dehydrogenase		
subunits ND1–6	−	+
RNA maturases[b]	+	−
Intron transposition endonuclease	+	−
RNA component RNase-P-like enzyme	+	−
Ribosome		
large and small rRNAs	+	+
ribosome-associated protein	+	−
tRNAs	24	22
Reading frames still unidentified[c]	3	−

[a] In mammalian mtDNAs, the gene encoding ATP synthase subunit 6 is preceded by and partially overlaps the short coding sequence A6L, that displays weak sequence similarity to the yeast ATP synthase subunit 8 gene (*aap1*). Both reading frames are expressed [6] (and references therein). In rat liver, the product of A6L has recently been identified as chargerin II, a hydrophobic protein associated with the F_0 component of ATP synthase [7].

[b] Intron-encoded reading frames for which in many cases genetic evidence indicates an involvement in RNA splicing.

[c] The existence of two short, additional reading frames with lengths of 243 bp and 140 bp has been noted by de Zamarocny and Bernardi [8].

bear some resemblance to the RNA maturase family [9]. Genes notably absent, but present in mtDNAs from sources as diverse as mammals, plants, fungi and trypanosomes, are those for subunits of the complex-I type of mitochondrial NADH:Q reductase. The absence correlates with the fact that *S. cerevisiae* appears to lack this type of NADH dehydrogenase [10, 11]. Since a compilation of all published DNA sequence data covers an estimated 92% of the genome, with few large gaps remaining [8], it seems unlikely that other genes remain to be discovered.

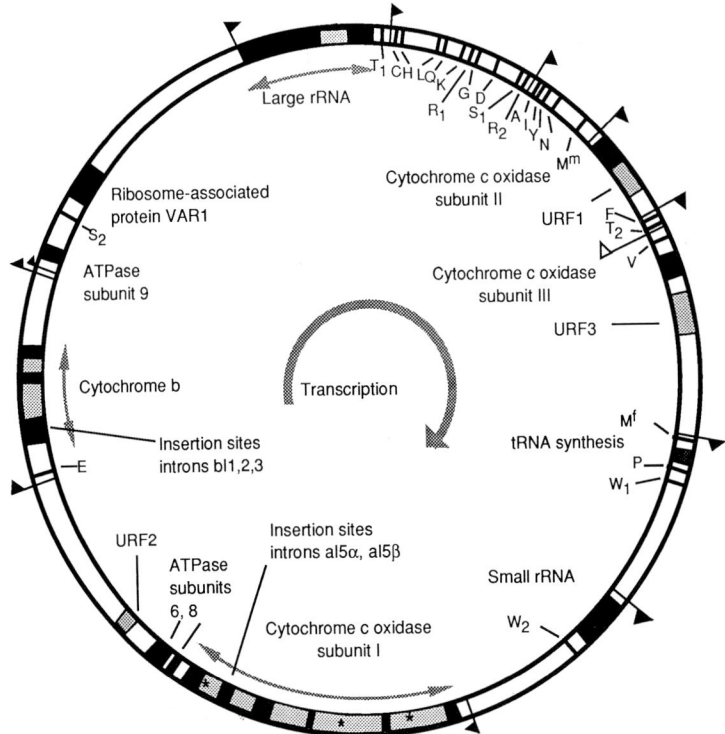

Fig. 1. *Yeast mitochondrial DNA*. The organization of mtDNA in *S. cerevisiae* D273-10B is shown, with identified genes being depicted as black bars and unassigned reading frames (URFs) as dark-grey shaded areas. URFs2 and 3 in fact consist of a series of overlapping reading frames linked by −1 or +1 frameshifts (see section on mitochondrial translation for discussion). For ease of presentation these coding sequences are shown as continuous areas on the map. Introns, where present, are shaded light grey and those of the group II type are additionally indicated by an asterisk. D273-10B contains a medium-sized mtDNA, which lacks a number of the introns present in the genes for apocytochrome *b* and coxI in so-called long strains (see [12, 16] for review). Insertion sites of the missing introns are indicated. In the case of tRNA-coding sequences, identification is by means of the cognate amino acid, using the one-letter code and subscripts to indicate iso-acceptors where necessary. For all genes but one (tRNA$_2^{Thr}$), transcription is in a clockwise direction and sites of transcriptional-initiation identified by *in vitro* capping of mitochondrial RNAs are indicated by (▶). In the case of the gene for tRNA$_2^{Thr}$ located of the opposite strand, ▷ indicates the position of the conserved nonanucleotide motif

As Fig. 1 shows, three genes are interrupted by introns. The unusual features of these (including the ability to self-splice *in vitro*) have been the subject of extensive review [12–15]. The exact structure of the split genes is strain-dependent, with certain introns being present in some strains but not in others. It is this variation that is mainly responsible for the differences in genome size of different strains. The gene order shown in Fig. 1 is conserved within the genus, but there is no obvious reason for this. Genes whose products are functionally related and required in stoichiometric amounts are widely separated and different orders pertain in other yeasts (see [16] and references therein).

B. NUCLEAR CONTROL OF MITOCHONDRIAL GENE EXPRESSION

1. Mitochondrial transcription and its regulation

Unlike mammalian mtDNAs (see [17] for review), yeast mtDNA contains 19 or 20 transcriptional initiation sites scattered around the genome. These were first identified by *in vitro* capping of RNAs with guanylyltransferase and subsequently confirmed by transcriptional studies *in vitro* using partially purified preparations of mitochondrial RNA polymerase [18–20]. Transcription starts within the sequence (A/T)TATAAGTA, with the last A being the site of initiation. Sequences outside this motif probably also contribute to promoter activity, as there are several sites in the genome, for which no corresponding capped transcripts are found [18, 21]. In addition, 'strong' and 'weak' promoter sites can be defined by their ability to compete for limiting amounts of RNA polymerase *in vitro*, with differences in strength being as much as 20-fold [22]. Sequences downstream of the initiation site at least in part responsible for this, there being a preference for A and T at positions +2 and +3 respectively of the stronger promoters [23, 24].

A single multicomponent RNA polymerase is responsible for transcription of coding sequences and for priming of DNA synthesis [25–27]. The enzyme can be dissociated chromatographically into two components, one of which is the product of the RP041 gene, a protein with molecular mass of 153 kDa. This is the catalytic subunit of the enzyme and, surprisingly, it displays regions of high sequence similarity to the DNA-directed RNA polymerases of bacteriophages T3 and T7 [28]. During transcription, this subunit is associated with the second component, a factor necessary for accurate transcriptional initiation [27, 29–31]. The existence of multiple factors with different specificities would clearly allow differential recognition of promoters, but whether this is in fact the case is not yet clear.

Many genes lack a transcriptional initiation site and the available evidence suggests that these form part of larger transcriptional units, with mature RNAs being generated by processing. The sequences co-transcribed in this way form rather mixed collections (Fig. 2) so that the advantages of this mode of organization are not immediately apparent. Pulse-

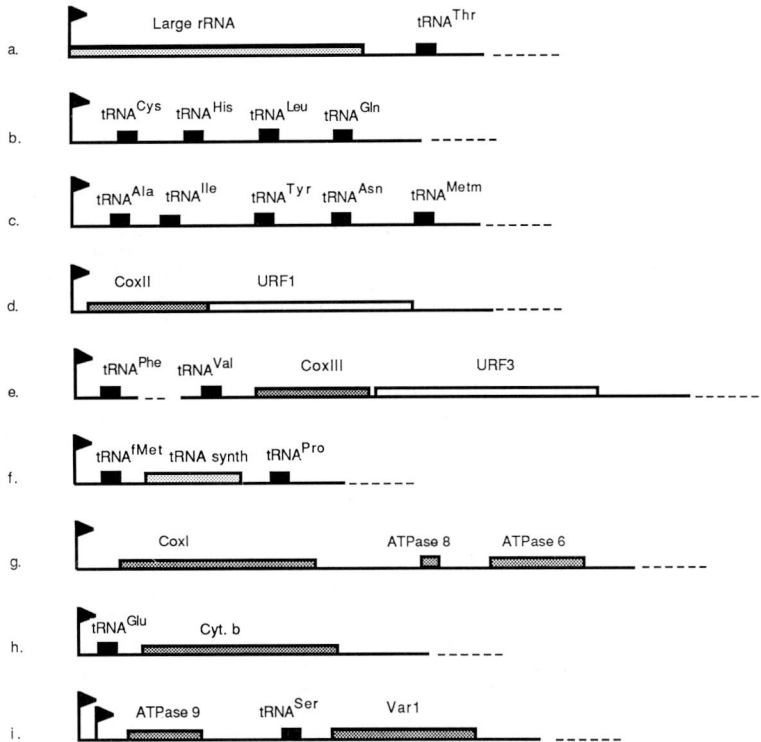

Fig. 2. *Transcriptional units in yeast mtDNA*. The figure shows the various gene clusters (not to scale), which are thought to belong to common transcription units, based on RNA capping data, the detection of multigenic precursor RNAs and intermediates in the processing of these: (a) [32, 33], (b) [33], (c) [21], (d) [34], (e) [35], (f) [36], (g) [37, 38], (h) [39], (i) [40]; see also [18, 22]. In all cases, the position of the 3′ end of the transcription unit is unknown due to uncertainty about the identity of the primary transcript. Several of the units contain potential sites of transcriptional-initiation which appear not to be used under normal conditions, but which give rise to low levels of transcripts in *petite* mutants. Sequence comparisons show that these sites lack the conserved features characteristic of 'strong' promoters (see text). The sites may possibly be used by wild-type cells in specific conditions or phases of growth

labelling reveals a dramatic fall-off in the level of synthesis of promoter-distal transcripts within some of the transcriptional units [22], suggesting that, just as in animal mitochondria [17], transcriptional attenuation may be used as a means of regulation. Two additional observations suggest that this is not the only form of control, however. First, for various transcripts the rate of synthesis predicted from pulse-labelling should in most cases lead to a 5–10-fold higher steady-state level than is actually found [41], indicating that differences in half-life are involved (see also section on RNA stability below). Second, comparison of steady-state RNA levels with measurements of the protein synthetic capacity of isolated mitochondria as an indirect assay for levels of translatable mRNAs [42] shows that transcripts and translation products sometimes attain different levels. For example, CoxI and cytochrome *b* are synthesized *in vitro* at a rate roughly similar to ATPase subunit 9, even though the steady-state levels of their mRNAs *in vivo* are lower (200–300 copies per cell compared with approximately 480 for the ATPase 9 mRNA). These differences are small and differences in protein turnover in isolated mitochondria might also have influenced the apparent rates at which particular proteins are labelled in this study. The findings do, however, lend initial support to the idea that controls on mRNA translation also exist (see below).

The repression of mitochondrial biosynthesis during growth on fermentable carbon sources is in part attributable to a decrease in the amount of transcription of mitochondrial genes. Mueller and Getz [41] have established that glucose repression uniformly decreases the level of individual mitochondrial transcripts by 3–6-fold. For the 21S rRNA, pulse-labelling experiments show that this decrease is due to a lower rate of synthesis. Since the amount of mtDNA decreases only about 2-fold under these conditions [43, 44], this decrease can only in part be ascribed to a reduction in available template. A decrease in the level of active mitochondrial RNA polymerase is probably responsible for the remaining decrease. Synthesis of the core RNA polymerase is in fact probably subject to catabolite control, steady-state levels of its mRNA being decreased 3–6-fold under conditions of glucose repression [28, 45].

The decreases in synthesis of various mitochondrial RNAs in response to catabolite repression are smaller in size than might be expected to account for the observed decrease in mitochondrial mass. This lack of stringent control may be due an inability of the mitochondrion to fine tune the expression of individual genes at the level of transcription. In addition, it may be part of a fail-safe mechanism for prevention of inadvertent loss of mtDNA. As Tzagoloff and Myers have pointed out [2], maintenance of the mitochondrial genome is dependent on mitochondrial translational activity. Decreased rates of mitochondrial transcription will reduce protein synthetic capacity, thus leading to loss of mtDNA and induction of cytoplasmic petite mutants.

2. Termination of transcription

By comparison with the detailed picture of events leading to the initiation of transcription, our knowledge of sequences signalling transcriptional termination is sketchy. The 3′ ends of many mRNAs coincide with the dodecamer motif

AAUAAUAUUCUU. This sequence is, however, primarily a site for endonucleolytic cleavage, as demonstrated by the fact that in the CoxI/ATPase transcription unit (Fig. 2), the 3′ end of the CoxI transcript directly abuts the 5′ end of the ATPase precursor RNA within the motif. Subsequent analysis of other multigenic transcription units has established a similar pattern, in which the RNAs arising from cleavage at this site have a stable 3′ terminus and further processing at the downstream 5′ terminus often occurs [38] (see also above).

A search of published mtDNA sequences reveals the presence of the dodecamer downstream of most protein-coding genes, including URFs 1 and 3 and two reading frames located within introns. In the case of the gene for cytochrome b, the mRNA is terminated by the dodecamer motif, but no transcripts have been detected in the downstream sequence that separates this gene from that for ATPase subunit 9. This suggests that, somewhat akin to the situation in animal mitochondria [46–48], the sequence might in fact serve a dual function in signalling both transcriptional attenuation and processing. As yet, however, a rapid turnover of downstream transcripts has not been rigorously excluded. The alternative that a high level of run-on transcription occurs, with stable 3′ ends being generated by processing, therefore remains perhaps more plausible [38].

3. RNA processing

Without exception, expression of the genes in yeast mtDNA requires some form of RNA processing, which is dependent on the action of nuclear-encoded enzymes. In addition to splicing, which is considered separately below, the modifications undergone include cleavage of individual transcripts from multigenic precursors, trimming at the 5′ or 3′ terminus and, in the case of tRNAs, -CCA addition and base modification.

As in bacteria, processing of tRNAs at their 5′ ends is dependent on the action of an enzyme that requires a specific RNA component. This RNA is a mitochondrial gene product encoded by the tRNA synthesis locus [49]. Similarity of the processing mechanism with that occurring in bacteria is suggested by the observations that the 5′ extensions of yeast mitochondrial pre-tRNAs can be accurately processed by *Escherichia coli* RNase P [50]. It has yet to be shown, however, that the RNA component of the yeast enzyme represents the catalytic centre.

Trimming of pre-mRNAs by cleavage at the sequence AAUAAUAUUCUU has been discussed above in connection with the generation of stable 3′ termini. The fact that this motif is found downstream only of mRNAs has led to suggestions that, besides signalling processing, the sequence may be a determinant of either mRNA stability or translatability [51]. Up to now, there has been no experimental system in which such suggestions can be tested, but this should be possible once micro-projectile techniques for transformation of mitochondria with mutated mtDNA sequences become more widely available [52, 53]. Controlled cleavage at the dodecamer could in itself be used as a means of modulating gene expression, as is shown by two examples. The first concerns regulation of the synthesis of the DNA endonuclease (fit1) encoded by and responsible for genetic transmission of the intron in the large rRNA gene. The dodecamer is present at the 3′ end of the coding sequences for this protein and, as Zhu et al. [51] have noted, a strain harbouring a mutant form of this motif is unable to carry out cleavage of RNA and is defective in intron transmission in crosses. These authors suggest that dodecamer cleavage is part of a pathway that generates functional mRNA for the fit1 protein from unspliced pre-rRNA, with the choice between splicing and cleavage perhaps being determined by accessibility of the substrate through its association with ribosomal proteins.

Controlled dodecamer cleavage may also be necessary for expression of URF1, the unassigned reading frame that lies downstream of the CoxII gene. The start of this URF overlaps the C-terminal part of the CoxII coding sequence, so that cleavage at the dodecamer positioned in the 3′-trailing sequence of CoxII transcripts will result in loss of expression [54, 55]. Synthesis of the URF1 protein may thus depend on factors that either prevent cleavage at this site, or perhaps allow cleavages at sites upstream, thus promoting efficient translational initiation.

Knowledge of sequences signalling processing at other sites is rather rudimentary. For the Var1 gene, the 5′ end of the major transcript is generated from a precursor that also contains sequences of ATPase subunit 9 and tRNA^{Ser} (Fig. 2). Processing is essential for the production of a functional mRNA [56, 57] and cleavage has been shown to occur at the heptanucleotide sequence AAUAUAA. This sequence is also frequently found at gene-internal positions, but only two of these are used to generate minor amounts of sub-genic RNAs of unknown function. How cutting at the remaining sites is prevented is unknown, but additional sequences and/or secondary structure seem likely to be involved.

4. RNA splicing

As mentioned above, three genes in yeast mtDNA are split. The introns within them can be divided into two main groups (I and II) by a number of criteria, including differences in their predicted secondary structures and the mechanism by which self-splicing occurs *in vitro*. Group I introns contain a set of conserved sequence elements that are involved in folding and help align exons. Self-splicing occurs via a transesterification mechanism, initiated by binding of a guanosine nucleotide to a site within the intron. Group II introns, in contrast, possess their own characteristic set of conserved features and self-splicing, although still basically a transesterification reaction, is initiated by nucleophilic attack of an intron-internal nucleotide on the 5′-exon/intron junction. Splicing is thus accompanied by accumulation of the excised intron in the form of a lariat which contains an unusual 5′-3′,2′-branched nucleotide.

For both types of intron, the picture we have is one of highly reactive RNAs, which, alongside their ability to splice, are capable of a variety of reactions involving the breakage and formation of phosphodiester bonds [58, 59]. However, despite their versatility in the test tube, several introns are not excised from precursor transcripts in cells which lack mitochondrial protein synthesis [14]. This suggests that, *in vivo*, RNA catalysis is dependent on the presence of functional RNA maturases, proteins encoded by the intron and synthesized by the mitochondrial translational machinery. How these proteins promote splicing is still not known. Speculations have included (a) stabilization of correct intron folding [60]; (b) prevention of aberrant side reactions [14] and (c) enhancement of the efficiency of RNA catalysis by electrostatic shielding and/or activation of reactive groups [61]. These ideas are open to experimental test, since genetically engineered versions of intron-encoded proteins can now be produced in both *E. coli* and yeast [62, 63]. Not unexpectedly, the proteins have been found to be capable of binding nucleic

Table 2. *Proteins required for splicing of mitochondrial introns*

Gene	Protein	Intron(s) affected	Reference
Mitochondrial DNA			
Intronic reading frames	RNA maturases	aI1,aI2,aI4,bI2,bI4	[1, 5, 12] (reviewed)
Nuclear DNA			
CBP2		bI5	[64]
MSS18		aI5b	[65]
MSS51		aI1,aI2,aI4,aI5	[66]
MSS116	RNA unwindase?	aI1,aI5a/b,bI1,bI2/3	[67]
MRS1		bI3	[68]
SUP-101		bI1	[69]
NAM1		groups I and II	[70]
NAM2	mt Leu-tRNA synthetase	aI4,bI4	[71]
CYT18	mt Tyr-tRNA synthetase	group I in *N. crassa*	[72]

acids with high affinity. More surprising, however, is their ability to promote DNA recombination and, in some cases, to act as highly specific DNA endonucleases. The relationship of these features to maturase action has not yet been established, but is clearly worthy of study.

Genetic studies have shown that besides the intron-encoded RNA maturases, a number of nuclear-encoded proteins are required for mitochondrial splicing (Table 2). Their involvement is surprising in view of the inherent simplicity of the self-splicing reaction *in vitro* and what might seem an already unnecessary dependence of this on RNA maturases. It is even more astonishing, however, that several appear to be specific to individual introns.

In two cases, proteins involved in splicing have been found to be mitochondrial aminoacyl-tRNA synthetases, or perhaps derivatives of these in association with another protein. In *Neurospora crassa*, the product of the nuclear *cyt18* gene is required for removal of group I introns from a number of transcripts. This is the mitochondrial tyrosyl-tRNA synthetase [72, 73]. In *S. cerevisiae*, the *nam2* gene encodes the yeast mitochondrial leucyl-tRNA synthetase [71]. As Fig. 4 shows, dominant mutations in *nam2* (*NAM2-1,2-6,2-7*) are capable of suppressing splicing deficiency resulting from changes in the RNA maturase encoded by intron bI4 in long versions of the gene for cytochrome *b* (corresponding to intron 1 in shorter versions of the gene; compare Fig. 3). In contrast to the broad action spectrum of the *cyt18* product in *Neurospora*, the *NAM2* product is apparently specific for only two introns (bI4 and aI4 in the genes for CoxI and cytochrome *b* respectively). In wild-type cells, the synthetase apparently promotes splicing via an interaction with the bI4 RNA maturase. In cells lacking the *NAM2*-bI4 maturase, the mutation allows the synthetase to activate the latent maturase encoded by the reading frame in aI4 [74]. As in *cyt18*, mutations responsible for altered splicing activity are located in the N-terminal part of the protein [72], a region likely to be involved in aminoadenylate formation rather than direct interaction with tRNA [75].

Akins and Lambowitz [72] have suggested that the *cyt18* product recognizes the introns on which it acts by way of their resemblance to tRNA. This cannot be formally ruled out, but the mutational data mentioned above make this less likely. In the case of *NAM2*, the dependence of action of the synthetase on a mitochondrially-coded RNA maturase could even mean that the synthetase makes no contact with an intron at all.

Of the remaining proteins, that encoded by the *CBP2* gene offers perhaps the best prospects for an understanding of its

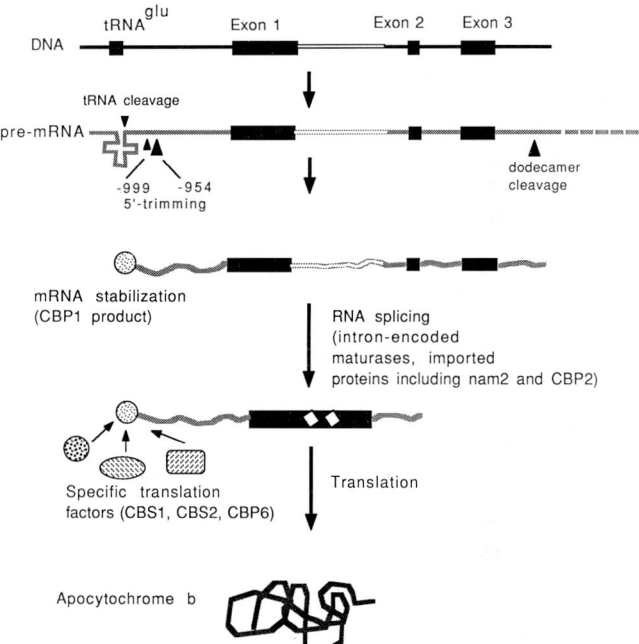

Fig. 3. *Nuclear genes required for the synthesis of apo-cytochrome* b. Steps in the synthesis, maturation and translation of the mRNA for apocytochrome *b* are shown schematically for the 'short' version of the gene in the *S. cerevisiae* strain D273-10B. This gene contains only two introns, the first of which corresponds to bI4, the fourth intron in strains with long versions of the gene. Solid blocks depict exons; the lines between them are introns (not to scale). The RNA maturase coding sequences in intron 1 are shown as a white bar. The characterization of *pet* mutants deficient in one or more of the steps shown has led to the identification of nuclear-encoded, imported proteins whose action is specific to this mRNA. Alongside these, other proteins, involved in cleavage downstream of tRNA^{Glu} [80] and at the 3' end of the mRNA (cf. [51]), in splicing and translation (see Tables 2 and 3) probably act on several other mitochondrial transcripts

mode of action. The protein is required for the splicing of only a single intron (bI5 in the gene for cytochrome *b* [76]) and has been purified on the basis of its ability to interact with this group I intron, reducing the dependence of self-splicing activity of the RNA on GTP and Mg^{2+} ions [64].

For other proteins listed in Table 2, it is unfortunately still not known whether they promote splicing by interacting with RNA, or even whether they enter mitochondria at all. Several

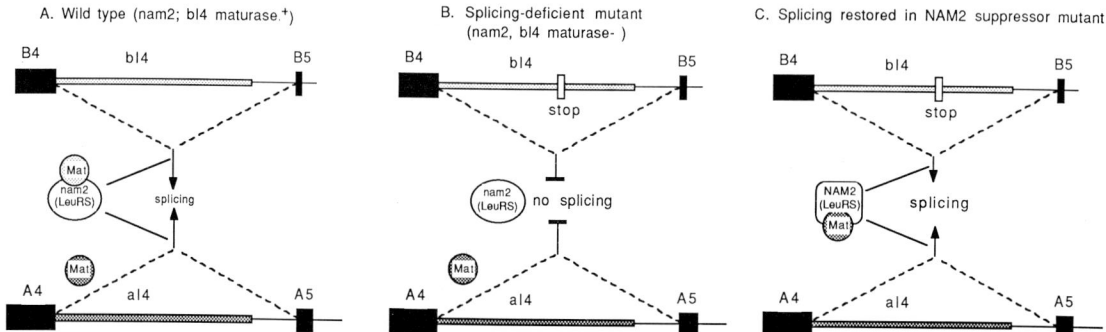

Fig. 4. *Helping RNA catalysis along*. Splicing of the group I introns bI4 and aI4 is likely to be an RNA-catalyzed process. *In vivo*, however, splicing both introns is dependent on the presence of a functional RNA maturase encoded by intron bI4 and on the product of the *nam2* gene, the mitochondrial leucyl-tRNA synthetase (A). Cells lacking bI4 maturase are respiratory-deficient as the result of an inability to synthesize apocytochrome *b* and coxI (B). Respiratory-sufficiency of such cells can be restored by dominant suppressor mutations in the leucyl-tRNA synthetase (NAM2; C). These changes are thought to allow the synthetase to interact with and activate the latent RNA maturase encoded by intron aI4, thus allowing splicing to occur. See text and [71, 74] for discussion

complications have been encountered during their characterization. First, it is often difficult to disentangle direct effects of mutation on splicing from indirect effects resulting from impaired translation and the consequent reduction in synthesis of maturases. Second, like the *NAM2* and *cyt18* products, which as aminoacyl-tRNA synthetases have roles in mitochondrial protein synthesis, many of the proteins so far identified may have more than one function in mitochondrial biogenesis. For example, strains harbouring mutations in the *MSS18* and *MSS116* genes remain respiratory-deficient even when their mitochondrial genome lacks introns [65, 67, 77]. The discovery [67] that the *MSS116* gene product is a member of a protein family, which includes the initiation factor eIF4A and several other nucleic acid unwindases, provides a ready explanation for this phenomenon in terms of a dual role of the gene product in both translation and splicing.

An additional problem has been the comparatively high frequency with which yeast DNA sequences capable of restoring splicing activity to splicing-deficient mutants are picked up in transformation-complementation assays, in particular when the transforming plasmid is present in multiple copies. Part of the problem here may arise from the high sensitivity of the complementation assay to even a low level of restoral of mitochondrial function, but additional factors are probably also involved. For example, the mutant M1301 studied by Schweyen and his colleagues is completely inactive in intron excision because of a one-base deletion in an internal loop in stem 3 of the conserved core structure of the group II intron bI1 [69] (and references therein). The effects of this mutation are suppressed by the nuclear *SUP-101* mutation (Table 2). So far, however, none of the different nuclear sequences cloned from the suppressor strain may represent the authentic *SUP-101* gene, since restoration is observed only at high plasmid copy number [78]. *In vitro*, under ionic conditions normally used for self-splicing, the M1301 mutation is leaky and mutant RNA is partially active in intron excision. *In vivo*, such leakiness might be encouraged by over-production of any of a number of proteins capable of inducing subtle changes in mitochondrial energy balance and/or ion content. In a similar fashion, other nuclear mutations affecting splicing may do so indirectly via effects on the mitochondrial energy balance, or on the synthesis and import of proteins directly involved in the splicing reaction and the effects of such mutations could be suppressible in a variety of ways.

5. RNA stability

Mitochondrial transcripts differ in their relative stabilities, but in most cases the determinants of stability are poorly understood. One of the best characterized nuclear genes affecting mRNA stability, is that known as *CBP1*. Strains harbouring mutations in this gene are unable to synthesize cytochrome *b*, because they lack the mRNA for this protein [79]. The reason for this becomes apparent from examination of the steps leading to the maturation of transcripts of the cytochrome *b* gene in both wild-type and *cbp1* mutant strains. As discussed above and shown in Fig. 3, this gene is co-transcribed with the gene for $tRNA^{Glu}$ and initial steps in processing involve endonucleolytic cleavages at three sites [79]. The first of these lies immediately 3' of the tRNA, while the others lie 100 and 145 nucleotides downstream at positions -999 and -954 relative to the initiator AUG. Cleavage at the most downstream site generates the 5' end of the mature mRNA. In the *cbp1* mutant, all transcripts of the cytochrome *b* gene are present at greatly reduced level, while the amount of $tRNA^{Glu}$ is normal. S1-nuclease analysis shows that although two of the cuts required for mRNA maturation occur apparently normally, no transcripts with a 5' terminus and the -954 position can be detected. The *CBP1* product could thus be either a nuclease specifically responsible for this last cleavage, or it could be an RNA-binding protein capable of stabilizing the pre-mRNAs with 5' extensions until the final cut has taken place. Either way, the findings imply that cytochrome *b* mRNAs with unusual 5' termini are extremely unstable and as a consequence suffer prompt degradation.

Effects of the *cbp1* mutation can be phenotypically suppressed by rearrangements in mtDNA that lead to the fusion of the coding sequences of the cytochrome *b* mRNA, together with part of the 5' leader, to the 5'-terminal regions of other mitochondrial mRNAs [79]. Characterization of such suppressor strains shows that the *CBP1* product is necessary for stability of an mRNA even if it contains only the first 250 nucleotides of the normal 954-nucleotide leader [81], implying that this region is also the target of the CBP1 protein. Interestingly, phenotypic equivalents of nuclear *cbp1* mutants have been found and shown to be deletions of the leader between positions -975 and -64 [82]. Sequences governing the stability of the RNAs that contain them have, of course, been identified in both bacterial and eukaryotic cytoplasmic mRNAs [83]. The surprising aspects of the situation in *cbp1*

Table 3. *Sequence similarity of yeast CBP1 protein with a number of nucleic acid binding proteins*
Sequence similarities were detected initially by use of a FASTA screen [85] of the Swissprot and NBRF protein databanks (releases 14 and 11 respectively). Local alignments were checked using the programmes LFASTA and PLFASTA, while the statistical significance of the matches found was assessed using the programme RDF2 (ktup = 1; 200 shuffles). aa = amino acid residues

Protein	Region CBP1 aligned	Length (aa)	Identity %	z value (SD)
Yeast RPO3 large subunit	10–240	239	18.0	6.31
Yeast topoisomerase II	345–483	141	20.6	7.32
Influenza A polymerase P1	1–112	115	19.1	6.03
EMC virus genomic polyprotein[a]	353–428	77	27.3	7.09

[a] Region corresponding to VP1-P2A.

mutants are, first, that the effects appear to be specific to a single mRNA and, second, that sequences in the immediate surroundings of the −945 cut lack any obvious features that might distinguish them from the remainder of the 5′ leader.

So far, there is no direct information on how the *CBP1* product stabilizes the cytochrome *b* mRNA, but some clues can be gleaned from an examination of the amino acid sequence derived from DNA sequence of its gene [84]. Computer comparisons of sequences present in the Swissprot and NBRF databanks show that the N-terminal half of this 654-amino-acid, rather basic, protein possesses sequence similarity with a number of nucleic-acid-binding proteins. The most extensive alignments are obtained with parts of the yeast DNA topoisomerase II and the largest subunit of RNA polymerase III (Table 3). Since in other systems mRNA degradation is often initiated by rate-limiting endonucleolytic cleavage in stem-loop structures [83], it is an attractive possibility that the CBP1 protein protects the cytochrome *b* mRNA by binding to sequences that might otherwise form a target for endonuclease action. CBP1 could be yet a further case of molecular opportunism, in which the mitochondrion has recruited what was once a cellular RNA-binding protein to its own ends, just as certain aminoacyl-tRNA synthetases may have been recruited to expedite the excision of self-splicing introns [72].

6. Translation

mRNA selection

Since the transcripts of the major protein-coding genes are either cappable *in vitro*, or arise from endonucleolytic cleavage of a primary transcript, the mRNAs of yeast mitochondria are unlikely to be capped. In addition, unlike the mRNAs of mammalian mitochondria, they lack-post-transcriptionally added poly(A) sequences [86]. Their most unusual feature is, however, the presence of long, extremely A + U-rich, untranslated 5′-leader and 3′-trailer sequences. The leader of the cytochrome *b* mRNA, for example, is 954 nucleotides long, the trailer 108 nucleotides. The initiator AUG is seldom the first in the mRNA and there are no obvious consensus sequences that might distinguish it from others in the immediate surroundings. The small ribosomal RNA of yeast mitochondria also lacks the equivalent of a Shine-Dalgarno sequence [87]. As in many eukaryotic cytoplasmic mRNAs, the 5′ leader sometimes contains one or more short reading frames, which could in principle mediate translational control of downstream coding sequences.

Genetic studies have identified two types of mutations affecting translatability of specific mitochondrial mRNAs [88]. In the first type, mutations in the 5′-untranslated leaders

Table 4. *Nuclear proteins required for translation of individual yeast mitochondrial mRNAs*

Mitochondrial mRNA affected	Nuclear gene	Reference
CoxII	*PET111*, *PET112*	[90]
CoxIII	*PET494*, *PET54*, *PET122*[a]	[91]
Apocytochrome *b*	*CBP6*, *CBS1*[b], *CBS2*[c]	[92, 93]

[a] Formerly *PET55*.
[b] Formerly *MK2*.
[c] Possibly identical to *CBP7* [94].

of various transcripts lead to the accumulation of stable, but untranslatable mRNAs. The changes, which so far are all located at some distance upstream of the initiator AUG, presumably identify sites in the mRNA that act as targets for components of the translational-initiation apparatus [56, 89]. So far, however, the limited number of changes studied has not allowed the formulation of any general conclusions regarding sequence or structure requirements.

In mutants of the second type, untranslatable mRNA also accumulates, but this time as the result of changes in nuclear genes. The mutations listed in Table 4 are of this type, specifically affecting the translation of mRNAs for coxII, coxIII, cytochrome *b* and possibly coxI too [95]. The mutations characterized so far are recessive, thus defining positive-acting products. In a number of cases, it has been possible to show that these products are indeed located inside mitochondria [92, 96, 97]. Mitochondrial mutations that allow bypass of the translational block in a number of these *pet* mutants have been isolated and shown to be deletions of mtDNA, resulting in fusion of the non-coding 5′-leader sequences of one mRNA to the body of another [98–101]. Such rearranged molecules can only be maintained in heteroplasmic cells, alongside a wild-type mitochondrial genome. Their existence is strong support for the idea that the products of these *PET* genes interact directly with mRNA, rather than with the mitochondrially coded protein during its assembly into a respiratory complex. In addition, comparison of the positions of the fusion points in different suppressors has given useful information on the leader sequences recognized. In the case of pet494, pet54 and pet122 (formerly pet55), which are all required for the translation of CoxIII mRNA, the factors all appear to have the same target and may in fact bind to the messenger as members of a single complex. The mutations lead to a translational block of any chimaeric-mRNA-bearing sequences derived from the 5′ end (−610) and position −172

Table 5. *Sequence similarities between proteins required for translation of individual mitochondrial mRNAs*

Sequences of PET111 [91], PET54 (Costanzo, M. C., Seaver, E. C. and Fox, T. D., unpublished sequence data), CBS1 [93], and CBP6 [94] were compared with each other using the programmes FASTA, LFASTA [85] and the results tabulated as percentage identical residues with the length of the matched region given in brackets. PET494 [90] has been omitted from the table, as no significant similarity of the sequence was observed with any of the other sequences tested; nd = none detected

Protein (amino acids)	Identity to			
	PET111	PET54	CBS1	CBP6
	%			
PET111 (718)	—	nd	21.1 (71)	17.9 (151)
PET54 (293)		—	20.0 (90)	nd
CBS1 (177)			—	19.3 (119)
CBP6 (162)				—

Table 6. *Sequence similarities between proteins required for translation of individual mitochondrial mRNAs and eukaryotic initiation factors 2α and 2β*

All sequences were screened against a subset of protein sequences from the Swissprot databank, consisting of both pro- and eukaryotic initiation factors. Local alignments were found by use of the programme LFASTA [85] and their statistical significance tested by use of the programme RDF2 (ktup = 1; 200 shuffles)

Protein	Match found with	Length (amino acids)	Identity	z value
			%	
PET494	eIF-2α (human)	113	21.2	8.34
CBP6	eIF-2β (yeast)	56	30.4	5.47

of the CoxIII mRNA [97, 102]. Since such fusions move this region closer to the site of initiation, there is probably no strict spatial relationship between the site of binding and the site of initiation.

The most likely function for the proteins encoded by the genes listed in Table 4 is that of mRNA-specific initiation factors (cf. [103]). Support for this idea is given by results of sequence comparison, which reveals evidence of relationships to both prokaryotic and eukaryotic initiation factors. As shown in Tables 5 and 6, *CBP6*, *CBS1*, *PET54* and *PET111* products are similar to each and *CBP6* product itself shows similarity to the yeast eIF-2β, which is probably significant by the criteria suggested by Pearson and Lipman [85]. The *PET494* product, on the other hand, while showing no resemblance to other *PET* products so far characterized, is clearly related to mammalian IF-2α. The finding of similarity to subunits of eukaryotic IF-2 is of especial interest, since the function of this complex is to charge the small ribosomal subunit with GTP and Met-tRNA prior to association with mRNA [104]. In mammalian cells, eIF-2α controls the level of eIF-2 binding to tRNA, while eIF-2β plays an important role in the scanning mechanism that determines start-site selection [105]. Should the mitochondrial analogues of these proteins function in a similar way, then their binding to a mitochondrial ribosome could determine the subsequent choice of both mRNA and initiator-AUG codon. In further analogy with other steps in the eukaryotic initiation process, other *PET* products may also function within multi-subunit complexes, thus providing an answer to the hitherto puzzling question of why so many of them should be required for the translation of a single mRNA. Fortunately, the purified proteins should be available for study in the not too distant future and their ability to interact with specific mitochondrial mRNAs can be tested directly. Study of their function in translational initiation will, however, require the development of suitable *in vitro* assays using active yeast mitochondrial ribosomes.

Frameshifting and translational efficiency

Another way in which yeast mitochondria may modulate translational efficiency is by use of ribosomal frameshifting. As can be seen from Fig. 1, URFs 2 and 3 lie downstream of and are co-transcribed with the genes for CoxIII and ATPase 6 respectively. Like URF1, which lies downstream of the CoxII gene, both URFs display sequence similarity to the RNA maturases encoded by group I introns, but consist in fact of overlapping reading frames, which are related by +1 or −1 frameshifts with overlaps centred on short G+C-rich sequences. Three possibilities could account for these observations.

a) The URFs could be pseudo-genes resulting from insertion of GC clusters, which seem to behave as mobile sequence elements in yeast mitochondria. Insertion would have to be a relatively recent event to account for the present degree of sequence conservation and absence of other types of mutation.

b) A process resembling the RNA-editing mechanism found in trypanosome mitochondria [106, 107] could be responsible for correction of the frameshifts. There is so far no evidence for the involvement of RNA-editing in the expression of other yeast mitochondrial genes but, without direct sequence data on the relevant transcripts, this possibility cannot yet be formally ruled out.

c) The frameshifts may serve to restrict expression of the URF to a low level relative to that of the highly expressed gene upstream [54, 108] in a fashion similar to that proposed for other frameshifted genes in other systems [109]. Frameshifting in other sequence contexts has previously been observed in yeast mitochondria [110] and this possibility thus deserves further examination.

C. NUCLEAR GENE EXPRESSION AND MITOCHONDRIAL BIOGENESIS

1. Carbon source and oxygen control

Under conditions in which the yeast cell is dependent on its mitochondria for oxidative metabolism, growth and division of the cell is matched by an increase in mitochondrial mass. As in other organisms, newly synthesized components are added to pre-existing organelles in such a way that functional constancy is preserved. How is this achieved? We are probably still far from knowing a full answer to this question, but results so far show the existence of controls at many levels, including transcription, protein import and assembly. As discussed below, a number of them can be nicely illustrated by an examination of just a single segment of the mitochondrial respiratory chain, that involved in electron transport between ubiquinol and cytochrome *b*.

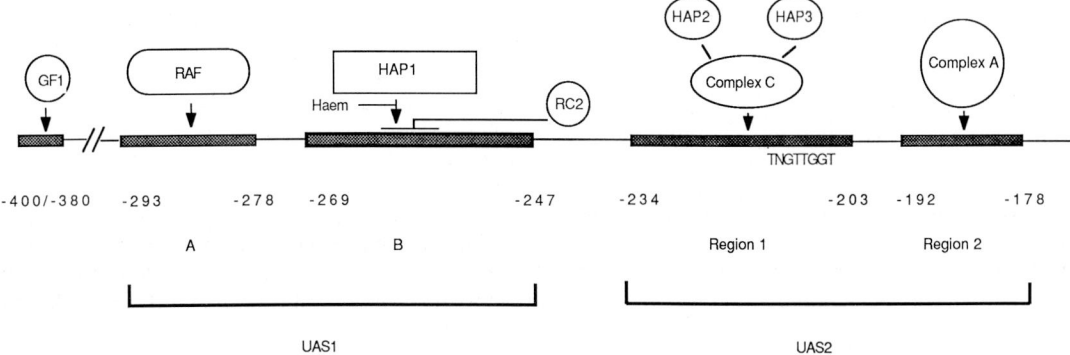

Fig. 5. *Organization of the upstream region of the gene for iso-1 cytochrome c (CYC1)*. Schematic representation (not to scale) of the 5' flank of the *CYC1* gene showing binding sites for the transcriptional activators HAP1−3 and for a number of other DNA-binding proteins, whose function has not yet been resolved. Distances are given in base pairs relative to the start site of RNA synthesis. Based on data presented in [112−114]. Data on the binding site for GF1, a factor in addition capable of binding to ARS and silencer elements, are taken from [145]. See also text for discussion

The enzyme mediating this electron transfer is the ubiquinol−cytochrome *c* reductase. As will be discussed more fully in a later section, this consists of at least nine subunits [11], of which only one (cytochrome *b*) is mitochondrially encoded. The remainder are, like cytochrome *c*, the acceptor of electrons from the complex, encoded by nuclear genes and imported from the cytoplasm. Steady-state levels of subunits of the complex and of cytochrome *c* vary in response to the need of the cell for mitochondrial function. For cytochrome *c*, it was shown at a very early stage that this response is governed primarily by changes in the rate of mRNA synthesis [111]. Subsequent analysis carried out mainly by Guarente and colleagues has given us the most detailed picture of the fine-structure of the promoter of the gene for this protein (*CYC1*) and how its transcriptional activity is regulated in response of the cell to oxygen and carbon source. As shown in Fig. 5, the *CYC1* gene is equipped with two upstream regions capable of activating transcription, namely UAS1 and UAS2.

UAS1 forms the target for binding by the transcriptional activator protein HAP1, whose task is to stimulate RNA synthesis under aerobic conditions. This protein is the product of the *CYP1* gene originally identified by Clavilier et al. [115] and its binding both to UAS1 [116, 117] and to the upstream activation of the gene for iso-2-cytochrome *c* (*CYC7*) is stimulated by haem [118−121], suggesting that oxygen control is in fact mediated by the level of intracellular haem. This makes sense, because the final stages in the synthesis of haem itself occur in the mitochondria and are dependent on oxygen. HAP1 continues to respond to haem even after passage over heparin−Sepharose [112], so it seems reasonable to assume that the protein is capable of binding haem directly. Support for this idea is also provided by examination of the sequence, which reveals the presence of several elements potentially capable of folding around haem, or a metal atom. It has been suggested that these elements could act as sensors, by responding to the redox state of the cell and thus mediating response to oxygen [122]. There is evidence, some of it indirect, that HAP1 control is also exerted on a large number of genes including those for non-haemoproteins [123, 124] and at least one non-mitochondrial protein [125]. The protein could thus play a key role in the coordinated synthesis of both mitochondrial and non-mitochondrial proteins during transitions from anaerobic to aerobic growth and vice versa.

Of interest for the mechanism of transcriptional activation is the finding that HAP1 is capable of binding with high selectivity to widely different DNA sequences and of responding differently to them [126]. For example, when bound to UAS1, the protein simulates transcription more than 300-fold and a major part of this increase is dependent on the presence of haem. When bound to the UAS upstream of the *CYC7* gene, however, stimulation is at least a factor 10 lower and haem dependence is marginal in comparison with *CYC1*. How this differential effect is achieved is not yet clear, since competition experiments using the two UAS sequences suggest that the affinity of HAP1 for both is similar and that the binding sites on the protein for each are either identical, or at least overlap closely [126]. The low degree of transcriptional activation of the *CYC7* gene is perhaps best explained by an inability of the UAS of this gene to promote conformational changes in the protein, which ultimately lead to a productive interaction with RNA polymerase (cf. [127]). Such an idea is consistent with the finding that in *CYP1-18* mutants, which display high transcriptional activity of the *CYC7* gene but low activity of *CYC1*, a serine → arginine change at position 64 of the protein has occurred. This change could influence UAS specificity by affecting the ability to form alternative and mutually exclusive Zn fingers [128].

UAS2, the second upstream sequence of *CYC1*, is required for activation of the gene in response to a shift to a non-fermentable carbon source [117]. Products of the genes *HAP2* and *HAP3* are both required for activation of UAS2 and probably act together as a single complex C (Fig. 5) [113, 114, 129, 130]. Mutational analysis shows that UAS2 in fact consists of two sub-domains, of which the first (region 1) mediates carbon-source induction, while the second (region 2) augments transcriptional activity via the binding of as yet uncharacterized factors [114]. The most critical part of region 1 appears to be the sequence TNATTGGT. A G → A transition at position 3 of this motif (−208 with respect to the start of the mRNA) increases transcription, but differentially dependent on carbon source (20-fold during growth on glucose; only 2-fold on lactate) and this has led to the suggestion [114] that shutdown of UAS2 during catabolite repression is due to limited binding of HAP2 and HAP3 under these conditions.

Although the response of UAS1 and UAS2 to the different HAP factors accounts in broad lines for the observed transcriptional behaviour of the *CYC1* gene, the final picture of regulatory mechanisms may be found to be considerably more complex. As pointed out above, a number of other proteins contribute to transcriptional activation of UAS2 and the action of these remains to be investigated. In UAS1, mutational analysis has also defined two domains, both of which are required for binding of HAP1 and for activity [131]. The first of these, region A, binds an as yet poorly defined protein called RAF (region A factor [112]) and methylation-interference footprint analysis suggests that protein–protein interactions between the two factors may contribute to the formation of a stable complex. Since C-terminally truncated versions of HAP1 bind to DNA and RAF without activating UAS1, both HAP1 and RAF apparently contribute to a domain responsible for transcriptional activation. Region B is part of a site that binds an additional factor called RC2 [132]. Cells deficient in haem lack RC2-binding activity and this cannot be restored by haem addition. HAP1 binds to region B, mutually exclusively with RC2, thus raising the interesting possibility that RC2 may have the function of blocking the action of UAS1 under conditions that UAS2 is active [112].

The action of HAP2 and HAP3 proteins is not confined to the *CYC1* gene. There is direct evidence that transcription of both *HEM1*, coding for the mitochondrial enzyme 5-aminolevulinate synthase and *COX4*, coding for subunit IV of cytochrome *c* oxidase, is reduced in *hap2* and *hap3* mutants [133]. In addition, the occurrence of sequences resembling the TNGTTGGT motif in the 5' flanks of genes for many mitochondrial proteins is an indirect indication that both proteins have many targets [134]. Just as for the CYC1 gene, however, regulation by HAP2 and HAP3 may be part of a complex set of controls, which together determine the final level of transcription and response to environmental changes. The *HEM1* gene, for example, displays constitutive expression, but this is in fact due to the balanced action of positive and negative regulatory elements responsive to carbon source and intracellular haem level [133]. Similarly, the apparent lack of response of transcription of the *COX4* gene to glucose repression may be due to the masking of HAP2/3 control by other factors [135]. Such interweaving of transcriptional controls is appearing as a theme common to the expression of many genes in yeast, with the full level of intricacy often only being exposed when the balance is disturbed by mutation [136]. The importance of a detailed dissection of both *cis*-acting and *trans*-acting elements therefore cannot be too highly stressed.

2. Mitochondrial biogenesis in relation to cell growth and division

Complex though the controls exerted by HAP and associated factors are, they cannot be the whole story. During budding, the daughter cell probably receives only a relatively small amount of mother cell mtDNA [137–140]. Should this distribution also reflect that of mitochondrial mass, then the cell must be able to monitor and respond to mitochondrial apportionment during growth and differentiation. The mechanism employed must be able not only to compensate for fluctuations in apportionment, but also respond to differences in growth rate and this last aspect in particular may involve interesting principles. When *S. cerevisiae* cells divide at slow growth rates, the bud is much smaller than the mother and must increase more in size after separation [141]. Measurements in fact show that the daughter cell has a longer G1 period in which this occurs [141]. At fast growth rates, however, separation occurs when the daughter cell has attained approximately the same size as the mother, so that there must be a mechanism that allows selective accumulation of mitochondrial mass and hence expression of both mitochondrial and nuclear genes in the daughter cell whilst it is still attached to the mother.

How this is achieved is not known, although analogy with other cell-cycle-regulated processes suggests several possibilities, including cell cycle control of the synthesis of transcriptional activators, asymmetric segregation of such activators and coupling of transcription to DNA replication (see [142] for review). Alongside these, some form of common regulation via attachment of groups of actively transcribed genes to some form of higher-order chromatin structure, such as a nuclear matrix or scaffold, also deserves consideration. The potential importance of a scaffolding network in transcriptional control via looping of DNA and physical clustering of active genes has been stressed by several authors (see [143, 144] and references therein). Such considerations may be especially relevant in the light of the findings by Dorsman et al. [145] that the 5'-flanking regions of genes coding for a number of nuclear-coded subunits of the yeast ubiquinol–cytochrome *c* reductase and of a number of other imported mitochondrial proteins contain binding sites for two abundant proteins also capable of recognizing genetic elements important in cell growth and division. The first of these (GF1) binds to elements important for optimal ARS function, while the second (GF2) recognizes a conserved element required for centromere function. GF1 strongly resembled the factors SBF-B and ABF1 previously identified by their ability to bind to ARS and silencer elements [146, 147], while GF2 may be identical to the centromere binding protein CP1 [148]. For these proteins too, roles in higher-order chromatin structure, or attachment to nuclear structures, have been speculated upon. Common attachment of yeast ARS and centromere sequences to the nuclear matrix in yeast has in fact recently been demonstrated, but the proteins mediating this attachment have not been identified [149].

D. A PATH FROM MITOCHONDRION TO NUCLEUS?

Although mitochondrial biogenesis might be expected to be dependent on a high level of nucleo-mitochondrial cross talk, the search for evidence of the mitochondrial contribution to such communication has so far been an elusive one. To what extent and how the mitochondrial genome exerts an influence on nuclear transcription and translation is still unclear and a matter for discussion. Cogent arguments against any major role for the mitochondrial genome in such regulation have been presented [2]. They are based on the observations that synthesis of many nuclear-coded components of mitochondrial respiratory complexes is not dependent on the presence of an intact mitochondrial genome and that the mitochondria of wild-type and cytoplasmic *petite* cells have essentially identical protein compositions.

On the other hand, there are several reported instances of alterations in the level of expression of nuclear genes in response to impaired mitochondrial function. In *Neurospora* and *Tetrahymena*, treatment of wild-type cells with chloramphenicol or ethidium bromide leads to the increased synthesis of a number of nuclear-encoded mitochondrial components, including methionyl-tRNA transformylase, RNA polymerase, DNA polymerase, at least one ribosomal protein and elongation factor G [150–153]. In yeast, steady-state levels of translatable mRNAs for nuclear-encoded subunits of ubiquinol-cytochrome c reductase were found to be much lower in a cytoplasmic *petite* mutant than in a wild-type strain growing under the same conditions [154]. More recently, levels of expression of a number of nuclear DNA sequences, not all of them directly related to mitochondrial biogenesis, were found to change in response to defects in, or loss of the mitochondrial genome [155, 156]. Early ideas that this modulation of nuclear gene expression is due to the action of a mitochondrial repressor protein which is exported to the cell nucleus [151, 157] have so far not received experimental support and, given the limited information content of most mtDNAs, seem unlikely. Alternative propositions involving regulation via metabolic changes induced as a result of impaired mitochondrial function are perhaps more plausible [155] and future efforts to delineate the elements of such control circuits will be of great interest.

E. IMPORTED PROTEINS: SHARING ENZYMES WITH THE REST OF THE CELL

Since more than 90% of all mitochondrial proteins are encoded by nuclear genes and synthesized in the cytoplasm, the final steps at which control on mitochondrial biogenesis can be exerted are those of import into the organelle and, for those proteins that form part of multi-subunit complexes, assembly. Able reviewers of the field of protein import abound [158–162], so that extensive coverage of the topic here is unnecessary.

One aspect of protein import of potential interest in the context of regulatory mechanisms concerns the growing list of enzymes that mitochondria share with the rest of the cell. An early and firmly held belief was that each compartment of the eukaryotic cell would contain only isoenzymes specific to it, but results have proved otherwise. Use is made of the appropriate addressing signals to direct the same protein to different destinations and different strategies are employed to control the relative amounts of protein sent to each.

One of the first documented example of this principle concerns the *HTS1* gene encoding the cytoplasmic and mitochondrial forms of the histidinyl-tRNA synthetase in yeast [163]. Both proteins are translated from mRNAs that are derived from the same coding sequences (Fig. 6). The mRNA for the mitochondrial enzyme is the longer of the two and its synthesis is initiated at a site upstream of two in-frame translational start sites, which lie 60 bp apart. Translation of this messenger from the AUG codon closest to the 5' end yields a protein bearing an N-terminal sequence with all the hallmarks of a mitochondrial cleavable presequence. The shorter message is initiated between the start sites. It therefore contains only the downstream AUG and the resulting protein lacks a presequence. A mutation destroying the upstream start site leads to a respiratory-deficient phenotype, but does not affect either the cytoplasmic histidinyl-tRNA synthetase, or cell viability. In contrast, mutations distal to the downstream AUG lead to loss of the cytoplasmic synthetase and lethality.

Other genes encoding products with a dual location in the cell include the *TRM1*, *TRM2*, *MOD5*, *LEU4*, *FUM1* and *VAS1* loci [164–169]. *TRM1*, *FUM1*, *LEU4* and *VAS1* resemble the histidinyl-tRNA synthetase gene in coding for long and short mRNAs, of which only the former is capable of directing the synthesis of a precursor protein with an N-terminal extension. Interestingly, the *TRM1* protein, which is involved in the post-transcriptional modification of both mitochondrial and nuclear-encoded tRNAs, has an amino-terminal extension which is not strictly required for import

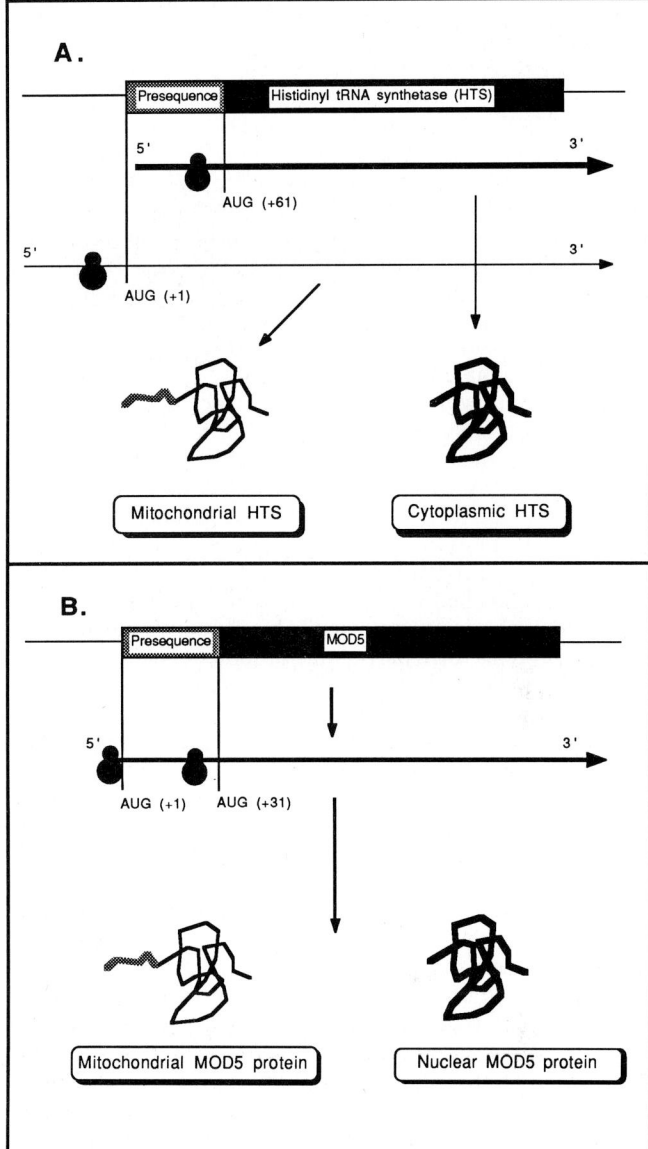

Fig. 6. *One-gene—one-protein—two-destinations.* The figure shows two different strategies used by the yeast cell to achieve differential synthesis of two proteins with a dual location in the cell. (A) Synthesis of mitochondrial and cytoplasmic forms of histidyl-tRNA synthetase encoded by *HTS1* gene by use of differential transcription to give two mRNAs differing at their 5' ends. (B) Synthesis of mitochondrial and nuclear forms of Δ^2-isopentenyl-pyrophosphate-tRNA isopentyl-transferase encoded by the *MOD5* gene from a single mRNA. In this case, proximity of the first AUG to the 5' end of the mRNA probably leads to more frequent initiation at the downstream site. Relative levels of synthesis are displayed schematically by differences in the thickness of the lines representing the mRNAs or the folded proteins. Based on data taken from [163] and [166]. See text for discussion

into mitochondria and is itself unable to target other proteins to the organelle. Its function in the mitochondrial version of *TRM1* may therefore be primarily to mask sequences responsible for addressing the protein to the nucleus.

MOD5, encoding Δ^2-isopentenyl-pyrophosphate-tRNA isopentenyltransferase displays an interesting variation on the theme of one-gene—one-protein—two-destinations (Fig. 6). Like *HTS1*, the 5' part of the coding sequence contains two translational initiation sites capable of directing the synthesis of two forms of the enzyme differing in length by 11 amino acids [166]. Unlike *HTS1*, the gene encodes a set of heterogeneous transcripts whose 5' ends all lie upstream of both sites, so that it is unlikely that different forms of the protein arise as a result of differential initiation of transcription. On the other hand, the leaders of the transcripts are short, with their 5' ends lying only 1–11 nucleotides upstream of the first AUG. This, together with the fact that the sequences flanking this first site show a poor match to the consensus sequence for optimal translational initiation [166], suggests that initiation of the synthesis of the longer (mitochondrially addressed) protein may be very inefficient compared with that of the nuclear form initiated at the downstream site.

The list of shared enzymes is probably far from complete. It is, however, worth speculating on the present value of this situation, since such sharing of enzymes clearly places limitations on their evolutionary flexibility. In the case of the aminoacyl-tRNA synthetases and tRNA-modification enzymes, for example, interaction must be maintained with very different substrates that are evolving at very different rates. In addition, shared enzymes are required to function in different cellular compartments in which different ionic conditions prevail. Why is sharing maintained? One attractive possibility is that the proteins form part of one or more regulatory circuits linking mitochondrial with cellular metabolism. In this respect, it may be no coincidence that most shared enzymes are involved in some way in translation, thus permitting coupling of the biosynthetic capacities of both cytosolic and mitochondrial systems.

F. ASSEMBLY

An as yet poorly understood area of nucleo-mitochondrial interactions is that of the role played by a number of the nuclear-encoded subunits in the catalytic activity of mitochondrial respiratory complexes. By comparison with their bacterial counterparts, mitochondrial versions of the respiratory enzymes are complex, with many additional subunits whose role in function is in the majority of cases ill-defined [11, 170, 171]. The availability of cloned genes for such subunits, coupled with the potential for their modification via site-directed mutagenesis and our ability to introduce the modified sequences into the yeast genome should, however, shortly put an end to this situation. An illustration of the value of this kind of approach comes from recent studies of the biosynthesis of the yeast ubiquinol–cytochrome *c* reductase which, as mentioned above, consists of nine subunits. Of the eight nuclear-encoded subunits, two carry prosthetic groups (cytochrome c_1 and the Rieske FeS protein) and the participation of these in electron transfer has been studied intensively. By comparison, much less is known of the roles played by the remainder.

Mutants lacking a functional gene for subunits II (40 kDa), VI (17 kDa), VII (14 kDa) and VIII (11 kDa) have been constructed in our laboratory and the effects of such deletions on assembly and function of the enzyme studied [172–176]. This analysis reveals that with the exception of the 17-kDa subunit VI, each of these subunits plays an important role in assembly, by stabilizing partially assembled complexes. As Table 7 shows, mutants lacking either the 14-kDaA or 11-kDa subunit contain severely reduced steady-state levels of the Rieske FeS protein, while the mitochondrially coded apocytochrome *b* is present in only trace amounts and holocytochrome *b* is undetectable. The mRNAs of the affected subunits are unaltered in level [175]

and, at present, the most likely explanation for the observed effects is that one of the main functions of subunits VII and VIII is to protect apocytochrome *b* and the FeS protein from proteolytic degradation. As discussed by Berden et al. [178], a similar function can be assigned to the two core proteins [44 and 40 kDa). Taken together with related observations made by Tzagoloff and his colleagues [179, 180], these findings lead to the proposal for assembly of the complex in the mitochondrial membrane as shown in Fig. 7, in which three subcomplexes can be distinguished. These associate with the Rieske FeS protein to form an enzymatically active complex. Two of the subcomplexes are intrinsically stable, while the third, the cytochrome *b* subcomplex consisting of apocytochrome *b* together with the 14-kDa and 11-kDa subunits, is not. The proteases responsible for this highly specific degradative process have yet to be characterized.

The additional subunits in the mitochondrial complexes may also serve to modulate catalytic activity in response to developmental needs or environmental changes. Direct experimental support for this idea is at present rather patchy, but one system where it could be studied further is yeast cytochrome *c* oxidase. Laboratory *S. cerevisiae* contain two genes for subunit V of this enzyme [181]. These are expressed differentially, with the *COX5a* gene being 25–55 times more active than *COX5b* in cells grown aerobically in glucose [135, 182, 183]. Expression of *COX5b* is, however, favoured when cells are grown aerobically in the absence of haem, or when they are shifted to anaerobic conditions. It is an intriguing possibility, therefore, that such differential expression could be advantageous if the products of the two genes confer functional differences on cytochrome *c* oxidase (e.g. altered K_m for O_2 binding, cf. [135–183]).

G. WHY mtDNA?

The scale of nuclear genetic investment in mitochondrial biogenesis and in the maintenance of the mitochondrial genetic system begs the question why mtDNA should need to be retained at all. According to the endosymbiont hypothesis, mitochondrial evolution has been marked by progressive transfer of genes to the nucleus. If so, why did this process stop? Possible answers are as follows.

a) The mitochondrial genetic system is a frozen accident [184]. The divergence of mitochondrial and nuclear genetic systems made genetic exchange with the nucleus more and more difficult until it finally became impossible, leaving a small set of genes isolated within the organelle.

b) Some proteins contain 'poison' sequences that are incompatible with their synthesis in soluble form in the cytoplasm and their transfer across the mitochondrial membrane [184]. They thus have to be made *in situ*. A variation on this theme is that certain proteins have to be made in the mitochondria, because they contain sequences that would otherwise be recognized by the machinery for co-translational translocation of proteins across the endoplasmic reticulum [185].

Of the two, the first is the less attractive. It takes no account of the fact that genetic exchanges are still possible between nucleus and mitochondrion, as evidenced by the existence of numerous reports on migratory DNA. Sequences homologous to mtDNA have found their way into the chromosomal DNA of various organisms [186–188]. In all cases, transfer seems to have been accidental. The transferred DNA is garbled and non-functional. Its presence does serve to illustrate, however, that genetic exchange is still possible.

Table 7. *Effects of subunit deficiencies on yeast ubiquinol–cytochrome c oxidoreductase*
The table summarizes the results of Western blot experiments, in which subunit-specific antibodies were used to quantify the steady-state levels of individual subunits in each of the subunit-lacking (subunit[0]) mutants. For ease of presentation, levels have been broadly classified as trace, low (< 20% wild-type level), or unaffected (> 70% wild-type level); see text and [178] for discussion

Mutant type	Activity	Subunit levels		
		trace	low	unaffected
I (11-kDa or 14-kDa protein[0])	absent	apocyt. b	11-kDa and 14-kDa proteins; FeS	cores I and II; Cyt c_1; 17-kDa and 7.2-kDa proteins
II (Core I[0] or II[0])	low[a]	none	11-kDa and 14-kDa proteins; Cyt. b; FeS	Cyt c_1; 17-kDa and 7.2-kDa proteins
III (17-kDa protein[0])	present	none	none	all subunits[b]
(FeS[0])	absent	none	none	all subunits

[a] For core I[0] only detectable with yeast cytochrome c at high ionic strength (225 mM), or horse cytochrome c at low ionic strength (33 mM) [177].

[b] With the possible exception of the 7.2-kDa subunit, the level of which has not yet been determined in the 17-kDa-protein[0] mutant.

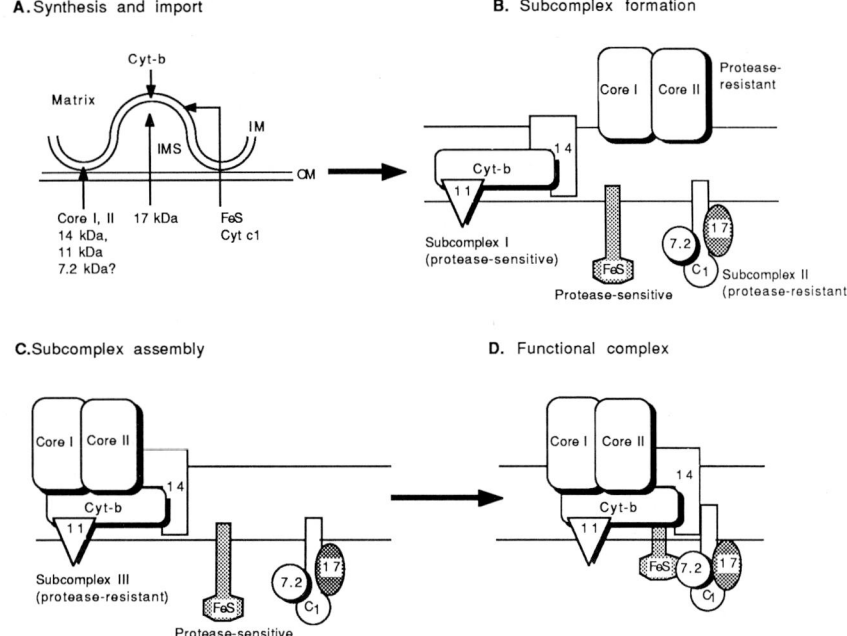

Fig. 7. *Model for assembly of the QH_2-cytochrome c oxidoreductase in yeast.* Studies with gene-disruption mutants have suggested [178] that the assembly of a functional ubiquinol–cytochrome c reductase occurs in four distinguishable phases, as shown in (A)–(D). In the first of these, individual subunits are synthesized in the cytoplasm and imported independently into the mitochondrion. Although not all subunits have been studied in detail, it seems probable that they attain their final locations on, or in the inner membrane by different routes (see [11] for discussion). Subsequent to addition of prosthetic groups to apocytochrome b, cytochrome c_1 and the Rieske FeS protein, subcomplexes are formed, as shown in (B). Of these, that containing b together with 14-kDa and 11-kDa subunits is sensitive to proteolytic degradation, unless stabilized by the FeS protein and the other subcomplexes shown. In the following stages, shown in (C) and (D), assembly of a functional complex occurs. The position of the 7.2-kDa subunit in the scheme is speculative. See text for further discussion

Of course, the present differences in genetic code between mitochondrion and nucleus form the current barrier to transfer of useful information, but these differences in themselves were probably not primarily responsible for prevention of transfer: the current view is that the variant genetic codes found in mitochondria do not represent remnants of a primitive code, but they have arisen in relatively recent evolutionary events [189, 190]. Moreover, the extent of information transfer appears to have been much greater in the case of mitochondria, which make use of variant genetic codes, than in chloroplasts, which do not.

The idea that some mitochondrial proteins have to be made *in situ* is more appealing, even though it may not apply equally to all proteins currently encoded by mtDNA. ATPase subunit 9, for example, is synthesized inside the mitochondrion in yeast, but in other organisms it is synthesized in the cytoplasm and imported into mitochondria [191, 192]. Moreover, by means of gene synthesis, or repeated cycles of *in vitro* mutagenesis on cloned mtDNA, it has been possible to transfer functional copies of mitochondrial genes to the nucleus and then demonstrate synthesis, import and function of the encoded proteins [63, 193–195]. The crucial test of the hypothesis will come, however, with the transfer of proteins like CoxI or cytochrome *b*, whose genes are mitochondrially located in all organisms so far examined.

H. CONCLUSIONS AND PROSPECTS

The overview given above should have convinced the reader that, despite considerable progress in our understanding of mitochondrial biogenesis, much still remains to be done. Areas that may provide new insights include the following.

1. Mitochondrial transformation

A frustrating limitation in our ability to test ideas concerning features of mitochondrial gene expression has been the lack of a transformation technique allowing introduction of manipulated DNA sequences into mitochondria. Recent findings suggest that this is now possible in yeast with a reasonable degree of success [52, 53]. The way is thus open to *in vivo* analysis of sequences involved in promoter function, mRNA translational efficiency and stability, and equally interesting, a detailed study of structure/function relationships in the mitochondrially coded components of respiratory enzymes.

2. Nucleo-mitochondrial interactions

The work described in this review has led mainly to the identification of nuclear gene products likely to play a role in the regulation of mitochondrial biogenesis and in outline, to basic ideas on how this regulation might be achieved. Efforts should now be directed at filling in this framework, so that we really can understand how the delicate balance of gene expression necessary for synthesis of this complex organelle is achieved and how this balance is maintained or modulated during the cell cycle and differentiation. In mammalian cells such studies are likely to have important spin-off in terms of our understanding of the mitochondrial myopathies, an important group of inherited diseases in which mitochondrial synthesis and/or function is impaired, often in a tissue-specific fashion.

3. Borrowed and shared proteins

Mitochondrial genomes are unconventional in the way they organize and express their genes and the study of these unusual features has given valuable insight into the problems of rRNA and tRNA design, ribosome structure, decoding, RNA splicing, mRNA selection and many other processes. In most instances, this typically mitochondrial gene expression is dependent on proteins supplied by the nucleus. Such proteins have possibly been recruited from cellular proteins with other primary functions, some of which may have been retained, others lost in the process of co-evolution with in most cases rapidly evolving mitochondrial genomes. The study of such borrowed or shared proteins, their evolutionary origins, the roles they fulfil in specific steps in gene expression in the nucleus, cell sap or mitochondrion may yield insight from a new angle into the evolution of mitochondrial genetic systems and into the way in which their present day dependence on the nucleus was attained.

4. Can we dispense with mtDNA?

Using information derived from the study of mitochondrial protein import, it should be possible to design addressing sequences capable of directing any protein to a specific location in the mitochondrion. A critical test of our understanding of the principles involved in the re-routing of proteins will be that of attempting to manoeuvre typical mitochondrial translation products such as CoxI and cytochrome *b* into their correct membrane topologies from the cytoplasm. Such attempts will pave the way to the more ambitious goal of constructing a eukaryotic cell in which all mitochondrial genes have been transferred to the nucleus. Achievement of this goal will provide an answer to the question of whether mtDNA can indeed be dispensed with and will considerably simplify the task of future reviewers.

I thank members of my laboratory for their discussion of many of the ideas presented here and for their comments on the manuscript. I am also grateful to Dr Tom Fox (Section of Genetics, Cornell University, Ithaca, USA) for making sequence data on *PET54* available before publication. Work carried out in my laboratory was in part supported by grants from the Netherlands Organization for the Advancement of Pure Research (NWO), under the auspices of the Netherlands Foundation for Chemical Research (SON).

REFERENCES

1. Dujon, B. (1981) in *The molecular biology of the yeast* Saccharomyces. *Life cycle and inheritance* (J. N. Strathern, E. W. Jones & J. R. Broach, eds) pp. 505–635, Cold Spring Harbor Laboratory.
2. Tzagoloff, A. & Myers, A. M. (1986) *Annu. Rev. Biochem. 55*, 249–285.
3. Dujon, B. (1983) in *Mitochondria 1983. Nucleo-mitochondrial interactions* (R. J. Schweyen, K. Wolf & F. Kaudewitz, eds) pp. 1–24, de Gruyter, Berlin.
4. Grivell, L. A. (1983) in *Mitochondria 1983. Nucleo-mitochondrial interactions* (R. J. Schweyen, K. Wolf & F. Kaudewitz, eds) pp. 25–45, de Gruyter, Berlin.
5. Wolf, K. (1987) in *Gene structure in eukaryotic microbes* (J. R. Kinghorn, ed.) pp. 41–67, IRL Press, Oxford.
6. Fearnley, I. M. & Walker, J. E. (1986) *EMBO J. 5*, 2003–2008.
7. Higuti, T., Negama, T., Takigawa, M., Uchida, J., Yamane, T., Asai, T., Tani, I., Oeda, K., Shimizu, M., Nakamura, K. & Ohkawa, H. (1988) *J. Biol. Chem. 263*, 6772–6776.
8. Zamaroczy, M. de & Bernardi, G. (1986) *Gene 47*, 155–177.
9. Hensgens, L. A. M., Bonen, L., de Haan, M., Van der Horst, G. & Grivell, L. A. (1983) *Cell 32*, 379–389.
10. de Vries, S. & Grivell, L. A. (1988) *Eur. J. Biochem. 176*, 377–384.
11. de Vries, S. & Marres, C. A. M. (1988) *Biochim. Biophys. Acta 895*, 205–239.
12. Grivell, L. A., Bonen, L. & Borst, P. (1983) *Horiz. Biochem. Biophys. 7*, 279–306.
13. Waring, R. B. & Davies, R. W. (1984) *Gene 28*, 277–291.
14. Tabak, H. F. & Grivell, L. A. (1986) *Trends Genet. 2*, 51–55.
15. Tabak, H. F. & Arnberg, A. C. (1986) *Oxf. Surv. Eukaryotic Genes 3*, 161–182.
16. Grivell, L. A. (1984) in *Genetic maps* (O'Brien, S. J., ed.) pp. 234–247, Cold Spring Harbor Press, Cold Spring Harbor, NY.
17. Clayton, D. A. (1984) *Annu. Rev. Biochem. 53*, 573–594.
18. Christianson, T. & Rabinowitz, M. (1983) *J. Biol. Chem. 258*, 14025–14033.
19. Osinga, K. & Tabak, H. F. (1982) *Nucleic Acids Res. 10*, 3617–3626.
20. Edwards, J. C., Levens, D. & Rabinowitz, M. (1983) *Cell 31*, 337–346.
21. Bordonné, R., Dirheimer, G. & Martin, R. P. (1987) *Nucleic Acids Res. 15*, 7381–7394.
22. Mueller, D. M. & Getz, G. S. (1986) *J. Biol. Chem. 261*, 11756–11764.
23. Biswas, T. K. & Getz, G. S. (1986) *Proc. Natl Acad. Sci. USA 83*, 270–274.

24. Wettstein-Edwards, J., Ticho, B. S., Martin, N. C., Najarian, D. & Getz, G. S. (1986) *J. Biol. Chem. 261*, 2905–2911.
25. Greenleaf, A. L., Kelly, J. L. & Lehman, I. R. (1986) *Proc. Natl Acad. Sci. USA 83*, 3391–3394.
26. Kelly, J. L. & Lehman, I. R. (1986) *J. Biol. Chem. 261*, 10340–10347.
27. Winkley, C. S., Keller, M. J. & Jaehning, J. A. (1985) *J. Biol. Chem. 260*, 14214–14223.
28. Masters, B. S., Stohl, L. L. & Clayton, D. A. (1987) *Cell 51*, 89–99.
29. Schinkel, A. H., Groot-Koerkamp, M. J. A., Touw, E. P. W. & Tabak, H. F. (1987) *J. Biol. Chem. 262*, 12785–12791.
30. Schinkel, A. H., Groot-Koerkamp, M. J. A. & Tabak, H. F. (1988) *EMBO J. 7*, 3255–3262.
31. Ticho, B. S. & Getz, G. S. (1988) *J. Biol. Chem. 263*, 10096–10103.
32. Locker, J. & Rabinowitz, M. (1981) *Plasmid 6*, 302–314.
33. Palleschi, C., Francisci, S., Zennaro, E. & Frontali, L. (1984) *EMBO J. 3*, 1389–1395.
34. Coruzzi, G., Bonitz, S. G., Thalenfeld, B. E. & Tzagoloff, A. (1981) *J. Biol. Chem. 256*, 12780–12787.
35. Thalenfeld, B. E., Hill, J. & Tzagoloff, A. (1983) *J. Biol. Chem. 258*, 610–615.
36. Martin, N. C., Miller, D. L., Underbrink, K. & Ming, X. (1985) *J. Biol. Chem. 260*, 1479–1483.
37. Simon, M. & Faye, G. (1984) *Mol. Gen. Genet. 196*, 266–274.
38. Osinga, K. A., De Vries, E., Van der Horst, G. & Tabak, H. F. (1984) *EMBO J. 3*, 829–834.
39. Christianson, T., Edwards, J. C., Mueller, D. M. & Rabinowitz, M. (1983) *Proc. Natl Acad. Sci. USA 80*, 5564–5568.
40. Zassenhaus, H. P., Martin, N. C. & Butow, R. A. (1984) *J. Biol. Chem. 259*, 6019–6027.
41. Mueller, D. M. & Getz, G. S. (1986) *J. Biol. Chem. 261*, 11816–11822.
42. McKee, E. E., McEwen, J. E. & Poyton, R. O. (1984) *J. Biol. Chem. 259*, 9332–9338.
43. Kelly, R. & Philips, S. L. (1983) *Mol. Cell. Biol. 3*, 1949–1957.
44. Goldthwaite, C. D., Cryer, D. R. & Marmur, J. (1974) *Mol. Gen. Genet. 133*, 87–104.
45. Wilcoxen, S. E., Peterson, C. R., Winkley, C. S., Keller, M. J. & Jaehning, J. A. (1988) *J. Biol. Chem. 263*, 12346–12351.
46. Etten, R. A. van, Bird, J. W. & Clayton, D. A. (1983) *J. Biol. Chem. 258*, 10104–10110.
47. Montoya, J., Gaines, G. L. & Attardi, G. (1983) *Cell 34*, 151–159.
48. Christianson, T. W. & Clayton, D. A. (1988) *Mol. Cell. Biol. 8*, 4502–4509.
49. Hollingsworth, M. J. & Martin, N. C. (1986) *Mol. Cell. Biol. 6*, 1058–1064.
50. Miller, D. L., Underbrink-Lyon, K., Najarian, D. R., Krupp, J. & Martin, N. C. (1983) in *Mitochondria 1983. Nucleo-mitochondrial interactions* (R. J. Schweyen, K. Wolf & F. Kaudewitz, eds) pp. 151–164, de Gruyter, Berlin.
51. Zhu, H., Macreadie, I. G. & Butow, R. A. (1987) *Mol. Cell. Biol. 7*, 2530–2537.
52. Johnston, S. A., Anziano, P. Q., Shark, K., Sanford, J. C. & Butow, R. A. (1988) *Science 240*, 1538–1541.
53. Fox, T. D., Sanford, J. C. & McMullin, T. W. (1988) *Proc. Natl Acad. Sci. USA 85*, 7288–7292.
54. Michel, F. (1984) *Curr. Genet. 8*, 307–317.
55. Bordonné, R., Dirheimer, G. & Martin, R. P. (1988) *Curr. Genet. 13*, 227–233.
56. Hibbs, A. R., Maheshwari, K. K. & Marzuki, S. (1987) *Biochim. Biophys. Acta 980*, 179–187.
57. Smooker, P. M., Wright, J. F., Linnane, A. W. & Lukins, H. B. (1988) *Nucleic Acids Res. 16*, 9081–9095.
58. Cech, T. R. (1987) *Science 236*, 1532–1539.
59. Van der Horst, G. & Tabak, H. F. (1987) *EMBO J. 6*, 2139–2144.
60. Davies, R. W., Waring, R. B., Ray, J. A., Brown, T. A. & Scazzocchio, C. (1982) *Nature 300*, 719–724.
61. Van der Veen, R., Kwakman, J. H. J. M. & Grivell, L. A. (1987) *EMBO J. 6*, 3827–3831.
62. Delahodde, A., Goguel, V., Becam, A. M., Creusot, F., Perea, J., Banroques, J. & Jacq, J. (1989) *Cell 56*, 431–441.
63. Banroques, J., Perea, J. & Jacq, C. (1987) *EMBO J. 6*, 1085–1091.
64. Gampel, A. & Tzagoloff, A. (1987) *Mol. Cell. Biol. 7*, 2545–2551.
65. Séraphin, S., Simon, M. & Faye, G. (1988) *EMBO J. 7*, 1455–1464.
66. Faye, G. & Simon, M. (1983) in *Mitochondria 1983. Nucleo-mitochondrial interactions* (R. J. Schweyen, K. Wolf & F. Kaudewitz, eds) pp. 433–439, de Gruyter, Berlin.
67. Séraphin, B., Simon, M., Boulet, A. & Faye, G. (1989) *Nature 337*, 84–87.
68. Kreike, J., Schulze, M., Ahne, F. & Lang, B. F. (1987) *EMBO J. 6*, 2123–2129.
69. Schmelzer, C. & Schweyen, R. J. (1986) *Cell 46*, 557–565.
70. Ben Asher, E., Groudinsky, O., Dujardin, G., Altamura, N., Kermorgant, M. & Slonimski, P. P. (1989) *Mol. Gen. Genet. 215*, 517–528.
71. Herbert, C. J., Labouesse, M., Dujardin, G. & Slonimski, P. P. (1988) *EMBO J. 7*, 473–483.
72. Akins, R. A. & Lambowitz, A. M. (1987) *Cell 50*, 331–345.
73. Garriga, G. & Lambowitz, A. M. (1986) *Cell 46*, 669–680.
74. Dujardin, G., Labouesse, M., Netter, P. & Slonimski, P. P. (1983) in *Mitochondria 1983. Nuclear-mitochondrial interactions* (R. J. Schweyen, K. Wolf & F. Kaudewitz, eds) pp. 233–250, de Gruyter, Berlin.
75. Blow, D. M. & Brick, P. (1985) in *Biological macromolecules and assemblies* (F. A. Jurnak & A. McPherson, eds) pp. 442–469, Wiley, New York.
76. McGraw, P. & Tzagoloff, A. (1983) *J. Biol. Chem. 258*, 9459–9468.
77. Séraphin, B., Boulet, A., Simon, M. & Faye, G. (1987) *Proc. Natl Acad. Sci. USA 84*, 6810–6814.
78. Koll, H., Schmidt, C., Wiesenberger, G. & Schmelzer, C. (1987) *Curr. Genet. 12*, 503–509.
79. Dieckmann, C. L., Koerner, T. J. & Tzagoloff, A. (1984) *J. Biol. Chem. 259*, 4722–4731.
80. Chen, J.-Y. & Martin, N. C. (1988) *J. Biol. Chem. 263*, 13677–13682.
81. Dieckmann, C. L. & Mittelmeier, T. M. (1987) *Curr. Genet. 12*, 391–397.
82. Dobres, M., Gerbl-Reiger, S., Schmelzer, C., Mueller, M. W. & Schweyen, R. J. (1985) *Curr. Genet. 10*, 283–290.
83. Brawerman, G. (1987) *Cell 48*, 5–6.
84. Dieckmann, C. L., Homison, G. & Tzagoloff, A. (1984) *J. Biol. Chem. 259*, 4732–4738.
85. Pearson, W. R. & Lipman, D. J. (1988) *Proc. Natl Acad. Sci. USA 85*, 2444–2448.
86. Groot, G. S. P., Flavell, R. A., Ommen, G. J. B. van & Grivell, L. A. (1974) *Nature 252*, 167–169.
87. Li, M., Tzagoloff, A., Underbrink-Lyon, K. & Martin, N. C. (1982) *J. Biol. Chem. 257*, 5921–5928.
88. Fox, T. D. (1986) *Trends Genet. 2*, 97–100.
89. Ooi, B. G., Lukins, H. B., Linnane, A. W. & Nagley, P. (1987) *Nucleic Acids Res. 15*, 1965–1977.
90. Strick, C. A. & Fox, T. D. (1987) *Mol. Cell. Biol. 7*, 2728–2734.
91. Costanzo, M. C., Mueller, P. P., Strick, C. A. & Fox, T. D. (1986) *Mol. Gen. Genet. 202*, 294–301.
92. Dieckmann, C. L. & Tzagoloff, A. (1985) *J. Biol. Chem. 260*, 1513–1520.
93. Forsbach, V. & Rödel, G. (1987) in *Cytochrome systems, molecular biology and bioenergetics* (Papa, S., Chance, B. & Ernster, L., eds) pp. 169–176, Plenum Press, New York.
94. Crivellone, M., Gampel, A., Murff, I., Wu, M. & Tzagoloff, A. (1987) in *Cytochrome systems, molecular biology and bioenergetics* (Papa, S., Chance, B. & Ernster, L., eds) pp. 67–77, Plenum Press, New York.
95. Kloeckener-Gruissem, B., McEwen, J. E. & Poyton, R. O. (1987) *Curr. Genet. 12*, 311–322.
96. Constanzo, M. C. & Fox, T. D. (1986) *Mol. Cell. Biol. 6*, 3694–3703.

97. Fox, T. D., Constanzo, M. C., Strick, C. A., Marykwas, D. L., Seaver, E. C. & Rosenthal, J. K. (1988) *Phil. Trans. R. Soc. Lond. B 319*, 97–105.
98. Mueller, P. P., Reif, M. K., Zonghou, S., Sengstag, C., Mason, T. L. & Fox, T. D. (1984) *J. Mol. Biol. 175*, 431–452.
99. Rödel, G., Korte, A. & Kaudewitz, F. (1985) *Curr. Genet. 9*, 641–648.
100. Rödel, G. (1986) *Curr. Genet. 11*, 41–45.
101. Rödel, G. & Fox, T. D. (1987) *Mol. Gen. Genet. 206*, 45–50.
102. Costanzo, M. C. & Fox, T. D. (1988) *Proc. Natl Acad. Sci. USA 85*, 2677–2681.
103. Benne, R. & Sloof, P. (1987) *Biosystems 21*, 51–68.
104. Pain, V. M. (1986) *Biochem. J. 235*, 625–637.
105. Donahue, T. D., Cigan, A. M., Pabich, E. K. & Castilho Valavicius, B. (1988) *Cell 54*, 621–632.
106. Benne, R., Van den Burg, J., Brakenhoff, J. P. J., Sloof, P., Van Boom, J. H. & Tromp, M. C. (1986) *Cell 46*, 819–826.
107. Feagin, J. E., Abraham, J. M. & Stuart, K. (1988) *Cell 53*, 413–422.
108. Séraphin, B., Simon, M. & Faye, G. (1987) *J. Biol. Chem. 262*, 10146–10153.
109. Craigen, W. J. & Caskey, C. T. (1987) *Cell 50*, 1–2.
110. Fox, T. D. & Weiss-Brummer, B. (1980) *Nature 288*, 60–63.
111. Zitomer, R. S., Montgomery, D. L., Nichols, D. L. & Hall, B. D. (1979) *Proc. Natl Acad. Sci. USA 76*, 3627–3631.
112. Pfeifer, K., Arcangioli, B. & Guarente, L. (1987) *Cell 49*, 9–18.
113. Olesen, J., Hahn, S. & Guarente, L. (1987) *Cell 51*, 953–961.
114. Forsburg, S. L. & Guarente, L. (1988) *Mol. Cell. Biol. 8*, 647–654.
115. Clavilier, L., Péré, G. & Slonimski, P. P. (1969) *Mol. Gen. Genet. 104*, 195–218.
116. Guarente, L. & Mason, T. (1983) *Cell 32*, 1279–1286.
117. Guarente, L., Lalonde, B., Gifford, P. & Alani, E. (1984) *Cell 36*, 503–511.
118. Prezant, T., Pfeifer, K. & Guarente, L. (1987) *Mol. Cell. Biol. 7*, 3252–3259.
119. Zitomer, R. S., Sellers, J. W., McCarter, D. W., Hastings, G. A., Wick, P. & Lowry, C. V. (1987) *Mol. Cell. Biol. 7*, 2212–2220.
120. Verdière, J., Creusot, F. & Guerineau, M. (1985) *Mol. Gen. Genet. 199*, 524–533.
121. Verdière, J., Creusot, F., Guarente, L. & Slonimski, P. P. (1986) *Curr. Genet. 10*, 339–342.
122. Creusot, F., Verdière, J., Gaisne, M. & Slonimski, P. P. (1989) *J. Mol. Biol. 204*, 263–276.
123. Myers, A. M., Crivellone, M. D., Koerner, T. J. & Tzagoloff, A. (1987) *J. Biol. Chem. 262*, 16822–16829.
124. Zitomer, R. S., Sellers, J. W., McCarter, D. W., Hastings, G. A., Wick, P. & Lowry, C. V. (1987) *Mol. Cell. Biol. 7*, 2212–2220.
125. Winkler, H., Adam, G., Mattes, E., Schanz, M., Artig, A. & Ruis, H. (1988) *EMBO J. 7*, 1799–1804.
126. Pfeifer, K., Prezant, T. & Guarente, L. (1987) *Cell 49*, 19–27.
127. Guarente, L. (1987) *Annu. Rev. Genet. 21*, 425–452.
128. Verdière, J., Gaisne, M., Guiard, B., Defranoux, N. & Slonimski, P. P. (1989) *J. Mol. Biol. 204*, 277–282.
129. Pinkham, J. L., Olesen, J. T. & Guarente, L. P. (1987) *Mol. Cell. Biol. 7*, 578–585.
130. Hahn, S. & Guarente, L. (1988) *Science 240*, 317–321.
131. Lalonde, B., Arcangioli, B. & Guarente, L. (1986) *Mol. Cell. Biol. 6*, 4690–4696.
132. Arcangioli, B. & Lescure, B. (1985) *EMBO J. 4*, 2627–2633.
133. Keng, T. & Guarente, L. (1987) *Proc. Natl Acad. Sci. USA 84*, 9113–9117.
134. Marse, A. C., De Haan, M., Bout, A. & Grivell, L. A. (1988) *Nucleic Acids Res. 16*, 5797–5811.
135. Wright, R. M., Trawick, J. D., Trueblood, C. E., Patterson, T. E. & Poyton, R. O. (1987) in *Cytochrome systems, molecular biology and bioenergetics* (Papa, S., Chance, B. & Ernster, L., eds) pp. 49–56, Plenum Press, New York.
136. Arndt, K. Styles, C. & Fink, G. R. (1987) *Science 237*, 874–880.
137. Dujon, B., Slonimski, P. P. & Weill, L. (1974) *Genetics 78*, 415–437.
138. Strausberg, R. L. & Perlman, P. S. (1978) *Mol. Gen. Genet. 163*, 131–144.
139. Sena, E. P., Welch, J. & Fogel, S. (1976) *Science 194*, 433–435.
140. Zinn, A. R., Pohlman, J. K., Perlman, P. S. & Butow, R. A. (1987) *Plasmid 17*, 248–256.
141. Pringle, J. R. & Hartwell, L. H. (1981) in *The molecular biology of the yeast* Saccharomyces. *Life cycle and inheritance* (J. N. Strathern, E. W. Jones, J. R. Broach, eds) pp. 97–142, Cold Spring Harbor Laboratory, New York.
142. Nasmyth, K. & Shore, D. (1988) *Science 237*, 1162–1170.
143. Gasser, S. M. & Laemmli, U. K. (1986) *Cell 46*, 521–530.
144. Gasser, S. M. & Laemmli, U. K. (1987) *Trends Genet. 3*, 16–22.
145. Dorsman, J. C., Van Heeswijk, W. C. & Grivell, L. A. (1988) *Nucleic Acids Res. 16*, 7287–7301.
146. Shore, D., Stillman, D. J., Brand, A. H. & Nasmyth, K. A. (1987) *EMBO J. 6*, 461–467.
147. Buchman, A. R., Kimmerly, W. J., Rine, J. & Kornberg, R. D. (1988) *Mol. Cell. Biol. 8*, 210–225.
148. Bram, R. J. & Kornberg, R. D. (1987) *Mol. Cell. Biol. 7*, 403–409.
149. Amati, B. B. & Gasser, S. M. (1988) *Cell 54*, 967–978.
150. Barath, Z. & Küntzel, H. (1972) *Proc. Natl Acad. Sci. USA 69*, 1371–1374.
151. Barath, Z. & Küntzel, H. (1972) *Nature 240*, 195–197.
152. Westergaard, O. & Lindberg, B. (1972) *Eur. J. Biochem. 28*, 422–431.
153. Kuiper, M. T. R., Akins, R. A., Holtrop, M., de Vries, H. & Lambowitz, A. M. (1988) *J. Biol. Chem. 263*, 2840–2847.
154. Van Loon, A. P. G. M., De Groot, R. J., Van Eyk, E., Van der Horst, G. T. J. & Grivell, L. A. (1982) *Gene 20*, 323–337.
155. Parikh, V. S., Morgan, M. M., Scott, R., Clements, L. S. & Butow, R. A. (1987) *Science 235*, 576–580.
156. Butow, R. A., Docherty, R. & Parikh, V. S. (1988) *Phil. Trans. R. Soc. Lond. B 319*, 127–133.
157. Williamson, D. H. (1968) in *Control of organelle development* (P. L. Miller, ed.) pp. 247–276, Cambridge University Press, London.
158. Schatz, G. (1987) *Eur. J. Biochem. 165*, 1–6.
159. Roise, D. & Schatz, G. (1988) *J. Biol. Chem. 263*, 4509–4511.
160. Eilers, M. & Schatz, G. (1988) *Cell 52*, 481–483.
161. Pfanner, N., Pfaller, R. & Neupert, W. (1988) *Trends Biochem. Sci. 13*, 165–167.
162. Pfanner, N., Hartl, F.-U. & Neupert, W. (1988) *Eur. J. Biochem. 175*, 205–212.
163. Natsoulis, G., Hilger, F. & Fink, G. R. (1986) *Cell 46*, 235–243.
164. Hopper, A. K., Furukawa, A. H., Pham, H. D. & Martin, N. C. (1982) *Cell 28*, 543–550.
165. Ellis, S. R., Hopper, A. K. & Martin, N. C. (1987) *Proc. Natl Acad. Sci. USA 84*, 5172–5176.
166. Najarian, D., Dihanich, M. E., Martin, N. C. & Hopper, A. K. (1987) *Mol. Cell. Biol. 7*, 185–191.
167. Beltzer, J. P., Morris, S. R. & Kohlaw, G. B. (1988) *J. Biol. Chem. 263*, 368–374.
168. Wu, M. & Tzagoloff, A. (1987) *J. Biol. Chem. 262*, 12275–12282.
169. Chatton, B., Walter, P., Ebel, J.-P., Lacroute, F. & Fasiolo, F. (1988) *J. Biol. Chem. 263*, 52–57.
170. Capaldi, R. A., Malatesta, F. & Darley-Usmar, V. M. (1983) *Biochim. Biophys. Acta 726*, 135–148.
171. Hauska, G., Hurt, E., Gabellini, N. & Lockau, W. (1983) *Biochim. Biophys. Acta 726*, 97–113.
172. Oudshoorn, P., Van Steeg, H., Swinkels, B. W., Schoppink, P. & Grivell, L. A. (1987) *Eur. J. Biochem. 163*, 97–103.
173. Maarse, A. C., Haan, M de, Schoppink, P. J., Berden, J. A. & Grivell, L. A. (1988) *Eur. J. Biochem. 172*, 179–184.
174. Schoppink, P. J., Hemrika, W., Reynen, J. M., Grivell, L. A. & Berden, J. A. (1988) *Eur. J. Biochem. 173*, 115–122.
175. Schoppink, P. J., Berden, J. A. & Grivell, L. A. (1989) *Eur. J. Biochem. 181*, 475–483.

176. Schoppink, P. J., de Jong, M. A., Berden, J. A. & Grivell, L. A. (1989) *Eur. J. Biochem. 181*, 681–687.
177. Schoppink, P. J., Hemrika, W. & Berden, J. A. (1989) *Biochim. Biophys. Acta 974*, 192–201.
178. Berden, J. A., Schoppink, P. J. & Grivell, L. A. (1988) in *Molecular basis of biomembrane transport* (F. Palmieri & E. Quagliariello, eds) pp. 195–208, Elsevier, Amsterdam.
179. Tzagoloff, A., Crivellone, M. D., Gampel, A., Muroff, I., Nishikimi, M. & Wu, M. (1988) *Phil. Trans. R. Soc. Lond. B. 319*, 107–120.
180. Crivellone, M. D., Wu, M. & Tzagoloff, A. (1988) *J. Biol. Chem. 263*, 14323–14333.
181. Cumsky, M. G., Ko, C., Trueblood, C. E. & Poyton, R. O. (1985) *Proc. Natl Acad. Sci. USA 82*, 2235–2239.
182. Trueblood, C. E. & Poyton, R. O. (1987) *Mol. Cell. Biol. 7*, 3520–3526.
183. Trueblood, C. E., Wright, R. M. & Poyton, R. O. (1988) *Mol. Cell. Biol. 8*, 4537–4540.
184. Borst, P. (1977) *Trends Biochem. Sci. 2*, 31–34.
185. Von Heijne, G. (1986) *FEBS Lett. 198*, 1–4.
186. Farrelly, F. & Butow, R. A. (1983) *Nature 301*, 296–301.
187. Jacobs, H. T. & Grimes, B. (1986) *J. Mol. Biol. 187*, 509–527.
188. Kemble, R. J., Mans, R. J., Gabay-Laughnan, S. & Laughnan, J. R. (1983) *Nature 304*, 744–747.
189. Fox, T. D. (1987) *Annu. Rev. Genet. 21*, 67–91.
190. Osawa, S. & Jukes, T. H. (1988) *Trends Genet. Sci. 4*, 191–198.
191. Gay, N. J. & Walker, J. E. (1985) *EMBO J. 4*, 3519–3524.
192. Viebrock, A., Perz, A. & Sebald, W. (1982) *EMBO J. 1*, 565–571.
193. Nagley, P., Farrell, L. B., Gearing, D. P., Nero, D., Meltzer, S. & Devenish, R. J. (1988) *Proc. Natl Acad. Sci. USA 85*, 2091–2095.
194. Nagley, P. & Devenish, R. J. (1989) *Trends Biochem. Sci. 14*, 31–35.
195. Banroques, J., Delahodde, A. & Jacq, C. (1986) *Cell 46*, 837–844.

Review

NMR studies of mobility within protein structure

Robert J. P. WILLIAMS

Inorganic Chemistry Laboratory, University of Oxford

(Received February 8, 1989) — EJB 89 0157

NMR studies of dynamics within structure have revealed that a quite new approach to protein structure and its relation to function is necessary. This approach requires the consideration in detail of the following:

1. Local movements of groups and small segments to allow fast recognition and fitting. The motion concerns on/off rates as well as binding. The observations affect surface/surface recognition, e.g. of antigen/antibody as well as of substrate and protein.

2. Somewhat larger interdomain or N- and C-terminal segments which allow rearrangement. Cases in point are the movement of segments in blood-clotting proteins or in histones.

3. Relative motion of helices in hinges. These actions are likely in such enzymes as kinases and P-450 cytochromes.

4. Relative motion of helices within domains (relative to other helices or sheets) in mechanical devices (triggers) e.g. in calmodulin.

5. General motion in random proteins. Examples extend from rubber-like proteins (entropy sensors), some glycoproteins, to proteins carrying peptide hormones to be generated only after hydrolysis.

6. Order → disorder transitions locally as in osteocalcin and metallothionine.

7. Swinging arm motions associated with special sequences such as $(Ala-Pro)_n$.

8. Of great interest is the power of NMR to look at proteins which are relatively large, up to 50 k Da proteins, and to isolate certain zones of interest. This needs careful temperature dependent studies and analysis of separated domains [72] as well as the use of a great variety of pulse sequences [15] and of nuclei other than protons.

9. In this article I have illustrated the different possibilities using work in my own group. This is done to lessen the burden of extensive review. I fully realise that the range of examples is now large. I would stress though that the production of the necessary technology was the endeavour of several of us within the Oxford Enzyme Group from 1970 to 1985, i.e. from 270 – 600 MHz Fourier-transform NMR spectroscopy.

10. While all of these features have been demonstrated by NMR methods there are parallel developments both using X-ray diffraction methods and theoretical approaches. All these procedures are changing the view of protein structure to one which incorporates dynamics all the way from conventional vibronic/rotational coupling to the disordered motions characteristic of random polymers. It is the understanding of dynamics that leads to an appreciation of function.

Correspondence to R. J. P. Williams, Inorganic Chemistry Laboratory, University of Oxford, South Parks Road, Oxford OX1 3QR, England

Abbreviations. FUR protein, iron-uptake regulatory protein; NOE, nuclear Overhauser effect; COSY, correlated spectroscopy; NOESY, nuclear Overhauser effect spectroscopy; Gla, γ-carboxyglutamate.

Note. While this article was being completed, a review was published in the European Journal of Biochemistry [76] which has summarised very effectively some of the problems faced by all users of NMR methods who attempt to describe either structure or dynamics of large flexible molecules. This review, on nucleic acid rather than protein structures, examined, in particular, both relaxation (especially NOE) and coupling constant data against the background of their time averaging. The present author has removed from his manuscript obviously overlapping passages including quantitative descriptions of both coupling constant and NOE averaging. The conclusions in the nucleic acid, the protein and the polysaccharide fields are similar. Great care is necessary in the description of the fluctuating ensembles of any large molecule. It is in the very nature of a biological macromolecule that it is not a static but a fluctuating structure. In fact it is the dynamics against the background of some low energy structure(s) which we must describe.

The physicochemical study of biology must be an analysis of motion within structure since a biological system is a dynamic system. The methods for the analysis of static structure are well described and we begin to have an understanding of the ways in which a variety of macromolecules fold. Our knowledge of the dynamics of these molecules is much less secure. The dynamics of interest are not so much small rotational/vibrational movements seen in crystals and measured by B-factors with time constants of $< 10^{-11}$ s. The most important dynamics are those with time constants on the order of a second down to 10^{-9} s. These are the time constants of enzyme action, of channel opening for electrical events, of fibre contraction in cell division, of muscle action, and so on. It is in this range of time constants that NMR is most powerful. This essay has, as its objective, not only a description of what we know today, but also of what it is that we need to know to explain many biological phenomena in terms of the motions of proteins within the structures defined largely by NMR in solution, even though many of these structures may be based partly on knowledge of structures in crystals.

I shall illustrate this review with recent NMR work of our own on some different types of proteins including cytochromes b [1] and c [2a], an enzyme (acylphosphatase) [3], a kinase (phosphoglycerate kinase) [4], two calcium binding proteins (calmodulin [5] and osteocalcin [6]), a domain of plasminogen (a kringle IV) [7] and some contractile proteins such as the myosin head group of muscle [8]. The object of these studies is to show the role of dynamics in protein functions so that the stress will be on the relationship:

composition and sequence : structure : dynamics : function [9, 10].

Using NMR, I shall show that the variety of dynamics which biology has put to use includes: (a) minor vibronic motions in electron transfer; (b) more extensive side chain motion and small segment motion in simple enzymes and at antigenic sites and here mainly on protein surfaces; (c) relative domain movement in complex proteins and enzymes based on hinge bending or loose connecting links; (d) segment/segment motion inside domains based largely on helix/helix motions. Here cooperativity and allosteric effects are implicated; (e) partial or even very considerable unfolding/folding switches of state. This is effectively a disorder/order reaction but it can be used in slow controls while movements in (d) are rapid.

It must be understood that other series of proteins could be used to illustrate the same points and the ones chosen here are simply those I know best.

INTRODUCTION TO THE STUDY OF DYNAMICS

In order to discuss dynamics we must start from a structure at a given time. Using X-ray crystallography we can give a picture of a protein trapped in a crystal and it will become possible to follow minor local slow movements in this trap. This is being attempted for the enzyme phosphorylase for example [11, 12]. Again, if biological systems supply ordered fibres as in muscle, it is possible to follow rate steps taking X-ray structure pictures at say 1 s^{-1}. Alternatively, one can forego the rate measurement and simply analyse the change of state in terms of the structural change, as Perutz has done for haemoglobin [13]. Here two structures are measured in quite different crystals. There is the danger that such studies will be treated as if they define a two-state switch (first-order phase change) in solution, but the objections to such an extrapolation are obvious and the process of molecular change is not likely to be so absolutely cooperative. The study of one molecule in two different crystal forms is, of necessity, the study of a first-order phase change [14]. Such a change is inherently slow and its nature in solution is likely to be very different.

Another way of finding a starting structure is to use NMR in solution. This has advantages and disadvantages. One advantage is that the structure is in the medium water and is of an isolated molecule. The major disadvantage is that the structure is a weaker interpretation of the experimental data than is the case for an X-ray diffraction analysis of a protein trapped in a lattice. In fact, the assumption that there is a structure in solution rather than an ensemble of structures could be wrong and it is not always easy to prove that it is not wrong. This is not a reflection on the fine work reported, particularly by Wüthrich [15], but it is a necessary warning. The NMR method is not secure except in the analysis of rigid molecules. It is very difficult to provide structures for dynamic molecules when using scattering, as in crystallography, but it is more difficult by far to provide structures using relaxation data based on transference of relaxation for molecules in solution where motions are less constrained. The best data NMR can provide at present are nuclear Overhauser (relaxation-transfer) effects (NOEs) see [15]. The reason for the uncertainty in the structure determination is clear in that the NOE involves two functions:

$$\text{NOE} = (\text{constants})(\text{relaxation times})(1/r^6), \quad (1)$$

where the relaxation term can involve several anisotropic local motions, dynamic relaxations, as well as the molecular tumbling time, and the second becomes a reported distance r between two protons from an ensemble average weighted by r^6. Such an average is full of problems as any simple calculation shows. The difficulties inherent in the nature of an NMR structure are that the structural data are always obained from effects which depend upon the dynamics. The only protein structures in solution, which can be well defined from previous knowledge of X-ray crystal structures together with NMR or directly today from NMR analyses, are the backbone and internal side chains of heavily cross-linked proteins, either through the formation of -S-S-, metal ion or H bonds (β-sheets). The molecules of greatest interest from the point of view of their dynamics are inherently those which are more mobile and are the very ones about which we know least of a possible starting structure. Yet NMR is the best tool which we have for their solution studies. We must work in a fairly rough and ready way, as I shall explain.

THE INFORMATION FROM NMR

Frequency shifts

As described elsewhere, NMR has a variety of ways of providing structural parameters [15]. I shall proceed from perturbations of the energy of transitions to perturbations of relaxation. The energy of an observed line leads to knowledge of its secondary shift from expectation based on the standard energy of the resonance of the same chemical group in an isolated small molecule in solution and where only solvent surrounds the group in question. From these secondary shifts, which are perturbations produced by induced or permanent magnetic dipoles in the macromolecule, e.g. phenyl rings, we obtain vectorial information. The simple case of an axial perturbed shift gives the formula:

$$\text{perturbed shift, } \Delta\delta_A = \text{constant} \cdot \frac{(3\cos^2\theta - 1)}{r^3}, \quad (2)$$

where r is the [ensemble average] distance from the centre of the molecular or radical magnetic dipole to the atomic nucleus in question and θ is the (ensemble average) angle between the alignment of the dipole and the direction from it to the atomic nucleus observed. (The more complicated case of a rhombic shift function is not fundamentally different.) In the case of a molecular dipole this is called a ring current (aromatic group) or a carbonyl shift. In the case of a radical, paramagnetic atom it is called a paramagnetic shift. For a rigid structure the first is temperature independent while the second is temperature dependent. While Eqn (2) gives us a way to describe structure, if we assume the constant above is truly constant, it also allows a description of dynamics in that we can follow the change in the perturbed shift with temperature both for diamagnetic and paramagnetic shifts. The analysis lets us

Table 1. *Proteins classified by mobility at 37°C*
The random coil is detected by the lack of dispersion of chemical shift, i.e. $\Delta\delta_A = 0$ in Eqn (2). Note that almost all enzymes are in the third column

Approximately random	Mobile	Relatively rigid
Tooth phospho-protein	calmodulins	trypsin inhibitor
Resilin	histones	trypsin
chromagranin A	kringles	cytochrome c
Many peptide hormones	coat proteins (Helical)	EGF
Apocytochrome c		toxins
Gla peptides		copper blue proteins
		acyl phosphatase

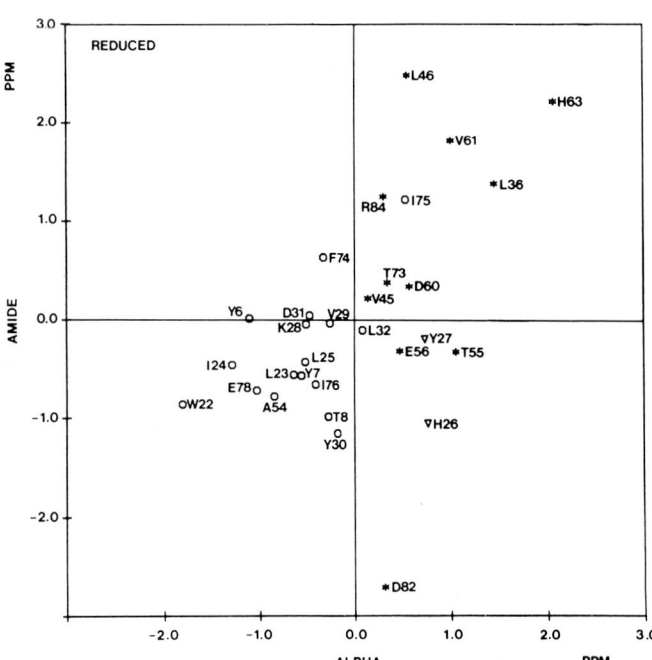

Fig. 1. *The relationship between the secondary chemical shift of α-CH protons and peptide NH protons of cytochrome b_5. The α-CH shift is downfield in a β sheet (right hand lower segment) and upfield in a helix (left hand upper segment). Circles and * represent points for β-sheet and α-helix protons respectively as shown by the crystal structure* [1]

know if there is a single rigid structure independent of temperature or not. If not the changes of secondary shift can be used to follow the rates of motion between two states.

Consider two states A and B with two values of the perturbed secondary shift $\Delta\delta_A$ and $\Delta\delta_B$ for one atom. If the rate of switch of state A→B is slow we see two different sharp NMR signals (slow exchange). If the rate is fast, then we see one sharp weighted average shift. At intermediate rates we see line broadenings of the two lines and shifts towards coalescence. Studies of temperature dependence can then lead to knowledge of the extreme structures and the rate constants and activation energies of exchange. Some examples of this method using NMR are the slow equilibration of states of slowly folding and unfolding proteins such as lysozyme [16]. A different case is the fast equilibrium of the states of calcium-free calmodulin [17]. From these simple data we know that there are two extreme classes of proteins, Table 1 [9, 18]. The proteins in the approximately rigid class are like lysozyme and have one major internal back-bone structure, but many smaller motions of the internal and external side chains as we shall show. The proteins in the mobile class are like calmodulin and have an ensemble of back-bone states in solution. This ensemble refers to the backbone/backbone relationships, i.e. the relationship of segments of the molecule to one another. In the extreme of the mobile class we reach the random coil protein. Note that before proceeding to a structure determination the temperature dependence of the one-dimensional NMR spectrum should always be reported. The character of the protein is then clear.

An example of this approach to a relatively rigid structure is shown in Fig. 1. Here the perturbed shift of the α-CH of many of the amino acids in reduced cytochrome b_5, which are virtually temperature independent, are plotted against the perturbed shift of their NH resonances [1]. The major perturbation is due to the peptide carbonyl group [19]. Those amino acids which are known to be in β-sheets give rise to the resonances in the lower left hand segment and those in the top right-hand segment are largely in α-helices. This structural division of the amino-acids based on the α-CH shift [and αCH/αCH cross-peaks (NOE)] in a folded protein is a very quick way into structure [19] and has been applied by us in the analysis of NMR spectra of a variety of proteins including calmodulins, cytochromes, iron-uptake regulatory (FUR) protein and colicin, all of which are α-helical, and the kringles and acylphosphatase which have larger β-sheet segments (see Fig.6B).

An example of the use of paramagnetic shifts is the analysis of the relatively rigid structure of oxidised cytochrome c in solution [20–23]. In collaboration with Dr D. Turner of Southampton University we have used here both ^{1}H and ^{13}C resonances. The proton resonances are virtually fully assigned today in several cytochromes c, see [2b, 2c]. The secondary paramagnetic shift is calculated from a diamagnetic blank. In Fig. 2 we plot the theoretical paramagnetic shift expected from the crystal structure and calculated from the theory of the rhombic paramagnetic shift (see [20, 21]) against the observed paramagnetic shift. The observations led not only to the general result that the structures in solution and crystal were closely similar but enabled us to see points of difference and to define the dynamics over a wide temperature range [22, 23]. Examples of ring current shifts will be analysed later in the article [24].

Chemical shift analysis can be pursued with great advantage in two-dimensional NMR spectra since this greatly increases resolution. An example from our own work is the use of two-dimensional NOE difference spectroscopy to find those resonances affected by site-specific mutagenesis on the basis of their changed secondary shifts [25].

It was necessary to give a quick introduction to this structural method in order to make statements such as: since the shifts of resonances of reduced cytochromes b and c are overwhelmingly temperature independent, their backbone structures are close to being fixed in terms of time averages. For some proteins this is far from being the case, e.g. calcium-free calmodulin.

At this stage it is necessary to give supporting evidence from other NMR experiments and experiments of quite a different kind for the conclusion that backbones of proteins have a wide range of mobilities.

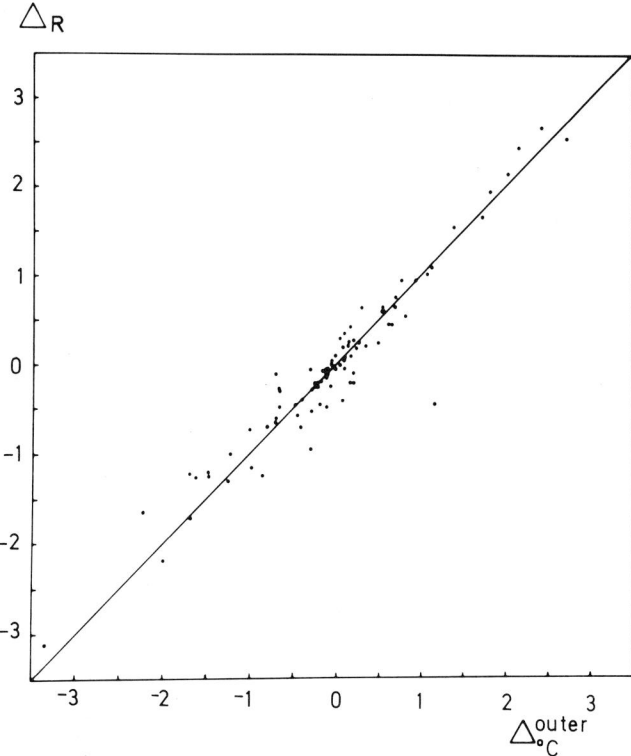

Fig. 2. *The relationship between the calculated Δ_{pc}^{outer} and the experimental, Δ_R, paramagnetic shift for proton resonances of cytochrome c [20]*

Table 2. *NH/ND exchange rates of peptide links*
The structure of colicin A is now known (see [77]). It is a remarkable example of a helical-switch protein. Colicin E has a similar structure as defined by NMR. FUR is the iron [Fe] uptake [U], regulatory [R] protein

Protein	Fast exchange	Slow exchange
Acyl phosphatase	only N-terminus	helices and β-sheet
Kringle IV	most of protein	small β-sheet
Snake toxins	small part of protein	β-sheet and cross-linked region
Calmodulin (-Ca^{2+})	most of protein	small β-sheet
FUR protein	most of protein	central helices
Colicin E.1	majority	central two helices

NH/ND exchange

The possible use of NH/ND exchange as a guide to mobility began with the experiments of Linderstrøm-Lang [26]. Unfortunately (for him) his work really produced telling results at the time when X-ray crystallographers dominated thinking about protein structure in crystals and perhaps unwittingly gave the biochemist the impression that they were like an NaCl crystal structure. What Linderstrøm-Lang showed was that proteins underwent differential fluctuations depending upon where you looked in the protein. He could not pinpoint NH groups to individual peptide bonds, but he could show that regions of proteins as well as different proteins, differed in mobility amongst themselves quite considerably. Today NMR assigns NH resonances specifically to individual residues. The examination of the first protein where this could be done, lysozyme, immediately permitted the analysis of the NH/ND exchange of specific tryptophans [27]. Subsequently it has proved possible to analyse NH/ND exchange of almost every amino acid, for example in the trypsin inhibitor, lysozyme, and so on [15, 16]. Without regard to detailed discussion of the mechanism of exchange, it has been found that proteins such as calcium-free calmodulin exchange almost all protons quite rapidly while proteins such as lysozyme exchange them but slowly, especially from the core, i.e. the β-sheet and α-helical elements (see Fig. 6B). In both cases local regions are differentiated. Table 2, compare Table 1, indicates that this is true for many proteins studied by us. The classification of proteins by their melting behaviour as shown by the temperature dependence of ring current shifts and by their NH/ND exchange rates leads to the same groupings, see Table 1. An additional frequent, if not invariable, finding is that the turns in a relatively rigid structure are more mobile than the extended secondary structure regions [3].

These regions are often those open to protease digestion. Again, domains exchange at different temperatures, as has been observed in phosphoglycerate kinase [4]. Once again two-dimensional NMR spectra vastly improves one-dimensional methods due to the increased resolution. It is also important to note too how much is gained in NH/ND exchange studies by the analysis of the pH and temperature dependence. Before passing on, attention must be drawn to the recent use of NH/ND exchange in the analysis of the folding of proteins. In some cases definite intermediates have been observed by using the temperature dependence of dynamics to trap certain states [28–30].

The data which NMR has provided about protein fluctuations is backed by quite other observations. Perhaps the most useful in the discussion of these statistical ensembles, which proteins in solution are, is calorimetry. In the hands of Privalov [31] in particular it has been shown that proteins like calcium free calmodulin are quite different from proteins like lysozyme. The first group of proteins take up heat over a wide and rather low temperature range, i.e. they have a gradual unfolding via a changing ensemble of states, like a glass, while lysozyme effectively melts like a crystal and takes up a latent heat associated with a phase change at a higher temperature. There may be a small number of partly unfolded states as well. Notice that this procedure examines the equilibrium ensemble and not the pathway of folding.

The confidence provided by the agreement between these observations and those of direct NMR studies led quickly to the conclusion that some proteins at room temperature were extreme random polymers which gave a virtual random-coil NMR spectrum, while others were more like crystalline protein molecules, at least internally, Table 1 [9]. There were other proteins and domains which had a construction between these extremes. We note that this is also true for certain types of small peptides, of polysaccharides [32] and of polynucleotides [33]. Two lines of development were then necessary. The first, which I shall take second, was better NMR procedures. The second, which I take first, was the possible value of knowledge of this variety of motions. My first concern is with the larger segmental motions, not with small internal or surface side chain motions to which I return later.

As an aside we note that a variety of other analyses of protein structure leads to similar conclusions. Practical methods include neutron diffraction, X-ray diffraction, Mössbauer and fluorescence spectroscopies, while the theoretical analysis of Karplus and Frauenfelder, for example, provide a protein model which likens a protein to a glass (see [6]). Theoretical descriptions go on to distinguish temperature ranges in which the protein behaves with simple harmonic elastic motions (low temperature) and grosser distortions resembling plastic materials at higher temperatures. The overall conclusion is that at room temperature even the most rigid protein molecule is not a crystallite but has many internal degrees of motional freedom. It is the purpose of this article to describe these motions in specific proteins studied by us.

The functional value of segmental motion

A random polymer, Table 3, is thought by most biochemists today to be useless. In fact it is extremely useful as an entropy sensor, an energy store as in a rubber, or as a flow aid. A perfect gas is an excellent substance for use in a device for measuring pressure or temperature by following its volume. $PV = RT$ is an equation of state not of structure. $\pi = RTc$ is a similar equation of state for a solute. In a similar way the volume in which a random coil polymer molecule is contained is related to the pressure and temperature. Hence a random coil polymer molecule can become a T, P (or volume, i.e. space) sensor. A charged random coil can be used to detect electric fields. When compressed it behaves like an ideal gas by reducing its configurational not its translational entropy [32, 34]. The ways in which this bulk sensor message can be relayed to a cell using a movable rigid helical domain in a membrane have been described [35]. Flow properties of molecules from adhesives to lubricants are based on such dynamics, but here the polymers which are used are frequently polysaccharides not proteins. This type of sensor has a graded response to bulk forces, is isotropic and is not suitable when a switch which operates over a very limited range of conditions is required. Some examples of loosely folded proteins are given in Table 3. Of course, the discussion applies to segments of molecules as well as to whole molecules.

By way of contrast a fixed, rigid, structure is a perfect device for precise chemical recognition by another device: die-in-mould fitting. Unfortunately, perfect fitting is likely to be accompanied by very slow docking. Specificity must be reduced in order to gain speed. The best way to do this is to make either the die or mould somewhat mobile, e.g. a key (rigid) fitting a lock (a little mobile). It will be more proficient if a fast response is required to relax the rigidity further making both die and mould mobile when we have hand-in-glove matching [9]. Not only are the hand and the glove mobile, but the hand in the glove together is a mobile entity. Selectivity can only be maintained in this device by using a large number of fingers. If the device operates in a cooperative way it is not only chemically selective but it can be made to operate with closer resemblance to a phase change of the Perutz kind [13].

Now, while in this discussion we have been examining mobility from a random coil towards a rigid structure, but usually staying a long way from either, we have not considered the possibility of introducing mobility in one dimension and not in others: anisotropic motions. This cannot be done with a random coil polymer. If we consider a set of loosely linked parallel rods then forces applied to one rod are transmitted to the others so that the rods slip relative to one another in particular directions. This is the essence of a mechanical machine. I shall maintain that biology is full of such molecular machines. We have to consider how to detect such motions, i.e. through the anisotropy of B-factors in crystals, by studies of oriented molecules or by defining motion closely by NMR. In general, the relaxation equations used in NMR are isotropic in form [see Eqn (1) which must be incorrect]. We must ask if motion is anisotropic in such proteins as those in Table 4. Helical proteins are composed of linked rods.

Table 3. *Loosely folded or random proteins*
These proteins may show local secondary structure but they do not fold in tertiary manner in the conventionally accepted sense, for example they resemble poly(glutamic acid) or polylysine. Note that there are no enzymes in this table

Protein	Function
Chromagranin A	unknown, peptide hormone carrier?
Osteocalcin	crystallisation inhibitor
Metallothionine	Zn buffering
Glucagon	hormone
Some RNA-binding proteins	structural support of ribosomes
Illexines	DNA binding protein
Phosvitin	heavily phosphorylated protein
Zinc fingers	zinc binding to connect to DNA

Table 4. *The functions of helical proteins*
Note the absence or low levels of tryptophan in these proteins. Tryptophan acts as a structural platform giving rigidity. Note that there is only one enzyme in this table, see Table 1

Class of protein	Protein	Function
Transducer	calmodulin (troponin C)	calcium signal relay to kinases
	FUR protein (compare DNA-binding proteins)	regulation of production of iron-chelating agents
	colicin	channel protein (open/shut) states
	c-peptide (ATP-synthetase)	proton binding relayed to ATP synthesis
Protective release	ferritin proteins	coat protein of ferritin
	histones	coat protein of DNA
	helical viral assembly proteins	coat proteins of viruses
Cooperative carriers and required-order enzymes	haemoglobin	
	haemerythrin	cooperative O_2 carriers
	haemocyanin	
	cytochrome *P*-450	required-order oxidase
Mechanical	myosin head	triggering of muscle

Segmental rearrangement need not be very fast. This is only demanded for such reactions as those of trigger proteins, e.g. calmodulins and channel pumps. Here the rates of reaction must be around 10^3 s^{-1}. For slower changes, e.g. those of regulation, the protein may change conformation at a slower rate. Here any switch between two conformers is possible provided that the protein is not over stable. Thus switches dependent on slow hormonal effects, e.g. those due to sterols, or dependent on changes in the environment, e.g. uptake regulation of metals such as zinc or iron, can be slow relative to switches dependent on hormones such as adrenaline linked to sodium, potassium, proton or calcium movements. We do not know very much about the structures of the latter group of regulatory proteins but it is thought that helical stretches bind into DNA grooves. There may well be a slow adjustment of a recognition centre based on a β-sheet which will drive changes in these helices allowing in the protein two sites of recognition, much as in calmodulins, but much slower to change.

We see that mobility has enormous potential whether it be that in a random coil [entropy sensor], that of the surface of a relatively rigid protein [fast recognition of good selectivity and useful in catalysis] or that of an adjustable set of anisotropic helical rods [a molecular mechanical machine]. My own NMR studies have been designed to reveal these possibilities, especially the last. I now turn to further methods of NMR analysis.

Correlation times

NMR studies of proteins in solution are made possible by the tumbling of the protein. It is usual and convenient (but clearly incorrect) to assume that the tumbling is isotropic when the tumbling is given one simple correlation time in Eqn (1) for example. Providing that this time is fast relative to the frequency of the measuring radiation the line width will be small and resolution good. There is a limitation on NMR methods to small rigid polymers, probably approximately < 20 kDa, if a full structure/mobility study is to be attempted, since larger proteins tumble slowly. Now it is observed that all the lines in a proton NMR spectrum of a protein are not of the same width and we are forced to conclude that there is differential motion [36—39]. Frequently it is the N and C termini of proteins which have sharper lines, but sometimes there are stretches in the middle of a sequence, and now and then large domains, which give rise to sharper resonances. The number of resonances in a given class is temperature dependent.

Some of the evidence appears to relate to random motion but in other cases the narrow line spectrum is found to be due to a particular short sequence which cannot move randomly but does move independently. Quite commonly now the sequence (XaaPro)$_n$, where Xaa is a second amino acid and the small part of the protein connected with this sequence has a mobility of its own. These findings have become the basis of ideas concerning the directed motion of swinging arms, group transfer, in enzymes [38, 39]. Here one short part of a protein or enzyme can flap about in constrained directions carrying with it either a chemical group or information about the state of the system. A second example is the freely moving arm in the muscle protein myosin, Table 5. The motion is to be compared with the up and down motion of a helical rod in a membrane. In a forthcoming publication Levine will show that these findings are general to certain groups of proteins [39].

Table 5. *Some XaaPro$_n$ sequences: possible swinging arms*
For details see [78]

Protein	Function
Ton B	transport in bacteria
Myosin light chains	mechanical energy transmission
β B.1 crystallin	not certain
Osteocalcin	bone growth control
Acyl-transfer protein	movement of acyl group

Extreme line narrowing is seen in random coil or fully extended proteins, Table 3. One such case was observed by NMR in the intact adrenal gland and we are puzzled about the function of the protein, chromagranin A, in the gland [40]. It may be a loose gel holding the vesicle intact but it may also be a large protein designed to be very easily hydrolysed so as to release peptide hormones. Other proteins examined by NMR are of this kind, so that they are open to rapid protease digestion to liberate signal peptides when they are rejected to the external solutions from the vesicles internal to a cell [18]. There is little to be gained from structural studies of such molecules.

Much information can be gathered then from the study of the line widths or relaxation times of various resonances. An example where the N and C termini of proteins have sharp resonances is in the domains of some kringles [7]. This immediately leads one to ask if it is true that inter-domain regions are mobile to allow protease digestion. Elsewhere, differential mobility has been observed in the inter-domain regions of kinases associated with hinge regions, e.g. of phosphoglycerate kinase, and these are now under detailed study, see Fig. 3 [4]. Finally, some regions, especially turns of compact proteins, show higher mobility, as seen from all or one of the following: (a) temperature dependence of chemical shift; (b) N$\underline{H}$/N$\underline{D}$ exchange; (c) sharper lines, longer T_2. From these data the association of antigenicity and mobile segments was postulated [9, 41]. However, the general point of mobility in loops, rather than in the α-helices or β-sheets, gives rise to the suggestion that in these regions may lie the points of communication between proteins via mechanical energy transfer, as well as the possibility of motion to allow reactants to move along reaction coordinates.

The mobility of local regions is reflected in two-dimensional NMR spectra in differential intensity patterns [42]. Thus, the sharpness of a NOE peak or the NOE intensity from a mobile part of a large protein is abnormally low, i.e. against expectation based on a single structure. The effect of local mobility in a protein on the two-dimensional correlation spectroscopy (COSY) experiment is to produce a set of broadened peaks due to fluctuating coupling constants, see below. For example both these effects have been observed in loop regions of acyl-phosphatase which were also shown to be more mobile than the α-helical and β-sheet regions from the faster rates of N$\underline{H}$/N$\underline{D}$ exchange [3]. These loops are the known antigenic regions. Even more mobile regions of this protein, an attached glutathione group and residues near the terminii, give sharp one-dimensional resonances, but poorly informative weak two-dimensional data. In two-dimensional work a lot of information about mobility can be gained from careful work over a range of temperature, by using a variety of time intervals in the pulse sequences and by using a variety of different pulse sequences. These advantages have not been fully explored as yet. (It would often be useful to have cross-

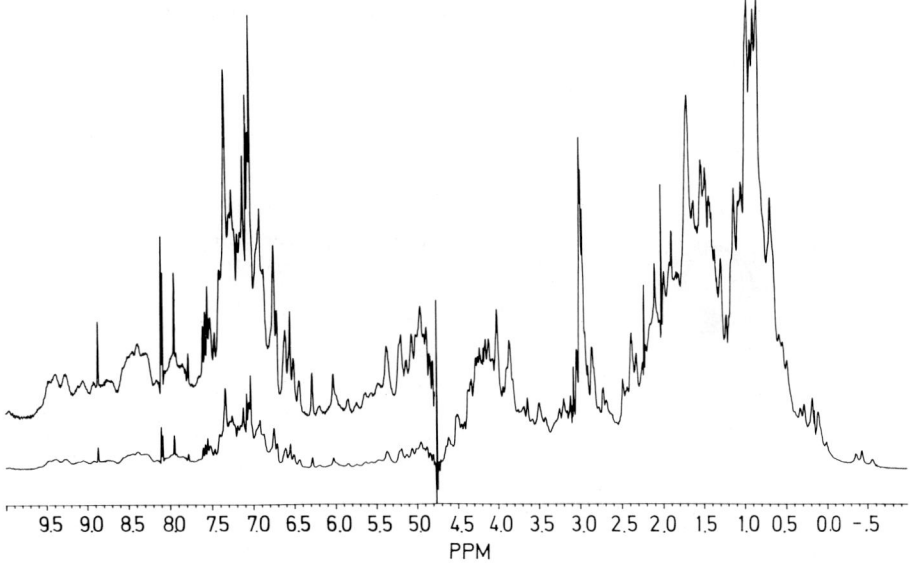

Fig. 3. *The one-dimensional spectrum of phosphoglycerate kinase showing the sharper lines of surface residues and of residues in the hinge region (see Fig. 9).* The scale for the aromatic region is increased to show the obvious differences in line widths for different resonances [4]

sections of the two-dimensional maps to show differential types of NOE, see Fig. 10B.)

A word of caution is repeated here. It is not easy to interpret relaxation data with full confidence. We have explained the reasons in a lengthy paper analysing the motion of side chains in a snake toxin [43, 44]. A general impression is easily obtained but an exact analysis, including rate constants, is extremely difficult.

Coupling constants and signal averaging

Side chain motion. The NMR spectral lines of individual protons are split by the fields of neighbouring protons due to the coupling, through-bond, between their magnetic spins. These are beautifully magnified in two-dimensional phase sensitive spectra. I shall not discuss the theory of these splittings nor their magnitudes which can often be fitted to the Karplus equation describing the splitting in terms of the dihedral angle between two C-H bonds on a frame CH-CH. The splittings of the coupling patterns give geometric information and therefore help structure determination as well as assignment. Clearly the description is only true for a rigid system. Motion will average the sizes of splittings and these averages may well be temperature dependent. If we consider the couplings within individual amino acids we can see if the groups rotate with respect to the bond between them in the side chains [9, 15, 44]. The data can be analysed in terms of rate constants and their activation energies, see footnote at the end of this article.

Now there is another way of observing side chain motion where a side chain is of low symmetry when represented as a rigid molecule in a field. Consider the phenyl ring of phenylalanine. In a fixed position in the field of a molecular magnet, e.g. another benzene ring, the left and right hand sides are not equivalent magnetically. There should be five single-proton signals. In fact, and very frequently, only three signals are seen: one single proton triplet (*para*) and one two-proton doublet (*ortho*) and one or two proton triplets (*meta*) showing that the benzene ring flips at a rate faster than the NMR time scale of 10^4 s^{-1}. When this was first seen, by three groups independently (see [15] for details), it was realised

Table 6. *Some slow-flipping rings in proteins*
Note that all these proteins have -S-S- or thioether crosslinks

Protein	Observation
Antibody	tyrosine immobilised on binding hapten
Osteocalcin	tyrosine immobilised on binding Ca^{2+}
Cytochrome *c*	five rings immobilised on haem incorporation
Trypsin inhibitor	one immobile tyrosine
Protein BDS-I	one immobile tyrosine
Kringle IV (plasminogen)	Tyr-9 immobile on binding substrate

Table 7. *Fast flipping rings in proteins*

Protein	Ring
Kringle IV	Tyr 40, Tyr 73, Phe 63
Lysozyme	Tyr 53, Phe 3, Phe 34
Cytochrome *c*	Phe 82, Phe 36
Acyl phosphatase	Tyr 11, Tyr 25, Tyr 91, Tyr 98, Phe 12, Phe 14, Phe 22, Phe 80
Cytochrome *b*	Tyr 6, Tyr 27, Tyr 30, Phe 35, Phe 58, Phe 74

that even so-called rigid proteins such as lysozyme must be breathing with considerable amplitude of their interiors. Interestingly some proteins breathe to a much smaller degree in some regions so that aromatic rings are found to be 'rigid'. Some examples of observations on 'rigid' and flipping aromatic rings, respectively, are given in Tables 6 and 7. As yet, there are no enzymes in Table 6. Do enzymes require more mobility than inhibitors or electron transfer proteins?

A very interesting case of two particular constrained aromatic rings is found in cytochrome *c* [2]. The N-terminal helix has a conserved phenylalanine, Phe-10, and the C-terminal helix has a semi-conserved aromatic at Tyr-97. They interact (Fig. 4), and both flip slowly. The N-helix is cross-linked

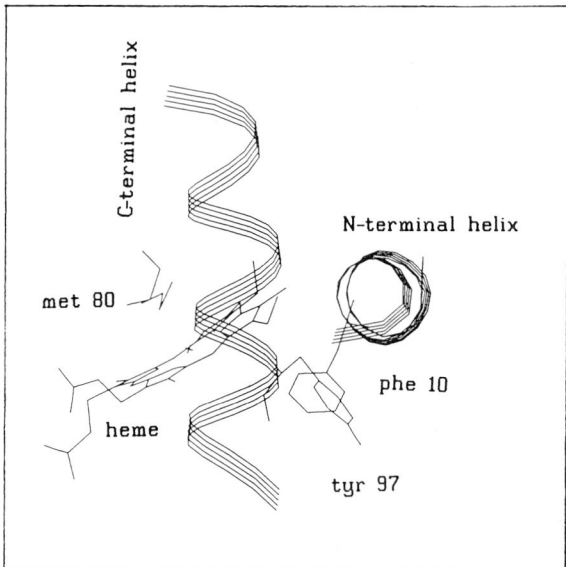

Fig. 4. *The relationship between the N- and C-terminal helices in cytochrome* c *with the two immobile aromatic residues, Phe-10 and Tyr-97 [2]*

to the haem via Cys-14, Cys-17 and His-18, while the C-helix is held by bulky aliphatic residues, known to be relatively immobile, against the haem. The helices are jammed in place. This observation is in marked contrast with the freely flipping aromatic rings which lie between helices of calmodulin [5]. As stressed above, cytochrome *c* is a segmentally rigid protein, but calmodulin is mobile. These differences are required for optimal function since cytochrome *c* is an electron transfer protein, very constrained mobility required (Fig. 5), while calmodulin is a trigger, considerable fast movement required. Another interesting example is the 'frozen' aromatic structure in calcium bound osteocalcin but not in the calcium free protein [6]. Inhibitor proteins are often rigid.

Now the analysis also applies to the asymmetric-top side chains of leucine and valine. Here the two methyl groups can be of very different secondary chemical shift or of very similar chemical shift. Only on the external faces of proteins have we observed rotating valine or leucine side chains. There is a difficulty in the interpretation of the data for asymmetric tops however.

The meaning of internal motions. In most proteins NMR shows that internal fast flipping rotation is found for all phenylalanine and tyrosine residues. In these same proteins NMR shows that no similar flipping is found apparently for leucine or valine tops. Rotation of a valine requires the development of a cavity replacing C-H by C-CH$_3$ of some 0.1-nm radius and rotating a benzene ring requires a similar cavity. To allow aromatic rings to flip between two states of equal energy therefore requires segment motion of the order of 0.2-nm at least 10^5 times/s and valines and leucines should flip equally rapidly. It may well be then that the valine and leucine flipping is just as fast but is not observed due to the potential energy assymetry associated with their tops, i.e. while a benzene ring rests equally in two orientations only one orientation is stable for an assymteric top. Thus, we only observe valine and leucine motions when these groups are on the surface of proteins. Notice that the dynamic generation of the above size of cavity is required to explain easy exchange of protons (N$\underline{H}$/N$\underline{D}$ exchange) via H$_2$O attack and fluorescence quenching by dioxygen. Structure determinations and the thermodynamic properties of structures need not be affected by such motions but the functional dynamic activity of a protein may be directly dependent on them.

From these observations and simulations of motion, Table 8, it has been found that while there are large internal fluctuations of some internal parts of some proteins these occur infrequently. The structure of the inside of most proteins is not greatly disordered, although it allows fast ring flipping. Hence, if we wish to discuss structure and the stability of the internal structure then we can use the X-ray or NMR coordinates and expect to get a meaningful understanding of the fold, especially the major units of secondary structure, helices and sheets.

Unfortunately we can make no such assumption about our understanding of the relationship between structure and function. The fact that all parts of very many proteins have fast internal fluctuations of up to say 0.1 nm and that fast external fluctuations of greater than 0.2 nm are common means that activity must be analysed in terms of dynamic not static structures. Activity is almost invariably associated with the infrequent passage of reactants over an energy barrier [but this is not true of simple electron transfer] and is a rare event. Partners, whether they are substrate and enzyme, or two proteins, or DNA/protein pairs, may well remain in particular energy wells for long periods (say 10^{-4} s) before a large fluctuation internally of greater than 0.1 nm occurs, when in 10^{-10} s, the system passes to a new resting energy well. It is not just the atoms of the small reactant molecules which move so as to form new compounds, but the atoms, even part of the interior of the large catalyst molecules move though in a cyclic pattern. Not only is this known from NMR data but all theoretical calculations of dynamics show it to be so. These motions are easily fast enough to match biological reactions. What is more important to note is that much of protein activity is related to a surface and not the interior, where motion is larger and faster.

Additionally the internal motion of some proteins is faster and much more extensive than others see Table 2. We shall see that the side chain motion of surface groups of proteins is much more rapid as well as extensive but here we run into a very real problem of interpretation of NMR data. Before we can provide a real dynamic map of a protein (Fig. 5) we must look at another major structural-dynamic approach to NMR data.

Relaxation perturbations

We have described structural and dynamic information from ring current, carbonyl and paramagnetic shifts, coupling constants, and we have noted the value of relaxation time data and N$\underline{H}$/N$\underline{D}$ exchange in differentiating different parts of proteins. There is one other source of structural information: the perturbation of relaxation which can be controlled in two ways: (a) paramagnetic relaxation probes [45], (b) nuclear Overhauser effects based on radiation-stimulated relaxation [15].

The first attempts at detailed structure of a protein relied on shift probes and paramagnetic relaxation probes using Mn^{2+}, Gd^{3+}, [Cr(CN)$_6$]$^{3-}$ and so on. The methods depended on the perturbation of relaxation of a proton at a distance *r* from the probe due to the slow electron relaxation time of the chosen metal ion [45]. The equation is of the form:

relaxation perturbation = $\Im$ (correlation times) $1/r^6$, (3)

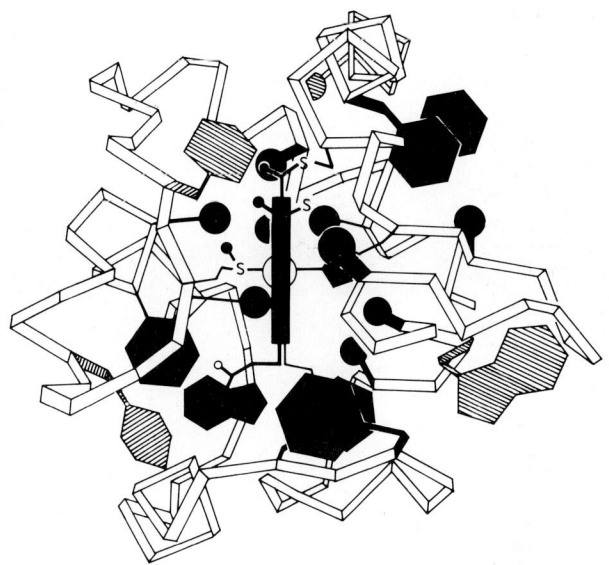

Fig. 5. *A mobility map of residues in cytochrome* c. It will be necessary to have such a map to understand functions of proteins since function is a property of mobility within structure. Phe-82 is the top left-hand side hatched residue. Hatched residues are more mobile than blocked-in residues [20]

where $\mathfrak{J}$ can be a complex function. The perturbation is measured as a linewidth change in the spectra using NMR difference spectroscopy. The combination of this type of probe with that from the use of a paramagnetic shift probe (fast electron relaxation time) plus coupling constant data was used to generate families of possible structures of complex small molecules in the range of molecular mass up to 1 kDa using computerized methods of search [46]. This is the general procedure (Burlesque) similar to that now used to obtain NMR structures of proteins but the probe relaxation is replaced by the NOE effect [15]. While the relaxation probe procedure gave excellent structural data for a limited region of lysozyme, it has not yet proved to be generally useful except in the discussion of metal ion and other binding sites on protein surfaces (see below). I believe that this method now merits detailed re-examination in the two-dimensional mode.

The second procedure has been developed in the two-dimensional NMR regime, mainly as a result of developmental work by Ernst and Freeman and their coworkers [15]. It is now the conventional approach to full structure determination of proteins of up to 100 amino-acids. Much of the credit for this development of two-dimensional method in proteins must go to Wüthrich [15]. Using this technique we for example determined an enzyme structure in solution for the first time (Fig. 6) [3]. The method is based on Eqn (3), but the perturbation is brought about by irradiating proton A, i.e. increasing its relaxation time, while observing the intensity of absorption of proton B, distance r away. (Remember r has to be treated as an ensemble average.) The spectral intensity changes due to changes in relaxation. The method is described by Wüthrich [15] and gives an excellent picture of the internal structure of relatively rigid proteins (Fig. 6A). The picture consists of an overlap of several (say ten) separately determined structures by computerized analysis. To some degree the spread of these structures, assuming a properly executed analysis, will reflect the mobility/disorder of the protein. The caveats about internal fluctuations are of little consequence while we discuss the internal structure, i.e. mainly the backbone secondary fold and the internal side chains of a cross-linked or β-sheet protein. The overall positional accuracy of the structure determination is often limited to about ±0.05 nm say and is usually worse than that of a good X-ray diffraction structure. All the same, the method is immensely valuable. This leaves the definition of the surface residues to be described. Here NOE and paramagnetic relaxation methods run into grave problems since there are too few observations for accurate structure determination and independent mobility of groups and small segments are involved. An NOE between two centres dies out for $r > 0.5$ nm. Since there may well be few neighbours in this distance range we find that the structural definition of much of the convex surface of proteins is too poor to be really useful in discussing function (see Note). Often where the surface is concave this is not so much the case and for example we can define by NMR the binding groove of the kringle IV of the blood-clotting domain from plasminogen quite well and fit the substrate, a lysine analogue, to the site (Fig. 7) (unpublished results). For much of the rest of the surface side chains on convex surfaces of proteins generally, and for the secondary structure of many loops and turns on the surface (usually), the NMR NOE method is not very powerful (see Fig. 6B). Undoubtedly the paramagnetic probe procedures are much more useful since the perturbation produced by them extends to 1.5 nm. As stated before, a good definition of the surface of lysozyme was achieved using them. The surface binding sites of ions have been defined too but there are now other problems [48]. Notice that use of the relaxation ratio method with paramagnetic probes removes the time dependence [45].

In solution the surface of a protein is strongly exposed to solvent. We may well be using an incorrect assumption when we search for the structure of it. It is more than likely that the surface is dynamic and only usefully constrained by the internal structure (Fig. 6B). For example we observe consistently that NMR signals from lysines especially ε-CH$_2$ groups are sharp, indicating very fast relaxation times and therefore freedom of motion. The interaction of a side chain with the solvent is as strong as that with neighbouring amino acids and the solvent fluctuates through many states at rates around 10^{-10} s^{-1}. Thus, the time of exchange is very short and often groups will find several energy minima which will be roughly equally occupied. There may be no meaningful singly occupied structure of a surface. The location of water molecules in crystals by X-ray crystallographic methods must be treated with caution unless close-to-full occupancy of the site is seen. Here the problem is of the structural and thermodynamic definition of the ground state. Given the problems faced by all methods and the fact that it is the surface of proteins which relates to binding and catalytic activity what can we say safely? We shall consider much studied examples. Always remember that the mobility of the surface has as great a functional value as the 'structure' of the interior. From here on in the article we take it that only the interior is defined structurally with certain dynamic properties, limited unless otherwise stated. We start from an analysis of the surface of a well-known protein.

THE SURFACE OF CYTOCHROME c

General

We have used the paramagnetic probe and the NOE method in attempts to uncover the nature of the surface of cytochrome c against the background of the very close corre-

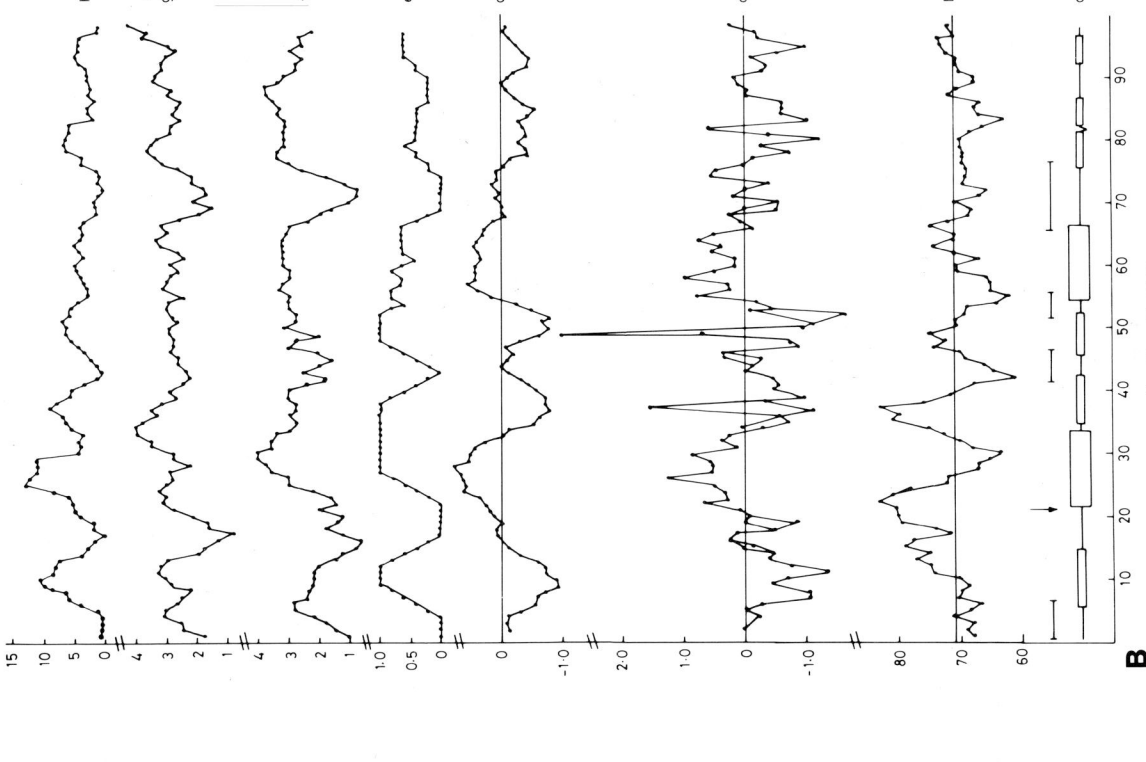

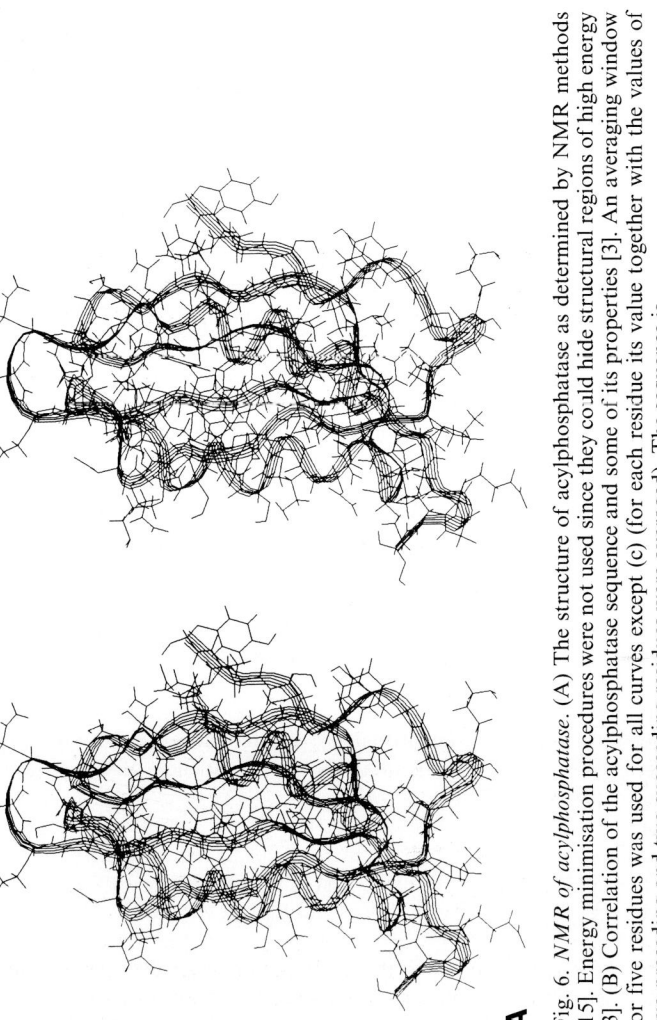

Fig. 6. *NMR of acylphosphatase.* (A) The structure of acylphosphatase as determined by NMR methods [15]. Energy minimisation procedures were not used since they could hide structural regions of high energy [3]. (B) Correlation of the acylphosphatase sequence and some of its properties [3]. An averaging window for five residues was used for all curves except (c) (for each residue its value together with the values of two preceding and two succeeding residues were averaged). The sequence is

AcSTARPLKSVDYEVFGRVQGVCFRMYAEDEARKIGVGWKNTSKGTVTGQ-
VQGPEEKVNSMKSWLSKVGSPSSRIDRTNFSNEKTISKLEYSNESVRYOH.

(a) Secondary structure: ☐, β-structure; ☐, α-helix; ——, loop; ∨, β-bulge. The bars indicate the continuous antigenic regions and the arrow the place of the glutathione attachment. (b) Hydrophobicity profile: hydrophobicity is given by larger numbers. (c) Secondary shift of the main chain CHα resonance (defined as the difference between the measured value and that of a residue in a peptide in a random coil conformation). (d) As (c) but using an averaging window of five residues and the values of the second CHα of glycines not included, compare (a). (e) Hydrogen exchange profile of the main chain amides. The residues with slowly exchanging amides were assigned a value of 1, the remaining residues 0. (f–h) NOE intensity profiles: the intensity of the cross-peaks in the two-dimensional NOE spectra were estimated as strong, medium or weak and assigned a value of 3, 2 or 1, respectively. (f) Main chain sequential NOE profile. The sum of the intensities of all sequential NOEs between CHα$_i$-NH$_{i+1}$, NH$_i$-NH$_{i+1}$ (or CH$_{2ai}$-CHα$_{i+1}$ for prolines) for each residue was used. (g) Intraresidue NOE profile. The sum of the intensities of the intraresidue NOEs (normalised over the number of magnetically non-equivalent hydrogens) was used for each residue. (h) Long-range NOE profile. The sum of the intensities of all non-sequential NOEs, normalised as for (g), was used for each residue

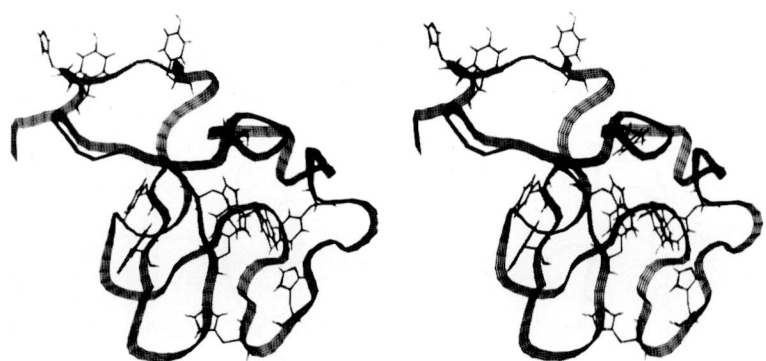

Fig. 7. *The structure of kringle IV of plasminogen by NMR methods.* The region between residue 5 and 10 cannot be defined due to the low intensity of resonances, see text [7, 47]. Aromatic side chains are shown

spondence between the internal structures generated by NMR and X-ray diffraction data (Fig. 5). The paramagnetic procedure is to treat $[Cr(CN)_6]^{3-}$, a relaxation probe, as a sphere which can bump into and roll over the surface of this positively charged protein even though it binds quite strongly [49]. Since a large number of protein proton resonances suffer relaxation perturbation, line broadening, we can inspect the line-intensity pattern changes with concentration trying to uncover binding sites and their strengths. In brief, the outcome of this study was to reveal that there were sites of considerable binding strength but that they were linked forming broad energy minima and we envisaged a rolling contact zone rather than a site. We have gone on to use this procedure in the analysis of protein/protein binding by competing cytochrome b_5 say with $[Cr(CN)_6]^{3-}$ for the surface of cytochrome c. This work is largely unpublished and is due to my collaborators, G. R. Moore and N. Veitch. An example of the results of this study relates to a particular 'surface' residue.

Phe-82 of cytochrome c

In a paper discussing the potential role of Phe-82 of cytochrome c in electron transfer reactions Salemme and co-workers have shown by theoretical analysis that Phe-82 can apparently move from its position close to the heme of the cytochrome some 1 nm into a position in the interfacial region of the cytochrome-c/cytochrome-b_5 complex [50]. This led to a discussion of the functional significance of the residue. Phe-82 is the hatched benzene ring at top left in Fig. 5. Now we have assigned the majority of proton resonances of both these proteins [1, 2] and we therefore know the whereabouts, using NOE and paramagnetic shifts, of Phe-82 not only in the structure of cytochrome c but also in the cytochrome-c/cytochrome-b_5 complex and we find no evidence for any change in its position as defined by such data. In the complex we do not believe that the Phe-82 moves from its position for more than say $1-5\%$ of its time, if that. Again it has been stated that Phe-82 is an amino acid important for electron transfer because of its aromaticity [51]. We (unpublished results) have been able to measure the rate of electron transfer in the complex using electron exchange NMR methods (line widths). We find a rate of about 10^3 s^{-1} after correction for the driving force. This is approximately the rate of transfer expected for a distance of some 1.5 nm. Moreover, this rate is little affected by mutation of Phe-82 to Gly. Finally we have no good reason from model experiments to expect any involvement of aromatic side chains of proteins in electron-transfer reactions which operate between redox couples at potentials close to 0.1 ± 0.2 V. It is necessary to re-examine the theoretical paper.

Our experiments also show that cytochrome-b_5/cytochrome-c pair are not in fixed association, their interaction is mobile, so that there is fast exchange between different sites. We, that is Dr G. Moore and myself, have been led to conclude that just as there are two types of proteins with (a) rigid segments and (b) mobile segments, so there are two types of protein/protein or domain/domain contacts which are (a) rolling contacts or (b) relatively rigid. The two allow very different time dependent functions. Some proteins are parts of organised systems but some, e.g. cytochrome c, are scavengers. I now give an example of such domain/domain interaction within cytochrome c and this analysis will permit us to go forward to other motions in proteins.

Groove opening [53, 54]

Apocytochrome c is a poorly structured protein. It folds readily around a porphyrin or a haem. The presence of the metal ion is not essential but the metal forms a further cross-link from His-18 to Met-80 through the iron atom, Fig. 8. This cross-link is not very stable. Using NMR, biological variants, and site-specific mutagenesis at a conserved residue, Phe-82, close to the weak link, Fe—Met80, in the region shown below the haem to the left in Fig. 5, we and others have investigated the stability and rates of the breaking of this bond in several ways. (a) The pK_a of the alkaline transition which allows a lysine, probably Lys-79, to replace Met-80 at the haem [51]. This is a slow reaction. (b) The temperature of the low-spin to high-spin switch, which is just a breaking of the Met80—Fe bond [23]. The switch is in fast exchange. (c) The binding and the kinetics of the reaction of CN$^-$ with the iron [54]. This is a substitution reaction, for Met-80, which has also been studied in the reaction with imidazole, azide and a cytochrome P-450 inhibitor drug [54]. All these reactions are slow.

This last study concentrates upon finding how one part of a protein can move away from another changing the solvation of parts of the surface. It is clear from all the data on all forms of oxidised cytochrome c that the Met80—Fe(III) bond is strained even in the native protein. In fact, there are series of cytochromes c, which are of high-spin, where the bond is not made at all, intermediate-spin and low-spin where it is a reasonably good bond. The strain in this methionine-iron bond is comparable to that in the blue copper proteins where the Cu(II) methionine bond is so strained as to be almost not a bond. Both strained situations have been called entatic states

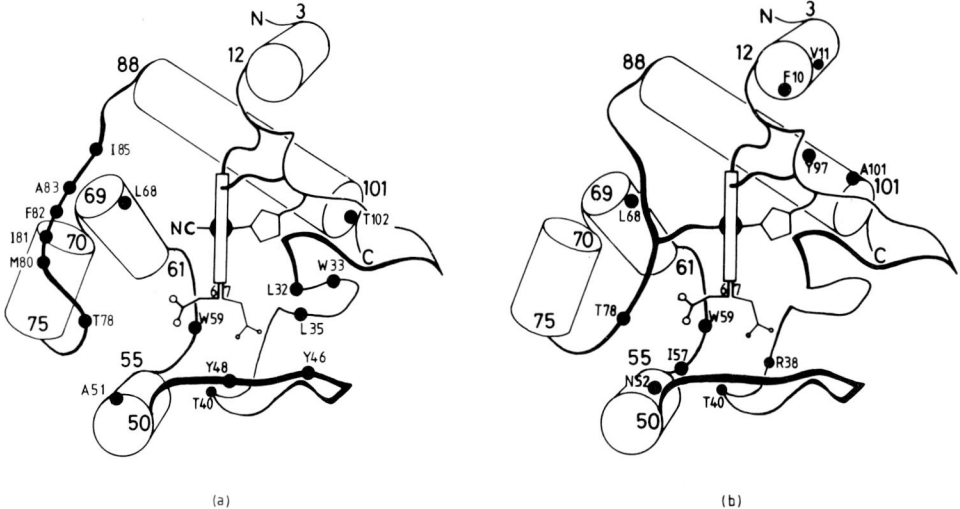

Fig. 8. *The difference in structure between the CN^- complex and cytochrome c itself illustrating the switch to a groove-open state [54]*

[55] and have benefit in electron transfer reactions in that they reduce the need to adjust the atoms forming the coordination sphere, while allowing small motions of the metal atom so as to assist electron transfer and hence optimise the electron-transfer potential and rate constant at the metal. On the whole, cytochromes *c* are very rigid except in this one region, as shown too by the fast spin-state equilibrium and the fast flip rate of Phe-82. The second feature of the strain is that it allows the small domain of the sequence from 78 – 90 to move away from the haem (Fig. 8).

We can observe, through NMR, parts of the assigned sequence of resonances of cytochrome *c* in residue 78 – 85. The residue at 78 is threonine and this residue is H-bonded to one propionate side chain of the heme. Its NMR spectrum shows that it is almost unaffected by any of the changes mentioned above. The behaviour of the surface residue 85, isoleucine in horse cytochrome *c*, is quite different and even a minimal mutation of Phe-82 to Tyr-82 causes a complete change in the conformation of residue 85 [25]. Note how site specific mutagenesis here is not site specific in a protein and contrast the effect of substitution in cytochromes *c* of serine (horse) by threonine (donkey) [41]. In passing, note too that it is the power of two-dimensional NOE NMR spectroscopy that allows an appreciation of the effects of mutations which cannot be obtained from a one- or two-dimensional COSY procedure since NOESY methods read out structure in a more sensitive way. In the case of cytochrome *c* or elsewhere, where paramagnetic probes can be used, there is also direct powerful structural fingerprinting in the two-dimensional chemical shift map.

(An interesting possibility arises when one considers the effect of site specific mutation. If the effect is purely a substitution then the effect is detected simply as a difference in local environments around the site. This should happen in stiff proteins. If the protein is more regionally flexible then the effect could be expected to be regional and to alter the dynamics as observed for example in a two-dimensional NOE spectrum (Fig. 9). We have compared the effects of site-specific mutations in cytochrome *c* with those in phosphoglycerate kinase [in the hinge region] and exactly these different effects are seen. NMR allows a very rapid analysis of these effects [25].)

We must now return to the particular experiment, the CN^- substitution on the iron for Met-80. We find by NMR that while Thr-78 remains attached to the heme propionate, Met-80 has moved its methyl group far from the haem and Phe 82 has also moved away (Fig. 8). The experiment indicates that it is the bonds between Thr-78 and Met-80 which take up the strain in forming the Met80 – iron link. Release of this strain allows rotation of the segment under the haem. This includes Lys-79 so that at high pH it may bind to the iron while Met-80 is rotated away from the haem. In fact the observed fold of cytochrome *c* is only conditionally the most stable at pH 7 and at pH 10.5 it is not so. (In cytochrome *f* it is highly probable that a lysine is bound instead in the pH 7 form [56].)

This example is given so as to direct attention to the generality of the motion of segments of proteins close to a surface such as the plane of a heme. This motion was first postulated as the cause of haemoglobin cooperativity [57] and the work of Perutz and his group showed it to be the case [13]. It is likely that similar motion allows the ordered reaction mechanism of cytochrome *P*-450. Recently we have pointed to the possibility that such a motion accounts for the proton pumping in the cytochrome oxidase and particle III electron transfer complex of mitochondria and in ion pumps of various kinds [58]. All these cases are probably examples of helix/helix motion (see below).

Loop antigenicity: cytochrome c and acyl phosphatase

While examining cytochrome *c* we became aware from a variety of NMR measurements [see above] that a particular part of the surface including a loop with a hydrophobic residue, Ile-57 (Fig. 9) had several states [23]. It was known that this was an antigenic site. Knowing that antibodies give rather rigid binding sites we postulated that generally antigenic sites would be regions which were relatively mobile. The idea was that a small rearrangement of the loosely folded part of the protein could be recognised in a highly selective and more powerful binding mode than if the antibody had to bind to the hydrophylic surface only. There has been some support and some dispute about the value of such an idea which can only apply to sequential antigenic sites of course [59]. We have

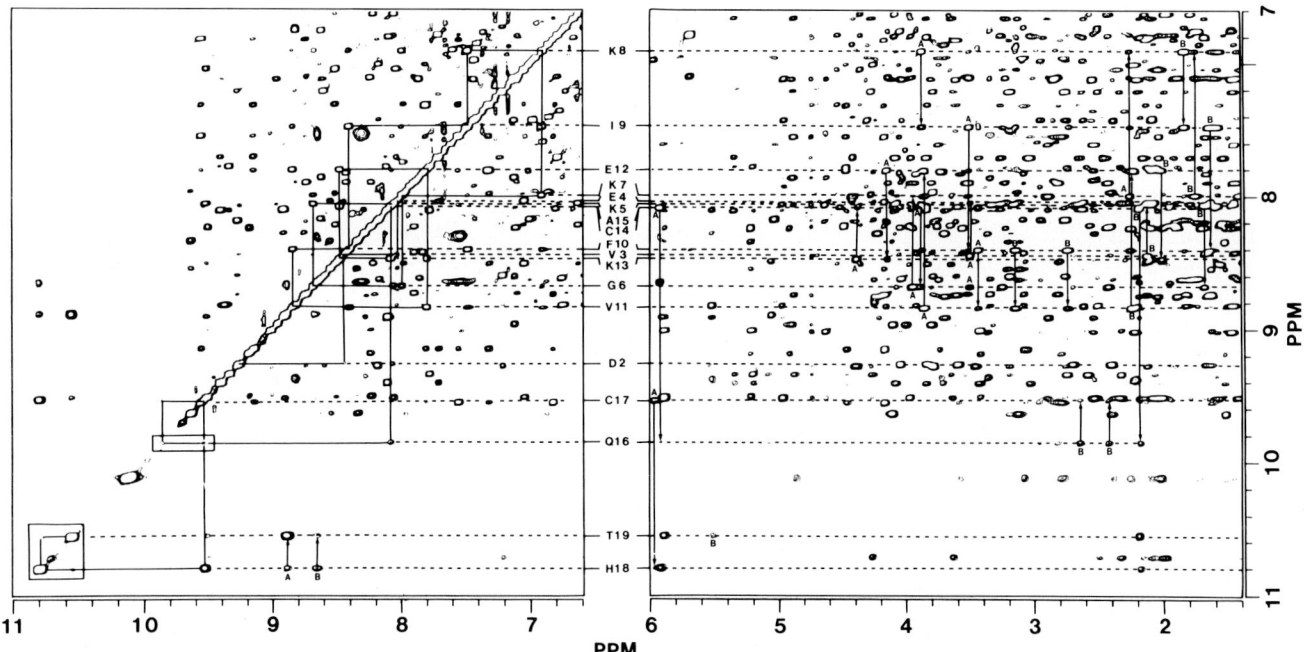

Fig. 9. *Part of the NOE spectrum of Fe(III) cytochrome* c. Apart from the structural data there is mobility information in the lineshapes, i.e. seen in spread of contours on one peak [42]. Assignments are given for the first helix. Since this work was finished [54] two excellent papers have appeared giving more details see [2b, 2c]

therefore determined the structure of a quite different protein, acylphosphatase, where the antigenic sites are well-known. We found that all the antigenic loops were of higher mobility than the secondary structural elements (Fig. 6B). Notice that there are no antigenic sites on any of the exposed surfaces of helical or β-sheet strands and that no part of them is mobile. The ideas are being explored by others in detail but it is to be remembered that fitting of complicated surfaces is far from easy if there is no mobility. A remarkable confirmation of the approach has come from the X-ray structure of the foot and mouth disease virus in which the antigenic site cannot be found in the structure (Stuart, D. I., personal communication).

Cytochrome c illustrates a case of surface mobility generally, of a small hinge region, of a loop region and a rearrangement of side-chains. We tackle next a protein in which there is larger hinge motion.

NMR STUDIES OF MOBILITY IN PROTEINS

Hinge proteins: phosphoglycerate kinase surfaces

Phosphoglycerate kinase is a large protein of 42 kDa, which has two domains. The C-terminal domain binds ATP and ADP and the N-terminal domain binds 3-phosphoglycerate and 1,3-diphosphoglycerate. The crystal structure of the yeast protein has been used together with NMR to understand the way in which the two domains move relative to one another about a hinge or connecting region between the domains [61–63]. From very early NMR observations (Fig. 3) it was known that addition of ATP caused relatively minor conformational changes while the (further) addition of 3-phosphoglycerate caused very considerable change, both in fast exchange on the NMR timescale of about 10^{-4} s. Even in the absence of sequence data three histidines, now known to be His-62, His-167 and His-170, were placed in the active-site binding region of 3-phosphoglycerate. The crystal structure and the NMR data agreed as to the way in which ATP was bound (Fig. 11); Fairbrother, W. et al., unpublished results. A series of papers by us will be published shortly relating to NMR studies of phosphoglycerate kinase) [61–63]. Further progress was delayed until both a sequence and a refined crystal structure were available. Even now the limited amount of information on the assignment of the 42-kDa protein that can be obtained from two-dimensional data (Fig. 10A) make confident discussion difficult. Assignment has been greatly helped however by using regions with different relaxation times, paramagnetic reagents, NOE methods and site-specific mutagenesis (Fairbrother, W. et al., unpublished results). It was found that there was a region around, and connected to, the hinge between the domains which gave relatively sharp lines (higher mobility) compared with the bulk of the protein, although one or two surface regions elsewhere are clearly more mobile e.g. around His-52 and His-53 (Fig. 10B). In effect, the large protein, which is far from spherical, has a slow anisotropic tumbling and lines from the interior of the two domains are too broad to resolve. (By way of contrast while the COSY spectra were indifferent the NOESY spectra are quite well resolved since NOE data improve with protein size.) A change in the value of the NMR method has resulted from two new approaches. First, the two domains have been split. The 20-kDa ATP-binding domain is intact and resonances from it are assignable since it gives a good two-dimensional spectrum. These resonances can then be superimposed on the intact 42-kDa protein. The 3-phosphoglycerate domain is damaged on splitting, it is more mobile, but by difference we know clearly which resonances belong to it. Second a series of site specific mutations have allowed confirmation of and new specific assignments in the hinge region. We are in a position to describe the properties of the hinge (Fairbrother, W. et al., unpublished results).

The first important point is that the hinge is not just the local region between the domains. On binding 3-phosphoglycerate there are small shifts running along helices on the

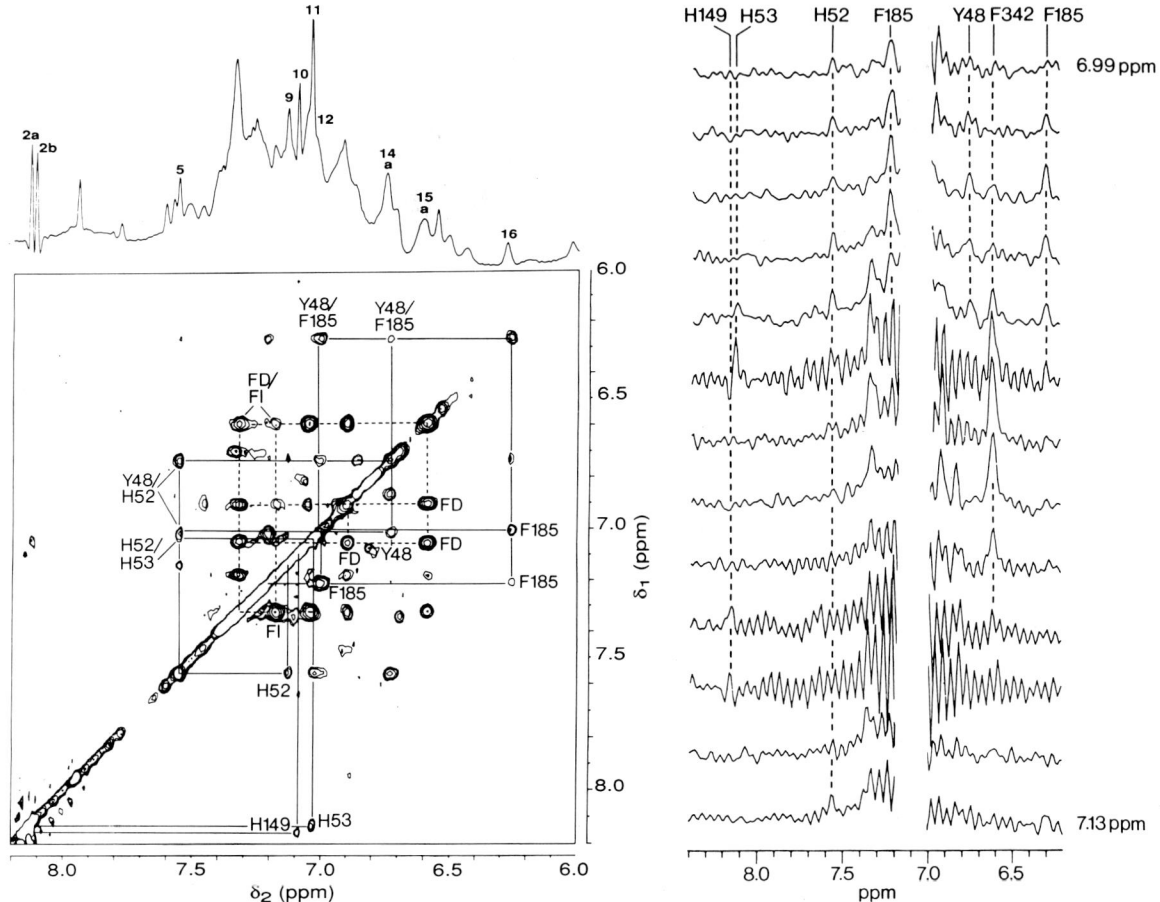

Fig. 10. *NOE spectroscopy of phosphoglycerate kinase.* (A) A two-dimensional NOESY proton NMR spectrum of phosphoglycerate kinase (molecular mass 42 kDa) [64]. The figure shows that it is the sharper lines of Fig. 3 which appear in the two-dimensional spectrum. Intensity of two-dimensional patterns need detailed exploration [42]. Lines 2a and 2b are of surface histidines, line 5 from a histidine of the hinge. Phenylalanines and tyrosines of the hinge are also shown. The aromatics from the core domain are not seen. (B) A series of cross-sections from a pure NOESY spectrum of the protein phosphoglycerate kinase [64]. The spectra are effectively one-dimensional spectra in stepped increments of 0.012 ppm through the aromatic region. Of the very large number of aromatic residues in the protein only a few are seen, see Fig. 9A, and those seen belong to mobile surface residues or residues in mobile loops or the more mobile hinge. NOESY spectra at this molecular mass, 42 kDa, reflect mobility as much as total structure

outside of the domains to quite distant parts of the protein (Fig. 11). The second important point is that many anions bind to the 3-phosphoglycerate site, but their effects on the protein conformation are different. We can only explain the observations in terms of a set of states of the hinge region, all in fast exchange, a few of which generate the catalytic condition of the enzyme. We have explored the effects of the site-specific mutation, not only on the average structure but also on the dynamics. Some ways in which this can be done by NMR are described in a recent paper [4, 25].

We believe that this study is a model study for the effects of domain/domain movement upon ordered helical structures, hinges. We are supposing that this is a general phenomenon which we relate to assemblies of domains or of proteins in such molecular machines as receptors and energy capture devices. Here the helical domains will be channel-forming systems in membranes or there will be a single helix acting like a lavatory chain to cause a cascade. In other cases we expect the link between domains to undergo order/disorder transitions providing an alternative mobility control without directional properties but this may be slow, see below.

Cross-linking and mobility

The evidence is accumulating that a major restriction on mobility in proteins is cross-linking. We have already referred to the cross-linking within a β-sheet, but this in itself provides no restriction on the side chains. It is easy to imagine that just as helices move relative to one another so sheets can move on one another (although their curvature makes this clumsy) but helices could move on sheets. A profound restriction on such motions would appear to be generated by chemical bond cross-linking. The best-known cross-link is the -S-S- bridge but we must not forget chemical cross-links in molecules such as collagens and components of cell walls. They clearly make for firm structures. NMR investigations of proteins such as lysozyme [65], neurotoxins [44], protease inhibitors [15] and phospholipase A_2 indicate little segmental mobility. A further type of cross-link is that due to metal ions such as calcium and zinc in many proteins and polysaccharides, and magnesium in polynucleotides, e.g. t-RNA. There are other cross-links due to iron in Fe_n/S_n proteins and other transition metal ions in other electron-transfer metalloproteins.

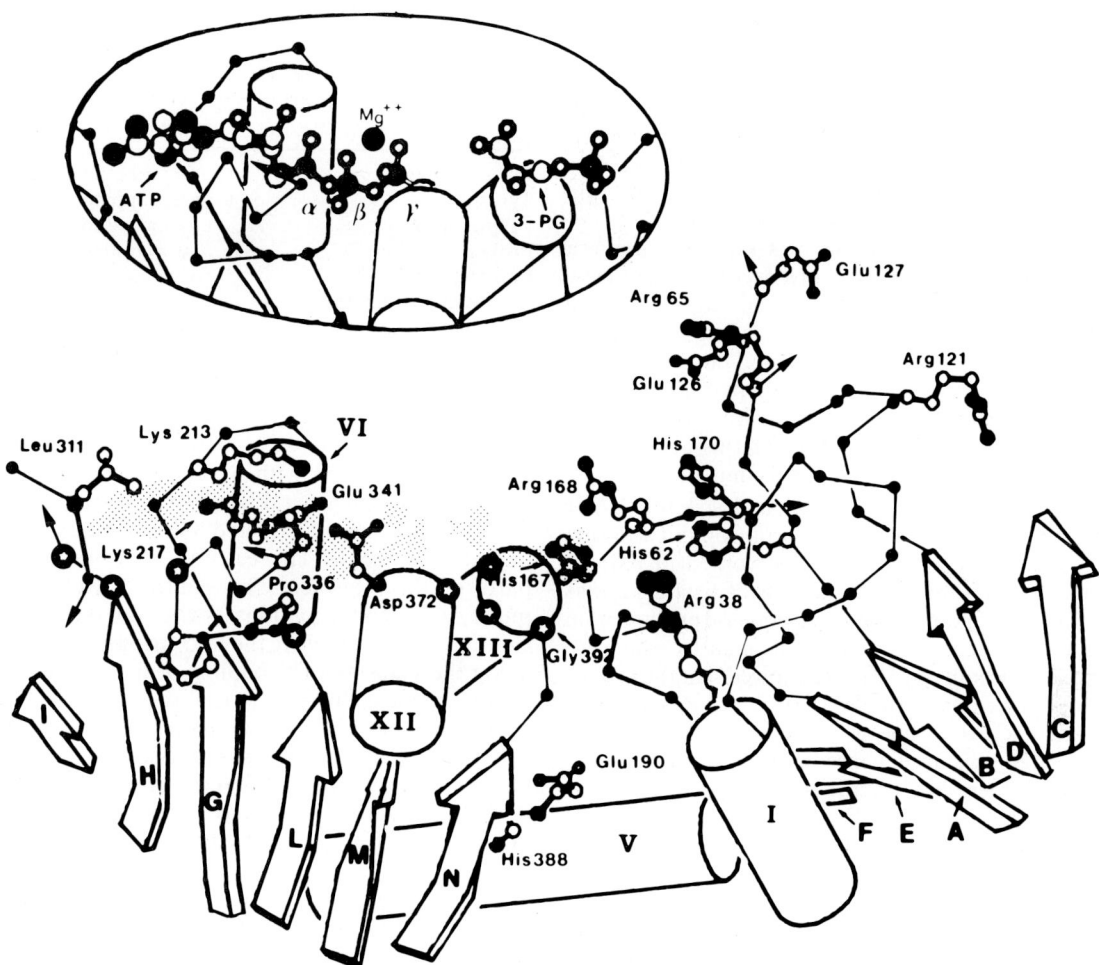

Fig. 11. *The structure of phosphoglycerate kinase determined by X-ray diffraction [61, 62].* The groove region is clearly shown

The cross-linking due to thio- bonds would seem to be very restrictive. All the cases of slowly flipping aromatic residues are in proteins which are cross-linked by -S- bonds Table 6. In the case of cytochrome *c* and osteocalcin this is supported by haem and calcium cross-linking, respectively. The implication is clear. Fast ring flipping implies segmental motion. It is then the case that there is segmental motion in many proteins. The use to which this is put will depend upon its degree and the whereabouts of the protein in the cell system.

We turn next to more mobile trigger proteins.

Trigger proteins: calmodulins and mobility within domains

The determination of the structure of parvalbumin by Kretsinger and his group [66] allowed the structure of troponin C and calmodulin to be quite well defined by NMR but due to the size of the protein only in two separated domains [5, 8]. There is a linker region. The crystal structures of calmodulin and troponin C are now available and apart from the confirmation of the outline conclusions from NMR there is much improved detail and a definition of the linker region. The first important conclusions from the X-ray structure and the NMR data are that the conformation change on binding calcium alters, but slightly, the site of calcium binding, but produces a knock-on effect in less than 10^{-3} s of considerable magnitude in the orientation of the two four-helix bundles in both domains and large rearrangements of loops. Control over this rearrangement is exerted by hydrophobic molecules [drugs] which bind to the surfaces of the helices [67]. The system then has two relay outlets: the loops at the ends of the helices and the sides of the helices. The overall picture is very reminiscent of the haemoglobin 'allosteric' switch which in solution is a similar multi-step helical switch process.

A problem has arisen in this work. NMR methods have not been able to show that the linker region between the two four-helical bundles exists. Whether this is a difficulty with NMR assignments or whether the region is disordered in solution is not known, although there is mounting evidence that the simple helix is not present at pH 7.

Intriguingly the NMR work of Campbell and his colleagues on small helical proteins which are to be found in membranes has shown that differential mobility can exist in helices themselves when they are broken by prolines [68]. We shall have to be very careful how we interpret any particular structure of a helical membrane protein which is known to be part of an ion or molecule pump.

Gross domain mobility: kringle-containing proteins

In 1976 I wrote an article on the dynamics of proteins [9]. In the course of the article I observed that certain parts of the proteins could not be observed or were poorly shown in structures derived from X-ray diffraction data. At that time [and still to some degree] it was not possible to distinguish between motion and disorder in crystals [14]. Today it is common to discuss such problems in separate terms of B-factors and disorder. Quite frequently however there is little indication of any structure for a small section of a protein, even in the crystals and here disorder is likely involving many conformations. A particularly interesting case is the structure of the fragment I of prothrombin where NMR and crystallographic data can be compared in some detail.

Prothrombin fragment I consists of a kringle domain, a Gla peptide domain and two large glycosyl units. Tulinsky and his collaborators [69] and Blake and his coworkers [70] have provided an X-ray structure of this fragment which shows a kringle domain. Much of the second domain is missing and the saccharide side chains are only partially defined. (The saccharides of nearly all glycoproteins are poorly defined [32].) The whole structure is discussed in terms of B-factors for the more rigid part and disorder for the missing part.

Simultaneously with this work we analysed the NMR spectrum of the N-terminal peptide of prothrombin and several kringles from different sources. We worked in parallel and often in collaboration with the groups of Esnouf, Llinas and Patthy [70, 71]. The N-terminal peptide of prothrombin contains γ-carboxylated glutamate, Gla, units to which we shall refer again. The joint NMR work showed that the kringle structure was similar in all kringles (Fig. 7) and its backbone was folded in much the same way for all of them despite considerable differences in sequence [69, 71]. Turning to mobility in solution there are several interesting observations from the NMR data. (a) The Gla peptide in the absence of calcium is virtually random except for structure imposed by the -S-S- bridge, as shown by chemical shift data. Calcium imposes structure which has been hard to define. (b) The N- and C- termini of the kringle fragments are not well structured. This remark applies to several kringles and in these regions very narrow resonance lines are often observed. (c) The kringle itself is a stable, but not a very stable, structure in the absence of binding ligands. The binding site of the ligands has been defined by both Llinas' group and our own. The ligand cross-links the kringle IV of plasminogen, for example between Arg-70 and Asp-54 and Asp-56. The kringle itself melts at about 45 °C. (d) The NMR structure determination is now highly refined and the back-bone is shown in a conventional manner in Fig. 7, but there are problems with the side chains of the structure. (e) There is a region of the protein around Tyr-9 (kringle IV of plasminogen numbering) which is not observable by NMR in any detail. There cannot be a well defined structure here in solution. It is probable that this Tyr-9 has an intermediate rate of flipping, i.e. approximately 10^4 s^{-1}. (f) There are minor peaks in the spectrum suggesting more than one orientation of some side chains which are in slow exchange. (g) The structure of the binding site is not rigid in that Phe-63 and Tyr-73 flip rapidly, as do all the other tyrosines except Tyr-9.

The overall impression of the structure of prothrombin and plasminogen is of one of differential domain mobility that it is designed to be such that only in the presence of calcium and ligand does it give a truly stable fold. It is notable that the secondary structure of the kringles only has small elements of a two-strand β-sheet according to the NMR data. Otherwise it is based on a set of single β-strands and turns. In the ligand free form most of the N$\underline{H}$ resonancs are in rapid N$\underline{H}$/N$\underline{D}$ exchange.

The interest in this dynamic molecule is related to its function. It is part of an extracellular control mechanism. Just as with osteocalcin [6], see below, the calcium binding is

Table 8. *Examples of possible side chain motions*
Many of these motions are seen in dynamic simulations but have not proved to be open to observation

Side chain	Motion
Phenylalanine	ring-flipping about β-CH$_2$
Tyrosine	ring-flipping about β-CH$_2$
Tryptophan	flapping about β-CH$_2$
Proline	cis $\rightleftarrows$ trans ring isomerism
Leucine/valine	asymmetric top rotation
Isoleucine/methionine etc.	asymmetric top rotation
Glutamate/aspartate	-CO$_2^-$ flip
Lysine/arginine	many rotamers possible

probably associated with a rather slow disorder $\rightarrow$ order change. Note the general similarity in structure of these two Gla peptides, i.e. [disordered Gla peptide] $\rightarrow$ [Ca^{2+}. Gla-peptide (ordered)]. The calcium cross-links rather remote parts of the sequence. This contrasts with the necessarily much faster conformational switch of the internal calcium calmodulin trigger where the calcium binds to a pocket formed along the sequence. The binding of the organic, lysine ligand to the kringle is different in that the site is largely preformed as a rather mobile surface region of hydrophobic character which has negative charges to one side and a positive charge to the other, i.e. in plasminogen kringle IV. It may well be that this structure is somewhat mobile even in the ligand-bound state. How can we judge this appropriately using NMR data?

A good and new example example has been given recently by Dobson and his coworkers [72]. They have shown that the domains of urokinase that are a protease (two domains), an EGF domain and a kringle show remarkably different mobilities in the intact protein. At the lowest temperatures there is more or less a single protein motion: a general tumbling. As the temperature is raised the domains are liberated in motion, but not chemically, assuming independent motion and they melt in an order, i.e. kringle, EGF then one of the protease domains [72]. The functional value of this differential mobility is not known. A more clear-cut example is given next.

(Before leaving this topic the mobility of the attached polysaccharides of such extracellular proteins as these is another area in which NMR is revealing the absence of simple single ground state structures of macromolecules [32].)

Partial domain unfolding

The extreme conformation change is an unfolding/folding reaction of such a protein as lysozyme. This article will not analyse the folding problem [27–29]. The use of smaller changes in folding is quite different. The most extreme example we have come across is the reaction of osteocalcin with calcium ions [or protons] at room temperature. The free protein has rather little structure except that imposed by the one -S-S- bridge. The folded protein is a set of four partially helical strands, cross-linked by calcium ions. Other examples are apocytochrome c and metallothionine which only fold around their cofactors (Table 3) [18].

Another, now small-scale, example is that of phosphorylase, the enzymatic reaction of which is known to be affected by the distant covalent binding of phosphate [11]. This may provide a general case of the effect of protein phosphorylation. The binding of phosphate itself causes a refolding of a small length of chain into an α-helical turn at the end of a preformed long helix. Transmission is via a shift of the long helix the other end of which is in contact with the active site. (We have observed above a similar movement of the helices in calmodulins and phosphoglycerate kinase.)

Two systems which have all the potentials of slow switches are osteocalcin and the zinc fingers. We have already described osteocalcin in the calcium free form, which it probably will be in the Golgi or vesicular systems of cells, as a disorder protein. It becomes ordered on exposure to the body fluids due to calcium binding and this transformation is slow. A possible parallel case is that of the zinc fingers. The zinc finger proteins in the absence of zinc are disordered, at least in small peptide form, and they become ordered on uptake of zinc as shown by an NMR study of Klevit and her coworkers [73]. The two cases generate positive surfaces, one of calcium (osteocalcin) the other of lysine and arginine (zinc fingers) in order to bind anionic phosphate surfaces of bone and DNA, respectively. However considerable motion of these surfaces, calcium movement and amino acid side-chain mobilities, will allow continuous adjustment of the surface interactions, a searching for the best fit. In the author's view this surface mobility will prove to be a common feature of protein interactions where the primary interaction energy is electrostatic, e.g. polyamines and histones with DNA, and cytochrome c with membranes (see above).

The segmental dynamics of proteins begins to fall into classes. We find fast motions in some trigger proteins, e.g. calmodulins. Here several states are of similar energy. We also find smaller fast motions in very many proteins but these are based on one stable ground state and a variety of types of slow segment changes which are probably connected to regulation and can be as extreme as complete unfolding. However proteins may also behave with a diversity of properties in different domains and linking segments.

Enzymes and mobility

Very few enzymes have been studied in detail by NMR since most of them exceed 20-kDa. It is, therefore, only possible to put forward points, deduced from smaller molecules, which are then open to test. The first point is that most enzyme sites are on concave surfaces, i.e. in pockets. This introduces some constraints on motion and pockets could be devised such that only a few necessary motions are allowed. Clearly specificity is enhanced both in binding and in the reaction pathway by this lowered mobility. It would be good to know it in detail. The second point is that the vast majority of enzymes are based on β-sheets, with or without attendant helices. This and cross-linking by -S-S- bridges or metal ions constrains segmental motion. It would be good to know how strong these limitations are. The third point is that active site residues are often carried on loops which by NMR methods we observe to be somewhat mobile. If this mobility assists activity then how does it do so? One class of enzyme, that which uses a preferred order mechanism, breaks the second rule but such enzymes require considerable conformational changes. They are generally based on helices (Table 9). Undoubtedly other enzymes are based on movement of loops (flaps) or of larger segments at hinges. NMR is beginning to get to grips with these problems but it will probably require much greater use of local structural knowledge using ^{15}N-, ^{13}C-, or D-labelled proteins.

The best set of assigned resonances of an enzyme of 'known' mechanism, now excluding electron transfer proteins

Table 9. *Mobility dependent functions of some helical proteins*

Protein	Function
Calmodulin	transfer of information
Cytochrome *P*-450	preferred-order enzyme reaction
Citrate synthase	preferred-order enzyme reaction
Haemoglobin	cooperative dioxygen uptake
Membrane ATPases	ion pumps
Bacteria rhodopsin	proton translocation

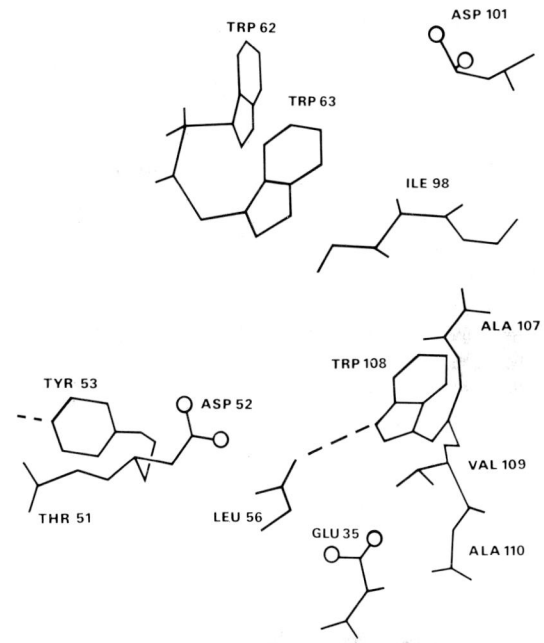

Fig. 12. *The active site region of lysozyme showing a fast-flipping Tyr-53 and a slow flipping Trp-62.* The active site remains mobile though mobility is reduced when inhibitors are bound [65]

Table 10. *Motions of side chains of lysozyme by NMR*
Very similar observations are available from a wide range of studies on other proteins. Note that the whole groove can also close somewhat: (a) in close proximity to the active site; (b) binding groove. I am grateful to Drs C. M. Dobson and C. Redfield [16] for their assistance with this table

Side chain	Location	Observation
His-15	surface	mobile on surface, sharp signals
Val-109(a)	surface	near equivalence of paramagnetic shifts on methyls
Trp-62(b)	surface	slow flapping motion (two signals)
Tyr-53(a)	surface	fast-flipping residue (signals averaged)
Tyr-20	interior	
Phe-3	surface	fast-flipping residue (signals averaged)
Phe-34	interior	
Lysines	surface	sharp signals from ε-CH_2

since they do not have a concave pocket, is for lysozyme (Fig. 12). In Table 10 we show that there is much motion in the region of the active site and that this motion is affected by the binding of inhibitors. The motions include more than the flapping of Trp-62, the flipping of Tyr-53 and Val-109.

There must also be motion around Trp-108 and Asp-52 if we are to explain the changes in NMR spectra on binding ligands and the NH/ND exchange rates. The general impression of the pocket and of the binding pocket of kringle IV (not an enzyme) is that there is considerable motion in such concave pockets. This makes good sense in that the catalyst is able to follow the changes in the substrate structure in a dynamic matching. These NMR results suggest that matching the transition state of the reaction by the enzyme ground state (Pauling postulate) is not desirable without the general possibility of dynamic matching of several states of the substrate/product pair while passing through the succession of barriers associated with intermediates and transition states. It must not be forgotten that enzyme turnover numbers are usually slow relative to ring flipping ($< 10^{-4}$ s). Another way of looking at this pocket mobility is to state that the solvent for the reaction, here the pocket of the enzyme, has constrained dynamics matching the movements of the reactant/product system. The importance of the solvent in solution kinetics is now well recognised and NMR measurements suggest that enzyme surfaces are constrained solvents.

The study of enzymes by NMR is still in its infancy. We know however that gross restrictions on side-chain mobility are not general. We know that enzymes are rarely built from helical segments alone. The evidence is accumulating that a β-sheet construct to supply a platform is generally required allowing only minor mobility of the active site.

CONCLUSION

The study of mobility in larger proteins has just begun. It is likely to be an analysis by NMR methods but the essential starting point in many cases will be an X-ray structure of the protein trapped in a crystal. Putting the data together will require great skill and probably many exercises in dynamic modelling. The scale of the motions involved may be quite small, as in extracellular enzymes, or it may be very considerable as in various kinases, ATPases, pumps and contractile systems. It is the very variety of NMR procedures alone which gives one reason to believe that we shall be able to understand functioning of molecular machinery in terms of composition (sequence): isolated structures: dynamics.

If we are to achieve our aim then, in the author's opinion, much more care is required in many studies of structure and mobility by NMR methods. The major problem is the use of NOE alone. For example it is necessary to report the failure to observe NOE when expected from the proposed structure. It is clear that relaxation times vary grossly within proteins such that some NOE signals expected from single structure representations do not appear or are of the wrong intensity. This occurs when the relaxation time is too short, too long or exchange processes interfere. The interiors of small, relatively rigid proteins are well defined but the essential surface structure of any protein cannot be obtained with confidence by the NOE methods since relaxation times are highly variable and there may be no one minimum energy structure. The findings of partial two-dimensional spectra for large proteins, e.g. in this article phosphoglycerate kinase, and of an incomplete two-dimensional spectrum for the kringle IV, both in the interior due to exchange and at the terminii, see also the acylphosphatase spectra (Fig. 6B), point out the problems. In fact, quite a good structure of a kringle was obtained by the combination of ring current shifts, paramagnetic shift and relaxation probes, which do not involve time dependence. The agreement between the analysis of the paramagnetic data and the interior parts of the X-ray structure of cytochrome c is impressive. Given that the assignment problem can be overcome in small proteins by NMR directly or with larger proteins from their crystal structures these time-independent methods merit reinspection as a help to the NOE method.

Here I wish to acknowledge a community: those who work in NMR of proteins and on whose work much of the above relies. I could not give all the references (see Abstracts XIII International Conference on Magnetic Resonance in Biological Systems (1988) at Madison, Wisconsin, USA) and I excuse myself on the grounds that I wished to make the article readable. Any original contributions refered to under my name below are due to intense collaboration with members of the Oxford Enzyme Group, now The Oxford Interdisciplinary Research Centre for Molecular Structure. They also remain nameless here.

REFERENCES

1. Veitch, N. C., Concar, D. W., Williams, R. J. P. & Whitford, D. (1988) *FEBS Lett.* 238, 49–55.
2a. Pielak, G. J., Boyd, J., Moore, G. R. & Williams, R. J. P. (1988) *Eur. J. Biochem.* 177, 167–177.
2b. Wand, A. J., DiStefano, D. L., Feng, Y., Roder, H. & Englander, S. W. (1989) *Biochemistry* 28, 186–194.
2c. Feng, Y., Roder, H., Englander, S. W., Wand A. J. & DiStefano, D. L. (1989) *Biochemistry* 28, 195–203.
3. Saudek, V., Williams, R. J. P. & Ramponi, G. (1989) *FEBS Lett.* 242, 225–232.
4. Wilson, H. R., Williams, R. J. P., Littlechild, J. A. & Watson H. C. (1988) *Eur. J. Biochem.* 170, 529–538.
5. Klevit, R. E., Dalgarno, D. C., Levine, B. A. & Williams, R. J. P. (1983) *Eur. J. Biochem.* 139, 109–114.
6. Williams, R. J. P. (1988) *Ciba Found. Symp.* 136, 198–201.
7. Esnouf, P., Lawrence, M. P., Mabbutt, B. C., Patthy, L., Pluck, N. & Williams, R. J. P. (1985) *Bull. Soc. Chim. Belg.* 94, 883–896 (and references therein).
8. Trayer, I., Levine, B. A. & Williams, R. J. P. (1989) in *Molecular mechanisms in muscle contraction* (Squire, J. M., ed.) Macmillan, London.
9a. Williams, R. J. P. (1978) *Angewandte Chemie Int. Ed. Engl. Suppl.* 16, 766–777.
9b. Williams, R. J. P. (1978) *Biol. Rev.* 54, 389–437.
9c. Williams, R. J. P. (1981) in *Mobility and migration of biological molecules* (Garland, P. B. & Williams, R. J. P., eds) Biochemical Society Press, London.
10. Blake, C. C. F., Grace, D. E. P., Johnson, L. N., Perkins, S. J., Phillips, D. C., Cassels, R., Dobson, C. M., Poulsen, F. M. & Williams, R. J. P. (1978) *Ciba Found. Symp.* 60, 137–185.
11. Prang, S. R., Acharya, K. R., Goldsmith, E. J., Stuart, D. I., Varvill, K., Fletterick, R. J., Madsen, N. B. & Johnson, L. N. (1988) *Nature* 336, 215–221.
12. Ringe, D. & Hajdu, J. (1987) *Nature* 329, 102–103.
13. Perutz, M. F. (1978) *Sci. Am.* 239, 68–73.
14. Huber, R. (1988) *Biochem. Soc. Trans.* 15, 1009–1021.
15. Wüthrich, K. (1986) *NMR of proteins and nucleic acids,* Wiley Interscience, New York.
16. Redfield, C. & Dobson, C. M. (1988) *Biochemistry* 27, 122–136.
17. Dalgarno, D., Drabikowski, R., Klevit, R., Levine, B., Scott, G. M. M. & Williams, R. J. P. (1983) in *Calcium-binding proteins* (de Bernard, B., ed.) pp. 83–91.
18. Williams, R. J. P. (1987) *Carlsberg Res. Commun.* 52, 1–30.
19. Dalgarno, D. C., Levine, B. A. & Williams, R. J. P. (1983) *Biosci. Rep. 3,* 443–452.
20. Williams, G., Clayden, N. J., Moore, G. R. & Williams, R. J. P. (1985) *J. Mol. Biol.* 183, 447–460.
21. Williams, G., Moore, G. R., Porteous, R., Robinson, M. N., Soffe, N. & Williams, R. J. P. (1985) *J. Mol. Biol.* 183, 409–446.

22. Campbell, I. D., Dobson, C. M. & Williams, R. J. P. (1985) *Biochem. J. 231*, 1.
23. Moore, G. R., Huang, Z.-X., Eley, C. G. S., Barker, G., Williams, G., Robinson, M. N. & Williams, R. J. P. (1983) *Faraday Discuss. Chem. Soc. 74*, 311–329.
24. Perkins, S. J. (1982) in *Biological magnetic resonance* (Berliner, L. J. & Reuben, J., eds) vol. 4, pp. 79–144, Plenum Press, New York.
25. Pielak, G. J., Atkinson, R. A., Boyd, J. & Williams, R. J. P. (1988) *Eur. J. Biochem. 177*, 179–185.
26. Linderstrøm-Lang, K. U. & Schellman, J. A. (1959) in *The enzymes 2nd ed.* (Boyer, P. D., Lardy, M. & Myrback, K., eds) vol. 1, pp. 443–480, Academic Press, New York.
27. Cassels, R., Dobson, C. M., Poulsen, F. M. & Williams, R. J. P. (1978) *Eur. J. Biochem. 92*, 81–97.
28. Udgaonkar, J. B. & Baldwin, R. L. (1988) *Nature 335*, 694–699.
29. Roder, H., Elore, G. A. & Englander, S. W. (1988) *Nature 335*, 700–704.
30. Evans, P. A., Dobson, C. M., Kautz, R. A., Hatfull, G. & Fox, R. O. (1987) *Nature 329*, 266–268.
31. Novokhatnyi, V. V., Kudinov, S. A. & Privalov, P. L. (1984) *J. Mol. Biol. 179*, 215–232.
32. Williams, R. J. P. & Fernandez, D. L. (1989) *R. Soc. Chem. Spec. Publ.*, in the press.
33. Hagerman, P. J. (1988) *Annu. Rev. Biophys. Chem. 17*, 265–286.
34. Doi, M. & Edwards, S. F. (1986) *The theory of polymer dynamics*, Clarendon Press, Oxford.
35. Egmond, M. R., Rees, D, Welsh, J. & Williams, R. J. P. (1979) *Eur. J. Biochem. 97*, 73–83.
36. Campbell, I. D., Dobson, C. M., Williams, R. J. P. & Wright, P. E. (1975) *FEBS Lett. 57*, 96–99.
37. Egmond, M. R., Verhagen, J. & Williams, R. J. P. (1978) *Biochim. Biophys. Acta 535*, 418–422.
38. Packmann, L. C., Perham, R. N. & Roberts, G. C. K. (1984) *Biochim. J. 217*, 219–227.
39. Evans, J. S., Levine, B. A., Trayer, I. P., Dorman, C. J. & Higgins, C. F. (1986) *FEBS Lett. 208*, 211–216.
40. Daniels, A. J., Williams, R. J. P. & Wright, P. E. (1978) *Neuroscience 3*, 573–585.
41. Moore, G. R. & Williams, R. J. P. (1980) *Eur. J. Biochem. 103*, 543–550.
42. Weiss, M. A., Eliason, J. L. & States, D. J. (1984) *Proc. Natl Acad. Sci. USA 81*, 6019–6023.
43. Hider, R. C., Drake, A. F., Inagaki, F., Williams, R. J. P., Endo, T. & Miyazawa, T. (1982) *J. Mol. Biol. 158*, 275–291.
44. Inagaki, F., Boyd, J., Campbell, I. D., Clayden, N. J., Hall, W. E., Tamiya, N. & Williams, R. J. P. (1982) *Eur. J. Biochem. 121*, 609–616.
45. Dwek, R. A., Richards, R. E., Morallee, K. G., Nieboer, E., Williams, R. J. P. & Xavier, A. V. (1971) *Eur. J. Biochem. 21*, 204–211.
46. Barry, C. D., North, A. C. T., Glasel, J. A., Williams, R. J. P. & Xavier, A. V. (1971) *Nature 232*, 235–239.
47. Reference deleted.
48. Arean, C. O., Moore, G. R., Williams, G. & Williams, R. J. P. (1988) *Eur. J. Biochem. 173*, 607–615.
49. Williams, R. J. P., Moore, G. R. & Williams, R. J. P. (1987) in *Progress in bioorganic chemistry and molecular biology* (Ovchinnikov, Yu. A., ed.) pp. 31–39, Elsevier, Amsterdam.
50. Wendoloski, J. J., Matthew, J. B., Weber, P. C. & Salemme, F. R. (1987) *Science 238*, 794–797.
51. Liang, N., Mauk, A. G., Pielak, G. J., Johnson, J. A., Smith, M. & Hoffman, B. M. (1988) *Science 240*, 311–313.
52. Reference deleted.
53. Pielak, G. J., Concar, D. W., Moore, G. R. & Williams, R. J. P. (1987) *Protein Eng. 1*, 83–88.
54. Gao, Y., Williams, G., Pielak, G. J. & Williams, R. J. P. (1989) *Eur. J. Biochem. 182*, 57–65.
55. Vallee, B. L. & Williams, R. J. P. (1968) *Proc. Natl Acad. Sci. USA 59*, 498–503.
56. Gadsby, P. M. A., Peterson, J., Foote, N., Greenwood, C. & Thomson, A. J. (1987) *Biochem. J. 246*, 43–54.
57. Banerjee, R., Alpert, Y., Leterrier, F. & Williams, R. J. P. (1969) *Biochemistry 8*, 2862–2869.
58. Williams, R. J. P. (1987) *FEBS Lett. 226*, 1–7.
59. Synthetic peptides as antigens (1986) *Ciba Found. Symp. 119*.
60. Reference deleted.
61. Blake, C. F. & Evans, P. R. (1974) *J. Mol. Biol. 84*, 585–601
62. Watson, H. C., Walker, N. P. C., Shaw, P. J., Bryant, T. W., Wendell, P. L., Fothergill, L. A., Perkins, R. E., Convoy, S. C., Dobson, M. J., Tuite, M. F., Kingsman, A. J. & Kingsman, S. M. (1982) *EMBO J. 1*, 1635–1640.
63a. Tanswell, P., Westhead, E. W. & Williams, R. J. P. (1976) *Eur. J. Biochem. 63*, 249–258.
63b. Wilson, H. R., Williams, R. J. P., Littlechild, J. A. & Watson, H. C. (1988) *Eur. J. Biochem. 170*, 529–538.
64. Reference deleted.
65. Redfield, C. & Dobson, C. M. (1988) *Biochemistry 27*, 122–136.
66. Kretsinger, R. H. (1976) *Annu. Rev. Biochem. 45*, 239–266.
67. Dalgarno, D. C., Klevit, R. E., Levine, B. A., Scott, G. M. M., Williams, R. J. P., Gergely, Z., Grabarek, Z., Leavis, P. C., Grand, R. A. J. & Drabikowski, W. (1984) *Biochim. Biophys. Acta 791*, 164–172.
68. Moody, M. F., Jones, P. T., Carver, J. A., Boyd, J. & Campbell, I. D. (1987) *J. Mol. Biol. 193*, 759–774.
69. Park, C. H. & Tulinsky, A. (1986) *Biochemistry 25*, 3977–3982.
70. Williams, R. J. P., Esnouf, P., Lawrence, M. & Cederholm-Williams, S. A. (1986) *FEBS Lett. 209*, 111–116.
71a. De Marco, A., Pluck, N. D., Banyai, L., Trexler, M., Laursen, R. A., Patthy, L., Llinas, M. & Williams, R. J. P. (1985) *Biochemistry 24*, 748–753.
71b. Mabbutt, B. C. & Williams, R. J. P. (1988) *Eur. J. Biochem. 170*, 539–548.
72. Ostwald, R. E., Bogusky, M. J., Bamberger, M., Smith, R. A. G. & Dobson, C. M. (1989) *Nature 337*, 579–582
73. Parraga, G., Horvath, S. J., Eisen, A., Taylor, W. E., Good, L., Young, E. T. & Klevit, R. E. (1988) *Science 241*, 1489–1492.
74. Wand, A. J., DiStefano, D. L., Feng, Y., Roder, H. & Englander, S. W. (1989) *Biochemistry 28*, 186–194.
75. Feng, Y., Roder, H., Englander, S. W., Wand, A. J. & DiStefano, D. L. (1989) *Biochemistry 28*, 195–203.
76. Van de Ben, F. J. M. & Hilbers, C. M. (1988) *Eur. J. Biochem. 178*, 1–38.
77. Parker, M. W., Pattus, F., Trucker, A. D. & Tsernoglou, D. (1989) *Nature 337*, 93–96.
78. Barr, G. C., Brewer, S., Dorman, C. J., Evans, J. S., Higgins, C. F., Levine, B. A., Tolley, I. M., Trayer, I. P. & Wormald, M. R. (1989) *J. Mol. Biol.*, in the press.

Review

Dehydrogenases for the synthesis of chiral compounds

Werner HUMMEL and Maria-Regina KULA

Institut für Enzymtechnologie der Heinrich-Heine-Universität Düsseldorf in der KFA Jülich

(Received December 28, 1988/April 10, 1989) — EJB 88 1517

CONTENTS

1. Introduction
2. Amino acid dehydrogenases
2.1. General porperties
2.2. L-Alanine dehydrogenase
2.3. L-Leucine dehydrogenase
2.4. L-Phenylalanine dehydrogenase
3. Hydroxy acid dehydrogenases
3.1. General properties
3.2. 2-Hydroxy acid dehydrogenases
4. Alcohol dehydrogenases
5. Diketone reductases
6. Enzymes for coenzyme regeneration
7. Conclusions

1. INTRODUCTION

Redox reactions are important steps in the metabolism and energy conversion of living cells. Besides the enzymes necessary, a number of coenzymes are involved in such reactions: ferrodoxins, lipoic acid, NAD(H), NADP(H), flavins and cytochromes. The coenzymes differ in their redox potential, in the binding constants and the mode of regeneration [1]. NADH/NADPH are the most frequently encountered coenzymes. In general they dissociate easily and need a second reaction with another metabolite for regeneration. These properties are one of the means by which nature directs the flow of intermediates in response to biosynthetic needs. Because of the spectral properties of the coenzyme moiety, dehydrogenases have been extensively studied in the past [2] and have found widespread applications in clinical and food analysis [3].

Two industrial fermentation processes rely on dehydrogenases for the final product formation. In the production of glutamate from molasses and ammonia by *Corynebacterium glutamicum*, glutamate dehydrogenase is the key enzyme converting 2-oxoglutarate accumulated in this strain by a defect in 2-oxoglutarate dehydrogenase to L-glutamate. In the reductive amination NADPH serves as coenzyme, it is regenerated in the metabolic sequence leading from hexoses to 2-oxoglutarate mainly by the isocitrate dehydrogenase reaction, which provides a very close stoichiometric and metabolic link in the biosynthesis of glutamate by *Corynebacterium glutamicum*. The ammonia fixation by glutamate dehydrogenase is an essential step in nitrogen metabolism, since the amino group is transferred by transamination to all other amino acids and serves as nitrogen precursor for purines, pyrimidines and other building blocks.

Also lactic acid production by various strains of *Lactobacillus* utilize an NADH-linked lactate dehydrogenase to reduce pyruvate stereospecifically to L- or D-lactate, respectively. In this case NADH is regenerated in the conversion of hexoses to pyruvate.

The stereospecific reduction of carbonyl groups is of interest for the production of various chiral compounds such as hydroxy acids, amino acids or alcohols from prochiral precursors [4–7]. Such products have a high economic value and find direct applications in food and feed, or serve as building blocks in the synthesis of therapeutics, herbicides, insecticides etc. To develop enzyme-catalyzed syntheses of such chiral compounds further, additional dehydrogenases with a suitable substrate range as well as efficient coenzyme regeneration systems are required. While the protein moiety of a dehydrogenase acts solely as a catalyst, coenzymes are consumed in stoichiometric relations with the product generated. In the processes discussed above, the necessary reduction of the $NAD^+/NADP^+$ to NADH/NADPH is coupled to the biosynthesis of the precursor oxoacids from cheap carbon sources. Such an approach is not always possible or desirable. For example, if the precursor of interest does not occur in the metabolism, or is consumed rapidly in other directions, or is not taken up efficiently by living cells. These difficulties can be resolved working with isolated enzymes as catalysts. Because stereoselectivity of enzymatic reactions is usually strictly preserved, the wider substrate range of catabolic enzymes should allow the design of new artificial pathways leading to the desired product in combination with a suitable regeneration system. Glucose co-metabolism can be replaced by different reactions; Simon et al. studied 'electromicrobial' or 'electroenzymatic' reductions utilizing artificial mediators like methylviologen that can regenerate NADH or replace it in some cases. These reactions are coupled with an electrochemical reduction of the mediator (for a review see [8]). These regeneration methods are difficult to carry out on a preparative scale [8]. They also investigated the regeneration of NADH with H_2

Correspondence to W. Hummel, Institut für Enzymtechnologie der Heinrich-Heine-Universität Düsseldorf in der KFA Jülich, P. O. Box 2050, D-5170 Jülich, FRG

Abbreviations. ADH, alcohol dehydrogenase; YADH, yeast alcohol dehydrogenase; *Tbr* ADH, *Thermoanaerobium brockii* alcohol dehydrogenase; FDH, formate dehydrogenase; PheDH, L-phenylalanine dehydrogenase; HmpDH, 2-hydroxy-4-methylpentanoate dehydrogenase 2-hydroxyisocaproate dehydrogenase; LeuDH, L-leucine dehydrogenase; AlaDH, L-alanine dehydrogenase.

and cells of *Clostridium* species containing a highly active hydrogenase activity. Enantioselective reductions of 2,3-unsaturated carboxylic acids, ketones (both with cells of *Clostridium* species) and 2-oxo-carboxylic acids (*Proteus*) were published (reviewed in [8]). The enzymatic oxidation of formate to CO_2 by formate dehydrogenase offers a further favorable method for the regeneration of NADH. In this way, product isolation is also simplified. In an enzyme-membrane reactor [9] catalysts from different origin can be easily combined and their relative activities predetermined by the operator. Such systems are amenable to detailed kinetic analysis and optimization by reaction engineering concepts. Basic features of the enzyme-membrane reactor have been described recently [9] as well as detailed reaction engineering studies of coupled dehydrogenase systems [10, 11]. In the following sections, we review NAD(P)H-dependent dehydrogenases potentially useful for the production of fine chemicals and strategies available for coenzyme regeneration. Progress, especially in NADH regeneration, has led to the search for and characterization of a number of new dehydrogenases. Despite the high price of NADH, of which every biochemist and enzymologist is aware, NADH-dependent enzyme-catalysed synthesis is about to enter production scale and industrial practice.

2. AMINO ACID DEHYDROGENASES

2.1. General properties

There are several L-amino acid dehydrogenases known which depend on nicotinamide coenzymes. They catalyze reactions according to the following scheme:

$$\text{R-CH(NH}_2\text{)-COOH} + \text{NAD}^+ + \text{H}_2\text{O} \rightleftharpoons \text{R-CO-COOH} + \text{NH}_4^+ + \text{NADH}.$$

The participation of NAD(P) makes these enzyme systems a valuable tool for analysis of L-amino acids or their corresponding oxo acids. By reductive amination of the oxo acid L-amino acids can be obtained in nearly quantitative yield because the equilibrium of the reaction favours amino acid formation. It appears technically feasible to carry out such reactions also on a large scale. Economic considerations show, however, that the enzyme-catalyzed production of L-amino acids competes with the fermentation or biotransformation routes, which employ cheap substrates such as molasses or methanol and ammonia and highly producing strains optimized by classical mutation methods or genetic engineering. This is true especially for the production of L-glutamate, where concentrations of >100 g/l can be obtained and carbon conversion rates of about 90%. As a rule, advantageous use of amino acid dehydrogenases can be expected in special cases, for example to produce unnatural amino acids such as 2-amino-3,3-dimethyl butanoic acid by leucine dehydrogenase [12] or to synthesize ^{15}N-labelled amino acids introducing the isotope via the [^{15}N]ammonium salt [13]. So far, the corresponding NAD-dependent D-amino acid dehydrogenases are not known. Therefore, enzyme-catalyzed synthesis of amino acids by reductive amination is restricted to the L-enantiomers as products.

2.2. L-Alanine dehydrogenase

Alanine dehydrogenase catalyzes the reversible deamination of L-alanine to pyruvate. The enzyme occurs in spores [14, 15] and vegetative cells of various bacteria [16–20]. Some strains of *Bacillus* especially reveal high levels of alanine dehydrogenase and some of these have been used for the preparation of the enzyme. AlaDH is a key enzyme in the degradation of L-alanine, the resulting pyruvate can easily be metabolized via the tricarboxylic acid cycle. Like leucine dehydrogenase, alanine dehydrogenase in spores seems to be responsible for generation of energy during sporulation.

2.2.1. Production of alanine dehydrogenase. An extended screening for the distribution of alanine dehydrogenase in bacteria is described by Ohashima and Soda [21], revealing high enzyme activities in *Bacillus sphaericus, B. aneurinolyticus, B. circulans* and *B. cereus*. The specific activity in the crude extract ranges between 0.15–0.41 units/mg. Enzyme purification procedures are described for alanine dehydrogenase from *B. subtilis* [22–24], *B. cereus* [25] and *B. sphaericus* [21]. The latter preparation starts from 2 kg (wet cells) and results, after seven purification steps, in a homogeneous enzyme preparation (157 units/mg). Leucine dehydrogenase, which is commonly present in the crude extract of *Bacillus*, can be separated by a DEAE-cellulose chromatography step.

2.2.2. Biochemical characterization of alanine dehydrogenase. The enzyme from *B. sphaericus* [21] has a molecular mass of 230 kDa (gel filtration) and is composed of six subunits (38 kDa). The enzyme oxidizes quite specifically L-alanine (K_m 18.9 μM), only slight activity was found for L-2-aminobutyrate, L-serine, L-norvaline, and L-valine. The substrate specificity for oxo acids is summarized in Table 1. The pH optimum for the reductive amination is around 9.0 and between 10.0–10.5 for the oxidative deamination. In contrast to alanine dehydrogenase from *Bacillus subtilis* [23], the *B. sphaericus* enzyme is not inhibited by several D-amino acids such as D-alanine or D-cysteine.

2.2.3. Applications of alanine dehydrogenase for the synthesis of L-amino acids. Due to the high substrate specificity, alanine dehydrogenase can be applied only for the synthesis of L-alanine. However, in spite of efficient coenzyme regeneration techniques, this would be an uneconomic route for L-alanine synthesis compared to the two-enzyme process starting with fumaric acid [26, 27]. The synthesis of [^{15}N]-

Table 1. *Substrate specificity of L-alanine dehydrogenase from B. sphaericus [21] for reductive amination and oxidative deamination reaction*

Reaction	Substrates	V	K_m
		mol	mM
Reductive amination	pyruvate	910	1.7
	2-oxobutyrate	933	11
	2-oxovalerate	184	23
	3-hydroxypyruvate	35	80
	glyoxylate	31	12
	2-oxo-3-methylbutanoate	5.5	11
Oxidative deamination	L-alanine	188	18.9
	L-2-aminobutyrate	4.1	330
	L-serine	2.6	39
	L-norvaline	0.26	14
	L-valine	0.13	20

alanine from pyruvate has been reported using alanine dehydrogenase and a coenzyme-regeneration system [28]. Wandrey et al. used the alanine dehydrogenase reaction to study a continuous process with three enzymes yielding L-alanine from D,L-lactate [29]. A space-time yield of $134 \text{ g} \cdot \text{l}^{-1} \cdot \text{d}^{-1}$ could be reached, producing 184 mmol L-alanine from 400 mmol D,L-lactate. Later, similar substrate-coupled cofactor-regeneration systems were studied for the production of L-leucine and L-phenylalanine from their corresponding racemic 2-hydroxyacids [11].

2.3. L-Leucine dehydrogenase

In strains of Bacillus high levels of leucine dehydrogenase are found, besides alanine dehydrogenase. It has been suggested that, by the action of this enzyme, growing cells are supplied with precursors for branched-chain fatty acids [30].

Several analytical applications of leucine dehydrogenase are described: the determination of branched-chain amino acids and their oxo analogs [31], and, in coupled assays, the determination of leucine aminopeptidase activity [32]. A possible therapeutic application of leucine dehydrogenase is indicated by the finding of Oki et al. [33] that crystalline leucine dehydrogenase reveals antineoplastic activity against Ehrlich ascites carcinoma in vivo.

2.3.1. Production of leucine dehydrogenase. Results of extensive screening procedures for leucine dehydrogenase activity among strains of Bacillus have been described repeatedly. Highly active mesophilic strains belong to B. sphaericus and B. cereus, from which the enzyme have been purified and characterized [34–36]. Although these enzymes originate from mesophilic organisms, all of them show remarkably high thermal stability and low deactivation rates during operation. Even more stable enzyme preparations can be obtained from thermophilic strains; B. sphaericus, B. stearothermophilus and three strains of Bacillus species with highly active leucine dehydrogenases have been found in a screening among thermophilic organisms [37, 38]. Rather low levels of leucine dehydrogenase occur in several strains of the thermophilic anaerob Clostridium thermoaceticum [39]. After gene cloning and expression in Escherichia coli, the E. coli extract showed a 900-fold higher leucine dehydrogenase activity than the extract from C. thermoaceticum. Table 2 summarizes the highly active strains derived by several screening procedures. The relative thermostability of the enzyme can be utilized for the purification. A large-scale isolation of the enzyme from B. cereus starting with 30 kg cells has been described by Schütte et al. [40]. Heat treatment of the crude extract and liquid-liquid extraction with aqueous two-phase systems resulted in a 16-fold enrichment and gave a product sufficient for technical applications. After three further chromatographic steps, a homogeneous enzyme preparation could be obtained with an overall yield of 48% and an 95-fold enrichment [40].

2.3.2. Biochemical characterization of leucine dehydrogenases. As summarized in Table 3, four leucine dehydrogenases from different bacterial sources have been purified and characterized. They have similar enzymological properties, although differences concerning the molecular mass, the temperature stability and the pH optimum point to a different protein structure. In contrast to the leucine dehydrogenases from B. sphaericus and B. stearothermophilus, the enzymes from B. cereus [41] and B. caldolyticus [42] are composed of eight identical subunits. For the B. cereus enzyme the octameric structure was confirmed by electron microscopy [41]. The kinetic data inportant for the application of leucine dehydrogenase are nearly identical for the enzymes described

Table 2. *Strains with highly active leucine dehydrogenase obtained by different screening procedures*

Strain	Temperature requirement	Enzyme activity U/mg	Reference
Bacillus sphaericus	mesophilic	0.42	[31]
Bacillus cereus	mesophilic	0.25	[40]
Bacillus stearothermophilus	thermophilic	0.15	[38]
Bacillus spec.	thermophilic	0.47	[37]
Bacillus spec.	thermophilic	0.72	[37]
Bacillus sphaericus	thermophilic	0.67	[37]
Clostridium thermoaceticum	thermophilic	0.01	[39]
E. coli C 600 – pJCD 242	mesophilic	9.4	[39]

Table 3. *Biochemical data of several leucine dehydrogenases*
The specific activity of the highly purified enzyme was measured for the reductive amination of 2-oxo-4-methylpentanoate, as was the pH optimum. Molecular mass was estimated by gel filtration

Parameter	Value for enzyme from			
	Bacillus sphaericus	B. cereus	B. stearothermophilus	Clostridium thermoaceticum
Reference	[34]	[40]	[38]	[39]
Specific activity (U/mg)	150	210	120	115
pH optimum	9.5	9.0–9.2	9.7	9.5
K_m (mM) for leucine	1.0	1.5	4.5	8.5
valine	1.7	2.5	3.9	6.8
isoleucine	1.8	1.0	1.4	3.9
2-oxo-4-methylpentanoate	0.31	0.45	–	0.8
2-oxo-3-methylbutanoate	1.40	2.1	–	4.0
2-oxo-3-methylpentanoate	–	9.0	–	–
K_m (µM) NADH	35	34	–	25
NAD	390	340	490	610
Molecular mass (kDa)	245	310	300	350
Subunits	6	8	6	6
Stable for 30 min at		50°C	65°C	75°C

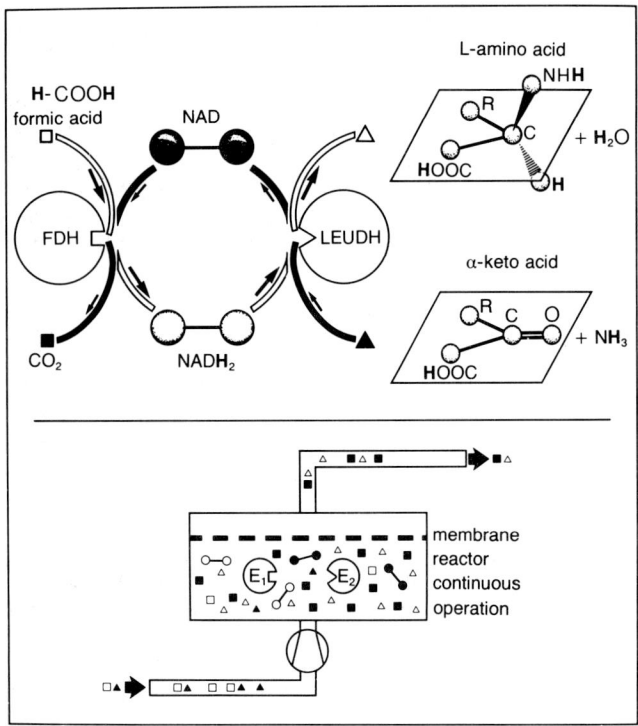

Fig. 1. *Continuous production of L-amino acids in a membrane reactor with leucine dehydrogenase (LeuDH, E_1) for the reductive amination of the oxo acid and formate dehydrogenase (FDH, E_2) for NADH regeneration.* (From [11])

except for the temperature stability. This higher stability can be advantageous in continuous processes, since the half-life of the catalysts influence the economy of the process. Besides a higher activity at higher temperatures for enzymes from thermophiles, they also show longer life at low temperatures. Ohshima et al. compared the continuous leucine production applying both a mesophilic and a thermophilic enzyme [43].

2.3.3. Use of leucine dehydrogenase for the synthesis of aliphatic L-amino acids. Several aliphatic L-amino acids have been prepared using leucine dehydrogenase in an enzyme-membrane reactor as shown in Fig. 1. For the continuous synthesis of L-leucine in an enzyme-membrane reactor, two routes concerning the coenzyme regeneration have been investigated: NADH regeneration from formate and the substrate-coupled route starting with the racemic mixture of D,L-2-hydroxy-4-methylpentanoate [11]. As emphasized in Table 4, the regeneration via formate/formate dehydrogenase offers some advantages in respect to the conversion and the space-time yield. In the substrate-coupled route the chemically more stable hydroxyacids can replace more expensive oxo acids. The substrate-coupled route appears attractive for the production of L-methionine from the racemic hydroxy acid analogues, which are cheaply available [44]. The reductive amination of trimethylpyruvate by leucine dehydrogenase seems to be economically feasable and scaling up this process to the 10-kg range has been successfully carried out. This optically active compound containing a bulky *tert*-butyl group is employed as a chiral inductor for chemical snythesis, for example of amino acids, according to Schöllkopf.

Table 4. *Continuous production of L-amino acids with the aid of dehydrogenases in an enzyme-membrane reactor*

Amino acid dehydrogenase	Enzyme for regeneration	Precursor[a]	Product	Product concn	Conversion	Space-time yield	Enzyme consumption	Reference
				mmol/l	%	g l^{-1} d^{-1}	U/kg	
Leucine DH	formate DH	oxomethyl-pentanoate	L-leucine	80	80	250	300 (LeuDH) 300 (FDH)	[11]
Leucine DH	D-HmpDH[b] L-HmpDH	DL-hydroxy-methylpentanoate (oxomethyl-pentanoate)	L-leucine	70	70	72	730 (LeuDH) 350 (D-HmpDH) 650 (L-HmpDH)	[12]
Leucine DH	D-HmpDH L-HmpDH	DL-hydroxy-methionine (oxomethyl-pentanoate)	L-methionine	240	60	143		[11]
Leucine DH	formate DH	trimethylpyruvate	L-Me$_3$-leucine	425	85	640	1000 (LeuDH) 2000 (FDH)	[12]
Alanine DH	L-lactate DH D-lactate DH	DL-lactate (pyruvate)	L-alanine	184	46	134	4700 (L-LacDH) 2600 (D-LacDH)	[11]
Phenylalanine DH	formate DH	phenylpyruvate	L-phenyl-alanine	114	95	456	1500 (PheDH) 150 (FDH)	[53]
Phenylalanine DH	D-HmpDH L-HmpDH	DL-phenyllactate	L-phenyl-alanine	22	43	28	—	[115]
Phenylalanine DH + ACA acylase	formate DH	acetamido-cinnamate	L-phenyl-alanine	70	88	277	1170 (acylase) 1770 (PheDH) 400 (FDH)	[54]

[a] Compounds in parentheses are necessary in trace amounts.
[b] HmpDH = 2-hydroxy-4-methylpentanoate dehydrogenase 'HicDH' 2-hydroxyisocaproate dehydrogenase.

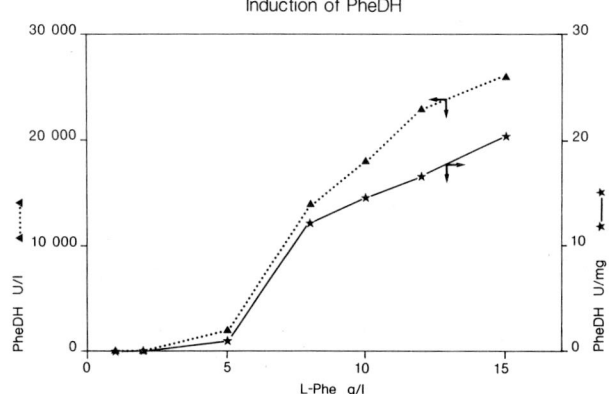

Fig. 2. *Induction of phenylalanine dehydrogenase of* Rhodococcus *species by* L-*phenylalanine.* (∗——∗) Specific activity; (▲····▲) volume activity

Table 5. *Comparison of phenylalanine dehydrogenase from* Brevibacterium *and* Rhodococcus *species*

Parameter	Brevibacterium	Rhodococcus
Microbiological data:		
enzyme yield (U/l) after addition of 1% of		
L-phenylalanine	210	15 200
L-histidine	120	1 800
L-phenylalaninamide		3 500
L-isoleucine	0	0
D-phenylalanine	204	0
DL-phenylalanine	214	0
Enzymological data:		
pH optimum		
reductive amination	9.0	9.25
oxidative deamination	10	10
K_m (mM)		
phenylpyruvate	0.11	0.16
p-hydroxy-phenylpyruvate	0.24	2.4
indolepyruvate	8.0	7.7
2-oxo-4-methylmercaptobutyrate	3.0	2.1
V_{max} (relative to phenylpyruvate)		
phenylpyruvate	100	100
p-hydroxy-phenylpyruvate	96	5
indolepyruvate	24	3
2-oxo-4-methylbutyrate	59	33
K_m (µM) NADH	47	130
K_m (mM) NH_4^+	431	387
Stability		
stored at 4°C ($t_{1/2}$)	4–8 h	10 d
deactivation (%/d) under operation	26	5
Reference	[51]	[46]

2.4. L-Phenylalanine dehydrogenase

None of the known glutamate, alanine of leucine dehydrogenases from different bacterial sources accepts 2-oxo acids possessing an aromatic side chain as substrate, although this type of compound was assayed routinely in the course of alanine or leucine dehydrogenase screening. The rising demand for L-phenylalanine as a building block of the dipeptide sweetener Aspartame has especially stimulated the development of several enzymatic routes for its synthesis. A suitable enzyme, L-phenylalanine dehydrogenase, performes the reductive amination of phenylpyruvate and was originally described by Hummel et al. [45] in a strain of *Brevibacterium* species. Later, the enzyme was found in *Rhodococcus* sp. [46], *Sporosarcina ureae* [47] and *Bacillus* sp. [48].

2.4.1. Isolation of phenylalanine dehydrogenase. Recently, we described a method for the selective enrichment of phenylalanine-dehydrogenase-producing microorganisms, resulting in a larger variety of bacteria with suitable dehydrogenases [49]. Especially those strains belonging to *Rhodococcus* sp. show remarkably high specific and volume activity and, in contrast to the enzyme from *Brevibacterium,* the *Rhodococcus* enzyme is relatively stable. The oxidative deamination catalyzed by phenylalanine dehydrogenase is the first step in the degradation of L-phenylalanine. The enzyme can be induced to yield 25 units/mg (in the crude extract) by addition of 10 g L-phenylalanine/l medium. This concentration has been optimized for growth in batch cultures as given in Fig. 2. The time curve of the enzyme production during batch fermentation reveals that the enzyme is deactivated or degraded after prolonged cultivation. A high enzyme yield can be obtained by harvesting the batch just after exhaustion of the L-phenylalanine. A frequent or continuous analysis of L-phenylalanine, e. g. by the flow injection analysis technique, therefore proved to be the best way to obtain high enzyme levels [50]. Concerning the purification of the enzyme, for technical applications it is sufficient to apply two enrichment steps after cell disintegration: extraction into the top phase of an aqueous two-phase system composed of poly(ethylene glycol), potassium phosphate and sodium chloride, followed by a second extraction to shift the enzyme into the bottom phase. A DEAE-cellulose chromatography step results in an enzyme preparation of more than 1000 units/mg. Due to the marked instability of the *Brevibacterium* enzyme in the crude extract, it is difficult to obtain and study the purified protein. Stored at 4°C, the half-life of the enzyme in the crude extract ranges between 4–8 h. Under operational conditions, the deactivation proved to be lower, about 25% per day. Thus, the stability was sufficient for the continuous production of L-phenylalanine. The presence of the coenzyme, substrate, product and/or poly(ethylene glycol) seems to be responsible for the stabilization of the enzyme.

2.4.2. Biochemical characterization of phenylalanine dehydrogenase. The two phenylalanine dehydrogenases derived from *Brevibacterium* and from *Rhodococcus* have been characterized extensively. Table 5 summarizes the kinetic data which are important for the application of the enzyme. The main differences between the two enzymes concern the specific activity, the activity against *p*-hydroxyphenylpyruvate, and the stability.

2.4.3. Application of phenylalanine dehydrogenase. Both the phenylalanine dehydrogenase from *Brevibacterium* and that from *Rhodococcus* have been used successfully for the continuous production of L-phenylalanine in a membrane reactor. With phenylpyruvate as substrate, nearly complete conversion can be reached. The activity of both enzymes was examined over a period of 12 days (d) and showed a mean productivity of 37.4 g · l^{-1} · d^{-1} applying the enzyme from *Brevibacterium* [51] and of 456 g · l^{-1} · d^{-1} for the enzyme from *Rhodococcus* [46]. Because of the instability of phenylpyruvate in aqueous solutions and its relatively high cost, two alternative routes have been studied. One starts from the racemic mixture of

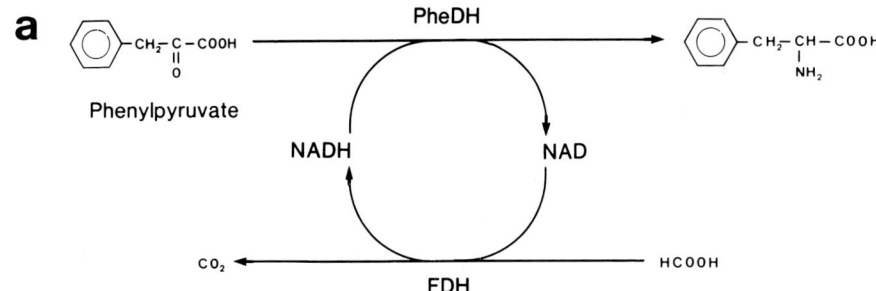

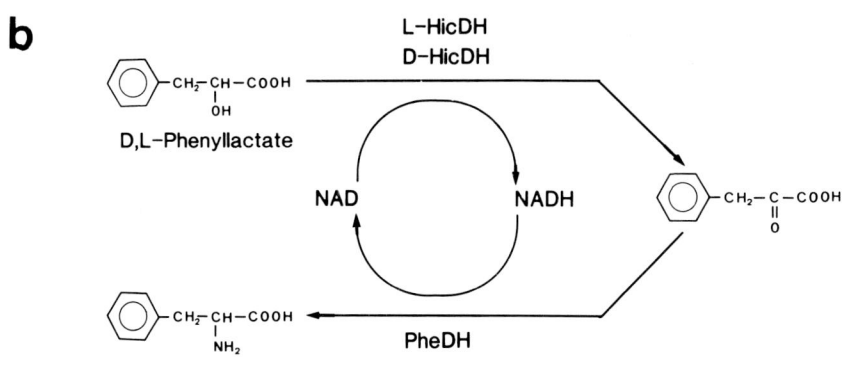

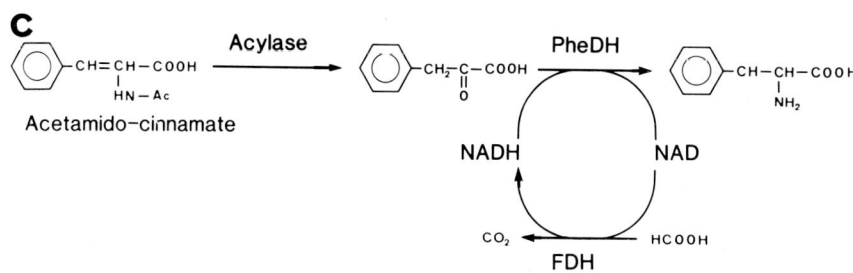

Fig. 3. *Enzymatic routes for the preparation of L-phenylalanine.* (a) Reductive amination of phenylpyruvate by phenylalanine dehydrogenase (PheDH) with simultaneously NADH regeneration using formate dehydrogenase (FDH). (b) Oxidation of DL-phenyllactate with D- and L-2-hydroxy-4-methylpentanoate dehydrogenase (HicDH) and simultaneous reductive amination of the *in situ* formed phenylpyruvate with phenylalanine dehydrogenase (PheDH). NADH is 'substrate-coupled' regenerated from phenyllactate. (c) *In situ* formation of phenylpyruvate by enzymatic deacetylation of acetamidocinnamic acid (Acylase) followed by simultaneous reductive amination with phenylalanine dehydrogenase (PheDH)

phenyllactate, the other from acetamidocinnamic acid (Fig. 3). In both routes, phenylpyruvate is formed *in situ* and converted simultaneously by the action of phenylalanine dehydrogenase to L-phenylalanine. The conversion of D,L-phenyllactate into the oxo acid can be achieved utilizing the side reaction of two enzymes, D- and L-2-hydroxy-4-methyl-pentanoate dehydrogenase (2-hydroxycaproate dehydrogenase). NADH is regenerated continuously by the substrate oxidation. The kinetic properties of the enzyme involved in the cyclic reaction make this approach unfavourable. Acetamidocinnamic acid is another stable precursor of phenylpyruvate. Deacetylation results in an unstable enamine-imine derivative, which hydrolysis spontaneously to yield phenylpyruvate. The deacetylation can be accomplished enzymatically by an acylase isolated from a strain of *Brevibacterium* sp. [52–54]. In this route, formate dehydrogenase is necessary for coenzyme regeneration. Table 4 summarizes the results of the continuous production of L-phenylalanine by the different enzyme-catalysed routes.

3. HYDROXY ACID DEHYDROGENASES

3.1. General properties

NADH-dependent hydroxy acid dehydrogenases allow the synthesis of chiral hydroxy acids from their corresponding oxo acids.

2-Oxo acid + NADH / H^+ ⇌ 2-Hydroxy acid + NAD^+.

These optically active products are valuable compounds for the production of various pharmaceuticals such as semisynthetic penicillins [55], cephalosporins [56] and anti-obesity compounds [57]. In contrast to amino acids, which can usually be synthesized by fermentation or biotransformation, hydroxy acids other than lactate are not known as primary or secondary metabolites of microorganisms. However, chemical synthesis of the racemate, followed by its resolution using crystallization, chromatography or microbial conversions [58–61], have been described. Only for D-mandelic acid has the biotransformation of benzoylformate by a fermentative *Leuconostoc dextranicus* been published [62]. The D- and L-lactic acids are usually produced by fermentation with such high yields and low costs that enzymatic processes are unattractive. Because of the very narrow substrate specificity of D- and L-lactate dehydrogenase for pyruvate, these enzymes have no importance for preparative applications. Two general approaches are possible to obtain 2-hydroxy acid dehydrogenases with altered substrate range: conventional screening or protein engineering. A first example of the latter is given by Clarke et al. [63]. Using site-directed mutagenesis on the lactate dehydrogenase gene from *Bacillus stearothermophilus*, they changed the substrate specificity from the lactate/pyruvate pair to the oxaloacetate/malate pair by three amino acid substitutions in the substrate binding site. Certainly, such rational attempts will be intensified in the future. Screening among certain genera during the last few years has revealed several new 2-hydroxy acid dehydrogenases which accept various straight and branched-chain aliphatic and aromatic substrates. These enzymes are called 2-hydroxy-4-methylpentanoate ('hydroxyisocaproate') or 2-hydroxyphenylacetate (mandelate) dehydrogenase with reference to the substrate with the lowest K_m value and will be described below. The enzymes have been found in many strains of Lactobacillaceae besides NADH-dependent or NADH-independent lactate dehydrogenases. At present the physiological function of these enzymes is unclear. The presence of lactate dehydrogenase at high levels in the fermentative Lactobacilli is obviously necessary, because this reaction enables the organisms to regenerate NAD.

3.2. 2-Hydroxy acid dehydrogenases

Most lactate dehydrogenases accept only pyruvate and to a minor degree the next homologue 2-oxobutyrate. Recently, we described a D-lactate dehydrogenase from *Lactobacillus confusus* [64], which also converts phenylpyruvate, (K_m: pyruvate 0.68 mM, oxobutyrate 7.2 mM, phenylpyruvate 3 mM). This stable enzyme can be obtained in relatively high yield and may be applied for the synthesis of D-phenyllactate. For the synthesis of other chiral 2-hydroxy acids, the newly discovered NAD-dependent hydroxy acid dehydrogenase are available.

3.2.1. Production of 2-hydroxy acid dehydrogenases. The choice of an appropriate organism is the first step of the enzyme production and isolation. Four different enzymes are known today, which were isolated and characterized from Lactobacillaceae: the L-2-hydroxy-4-methylpentanoate dehydrogenase from *Lactobacillus confusus* [65], the D-2-hydroxy-4-methylpentanoate dehydrogenase from *Lactobacillus casei* [66] and the D-mandelate dehydrogenases from *Lactobacillus curvatus* [67] and *Streptococcus faecalis* [68]. Extensive screening procedures among *Lactobacillus* strains showed no systematic occurrence of such an enzyme activity. The results of a screening for NAD- and NADP-dependent 2-hydroxy acid dehydrogenases among strains of the *Lactobacillus* subgenus *Thermobacterium* are given in Table 6. In six of the eight strains tested the presence of the desired activity was demonstrated. Perhaps the inactive strains possess

Table 6. *Screening for 2-hydroxyacid dehydrogenase activity among the subgenus* Thermobacterium *of the genus* Lactobacillus

Species	NAD-dependent activity against		
	oxomethyl-pentanoate	phenyl-pyruvate	benzoyl-formate
	U/mg		
L. acidophilus	0	0	0
L. helveticus	0.08	0.02	0.03
L. bulgaricus	0.11	0.20	0.17
L. lactis	1.03	0	0
L. delbrueckii	0.06	0.06	0.08
L. leichmannii	0.12	0.22	0.18
L. salivarius	0.07	0.21	0.01
L. jensenii	0	0	0

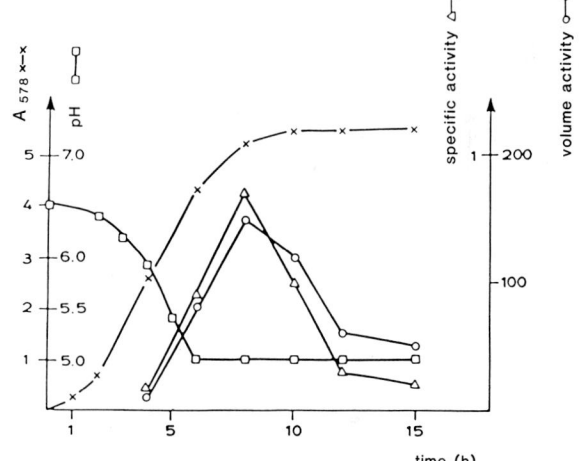

Fig. 4. *Production of D-2-hydroxy-4-methylpentanoate dehydrogenase during growth of* Lactobacillus casei *on a 10-l scale.* (△——△) Specific activity; (○——○) volumetric activity

hydroxy acid dehydrogenase which depend on other coenzymes. Harvesting of batch cultures for the production of hydroxy acid dehydrogenase is a crucial step. This is demonstrated in Fig. 4, presenting a time curve for the enzyme production, typically for all the hydroxy acid dehydrogenases. The specific activity (units/mass protein) and content (units/volume of culture) of D-2-hydroxy-4-methylpentanoate dehydrogenase from *Lactobacillus casei* increases during the growth phase of the organism followed by a rapid deactivation or degradation. In order to obtain a maximum enzyme content of the cells, harvest of the culture has to be carried out within a very small time interval. At present, enzyme production cannot be followed directly on-line to determine the best harvest point. Hence, harvesting was carried out after defined times, a procedure which requires a high standardization of inoculation and treatment of the medium. For large-scale isolation and enrichment of the enzymes, extraction in aqueous two-phase systems has been applied, followed by ion-exchange chromatography. These enrichment steps are sufficient for an enzyme preparation suitable for technical applications. In the case of mandelate dehydrogenase from *L. curvatus*, these steps alone yield a nearly homogeneous protein (700-fold enriched, 2100 U/mg). The other enzymes have been purified further by applying affinity chromatography on Blue

Table 7. *Biochemical data for 2-hydroxy acid dehydrogenases*

Parameter	Lactobacillus confusus	Lactobacillus casei	Streptococcus faecalis	Lactobacillus curvatus
Chirality of product	L	D	D	D
pH optimum, reduction	7.0	5.5–7.0	4.5[a]	6.0[a]
oxidation	8.0–8.5	8.0–9.0	9.2[a]	8.0[a]
Specificity activity (U/mg)[b]	479	110	897	2122
K_m for oxoacids (µM):				
2-oxo-4-methylpentanoate	60	60	700	90
2-oxo-3-methylpentanoate	500	2200	500	95
2-oxopentanoate	100	110	3300	170
phenylpyruvate	120	150	7000	150
K_m for hydroxy acids (mM):				
DL-2-hydroxy-4-methylpentanoate	0.7	1.4		1.1
DL-2-hydroxypentanoate	1.8	0.7		13.0
DL-phenyllactate	0.65	1.1		2.5
K_m NADH (µM)	33	10	35	36
K_m NAD (µM)	330	500		210
Molecular mass (kDa)	125	74	72	60
Subunits	4	2	2	2
Reference	[65]	[66]	[68]	[67]

[a] Determined with benzoylformate or DL-mandelate, respectively.
[b] Highest specific activity; more than 95% pure.
[c] Determined by gel filtration.

Sepharose or 5′AMP-Sepharose (L-2-hydroxy-4-methyl-pentanoate dehydrogenase from *L. confusus*) or ion-exchange chromatography on Amberlite CG 50 (D-2-hydroxy-4-methylpentanoate dehydrogenase from *L. casei*).

3.2.2. Biochemical characterization of hydroxy acid dehydrogenases. The main biochemical parameters important for the application of these enzymes are listed in Table 7. It is remarkable that pyruvate and lactate show very high K_m values if they are converted at all. Hence, it is certain that 2-hydroxy acid dehydrogenase activity is not a side reaction of a lactate dehydrogenase. In the case of *L. confusus*, both enzymes have been separated and characterized. The resulting products are of opposite chirality, D-lactate from D-lactate dehydrogenase and L-lactate from L-hydroxymethylpentanoate dehydrogenase. The analysis of the protein structure and kinetic data of the known enzymes reveals marked differences. The enzyme from *L. confusus* is unique in its L-specificity. In addition, only L-hydroxymethylpentanoate dehydrogenase exhibits a tetrameric structure while all the D-hydroxy acid dehydrogenases described are dimers of identical subunits. Despite extensive screening efforts, no other L-hydroxy acid dehydrogenase producer has been found as yet. The enzyme from *L. casei* shows unexpected differences regarding the activity against the closely related 2-oxo-4-methylpentanoate ('oxoleucine') and 2-oxo-3-methylpentanoate ('oxoisoleucine'). The latter is accepted with a 37-fold weaker affinity, whereas the enzymes from other sources show similar K_m values. The enzyme from *L. casei* is remarkably stable during storage at different pH values, more than 80% of the activity could be recovered between pH 3.3 and 9.7. The enzymes differ with regard to the sensitivity against 0.1 mM Hg^+ or Hg^{2+}. The enzymes from *Streptococcus faecalis* and *L. casei* were found to be nearly insensitive against a pretreatment with Hg ions, whereas the other D-specific enzymes are completely deactivated. For the *L. casei* enzyme it could be confirmed meanwhile that the enzyme contains no cysteine [68a].

4. ALCOHOL DEHYDROGENASES

Alcohol dehydrogenase is widely distributed in nature and has been found in many microorganisms, plant and animal tissues [69].

$$R-\underset{\underset{O}{\|}}{C}-R' + NAD(P)H + H^+ \rightleftharpoons R-\underset{\underset{OH}{|}}{CH}-R' + NAD(P)^+$$

Depending on the biological source, alcohol dehydrogenases show different substrate specificities; NAD and NADP-dependent enzymes are known. Preparative applications are limited by some enzyme instability and the narrow substrate specificity of most of the known alcohol dehydrogenases [70–72]. Two commercially available, rather inexpensive, alcohol dehydrogenases are often used by organic chemists for laboratory-scale synthesis, one isolated from yeast (YADH) and the other from horse liver (HLADH). Due to the limited stereoselectivity of both enzymes, direct reduction of the carbonyl function is seldom described; as a rule, these enzymes are used for resolution of the racemic hydroxy compound, followed by separation of the unreacted optically active compound. An example for the application of HLADH is the preparation of (1S,2R)-*cis*-2-carboxymethyl-3-cyclopenten-1-ol lactone [73], an interesting starting compound for prostaglandin synthesis (Fig. 5). After oxidation of the (1R,2S)-enantiomer from the racemic mixture by HLADH, the unreacted (1S,2R)-alcohol can be isolated and chemically oxidized.

HLADH is probably the best investigated enzyme for such applications. Acyclic, mono- bi- and even tetracyclic (steroid) structures are accepted by this enzyme. Based on a large number of substrates tested, Prelog [73] developed the 'diamond lattice' model for this enzyme, which enables the confident prediction of the reactivity and stereochemistry of a new substrate. For cyclic ketones, this model has subsequently been slightly altered [74, 75]. Only acyclic secondary alcohols

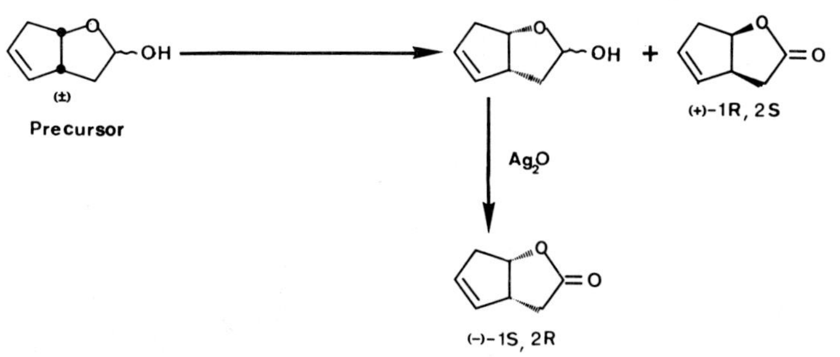

Fig. 5. *HLADH-catalyzed preparation of (1S, 2R)-cis-2-carboxymethyl-3-cyclopenten-1-ol lactone, a starting compound for prostaglandin synthesis* [73]

are poor substrates. An elegant way to circumvent this limitation is described by Jones [76, 77], who converted the acyclic substrate into a heterocyclic derivative, followed by subsequent enzymatic transformation to the desired acyclic product.

YADH is of limited use for the preparation of chiral products because of its rather narrow substrate specificity [69]. One of the substituents of the substrate carbonyl groups must necessarily be either hydrogen (aldehyde) or a methyl group. Hence, cyclic compounds are excluded as substrates for YADH.

For preparative applications YADH and HLADH have several disadvantages: they are unstable at temperatures above 30 °C [75], quite sensitive to organic solvents, and tend to loose their activity during immobilization. Therefore, many other alcohol dehydrogenases have been identified and tested for preparative applications. The groups of Prelog and Dutler investigated appropriate enzymes from *Curvularia falcata*, *Mucor javanicus*, and pig liver [78, 79]. The *Curvularia* enzyme has been studied with six-membered ring ketones and alcohols as substrates. As with HLADH, (S)-alcohols are formed, whereas the *Mucor* enzyme exhibits the opposite stereospecificity, accepting a rather broad spectrum of structures [80].

In recent years alcohol dehydrogenases have been isolated from thermophilic microorganisms. These enzymes are especially interesting because not only are they active at higher temperatures, but they are found to be more stable at moderate temperatures and in the presence of organic solvents [81, 82]. The group of Rossi reported an NAD-dependent alcohol dehydrogenase isolated from the archaebacterium *Sulfolobus solfataricus* [83]. After a 280-fold purification, the enzyme had a specific activity of 3.9 U/mg. It has a broad substrate specificity that includes primary and secondary alcohols and ketones of linear and cyclic structure; anisaldehyde shows the lowest K_m value. The thermal stability can be demonstrated by the half-life of the enzyme, which is 20 h at 60 °C and 5 h at 70 °C.

An NADPH-dependent alcohol dehydrogenase has been isolated from *Thermoanaerobium brockii* (*Tbr*ADH) by Lamed and Zeikus [84]. After four purification steps, the enzyme has a specific activity of 100 U/mg (at 40 °C). The activity was not influenced by heating at 65 °C for 70 min. The highest activity was found for the conversion of secondary alcohols; the activity decreased with acetaldehyde, linear ketones, cyclic ketones or primary alcohols as substrates.

Meanwhile, several applications have been described using this enzyme for synthetic purposes, e.g. the gram-scale conversion of NADP to NADPH applying 0.3% 2-propanol as the reductant with a coenzyme recycling number of 20 000 [85] or the reduction of the 2-ketone to the chiral (S)-(+)-2-pentanol in the presence of 2-propanol [85], erroneously assigned as (R)-(+)-2-pentanol [86]. Ketones possessing a C_5 chain (such as 2-pentanone) or larger give the (S)-alcohol with excellent optical yields [86]. This reversal of the enzyme stereospecificity has also been observed for other enzymes [87, 88] and can be explained by the structure of the active site in the enzyme. A small and a larger site is postulated; in the case of 2-butanone, for example, the ethyl group occupies the 'large' site, but in 3-heptanone the same group is forced into the 'small' site resulting in an inverted selectivity (Fig. 6).

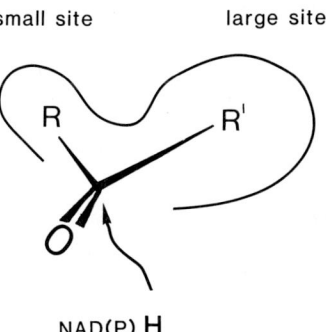

Fig. 6. *Scheme of the active site of* Thermoanaerobium brockii *alcohol dehydrogenase*. (Redrawn from [90])

Diketones are further interesting substrates for alcohol dehydrogenases. Ketoesters with 1,5-located carbonyl function can be reduced by the *Tbr*ADH enzyme, optically pure (S)-lactones are obtained spontaneously under reaction conditions out of the 5-hydroxyester [89]. Chiral aliphatic chloroalcohols are valuable building blocks that can be prepared from the corresponding chloroketones with *Tbr*ADH [89]. *Tbr*ADH can be immobilized successfully. Coupling to crosslinked agarose via cyanogen bromide as well as to epoxide-containing resins resulted in active enzyme preparations that showed no apparent loss in activity during 30 days of operation [90].

In the case of *Tbr*ADH, the dependence on NADPH is no limitation, because the coenzyme can be regenerated simultaneously by the same enzyme employing an excess of 2-propanol. Concentrations up to 30% isopropanol in water have been tested and resulted in only partial loss of activity after 34 h at 50 °C. Detailed kinetic analysis of these processes is still missing.

Table 8. *Substrate specificity of acetoin dehydrogenase from* Lactobacillus kefir
Inactive substrates are 3-oxobutyric ethyl ester, acetoin, benzoylformate, and acetophenone. V_{max} values are given relative to diacetyl = 100

Substrate	V_{max}	K_m
		mM
2,3-Butandione (diacetyl)	100	310
2,3-Pentandione	550	
2,4-Pentandione	8	50
2,5-Hexandione	7	26
Pyruvate	104	0.057
Pyruvate ethyl ester	109	2.4
1-Phenyl-1,2-propandione	12	
Phenylpyruvate	17	0.48

5. DIKETONE REDUCTASES

The selective reduction of diketones to hydroxy-ketones is difficult to achieve.

$$R\text{-}\underset{\underset{O}{\|}}{C}\text{-}\underset{\underset{O}{\|}}{C}\text{-}R' + NAD(P)H + H^+ \rightleftharpoons R\text{-}\underset{\underset{O}{\|}}{C}\text{-}\underset{\underset{OH}{|}}{C}\text{-}R' + NAD(P)^+ .$$

Commonly diacetyl-reducing enzymes carry out the conversion of the substrates up to 2,3-butandiol as is known for the enzyme activities from *Streptococcus diacetilactis* [91] or from *Klebsiella pneumonia* [92, 93]. For acetoin, the transformation of pyruvate to (−)-acetoin by acetone-dried cells of *K. pneumoniae* with a low product yield (0.55 g/l) is described [94]. An enzymatic route was described by Lee and Whitesides [95], utilizing the broad substrate specificity of commercial available glycerol dehydrogenase (from *Cellulomonas* sp. or *Enterobacter aerogenes*) for the enantio-selective oxidation of diols. NAD was regenerated with diaphorase, methylene blue and dioxygen. Probably due to product inhibition, the reaction stopped when 30% of the *meso*-substrate had been consumed.

Recently, we carried out a screening procedure among strains of the genus *Lactobacillus* in order to obtain NADH-dependent enzymes which convert diacetyl to acetoin. *L. kefir* was found to be a suitable strain. The heat stability of the enzyme can be exploited for enrichment. After two chromatographic steps, a specific activity of 1061 U/mg could be obtained. Considering the K_m values for different substrates, the enzyme probably acts as a D-lactate dehydrogenase in cells, but is able to convert a broad variety of 1,2- and 1,3-diketones, often with nearly the same reaction velocity as with pyruvate (Table 8). For acetoin, the enantiomeric specificity could be determined by chiral GC (Fig. 7).

6. ENZYMES FOR COENZYME REGENERATION

The majority of the presently applied oxidoreductases depends on nicotinamide coenzymes. The use of these enzymes as catalysts for large-scale enzyme-catalysed synthesis of chiral products requires efficient procedures for *in situ* regeneration of the coenzyme. Isolation and separate regeneration is much too expensive. In addition, a continuously operating process is desirable for industrial applications to utilize the expensive coenzyme and enzymes extensively as long as they are active.

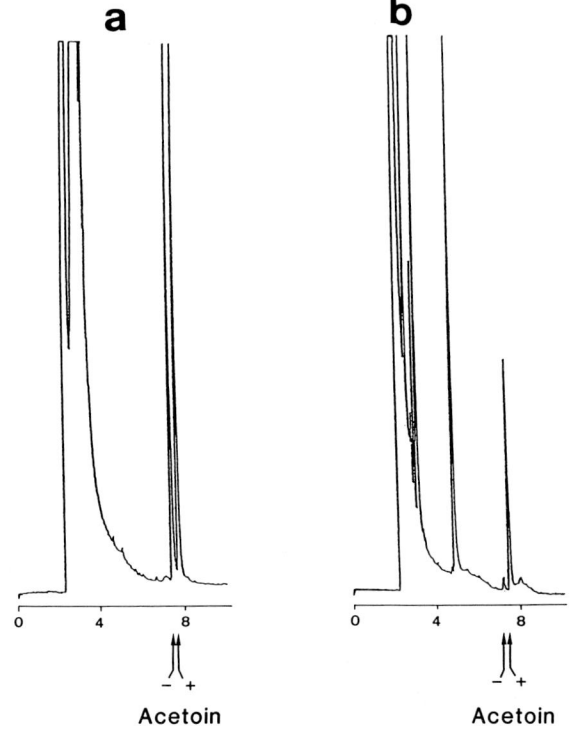

Fig. 7. *Chiral gas chromatographic analysis of acetoin.* (a) Racemic mixture; (b) product of the enzyme-catalyzed conversion (purified acetoin dehydrogenase from *Lactobacillus kefir*)

Enzyme-membrane reactors can advantageously be employed for multienzyme systems performing the catalysis in homogeneous phase and retaining the enzyme and (modified) coenzyme in the reactor by an ultrafiltration membrane. Since the native coenzyme would be washed through the membrane together with products and unconverted substrates, it is covalently linked usually to a water-soluble polymer through the exocyclic amino group of the adenine ring. In this way a retention ≥99.9% by an ultrafiltration membrane can be achieved. Poly(ethylene glycol) appears to be the best choice as water-soluble polymer. An efficient chemical synthesis of the modified coenzyme has been described [96]. Limitations of this strategy may arise due to altered affinities of the coenzyme derivatives towards the dehydrogenases. In most cases studied so far K_m values varied within one order of magnitude, V_{max} was even less effected. The observation that the modified coenzyme gives 1.5-fold higher activity with formate dehydrogenase is surprising. On the negative side, glucose dehydrogenase from *Bacillus* species exhibits more than 1000-fold higher K_m values, which makes this enzyme unattractive for coenzyme regeneration in an enzyme-membrane reactor.

For the selection of an appropriate regeneration system, the following requirements have to be taken into consideration. The reaction should be thermodynamically favoured, neither the co-product nor the unreacted co-substrate of the regeneration reaction should interfere with the enzyme or with the isolation method for the desired product. Regarding the coenzyme, by-product formation should be negligible, a demand that rules out most chemical methods.

Fig. 8 summarizes principles approaches towards the regeneration of a coenzyme, demonstrated here for the case of NAD(P)H-dependent reactions. Method one can be applied in general; it requires a second enzyme and a second substrate.

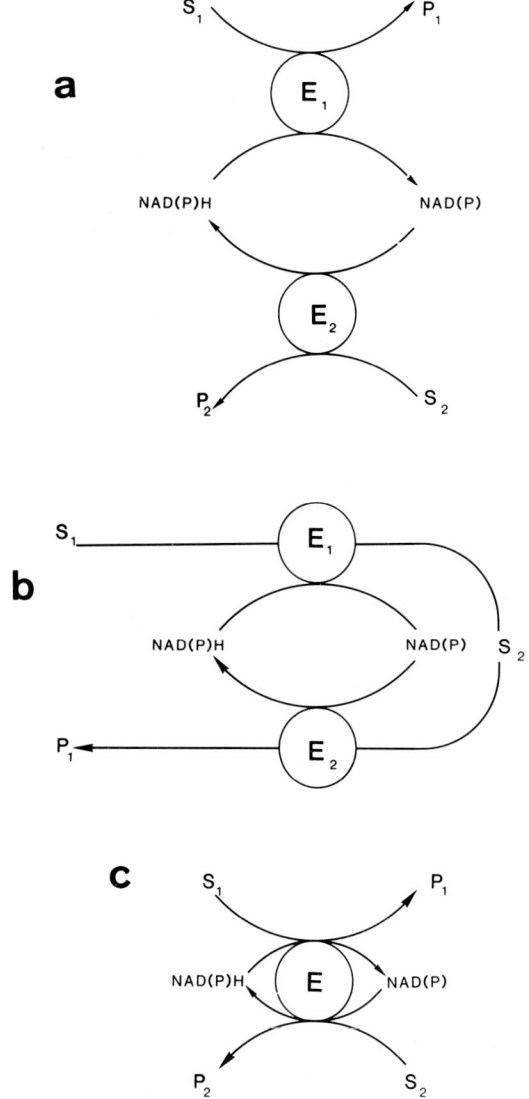

Fig. 8. *Different method for regeneration of a coenzyme.* (a) Regeneration applying a second enzyme (E_2) and a second substrate (S_2). (b) 'Substrate-coupled' [11] regeneration of the coenzyme. The substrate (S_2) of the desired product (P_1) is formed *in situ* from the precursor S_1; both reactions (enzyme E_1 and enzyme E_2) depend on the same coenzyme. (c) 'Coupled substrate approach' [11], utilizing the same enzyme (E) for the reduction of the substrate S_1 to the desired product P_1 as well as for the coenzyme regeneration [NAD(P)H] oxidizing S_2 to P_2.

As a general NADH-regenerating enzyme, we utilize formate dehydrogenase from *Candida boidinii* [97, 98]. The enzyme is expressed at high levels during growth on methanol. The equilibrium of the reaction catalysed by formate dehydrogenase strongly favours CO_2 formation and reduction of NAD^+ to NADH. This property simplifies kinetic modelling of coupled reactions. Besides, formate is a cheap co-substrate and neither formate nor CO_2 influence enzyme or product stability. Commercial formate dehydrogenase is still quite expensive on an activity basis but it can be isolated easily and cheaply on a large scale [99, 100]. Therefore, the present price of the enzyme does not reflect the ultimate cost of the technical catalyst. NADH regeneration by formate dehydrogenase has reached the most advanced stage of development. Up to 600 000 mol product have been produced/mol coenzyme lost in the process [51]. Formate dehydrogenase from *C. boidinii* does not accept NADP as substrate, however. For NADPH regeneration, glucose-6-phosphate dehydrogenase from *Leuconostoc mesenteroides* [101] as well as glucose dehydrogenase from *Bacillus cereus* [102, 103] have been investigated successfully. Both enzymes are rather inexpensive, highly active (>250 U/mg) and stable. The reactions proved to be nearly irreversible due to the hydrolysis of the gluconolactone or 6-phosphogluconolactone, respectively. Applications for gram-scale enzyme-catalyzed synthesis have been reported utilizing immobilized enzymes and native coenzyme, but continuous processes have not yet been developed. Besides, glucose 6-phosphate is too expensive as co-substrate in most cases, even for synthesis on a molar scale.

Alcohol dehydrogenases are not particularly attractive, in general, as regenerating enzymes since ethanol but especially acetaldehyde accumulating to rather high concentrations deactivate enzymes rapidly. The specificity of chemical methods for NAD(P)H regeneration is not sufficient [104, 105].

The reverse reaction, the regeneration of the oxidized coenzymes, is needed for the synthesis of ketones or the resolution of racemic hydroxy compounds. It is a more difficult problem due to the fact that oxidation reactions often show severe product inhibition and the equilibrium of most dehydrogenase reactions in the oxidation direction are unfavourable. A nonenzymatic method for the regeneration of NAD is described by Jones et al. [106, 107] adding flavin mononucleotide (FMN) and O_2 as the ultimate oxidizing agent. Ketones on a gram-scale have been prepared by this method [108]. An improvement of this reaction is the procedure of Drueckhammer et al. [109], employing the comercially available FMN reductase from *Photobacterium fischerei*. Catalase is added to destroy peroxides which are formed during the oxidation of $FMNH_2$. Considering the oxidation of NADH, addition of 10 mg FMN reductase/l increased the productivity from 17 to 400 mmol/$l^{-1} d^{-1}$ under otherwise identical conditions [109]. A detailed comparison of three methods for the regeneration of NAD^+ in gram-scale preparations has been presented by Lee and Whitesides [110]; reactions have been carried out using the 2-oxoglutarate/glutamate dehydrogenase system, the methylene blue/diaphorase/O_2 method and ferricyanide/diaphorase. The use of the secondary alcohol dehydrogenases in the presence of acetone for the reoxidation of the coenzyme has not yet been described but appears possible.

The second method, the substrate-coupled regeneration, is characterized by the fact that the substrate for the desired reaction emerges as product from a preceding reaction, both utilizing the same coenzyme. To catalyze both halves of the cycle efficiently, appropriate enzymes must be available. Especially in the case of unstable precursors (S_2 in Fig. 8), this kind of *in situ* liberation is advantageous. Wandrey et al. developed such a three-enzyme process for the conversion of the corresponding racemic hydroxy acid into L-methionine, L-alanine or L-leucine, respectively [11]. The only by-product of this reaction is water. It should be noted that there is a stoichiometric link between amino acid and NAD^+ formation. Therefore escape of the intermediate oxo acid across the membrane before reductive amination will eventually stop continuous coenzyme cycling. This can be prevented by supplying enough oxo acid with the substrate stream to compensate for wash out and keep the coenzyme cycling.

The third method utilizes the same enzyme simultaneously for the formation of the desired product and for the coenzyme regeneration. Lamed et al. [90] applied it to reactions catalyzed by alcohol dehydrogenase (*Tbr*ADH) in which excess isopropanol (S_2 in Fig. 8) is oxidized to acetone ('coupled substrate approach'). This kind of process becomes possible due to the stability of *Tbr*ADH in organic solvents. In principle, the coenzyme remains in the same active site and acts more as a prosthetic group with alternating oxidation and reduction cycles. Transport of the coenzyme between different enzymes as in method 1 or 2 is not necessary in this case, because oxidation and reduction of the coenzyme occur at the same site. Hence, it will only be a small step to consider a coupling of the coenzyme directly to the enzyme with the chance to increase its catalytic activity and to reduce the amount of the cofactor [110a].

7. CONCLUSIONS

Dehydrogenase can be considered as the first group of coenzyme-dependent enzymes that are applicable for synthetic organic chemistry on a technical scale. Due to the general regeneration technique via formate dehydrogenase, NADH-dependent reduction reactions leading to valuable optically active compounds can easily be scaled-up and also appear economically feasable [111]. The continuously operating membrane reactor offers the advantage that a long-term use of the different enzymes is possible and limited only by their thermal stability. NADPH-dependent reactions and oxidations requiring NAD(P)-regeneration techniques are well investigated but only used on a laboratory scale. Product isolation and the regeneration steps are more complicated and expensive. Fortunately, a variety of NADH-dependent enzymes are available catalyzing a broad substrate spectrum; for example, the D-2-hydroxy-4-methylpentanoate dehydrogenases from *Lactobacillus* species or the acetoin dehydrogenase from *L. kefir*. Specific enzyme screening procedures [112] and efficient large-scale enzyme purification steps [113, 114] enable the preparation of further enzymes useful for stereoselective synthesis.

We would like to thank Prof. Wandrey for fruitful and stimulating discussions. Work of the authors referred to in this review was supported within the biotechnology program of the *Bundesministerium für Forschung und Technologie* (PTB 03 8409) and by Degussa AG.

REFERENCES

1. Wang, S. S. & King, C.-K. (1979) *Adv. Biochem. Eng.* 12, 119 – 146.
2. Boyer, P. D. (ed.) (1975) *The enzymes*, 3rd edn., vol. XI, Academic Press, New York.
3. Bergmeyer, H. U. (ed.) (1985) *Methods in enzymatic analysis*, Verlag Chemie, Weinheim.
4. Findeis, M. A. & Whitesides, G. M. (1984) *Annu. Rep. Med. Chem.* 19, 263 – 272.
5. Whitesides, G. M. & Wong, C.-H. (1983) *Aldrichimica Acta* 16, 27 – 34.
6. Jones, J. B. (1985) in *Asymmetric synthesis* (Morrison, J. D., ed.) Academic Press, New York.
7. Whitesides, G. M. & Wong, C.-H. (1985) *Angew. Chem. Int. Eng. Ed.*, 617 – 638.
8. Simon, H., Bader, J., Günther, H., Neumann, S. & Thanos, J. (1985) *Angew. Chem.* 97, 541 – 555.
9. Kula, M.-R. & Wandrey, C. (1987) *Methods Enzymol.* 136, 9 – 21.
10. Wichmann, R., Wandrey, C., Bückmann, A. F. & Kula, M.-R. (1981) *Biotechnol. Bioeng.* 23, 2789 – 2802.
11. Wandrey, C. (1986) in *Enzymes as catalysts in organic synthesis* (Schneider, M. P., ed.) pp. 263 – 284, D. Reidel, Dordrecht.
12. Wandrey, C. & Bossow, B. (1986) *Biotechnol. Bioind.* 3, 8 – 13.
13. Greenaway, W. & Whatley, F. R. (1975) *J. Labelled Cpd.* 11, 395 – 400.
14. O'Conner, R. J. & Halvorson, H. (1961) *Arch. Biochem. Biophys.* 91, 290 – 299.
15. Nitta, Y., Yasuda, Y., Tochikubo, K. & Hachisuka, Y. (1974) *J. Bacteriol.* 117, 588 – 592.
16. Hong, M. M., Shen, S. C. & Braunstein, A. E. (1959) *Biochim. Biophys. Acta* 36, 288 – 289.
17. Zink, M. W. & Sanwal, B. D. (1962) *Arch. Biochem. Biophys.* 99, 72 – 77.
18. McCowen, S. M. & Phibbs, P. V. Jr (1974) *J. Bacteriol.* 118, 590 – 597.
19. Germano, G. J. & Anderson, K. E. (1968) *J. Bacteriol.* 96, 55 – 60.
20. Holmes, P. K., Dundas, I. E. & Halvorson, H. O. (1965) *J. Bacteriol.* 90, 1159 – 1160.
21. Ohashima, T. & Soda, K. (1979) *Eur. J. Biochem.* 100, 29 – 39.
22. Yoshida, A. & Freese, E. (1964) *Biochim. Biophys. Acta* 92, 33 – 43.
23. Yoshida, A. & Freese, E. (1965) *Biochim. Biophys. Acta* 96, 248 – 262.
24. Lebeault, J. M., Zévaco, C. & Hermier, J. (1970) *Bull. Soc. Chim. Biol.* 52, 1073 – 1088.
25. McCormick, N. G. & Halvorson, H. O. (1964) *J. Bacteriol.* 87, 68 – 74.
26. Nishida, Y., Sato, T., Tosa, T. & Chibata, I. (1979) *Enzyme Microb. Technol.* 1, 95 – 99.
27. Jandel, A.-S., Hustedt, H. & Wandrey, C. (1982) *Eur. J. Appl. Microbiol. Biotechnol.* 15, 59 – 63.
28. Wandrey, C. (1984) *Forum Mikrobiologie 7, Sonderheft Biotechnologie*, 2789 – 2802.
29. Wandrey, C., Fiolitakis, E., Wichmann, R. & Kula, M.-R. (1984) *Ann. NY Acad. Sci.* 443, 91 – 94.
30. Obermeier, N. & Poralla, K. (1976) *Arch. Microbiol.* 109, 59 – 63.
31. Ohshima, T., Misono, H. & Soda, K. (1978) *Agric. Biol. Chem.* 42, 1919 – 1922.
32. Takamiya, S., Ohshima, T., Tanizawa, K. & Soda, K. (1983) *Agric. Biol. Chem.* 47, 893 – 895.
33. Oki, T., Shirai, M., Ohshima, T., Yamamoto, T. & Soda, K. (1973) *FEBS Lett.* 33, 286 – 288.
34. Ohshima, T., Misono, H. & Soda, K. (1978) *J. Biol. Chem.* 253, 5719 – 5725.
35. Hummel, W., Schütte, H. & Kula, M.-R. (1981) *Eur. J. Appl. Microbiol. Biotechnol.* 12, 22 – 27.
36. Zink, M. W. & Sanwal, B. D. (1962) *Arch. Biochem. Biophys.* 99, 72 – 77.
37. Ohshima, T., Wandrey, C., Sugiura, M. & Soda, K. (1985) *Biotechnol. Lett.* 7, 871 – 876.
38. Ohshima, T., Nagata, S. & Soda, K. (1985) *Arch. Microbiol.* 141, 407 – 411.
39. Shimoi, H., Nagata, S., Esaki, N., Tanaka, H. & Soda, K. (1987) *Agric. Biol. Chem.* 51, 3375 – 3381.
40. Schütte, H., Hummel, W., Tsai, H. & Kula, M.-R. (1985) *Appl. Microbiol. Biotechnol.* 22, 306 – 317.
41. Lünsdorf & Tsai, H. (1985) *FEBS Lett.* 193, 261 – 266.
42. Kärst, U., Schütte, H., Baydoun, H. & Tsai, H. (1987) *Proc. 4th Eur. Congr. Biotechnol.* 2, 220 – 223.
43. Ohshima, T., Wandrey, C., Kula, M.-R. & Soda, K. (1985) *Biotechnol. Bioeng.* 27, 1616 – 1618.
44. Tichy, S., Vasić-Rački, Dj., Schütte, H., Talsky, G., Wandrey, C. (1987) *Chem. Biochem. Eng. Q1*, 25 – 30.
45. Hummel, W., Weiß, N. & Kula, M.-R. (1984) *Arch. Microbiol.* 137, 47 – 52.
46. Hummel, W., Schütte, H., Schmidt, E., Wandrey, C. & Kula, M.-R. (1987) *Appl. Microbiol. Biotechnol.* 26, 409 – 416.
47. Asano, Y. & Nakazawa, A. (1985) *Agric. Biol. Chem.* 49, 3631 – 3632.
48. Asano, Y. (1987) *J. Biol. Chem.* 262, 10346 – 10354.

49. Hummel, W., Schmidt, E., Schütte H. & Kula, M.-R. (1987) in *Biochemical engineering* (Chmiel, H., Hammes, W. P. & Bailey, J. E., eds) G. Fischer, Stuttgart, New York.
50. Nalbach, U., Schimenz, H., Stamm, W., Hummel, W. & Kula, M.-R. (1988) *Anal. Chim. Acta 213*, 55–60.
51. Hummel, W., Schmidt, E., Wandrey, C. & Kula, M.-R. (1986) *Appl. Microbiol. Biotechnol. 25*, 175–185.
52. Hummel, W., Schmidt, E., Schütte, H. & Kula, M.-R. (1987) *Proc. 4th Eur. Congr. Biotechnol.*, 171–172.
53. Hummel, W., Schütte, H., Schmidt, E. & Kula, M.-R. (1987) *Appl. Microbiol. Biotechnol. 27*, 283–291.
54. Schmidt, E., Hummel, W. & Wandrey, C. (1987) *Proc. 4th Eur. Congr. Biotechnol.* 189–191.
55. Furlenmeier, A., Quitt, P., Vogler, K. & Lanz, P. (1976) US Patent 3957758.
56. Bickel, H., Kocsis, K. & Peter, H. (1975) Ger. Offen. 2514912.
57. Mills, J., Schmiegel, K. K. & Shaw, W. N. (1983) US Patent 4391826.
58. Kesslin, G. & Kelly, K. W. (1982) US Patent 4322548.
59. Muto, M. (1984) Eur. Pat. Appl. EP 98 707.
60. Mori, K. & Akao, H. (1980) *Tetrahedron 36*, 91–99.
61. Sawada, T., Ogawa, M., Ninomiya, R., Yokose, K., Fujiu, M., Watanabe, K., Suhara, Y. & Maruyama, H. (1983) *Appl. Environ. Microbiol. 45*, 884–891.
62. Kanegafuchi Chemical Ind. (1982) JP 57,198,096.
63. Clarke, A. R., Smith, C. J., Hart, K. W., Wilks, H. M., Chia, W. N., Lee, T. V., Birktoft, J. J., Banazak, L. J., Barstow, D. A., Atkinson, T. & Holbrook, J. J. (1987) *Biochem. Biophys. Res. Commun. 148*, 15–23.
64. Hummel, W., Schütte, H. & Kula, M.-R. (1983) *Eur. J. Appl. Microbiol. Biotechnol. 18*, 75–85.
65. Schütte, H., Hummel, W. & Kula, M.-R. (1984) *Appl. Microbiol. Biotechnol. 19*, 167–176.
66. Hummel, W., Schütte, H. & Kula, M.-R. (1985) *Appl. Microbiol. Biotechnol. 21*, 7–15.
67. Hummel, W., Schütte, H., Kula, M.-R. (1988) *Appl. Microbiol. Biotechnol. 28* 433–439.
68. Yamazaki, Y. & Maeda, H. (1986) *Agric. Biol. Chem. 50*, 2621–2631.
68a. Kallwaß, H. (1988) Thesis, Technical University Braunschweig.
69. Sund, H. & Theorell, M. (1963) in *The enzymes* (Boyer, P. D., Lardy, H. & Myrback, K., eds) 2nd edn, vol. VII, pp. 25–83, Academic Press, New York.
70. Lok, K. P., Jakovak, I. J. & Jones, J. B. (1985) *J. Am. Chem. Soc. 107*, 2521–2526.
71. Legoy, M. D., Kim, H. S. & Thomas, D. (1985) *Process Biochem. 20*, 145–148.
72. Keinan, E., Seth, K. K. & Lamed, R. (1986) *J. Am. Chem. Soc. 108*, 3474–3480.
73. Prelog, V. (1984) *Pure Appl. Chem. 9*, 119–130.
74. Irwin, A. J. & Jones, J. B. (1977) *J. Am. Chem. Soc. 99*, 1625–1630.
75. Jones, J. B. & Beck, J. F. (1976) *Techn. Chem. 10*, 107–401.
76. Takemura, T. & Jones, J. B. (1983) *J. Org. Chem. 48*, 791–799.
77. Davies, J. & Jones, J. B. (1979) *J. Am. Chem. Soc. 101*, 5405–5410.
78. Rétey, J. (1963) Thesis 3409, ETH Zürich.
79. Hochuli, E. (1974) Thesis 5284, ETH Zürich.
80. Lemière, G. L. (1986) in *Enzymes as catalysts in organic synthesis* (Schneider, M. P., ed.) pp. 19–34, D. Reidel Publishing Company, Dordrecht.
81. Fontana, A. (1984) in *Theromophilic enzymes and their potential use in biotechnology*, pp. 221–232, Dechema, Weinheim.
82. Friedmann, S. M. (ed.) (1978) *Biochemistry of thermophiles*, Academic Press, New York.
83. Rella, R., Raia, C. A., Pensa, M., Pisani, F. M., Gambacorta, A., de Rosa, M. & Rossi, M. (1987) *Eur. J. Biochem. 167*, 475–479.
84. Lamed, R. & Zeikus, J. G. (1981) *Biochem. J. 195*, 183–190.
85. Lamed, R., Keinan, E. & Zeikus, J. G. (1981) *Enzyme Microb. Technol. 3*, 144–148.
86. Keinan, E., Hafeli, E. K., Seth, K. K. & Lamed, R. (1986) *J. Am. Chem. Soc. 108*, 162–169.
87. Zhou, B., Gopalan, S., van Middelsworth, F., Shieh, W. R. & Sih, C. J. (1983) *J. Am. Chem. Soc. 105*, 5925–5926.
88. Sabbioni, G., Shea, M. L. & Jones, J. B. (1984) *J. Chem. Soc. Chem. Commun.*, 236–238.
89. Keinan, E., Seth, K. K. & Lamed, R. (1986) *J. Am. Chem. Soc. 108*, 3474–3480.
90. Keinan, E., Seth, K. K. & Lamed, R. (1987) *Ann. N. Y. Acad. Sci. 501*, 130–149.
91. Seitz, E. W., Sandine, W. E., Elliker, P. R. & Day, E. A. (1963) *Can J. Microbiol. 9*, 431–439.
92. Bryn, K., Hetland, O. & Stormer, F. C. (1971) *Eur. J. Biochem. 18*, 116–119.
93. Hetland, O., Olsen, B. R., Christensen, T. B. & Stormer, F. C. (1971) *Eur. J. Biochem. 20*, 200–205.
94. Berl, S. & Bueding, E. (1951) *J. Biol. Chem. 191*, 401–411.
95. Lee, L. G. & Whitesides, G. M. (1986) *J. Org. Chem. 51*, 26–36.
96. Bückmann, A. F., Morr, M. & Kula, M.-R. (1987) *Biotechnol. Appl. Biochem. 9*, 258–268.
97. Schütte, H., Floßdorf, J., Sahm, H. & Kula, M.-R. (1976) *Eur. J. Biochem. 62*, 151–160.
98. Shaked, Z. & Whitesides, G. M. (1980) *J. Am. Chem. Soc. 102*, 7104–7108.
99. Kroner, K. H., Schütte, H., Stach, W. & Kula, M.-R. (1982) *J. Chem. Tech. Biotechnol. 32*, 130–137.
100. Cordes, A. & Kula, M.-R. (1986) *J. Chromatogr. 376*, 375–384.
101. Wong, C.-H. & Whitesides, G. M. (1981) *J. Am. Chem. Soc. 103*, 4890–4899.
102. Wong, C.-H. & Drueckhammer, D. G. (1985) *Bio/Technology 3*, 649–651.
103. Wong, C.-H., Drueckhammer, D. G. & Sweers, H. M. (1985) *J. Am. Chem. Soc. 107*, 4028–4031.
104. Jones, J. B. (1980) in *Enzymic and nonenzymic catalysis* (Dunnill, P., Wiseman, A. & Blakebrough, N., eds) pp. 58–83, Ellis Harwood Ltd, Chichester.
105. Baricos, W., Chambers, R. & Cohen, W. (1975) *Enzyme Technol. Dig. 2*, 39–53.
106. Jones, J. B. & Taylor, K. E. (1976) *Can. J. Chem. 54*, 2969–2973.
107. Jones, J. B. & Taylor, K. E. (1976) *Can. J. Chem. 54*, 2974–2980.
108. Jakovac, I. J., Goodbrand, H. B., Lok, K. P. & Jones, J. B. (1982) *J. Am. Chem. Soc. 104*, 4659–4665.
109. Drueckhammer, D. G., Riddle, V. W. & Wong, C.-H. (1985) *J. Org. Chem. 50*, 5387–5389.
110. Lee, L. G. & Whitesides, G. M. (1985) *J. Am. Chem. Soc. 107*, 6999–7008.
110a. Mosbach, K. (1984) *Ann. N. Y. Acad. Sci. 434*, 239–248.
111. Wandrey, C. (1987) *Proc. 4th Eur. Congr. Biotechnol. 4*, 171–188.
112. Schütte, H., Hummel, W. & Kula, M.-R. (1985) *Anal. Biochem. 151*, 547–553.
113. Hustedt, H., Kroner, K. H. & Kula, M.-R. (1985) in *Partitioning in aqueous two-phase systems: Methods, uses and application to biotechnology* (Walter, H., Brooks, D., Fisher, D., eds) pp. 529–587, Academic Press, Orlando.
114. Kula, M.-R. (1987) in *Protein purification: micro to macro*, pp. 99–115, Alan R. Liss, New York.

Review

Chemical model systems for drug-metabolizing cytochrome-*P*-450-dependent monooxygenases

Daniel MANSUY, Pierrette BATTIONI and Jean-Paul BATTIONI

Laboratoire de Chimie et Biochimie Pharmacologiques et Toxicologiques, Unité Associée au Centre National de la Recherche Scientifique en Développement Concerté avec l'Institut National de la Santé et de la Recherche Médicale, no. 400, Université René Descartes, Paris

(Received January 18, 1989) — EJB 89 0077

Monooxygenases are widely distributed enzymes which catalyze dioxygen activation using two electrons and two protons, with the insertion of one oxygen atom from O_2 into a substrate and the formation of water (Eqn 1).

$$RH + O_2 + 2e^- + 2H^+ \longrightarrow ROH + H_2O \quad (1)$$

A great number of these monooxygenases contain a heme protein called cytochrome *P*-450 which is the site of dioxygen activation. These cytochrome-*P*-450-dependent monooxygenases are involved in many steps of the biosynthesis and biodegradation of endogenous compounds such as steroids, fatty acids, prostaglandins and leukotrienes. They also play a key role in the oxidative metabolism of exogenous compounds such as drugs and other environmental products allowing their elimination from living organisms. Because of their wide distribution in living organisms and their very important role in biochemistry, pharmacology and toxicology, cytochromes *P*-450 have been the subject of many studies during the last 20 years. Thus, much is known about the structure and function of cytochromes *P*-450 (for recent reviews, see [1–4]) and more than 60 cytochromes *P*-450 from various origins have been sequenced [5].

However, because of the high molecular mass of cytochromes *P*-450 (about 50 kDa), it is still difficult to determine the detailed mechanism of substrate oxidations and the nature of the iron intermediates involved in these processes. In particular, it is difficult to determine the molecular structure of the iron-metabolite complexes formed from time to time during the metabolism of some classes of substrates. A possible way to solve these problems is to use biomimetic chemical systems involving iron-porphyrins.

In fact, iron-porphyrin model systems for cytochromes *P*-450 may be used for three main purposes. The first is to determine the nature of the iron complexes involved as intermediates in the catalytic cycle of dioxygen activation and substrate hydroxylation by cytochrome *P*-450. Model iron-porphyrin complexes for all these intermediates except one have been prepared and completely characterized by various spectroscopic techniques. Their structures have been established by X-ray analysis. The spectroscopic characteristics of these model complexes have been found to be very similar to those of the corresponding enzymatic complexes and the use of these models has largely contributed to our present detailed knowledge of the catalytic cycle of cytochrome *P*-450. This first interest of the use of models has been described in previous reviews (see for instance [6–9]) and will not be treated in this article.

A second objective of the use of chemical models for cytochromes *P*-450 is to understand as much as possible about the mechanisms of cytochrome *P*-450 during these reactions. In particular, the preparation and complete characterization of models for the cytochrome-*P*-450—iron—metabolite complexes formed during the metabolism of some drugs or exogenous compounds should allow us to determine the chemical functions responsible for the formation of such inhibitory complexes which have important consequences in pharmacology and toxicology.

A third objective in this field is to build up catalytically active chemical systems able to reproduce the main reactions catalyzed by cytochromes *P*-450. This is an important challenge in the field of homogeneous catalysis for the discovery of new catalysts for selective hydroxylation of alkanes or aromatic compounds under mild conditions. This is also an important challenge in the field of drug metabolism, as an efficient chemical model catalyst for cytochrome *P*-450 would allow one to prepare large amounts of oxidized metabolites of a given drug or of a given agrochemical, for instance. Moreover, the use of such a model catalyst would allow one to predict to a certain extent not only the pattern of the oxidized stable metabolites of a drug which are of interest in pharmacology, but also the possible formation of reactive metabolites which are of interest in toxicology.

The use of chemical models for the second and third objectives indicated above will be the subject of sections II and III of this article. The first section will be devoted to a brief survey of what is known on the main characteristics of the enzymatic systems that can be reproduced by chemical models.

I. THE MAIN CHARACTERISTICS OF THE REACTIONS CATALYZED BY CYTOCHROMES *P*-450: WHAT IS TO BE MIMICKED?

I-1. NATURE OF THE ACTIVE SITE OF CYTOCHROMES *P*-450

The active site of all cytochromes *P*-450 contains iron(III)-protoporphyrin IX. The iron is linked to the apoprotein by a

Correspondence to D. Mansuy, Laboratoire de Chimie et Biochimie Pharmacologiques et Toxicologiques, Unité Associée au CNRS en Développement Concerté avec l'INSERM, no. 400, Université René Descartes, 45 rue des Saints Peres, F-75270 Paris, France

Abbreviations. TMP, tetramesitylporphyrin; TPP, *meso*-tetraphenylporphyrin; TDCPP, tetra-2,6-dichlorophenylporphyrin.

cysteine residue. The existence of this cysteinate-iron bond in cytochromes *P*-450 from various sources was deduced from several spectroscopic studies [1–9]. It was recently confirmed by X-ray analysis of a bacterial cytochrome *P*-450 from *Pseudomonas putida* grown on camphor (*P*-450$_{cam}$) [10, 11]. This cytochrome catalyzes the hydroxylation of camphor in position 5-*exo*. The X-ray structure of its complex with camphor gives a clear picture of the active site, with the iron axially bound to the sulfur atom of cysteine 357. The camphor molecule is maintained close to the heme mainly by a hydrogen bond of the camphor oxo group with a tyrosine residue of the protein [10]. In this *P*-450 – substrate complex, the C-H bond in position 5-*exo* of camphor is very well located to receive an oxygen atom from the iron. So far, this is the only cytochrome *P*-450 for which X-ray data are available. However, from many other spectroscopic data obtained on other cytochromes *P*-450, it appears that the cysteine – iron-porphyrin part of the active site is always present. The various cytochromes *P*-450 should differ by the nature of the binding of their substrates to the protein active site. Ionic and hydrophobic interactions should be involved in addition to hydrogen bonds.

I-2. CATALYTIC CYCLE OF CYTOCHROME *P*-450
(FIG. 1).

In its resting state, two forms of cytochrome *P*-450 are in equilibrium: a hexacoordinate low-spin iron(III) complex bearing two axial ligands, a cysteine and presumably an OH-containing residue, and a pentacoordinate high-spin iron(III) complex with the cysteine as only axial ligand. The binding of a substrate which occurs, in general, on a protein-binding hydrophobic site close to the heme, leads to a shift of this equilibrium towards the pentacoordinate state. The high-spin enzyme-substrate complex is then reduced by one electron coming from NADPH via an electron transfer chain. The high-spin pentacoordinate ferrous complex derived from this step is able to bind many ligands such as CO, isocyanides, nitrogenous bases, phosphines and dioxygen. Binding of dioxygen to cytochrome *P*-450 Fe(II) leads to a relatively stable hexacoordinate low-spin complex. Model iron-porphyrin complexes for these four intermediates of the cytochrome *P*-450 catalytic cycle have been prepared and completely characterized by X-ray analysis (for reviews, see [6–9]). This has largely contributed to providing a detailed view of the variations of the coordination and spin state of the iron during the catalytic cycle of cytochrome *P*-450. The intermediate of the catalytic cycle, which is directly responsible for the transfer of an oxygen atom to substrates, is derived from a one-electron reduction of the ferrous-dioxygen intermediate. Unfortunately, this oxidizing species has a lifetime so short that, to date, it could not be studied by any spectroscopic technique. Our knowledge about its possible nature and about the final steps of the catalytic cycle is based on indirect evidence obtained from studies on the characteristics of the oxidation reactions, and on comparisons with better known active oxygen complexes derived from other hemoproteins or from iron-porphyrins. From all these data, the most likely mechanism for O_2 activation by cytochrome *P*-450 involves (a) a heterolytic cleavage of the O-O bond of a possible Fe(III)-O-O-H intermediate formed by one-electron reduction of the Fe(II)-O_2 complex, (b) the formation of a high-valent iron-oxo complex derived formally from a two-electron oxidation of the ferric state and the binding of an oxygen atom to the iron, and (c) the transfer of the oxygen atom of this iron-oxo complex to the substrate. Accordingly, single-oxygen-atom donors such as C_6H_5IO, H_2O_2 or $NaIO_4$ can replace O_2 and NADPH for the cytochrome-*P*-450-catalyzed oxidations of many substrates [1–4].

The use of iron-porphyrin models has been of great help in understanding the possible nature of the intermediate high-valency iron complexes involved in the catalytic cycle of peroxidases, catalases and cytochromes *P*-450. Actually, a high-valency porphyrin-iron-oxo complex corresponding formally to an Fe(V)=O structure has been prepared by reacting metachloroperbenzoic acid with Fe(TMP)(Cl) [12] and studied by ^{1}H-NMR, EPR, Mössbauer and EXAFS spectroscopy [13] (TMP = tetramesitylporphyrin). All its characteristics are compatible with a (porphyrin-radical cation) Fe(IV)=O structure and similar to those of horseradish peroxidase compound I. Later, analogous high-valency iron-oxo complexes were obtained from other iron-porphyrins and oxidants [14] (and references therein).

I-3. NATURE AND MECHANISMS
OF THE MAIN REACTIONS
CATALYZED BY CYTOCHROME-*P*-450-DEPENDENT
MONOOXYGENASES

The main monooxygenation reactions performed by the cytochrome-*P*-450 active-oxygen species are indicated in Fig. 1. This includes the hydroxylation of C-H bonds, the epoxidation of alkene double bonds and of aromatic rings, and the transfer of an oxygen to nitrogen atoms of amines and sulfur atoms of thioethers. Simplified mechanisms which are most generally admitted for these reactions are given in Fig. 2. Hydroxylation of C-H bonds is believed to occur in two steps: (a) abstraction of a hydrogen atom by the high-valency iron-oxo intermediate having a free-radical-like reactivity, and (b) oxidation of the intermediate free radical by the Fe(IV)-OH species inside the active site with transfer of its OH ligand, leading to the hydroxylated substrate. Since the second step is fast enough, the intermediate free radical has no chance to escape from the active site. However, in some rare cases, it has enough time to invert its configuration or to isomerize itself before being oxidized by Fe(IV)-OH ([15, 16] and references cited in [16]).

Several mechanisms have been proposed for the oxidation of double bonds of alkenes or of aromatic rings by the active oxygen species of cytochrome *P*-450. Intermediate free radicals or cations formed by addition of the Fe=O species to the double bond would undergo an intramolecular reaction with formation of the corresponding epoxide (Fig. 2). The same intermediate species could also react with one pyrrole nitrogen atom of the heme explaining the formation of green pigments (*N*-alkylated hemes) during the *in vivo* oxidation of terminal alkenes by cytochromes *P*-450 (see following section) [17]. Interestingly, the high-valency iron-oxo model complex formed by reaction of Fe(III)(TMP)(Cl) with ArCO$_3$H was found to be able to transfer its oxygen atom to alkenes with an almost quantitative formation of the corresponding epoxide [12].

It is noteworthy that cytochromes *P*-450 are also able to catalyze the reductions of some particular substrates especially under low dioxygen pressure or under anaerobic conditions. These reductions, which are considerably less important than monooxygenation reactions from a physiological point of view, are performed by cytochrome *P*-450 in its ferrous state. This intermediate is able to transfer electrons to easily reducible compounds such as CCl_4, nitroaromatics,

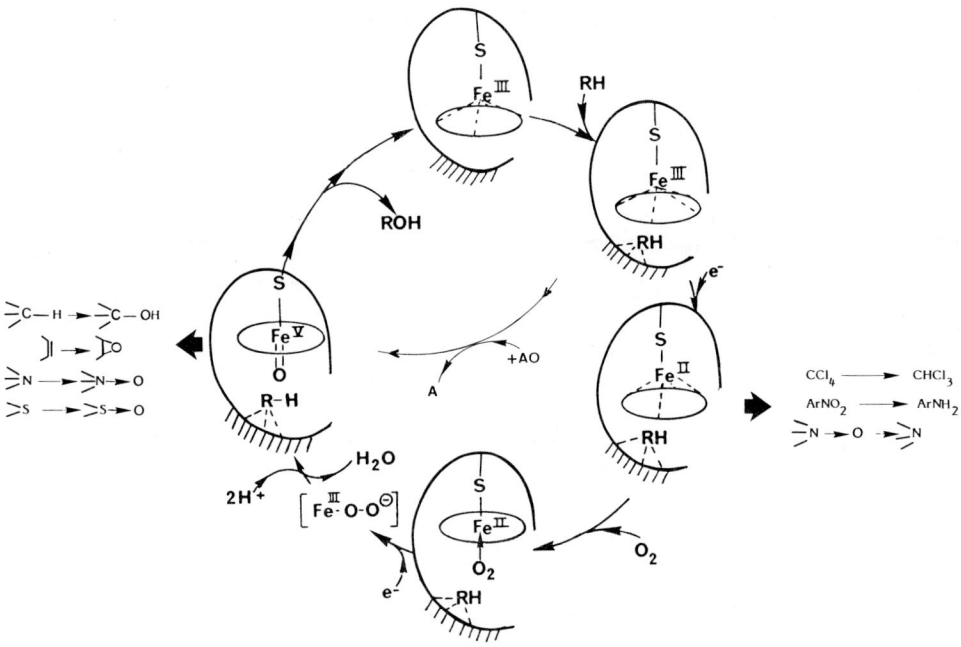

Fig. 1. *Catalytic cycle of cytochrome P-450: main oxidation and reduction reactions catalyzed*

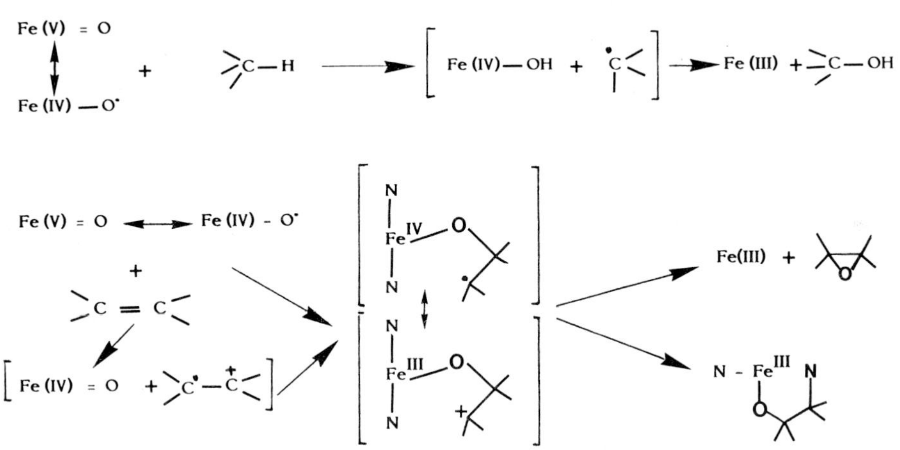

Fig. 2. *Generally admitted mechanisms for alkane hydroxylation and alkene (or arene) epoxidation catalyzed by cytochrome P-450*

tertiary amine oxides or arene oxides and to catalyze their reductions to $CHCl_3$, anilines, tertiary amines and arenes respectively [1–4] (Fig. 2).

I-4. CYTOCHROME *P*-450 CHARACTERISTICS THAT COULD BE REPRODUCED BY IRON-PORPHYRIN-MODELS

The various cytochromes *P*-450 which have been discovered and studied so far have several common characteristics, such as (a) the same cofactor (iron-protoporphyrin IX), (b) the same endogenous axial ligand of the iron (a cysteinate residue from the protein), (c) very similar catalytic cycles of dioxygen activation, and (d) in most cases, very similar mechanisms for the main classes of reactions that they catalyze (C-H bond hydroxylation, alkene epoxidation, aromatic ring hydroxylation, etc.). The various cytochrome *P*-450 isozymes differ mainly by their amino acid sequences and, in particular, by the nature of the amino acid residues of the polypeptide chain which is responsible for the binding of the substrate and for the positioning of this substrate relative to the iron catalytic site. Thus, it seems likely that the chemoselectivity of the oxidations catalyzed by cytochromes *P*-450 is mainly dependent on the cysteine-iron-porphyrin part which is common to all cytochromes *P*-450 and on the intrinsic reactivity of the active oxygen complex formed at the iron. In contrast, the substrate specificity and the regioselectivity of substrate oxidations by cytochromes *P*-450 appear mainly dependent on the nature of the amino acid residues from the protein which are in the active site and which may be very different in the various *P*-450 isozymes. Therefore, at least *a priori*, the likely objectives of the use of simple iron-porphyrin models are to mimic the formation of an active-oxygen species at the iron very similar to that found in cytochrome *P*-450 and to obtain a catalytic system having grossly the chemoselectivity of cytochromes *P*-450. However, it is clear that such simple iron-porphyrin models will not lead to the particular specific substrate recognition and oxidation regioselectivity due to the various protein active sites of cytochrome P-450 isozymes.

Simple iron-porphyrin models should be of great help, first, to study the detailed mechanisms of substrate oxidation, and, in particular to determine the structure of more-or-less stable reactive intermediates either free or bound to the iron; secondly to produce at the same time the various metabolites formed by oxidation of a substrate by the different P-450 isozymes.

Concerning this second point, it will be important to obtain catalytic rates and yields similar to those of cytochrome-P-450-dependent monooxygenases. Depending upon the nature of the substrate and of the cytochrome P-450 isozyme, these rates vary between 1 and 100 turnover/min, and the yields based on the reducing agent (NADPH or NADH) are most often high (between 50% and 100%).

II. THE USE OF HEME MODELS TO UNDERSTAND THE FORMATION OF IRON-METABOLITE COMPLEXES DURING THE METABOLISM OF SOME XENOBIOTICS

The formation of more or less stable cytochrome-P-450—iron-metabolite complexes has been found during the metabolim of various xenobiotics. In most cases, these complexes have been detected by ultraviolet — visible difference spectroscopy and possible structures have been proposed on the basis of several pieces of indirect evidence. Actually, it was always very difficult to determine clearly the structure of iron-metabolite complexes of such large hemeproteins. This section shows how the use of iron-porphyrin models, which are easily studied by various spectroscopic techniques including X-ray spectroscopy, was particularly helpful to determine the nature of these iron-metabolite complexes.

II-1. IRON-METABOLITE COMPLEXES CONTAINING AN IRON-NITROGEN BOND

Stable cytochrome-P-450—iron-metabolite complexes are formed during the oxidation of 1,1-dialkylhydrazines such as N-amino-piperidine by liver microsomes in the presence of NADPH and O_2 [18]. They are characterized by Soret peaks around 438 nm in their highest valency state (presumably ferric) and around 450 nm in their reduced state (presumably ferrous). It has been speculated that they were iron (III or II)-nitrene complexes, the R_2N-N amino-nitrene ligand being formed by in situ oxidation of the starting R_2NNH_2 by the active-oxygen species of cytochromes P-450 [18] (Eqn 2). The use of iron-porphyrin models led to an understanding of these reactions since it was found that the oxidation of a 1,1-dialkylhydrazine by PhIO or O_2 in the presence of an iron-porphyrin led to an iron-nitrene complex in high yields (Eqn 3) [19, 20]. This complex was isolated and completely characterized by several spectroscopic techniques (ultraviolet—visible, infrared, ^1H- and ^{13}C-NMR, EPR and X-ray spectroscopy). Its X-ray structure definitely proved that it contained an iron-nitrene bond. This showed that oxidation of a R_2NNH_2 derivative by the high-valency iron-oxo intermediate formed upon reaction between an iron-porphyrin and PhIO effectively led to an iron-nitrene complex. The results of these model studies strongly suggest that the iron-metabolite complexes observed during oxidation of 1,1-dialkylhydrazines by cytochrome P-450 are also iron-nitrene complexes.

$$\begin{matrix} R \\ R \end{matrix}\!\!>\!\!N-NH_2 \xrightarrow[NADPH, O_2]{P-450} [P-450-Fe(III) \leftarrow N-NR_2] \lambda \approx 438\text{nm}$$

$$-e^- \updownarrow +e^- \quad (2)$$

$$[P-450-Fe(II) \leftarrow N-NR_2] \lambda \approx 450\text{nm}$$

$$\begin{matrix}\bigcirc\\N-NH_2\end{matrix} + Fe(P)(Cl) \xrightarrow[\text{or PhIO}]{O_2} (P)Fe(II) \leftarrow N-N\begin{matrix}\bigcirc\end{matrix}$$

P = porphyrin   Fe-N = 180 pm   (3)
    when P = TpClPP

$$P-450Fe(III) + \begin{cases} RN(CH_3)_2 \\ \text{or } RNH_2 \\ \text{or } RNHOH \end{cases} \xrightarrow[O_2]{NADPH} [P-450Fe(II)-RNO] \xleftarrow[\text{or } S_2O_4Na_2]{NADPH} P-450Fe(III) + RNO_2 \quad (4)$$

$$RNHOH + Fe(III)(P)(Cl) \longrightarrow \left[(P)Fe(II) \leftarrow N\!\!<\!\!\begin{matrix}O\\R\end{matrix} \right] \quad (5)$$

Fe-N = 186 pm
when P = TPP, R = iPr

Very stable iron-metabolite complexes are also formed during the oxidative metabolism of various drugs containing an amine function such as amphetamines, propoxyphene, erythromycin and troleandomycin [21, 22]. Most of them are very stable so that cytochromes P-450 engaged in such complexes are strongly inhibited. Some of them, like those derived from the antibiotic troleandomycin, are very stable in vivo and have even been detected in the liver of humans treated with this antibiotic [23]. The strong inhibition of cytochromes P-450 due to the formation of these complexes would explain, at least in part, the adverse clinical effects observed when the antibiotic was associated with various drugs [24]. These complexes are clearly derived from metabolic oxidation of the amine function of the above-mentioned drugs, are P-450—Fe(II)-metabolite complexes characterized by Soret peaks around 455 nm, and are not stable when P-450 becomes ferric after oxidation by $Fe(CN)_6K_3$ for instance [21, 22]. Since they are equally formed during P-450-dependent oxidation of alkylhydroxylamines RNHOH and P-450-dependent reduction of the corresponding nitro compounds RNO_2 (Eqn 4), it has been proposed that they are derived from the binding of nitrosoalkane intermediates to P-450—Fe(II) [25—27]. Heme model studies completely confirmed this proposition, as reaction of various hydroxylamines RNHOH with Fe(III)-porphyrins, such as Fe(TPP)(Cl), (TPP = meso-tetraphenyl-porphyrin) led to new very stable complexes of the porphyrin-Fe(II)-RNO type [28]. These model complexes were also obtained by reduction of RNO_2 compounds by dithionite in the presence of iron-porphyrins [28] (Eqn 5). X-ray structure determination of one of them, Fe(TPP)(iPrNO)(iPrNH_2), showed that the nitrosoalkane ligand was bound to Fe(II) by its nitrogen atom [29]. Several spectroscopic studies on the isolated model complexes also underlined the considerable analogy between the Fe(II)-RNO and Fe(II)-O_2 bonds in porphyrin complexes [29] (RNO is isoelectronic to O_2). Thus, model studies have (a) established the existence of Fe(II)-RNO bonds, (b) given the precise structure of these bonds, and (c) showed their great stability, explaining why cytochromes P-450 engaged in such complexes are so strongly inhibited. It is noteworthy that the formation of such complexes is not limited to cytochromes P-450 and iron-porphyrins but also occurs when hemoglobins or myoglobins react directly with alkylhydroxylamines RNHOH or with nitroalkanes RNO_2 in the presence of dithionite [30].

II-2. IRON-METABOLITE COMPLEXES CONTAINING AN IRON-CARBON BOND

II-2-1. Complexes formed during reductive metabolism of halogenated compounds

Several compounds containing reactive carbon-halogen bonds such as CCl_4 are reduced by microsomal cytochromes P-450 in the presence of dithionite or NADPH under very low dioxygen pressures, with formation of P-450−iron-metabolite complexes characterized by Soret peaks between 450−486 nm. This is the case for CCl_4 and many other polyhalogenomethanes which lead to complexes exhibiting Soret peaks between 450−460 nm [31]. Other compounds, like benzyl halides [32] and the anaesthetic agent halothane ($CF_3CHClBr$) [33], give complexes characterized by even more unusually red-shifted Soret peaks (470−480 nm).

Model studies using simple iron-porphyrins have played a key role in our present understanding of the nature of these complexes. They showed that, in the presence of a reducing agent in excess, Fe(II)-porphyrins reacted with compounds containing a sufficiently reactive carbon-halogen bond with formation of a porphyrin-Fe(III)-σ-alkyl complex. Depending upon the nature of the starting halogenated compounds and of the reaction conditions, the σ-alkyl complex is either formed in steady-state concentrations allowing its study by some spectroscopic techniques (as found with benzyl halides [34]) or is stable enough to be isolated and studied by many spectroscopic techniques (as found with halothane [35]; Eqns 6 and 7). The σ-alkyl complexes formed by reduction of benzyl halides and halothane by Fe(II)(TPP) are low-spin ferric complexes at room temperature like other (TPP)Fe(III)−σ-alkyl complexes prepared by more conventional methods [36a] (for other reviews on porphyrin-Fe−σ-alkyl complexes see [36b, 36c]). In the case of the reduction of other compounds containing at least two halogen atoms on the same carbon, such as CCl_4, CBr_4, CCl_3CN or CCl_3CF_3, the reaction does not stop at the level of the σ-alkyl ferric complex. In these cases, the σ-alkyl intermediate complex would undergo a new reductive elimination of a halogen atom with eventual formation of a Fe(II)-carbene complex [37−39] (Eqn 8). These carbene complexes have been isolated and completely characterized, an X-ray structure being available for one of them, (TPP)Fe←CCl_2 [38]. They are all low-spin ferrous diamagnetic complexes [39].

Thus, model studies have definitely established the formation of ferric-σ-alkyl and ferrous-carbene complexes in the reduction of halogenated compounds by iron-porphyrins in the presence of a reducing agent. It is noteworthy that the Fe(TPP)(CCl_2) complex was the first-described dihalogenocarbene complex of a transition metal.

Coming back to the iron-metabolite complexes formed during reduction of halogenated compounds by cytochrome P-450, it is interesting to note that the complexes derived from benzyl halides and halothane have all their Soret peaks highly red-shifted (around 480 nm) and that the halothane-derived complex exhibits an EPR spectrum characteristic of a low-spin Fe(III) species [40]. These spectral data of the halothane-derived complex and its decomposition into P-450-Fe(III) and $CHCl = CF_2$ [41] are in complete agreement with the σ-alkyl Fe(III)-$CHClCF_3$ structure demonstrated in model studies (Eqns 9 and 10). In contrast, the P-450 complexes derived from CCl_4 and other polyhalogenomethanes exhibit less red-shifted Soret peaks (between 450−468 nm). The CCl_4-derived species seems to be a ferrous complex which is slowly transformed into P-450-Fe(II)(CO) [31]. These data are in agreement with the Fe(II)-CCl_2 carbenic structure found in model systems (Eqn 11). A low-spin diamagnetic structure similar to that of model complexes [Fe(TPP)(CCl_2)] would also explain the absence of EPR signals for the CCl_4-derived P-450−metabolite complex.

$$(P)Fe(II) + ArCH_2X \xrightarrow[-X^-]{+e^-} (P)Fe(III)-CH_2Ar \qquad (6)$$

$$(P)Fe(II) + CF_3CHClBr \xrightarrow[-Br^-]{+e^-} (P)Fe(III)-CHClCF_3 \qquad (7)$$

$$(P)Fe(II) + CCl_4 \xrightarrow[-Cl^-]{+e^-} [(P)Fe(III)-CCl_3] \xrightarrow[-Cl^-]{+e^-} (P)Fe(II) \leftarrow CCl_2 \qquad (8)$$

P = porphyrin $\qquad$ Fe−C = *183 pm* when P = TPP

$$P\text{-}450Fe(II) + ArCH_2X \xrightarrow{NADPH} [P\text{-}450Fe(III)-CH_2Ar] \; (\lambda_{max} \approx 480\,nm)$$

$$\downarrow H^+$$

$$P\text{-}450Fe(III) + ArCH_3 \qquad (9)$$

$$P\text{-}450Fe(II) + CF_3CHClBr \xrightarrow{NADPH} [P\text{-}450Fe(III)-CHCl-CF_3] \; (\lambda_{max} = 470\,nm)$$

$$\downarrow$$

$$P\text{-}450Fe(III) + CHCl = CF_2 + F^- \qquad (10)$$

$$P-450\text{Fe(II)} + CCl_4 \xrightarrow{\text{NADPH}} [P-450\text{Fe(II)} \leftarrow CCl_2] \ (\lambda_{max} \approx 460\text{nm})$$

$$\downarrow H_2O$$

$$[P-450\text{Fe(II)} \leftarrow CO] \tag{11}$$

$$(P)\text{Fe(III)} + RNHNH_2 \xrightarrow{\text{OX.}} [(P)\text{Fe(II)} \leftarrow NH=NR] \xrightarrow[-N_2]{\text{OX.}} [(P)\text{Fe(III)}-R] \tag{12}$$

P = porphyrin

$$P-450\text{Fe(III)} + RNHNH_2 \longrightarrow [P-450-\text{Fe(III)}-R] \ (\lambda_{max} \approx 480\text{nm}) \tag{13}$$

R = CH_3, C_2H_5, Ph

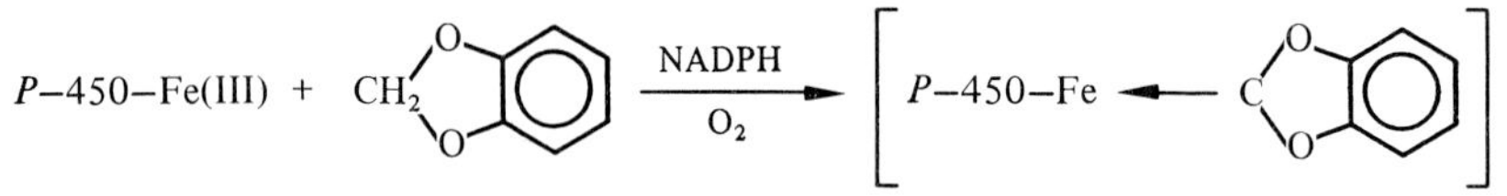

$$P-450-\text{Fe(III)} + CH_2\underset{O}{\overset{O}{\diagup\!\!\!\diagdown}}\!\!\bigcirc \xrightarrow[O_2]{\text{NADPH}} \left[P-450-\text{Fe} \leftarrow C\underset{O}{\overset{O}{\diagup\!\!\!\diagdown}}\!\!\bigcirc \right] \tag{14}$$

$$(P)\text{Fe(II)} + Cl_2C\underset{O}{\overset{O}{\diagup\!\!\!\diagdown}}\!\!\bigcirc \xrightarrow[-2\,Cl^-]{+2e^-} \left[(P)\text{Fe(II)} \leftarrow C\underset{O}{\overset{O}{\diagup\!\!\!\diagdown}}\!\!\bigcirc \right] \tag{15}$$

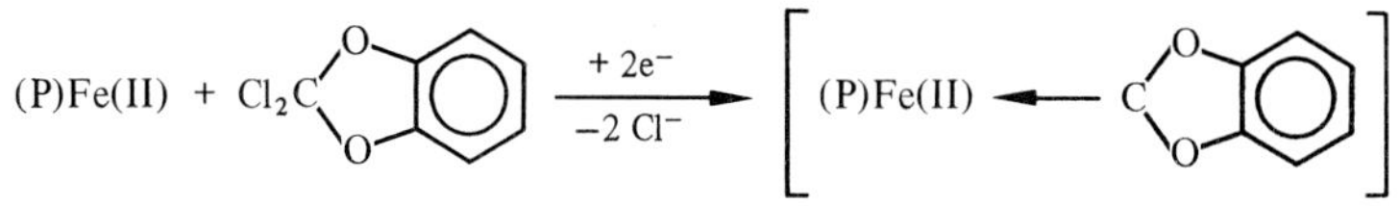

II-2-2. Complexes formed during oxidation of monosubstituted hydrazines

Cytochrome-P-450–iron-metabolite complexes, exhibiting spectral characteristics (unusually red-shifted Soret peaks around 480 nm) and reactivity similar to those derived from halothane or benzyl halide reduction, are formed upon oxidative metabolism of hydrazines $RNHNH_2$ both *in vitro* [42, 43] and *in vivo* [44]. In order to understand their structure and mode of formation, model studies between iron-porphyrins and hydrazines ($RNHNH_2$) in the presence of O_2 or other oxidizing agents have been performed. Eventual formation of porphyrin-Fe(III)–σ-alkyl (or aryl) complexes was demonstrated [45]. For some hydrazines, intermediate formation of Fe(II)–alkyldiazene complexes was observed (Eqn 12) [45]. It is likely that the 480-nm-absorbing P-450 complexes derived from hydrazines $RNHNH_2$ are also σ-alkyl (or aryl) Fe(III)–R complexes [43] (Eqn 13). This is in agreement with (a) the relative stability of the P-450 complex derived from $PhNHNH_2$ towards O_2 in contrast to the great instability of the CH_3NHNH_2-derived complex, since porphyrin Fe(III)–Ph complexes are much more stable towards O_2 than porphyrin-Fe(III)–alkyl complexes [45], and (b) the general formation of hemoprotein-Fe(III)–Ph complexes upon reaction of $PhNHNH_2$ with myoglobin, hemoglobin and catalase [46–50]. The myoglobin-Fe(III)–Ph structure was definitely established by X-ray spectroscopy [49].

II-2-3. Complexes formed during oxidation of benzodioxole derivatives

Benzodioxole derivatives such as piperonyl butoxide or safrole are well-known inhibitors of cytochromes P-450 both *in vitro* and *in vivo*. Inhibition by these compounds is due, at least in part, to the formation of very stable P-450–iron-metabolite complexes characterized by Soret peaks around 435 nm in their higher oxidation state and around 455 nm after reduction [21]. It has been proposed that they could be P-450–iron-carbene complexes formed by *in situ* oxidation of the CH_2 moiety of the benzodioxole ring [51] (Eqn 14). Model iron-porphyrin complexes containing the proposed iron-carbene bond have been prepared and characterized (Eqn 15) [52]. Their spectral properties are in agreement with those of the corresponding P-450 complexes.

In a general manner, the addition at low temperature of an alkylthiolate RS^- ligand to several isolated model porphyrin–iron-metabolite complexes, which should give an RS-Fe-metabolite complex, has led to visible spectra with red-shifted Soret peaks very similar to those of the corresponding P-450–iron-metabolite complexes. This was the case for the benzodioxole-derived carbene complex [52], the $PhNHNH_2$-derived σ-Ph complex [43] and the ferric σ-alkyl complex derived from halothane [40].

II-2-4. Complexes formed by interaction of diazo compounds with cytochrome P-450

It has been postulated that iron-carbene complexes could be formed upon reaction of diazo compounds with cytochrome P-450 [53]. Model studies of reactions between Fe(II) porphyrins and the diazo compounds $RCOCN_2CH_3$ (R = Ph or CH_3) led very recently to the isolation and complete characterization of porphyrin-Fe(II)-C(CH_3)COR carbene complexes [54] (Eqn 16).

$$(P)\text{Fe(II)} + N_2{=}C{<}^{CH_3}_{COR} \xrightarrow{-N_2} \left[(P)\text{Fe(II)} \leftarrow C{<}^{CH_3}_{COR} \right] \tag{16}$$

P = TpClPP, R = Ph or CH_3

II-2-5. About the importance of the formation of iron-carbon bonds in cytochrome P-450 reactions

Present evidence based on spectral and chemical properties of the aforementioned P-450-metabolite complexes and the data accumulated on iron-porphyrin model reactions, strongly indicate the importance of the organometallic chemistry of cytochrome P-450. During its reactions with substrates, complexes containing iron-alkyl (or aryl) or iron-carbene bonds appear to be formed. Some of them are stable and may be detected spectrally like those described in the previous paragraphs. However, many others could be formed as transient intermediates during the oxidation of various substrates. For instance, the formation of iron-carbene complexes upon metabolic oxidation of benzodioxole derivatives has been interpreted with the mechanism of Eqn (17) [55]. This mechanism involves the abstraction of a hydrogen of the CH_2 group by the high-valency iron-oxo species of P-450 and the combination of the intermediate carbon-centered free radical with the iron. This mechanism could be general for C-H bond hydroxylation, the most probable evolution of the intermediate high-valency σ-alkyl complex being a reductive elimination of the σ-alkyl and OH ligands that are in cis position on the iron, with eventual formation of the hydroxylated metabolite [55]. In the particular case of benzodioxole derivatives, there is another possible evolution of the intermediate σ-alkyl complex, as 1,2-elimination of water could lead to the observed stable iron-carbene complexes. Such a formation of transient iron-carbon bonds between P-450 iron and free radicals derived from substrate activation inside the active site of cytochrome P-450 could be a good way for the enzyme to control these free-radical intermediates. Therefore, iron-carbon bonds could also be formed during alkene epoxidation or aromatic ring hydroxylation (Eqns 18 and 19).

In that regard, a recent peculiar result of the epoxidation of propene by a purified cytochrome P-450 well illustrates the importance of this organometallic chemistry of cytochrome P-450. The deuterium–hydrogen exchange observed during the epoxidation of *trans*-(1-D)propene has been explained by the mechanisms indicated on Fig. 3 [56]. Its first step is the formation of a four-membered metallocycle after addition of

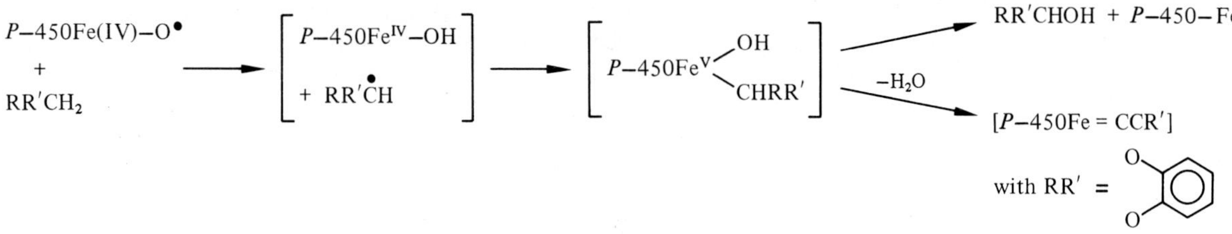

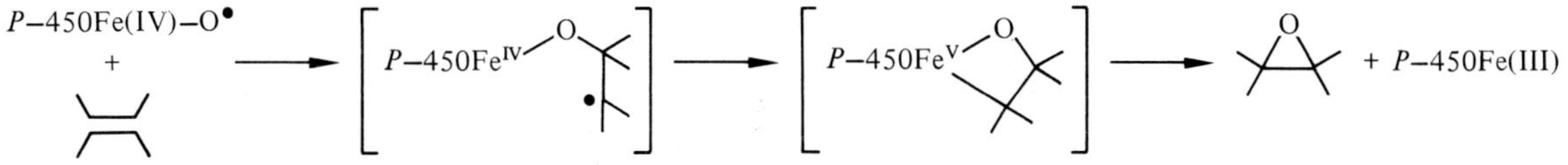

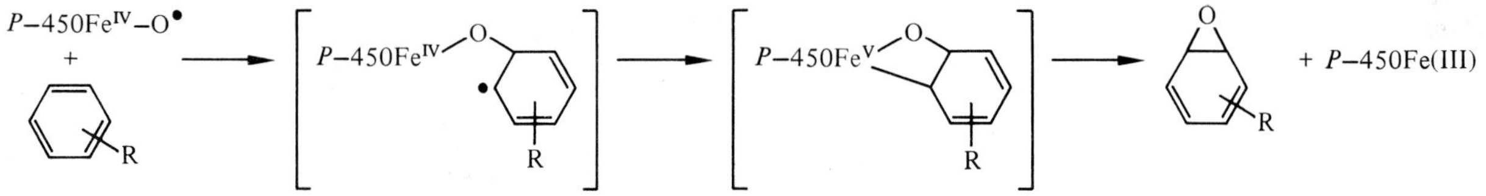

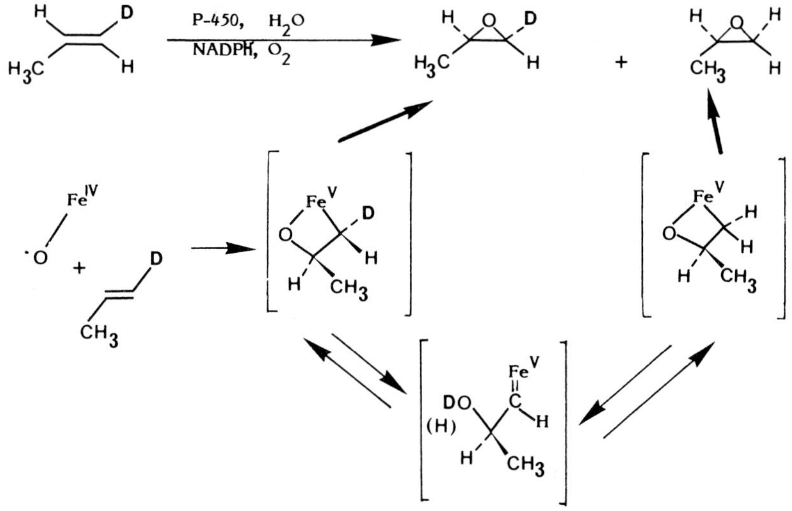

Fig. 3. *Postulated mechanism for the hydrogen-deuterium exchange observed during epoxidation of propene by cytochrome P-450 LM_2 (from [56])*

the iron-oxo species to the double bond and combination of the intermediate free radical with the iron as proposed in Fig. 3. Indirect evidence for the existence of such metallocycles in model iron-porphyrin systems has been published [57, 58]. This metallocycle could be in equilibrium with a carbene complex after α-elimination of D^+. After D/H exchange at the level of the OD(H) function and nucleophilic addition of OH(D) on the Fe=C double bond, it is also in equilibrium with the corresponding metallocycle where the deuterium atom has been replaced by a hydrogen atom. By reductive elimination, these two metallocycles lead to a mixture of the deuterated and non-deuterated epoxides. This mechanism, which implies the intermediate formation of iron-carbon σ bonds and iron-carbene bonds, is consistent with the results described in section II-2 and rationalizes experimental data that are difficult to explain otherwise.

II-3. THE USE OF MODEL SYSTEMS TO UNDERSTAND REACTIONS INVOLVING N-ALKYLATIONS OF CYTOCHROME P-450 HEME

During the *in vivo* metabolism of many compounds, green pigments coming from N-alkylations of the heme of cytochromes P-450 are formed [3, 17]. Iron-porphyrin model studies have contributed to our understanding of the participation of pyrrole nitrogen atoms in hemoprotein reactions.

II-3-1. The participation of pyrrole nitrogen atoms in some iron-porphyrin reactions

Iron-porphyrin complexes containing axial ligands such as σ-alkyls, carbenes or nitrenes are relatively stable when the formal oxidation state of the iron is low enough [III for the σ-alkyl complexes, II for the nitrene or carbene complexes: Fe(II) ← CRR′ or formally Fe(IV) = CRR′]. However, a one-electron oxidation of these complexes was found to lead to a migration of the axial ligand from the iron to a pyrrole nitrogen atom (Eqns 20–22). The corresponding final complexes, which are Fe(II)-N-alkyl complexes [59–62] when starting from σ-alkyl-Fe(III) compounds, and bridged carbene [Fe(III)-C-N] [63, 64] or bridged nitrene [Fe(III)-N-N] complexes [65] when starting from iron-carbene or iron-nitrene complexes, were isolated and completely characterized. These data clearly establish the great propensity of iron-porphyrin complexes to migrate their iron axial ligand to a pyrrole nitrogen when the iron is in a sufficiently high oxidation state. Such a participation of pyrrole nitrogens has also been found during reactions of other metalloporphyrins containing a metal different from iron (for a review, see [62]).

II-3-2. Formation of N-alkylated hemes during oxidation of terminal alkenes and alkynes

Green pigments accumulate in the liver of animals treated by various terminal alkenes or alkynes not too hindered in the vicinity of the double (or triple) bond [17]. In many cases, the structures of the pigments found after acidic extraction were determined: the pigments found after treatment by R-CH=CH$_2$ have an N-CH$_2$-CHOHR—protoporphyrin IX structure [17]. Fig. 4 shows a possible mechanism for alkene epoxidation and N-alkyl-porphyrin formation which takes into account the intrinsic properties of iron-porphyrins described in the previous paragraphs and particularly the propensity of a possible high-valency σ-alkyl metallocyclic complex to migrate its σ-alkyl ligand to a pyrrole nitrogen.

Model reactions for alkene oxidation using iodosoarenes as oxygen atom donors and iron-porphyrins as catalysts reproduced quite well the reactions observed with cytochromes P-450 [66–73]. With Fe(TDCPP)(Cl) (TDCPP = tetra-2,6-dichlorophenylporphyrin, see Fig. 8 for the formula) as catalyst, terminal alkenes RCH=CH$_2$ are oxidized to the corresponding epoxide and the catalyst is transformed into a ferric-N-alkylporphyrin whose alkyl group has a CH$_2$CHOHR structure identical to that found in cytochrome P-450 green pigments (Eqn 23) [67, 68]. The formation of the ferric metallocyclic Fe-O-C-C-N complex, expected from the mechanism proposed in Fig. 4, was demonstrated [67,70]. Interestingly, oxidation of the same alkenes, RCH=CH$_2$, by PhIO catalyzed by Fe(TPP)(Cl) or Fe(TpClPP)(Cl) (Fig. 8) leads to the formation of N-alkylporphyrins which should

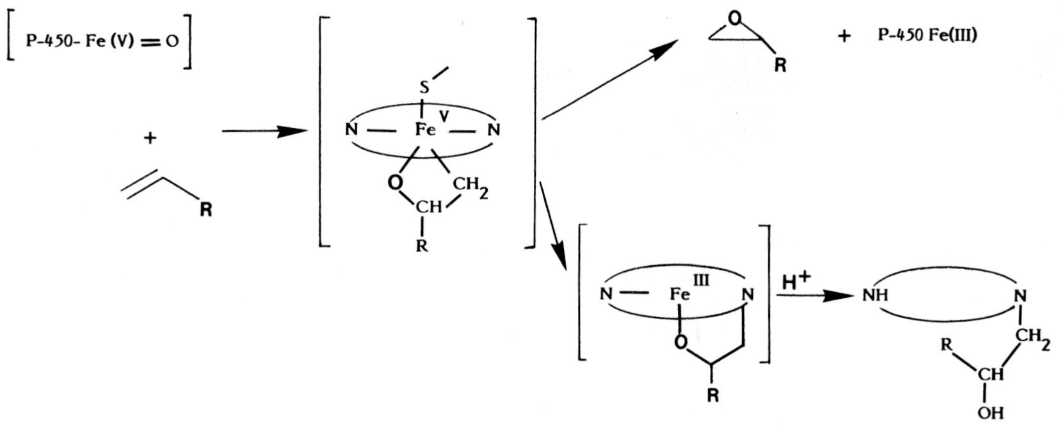

Fig. 4. *Possible mechanism for the formation of N-alkylated hemes during cytochrome-P-450-catalyzed oxidation of terminal alkenes*

result from the addition of the active Fe = O species to the less substituted alkene vinylic carbon (Eqn 24) [71] and not to the more substituted vinylic carbon as observed with P-450 or Fe(TDCPP)(Cl) as catalysts. Such branched N-alkylporphyrins derived from the binding of a pyrrole nitrogen to the more substituted vinylic carbon of terminal alkenes have not been isolated so far from corresponding P-450 reactions. Iron-porphyrin model studies on oxidations of various alkenes showed recently that N-alkylporphyrins were formed not only from terminal alkenes but also as transient intermediates during oxidation of 1,2-disubstituted alkenes such as norbornene [69, 72]. In the case of the oxidation of *trans*-hex-2-ene catalyzed by Fe(TpClPP)(Cl), the two possible isomeric N-alkylporphyrins with the N-CH(CH$_3$)-CHOHC$_3$H$_7$ and N-CH(C$_3$H$_7$)-CHOHCH$_3$ structures have been isolated and characterized [73]. Thus, on the basis of data reported so far, the formation of N-alkylporphyrins during alkene oxidation seems to be more general with iron-porphyrin catalysts than with cytochromes P-450 (Table 1). It is more general because it is not limited to unhindered terminal alkenes and because it leads to the two possible isomeric structures derived from the binding of a pyrrole nitrogen to the two vinylic carbons of the alkene molecule. The fact that the formation of branched N-CHR-CHR'OH-protoporphyrins has never been observed in P-450-catalyzed oxidation of alkenes could be due to the relative instability of the N-alkylporphyrins compared to N-CH$_2$-CHROH-porphyrins under the harsh conditions of their extraction from the biological medium. Further experiments are required to determine whether such branched N-alkylporphyrins are definitely not formed in P-450-catalyzed alkene oxidation.

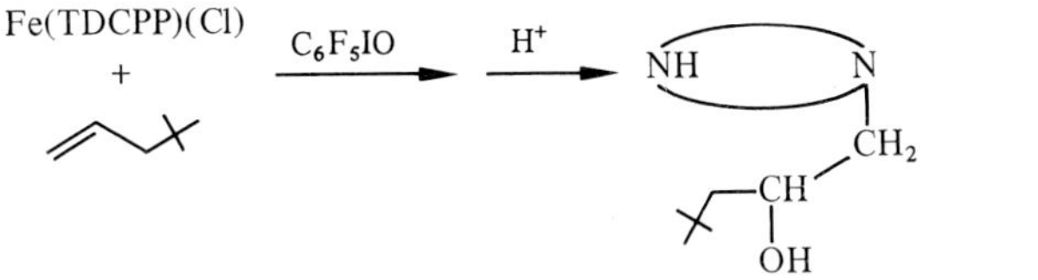

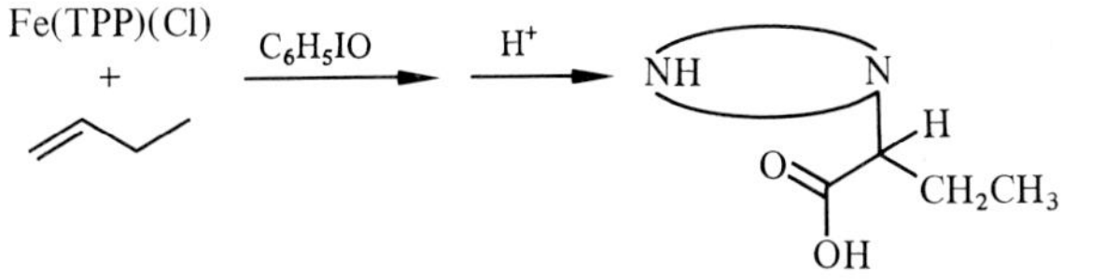

Table 1. *Structure of the N-alkylporphyrins isolated from oxidation of alkenes catalyzed either by cytochromes P-450 or by iron-porphyrins*

Alkene structure	Structure of N-alkylporphyrins isolated from alkene oxidation catalyzed by		
	P-450	Fe(TDCPP)(Cl)	Fe(TpClPP)(Cl)
R−CH=CH$_2$	CH$_2$CHOHR [17]	CH$_2$CHOHR [67−69]	CHR-COOH [71]
R−CH=CHR' (*trans*-hex-2-ene)	−[a]	−[b]	CHRCHOHR' + CHR'CHOHR [73]

[a] No green pigment detected so far from 1,2-dialkylethylenes.
[b] No N-alkylated porphyrin isolated so far, but transient formation of iron-N-alkylporphyrins detected during reaction of various iron-porphyrins with ArIO and other 1,2-disubstituted alkenes [69].

II-3-3. *Formation of N-alkylated hemes during oxidation of sydnones and during reaction of diazo compounds*

Some sydnones (Fig. 5) act as suicide substrates of cytochrome P-450 [53, 74]. Their metabolic oxidation leads to an N-alkylation of P-450 heme with formation of N-vinylprotoporphyrin and of N-(2-phenylethyl)- and N-(2-phenylvinyl)-protoporphyrin from sydnones bearing R = SPh and R = Ph substituents, respectively. Mechanisms indicated in Fig. 5 have been proposed for these reactions [53, 74]. They involve the intermediate formation of diazo compounds and their interactions with cytochrome-P-450-iron to give transient iron-carbene or iron−σ-alkyl complexes as precursors of N-substituted hemes. Recently, iron-porphyrin model reactions have been reported for several steps of these mechanisms. As indicated in section II-2-4, some diazoketones react with iron(II)-porphyrins to give iron-carbene complexes [54]. Their one-electron oxidation by Br$_2$ or K$_2$Cr$_2$O$_7$ leads to iron-N-alkylporphyrins derived from the migration of the carbene moiety to a pyrrole nitrogen (Fig. 6) [54]. Reac-

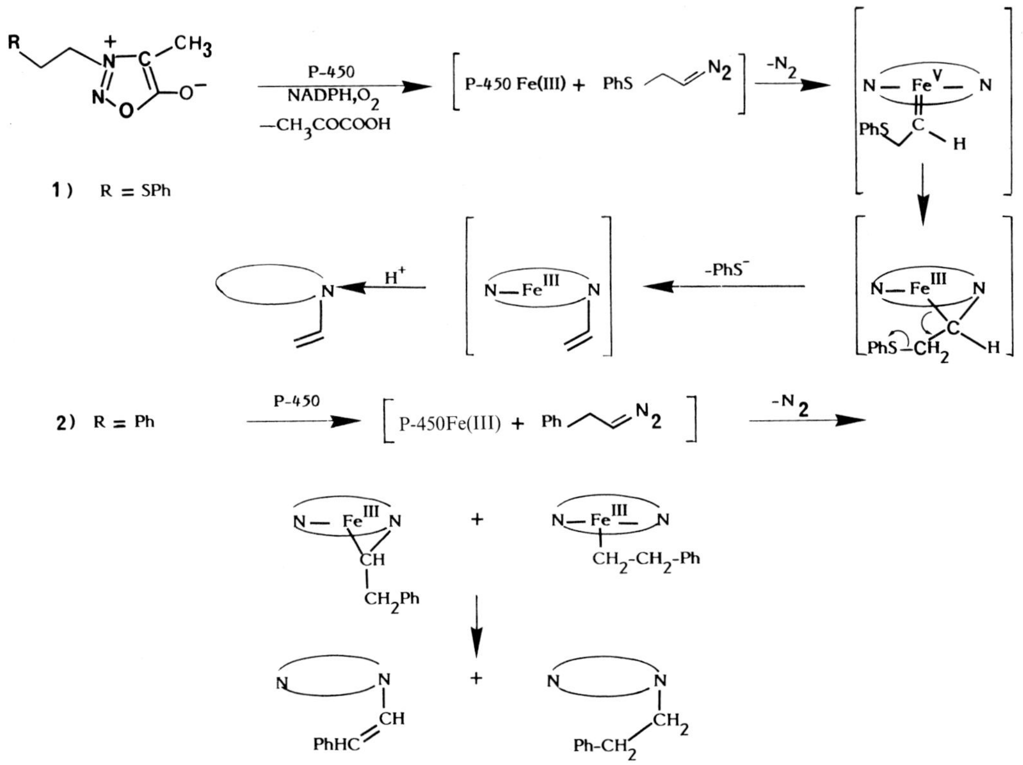

Fig. 5. *Postulated mechanisms for the formation of N-substituted hemes during cytochromes-P-450-catalyzed oxidation of sydnones (from [53, 74])*

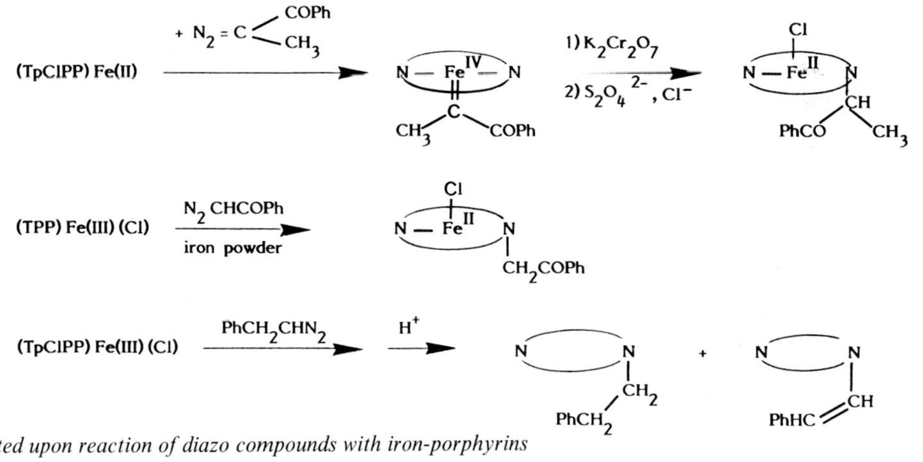

Fig. 6. *Complexes isolated upon reaction of diazo compounds with iron-porphyrins*

tions of PhCOCHN$_2$ [75], N$_2$CHCOOEt, N$_2$CHCF$_3$ and N$_2$CHCH$_2$Ph (I. Artaud, N. Grégoire, J. P. Battioni and D. Mansuy, unpublished results) with iron(III)-porphyrins lead directly to the corresponding iron-*N*-CH$_2$COPh-, -*N*-CH$_2$COOEt-, -*N*-CH$_2$CF$_3$- and -*N*-CH$_2$CH$_2$Ph- and -*N*-CH = CHPh-porphyrins (Fig. 6).

III. CATALYTICALLY ACTIVE MODEL SYSTEMS FOR CYTOCHROMES *P*-450: NEW CHEMICAL CATALYSTS FOR STUDYING AND PREDICTING THE OXIDATIVE METABOLISM OF XENOBIOTICS

III-1. PROBLEMS TO BE SOLVED TO BUILD UP EFFICIENT CATALYTICALLY ACTIVE MODELS

On the basis of the known characteristics of cytochrome *P*-450 reactions summarized in section I, an ideal model system would associate an iron-porphyrin, a thiolate ligand, a reducing agent and a proton donor, and O$_2$ itself as oxygen atom donor. However, the use of thiolate ligands appears difficult because of their rapid oxidation in strongly oxidizing media. Moreover, a serious problem in the design of chemical systems mimicking the long catalytic cycle of cytochrome *P*-450, using O$_2$ itself, is to avoid a too-fast reduction of the Fe(V) = O species by the reducing agent in excess. The separation between the Fe(V) = O species and NADPH ensured by the enzymatic system should be difficult to reach in a simple chemical system (for a detailed analysis of this particular problem, see [76]). Thus, it appears far easier to mimic the shortened catalytic cycle of cytochrome *P*-450 using oxygen atom donors (AO, Fig. 1). However, the level of difficulty in generating a high-valency iron-oxo complex upon reaction of an iron-porphyrin with an oxygen atom donor will depend on the nature of this oxidant. It is likely that oxidants containing only one oxygen atom linked to a leaving group, such as PhIO

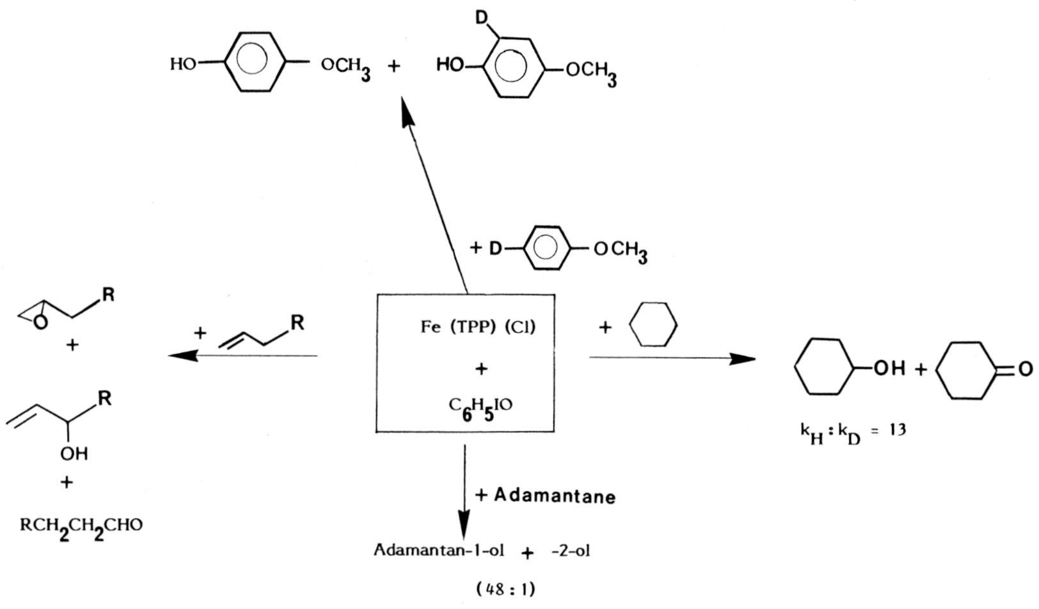

Fig. 7. *Examples of oxidations by PhIO catalyzed by iron-porphyrins*

or ClO$^-$, should transfer their oxygen atom more easily to metalloporphyrins than oxidants containing an O-O bond, such as alkylhydroperoxides or H$_2$O$_2$, for which two modes of cleavage of this O-O bond (homolytic and heterolytic) are possible. These considerations explain the order of presentation of the model systems in the following.

In the design of efficient metalloporphyrin-based model systems for cytochromes *P*-450, the following key properties must be obtained.

a) The involvement of high-valency metal-oxo intermediates leading to a chemoselectivity similar to that of cytochromes *P*-450, with the ability to hydroxylate aliphatic C-H bonds (with preferential attack of tertiary C-H bonds), to epoxidize double bonds, to hydroxylate aromatic rings, to perform the *N*-dealkylation of secondary and tertiary amines or the *O*-dealkylation of ethers, and to transfer an oxygen atom to nitrogen and sulfur atoms of amines and thioethers.

b) Rates around 1–100 turnover/min and yields based on the reducing agent (if O$_2$ itself is used as an oxygen atom donor) between 50–100%.

c) Good stability of the metalloporphyrin catalyst in the strongly oxidizing medium. This is a generally difficult problem in oxidation catalysis. In fact, reasonable stability of the catalyst is required if one wants to use such model systems for preparative purposes, such as the synthesis of hydroxylated metabolites from a drug at the level of gram amounts.

The purpose of this section is not to give a complete list of all the oxidations by Fe- or Mn-porphyrin-dependent model systems reported so far [76–79], but to show the developments in this field during this last decade and its potential applications.

III-2. MODEL SYSTEMS USING SINGLE OXYGEN ATOM DONORS LIKE IODOSYLARENES

As first shown by Groves et al. [77], iron-*meso*-tetraarylporphyrins are very good catalysts for the transfer of the oxygen atom of PhIO into many substrates in a manner very similar to cytochromes *P*-450 (for recent reviews see [78, 79]). For instance, alkane hydroxylation occurs with high isotopic effects (k_H/k_D of about 13 for cyclohexane [80]), retention of configuration of the C-H bond for hydroxylation of *cis*-decalin into 9-decalol [80, 81] and preferential reaction on tertiary C-H bonds [80] (Fig. 7). The four reactions observed upon cytochrome-*P*-450-dependent oxidation of alkenes, i.e. the stereospecific epoxidation of the double bond [82, 83], the hydroxylation of allylic C-H bonds [84], the minor formation of aldehydes RCH$_2$CHO from RCH=CH$_2$ [82, 85, 86] and the transformation of the iron-porphyrin catalyst into an *N*-alkyl-porphyrin (see section III-3), have also been reproduced by these model systems. Moreover, hydroxylation of aromatic hydrocarbons occurs with migration of the hydrogen (or deuterium) atom present at the hydroxylation site to the *ortho* position ("NIH shift") (72% retention of deuterium in *p*-methoxy-phenol formed by hydroxylation of (*p*-D)anisole [87]) (Fig. 7). From the various metalloporphyrins tested as catalysts in such systems, Fe(III)- and Mn(III)-porphyrins appeared as the best ones, Mn(III)-porphyrins being in many cases more resistant toward oxidative degradation than the corresponding Fe(III)-porphyrins. Unfortunately, simple Fe- or Mn-tetraarylporphyrins like Fe(TPP)(Cl) are oxidatively destroyed especially during oxidation of poorly reactive substrates such as alkanes. Major improvements of the catalytic activities of these model systems were obtained by using tetraarylporphyrins containing halogen-substituted aryl groups such as the pentafluorophenyl [87] and the 2,6-dichlorophenyl [88] groups. The iron-porphyrin, Fe(TDCPP)(Cl) (Fig. 8), catalyzes alkene epoxidation by C$_6$F$_5$IO with an initial rate as high as 300 turnovers/s [89], and more than 100000 mol epoxide are obtained/mol catalyst without appreciable destruction of this catalyst. More recently, the even more robust Fe [(tetrakis(2,6-dichlorophenyl)octabromoporphyrin] [Cl] complex was found to be able to hydroxylate norbornane with a 75% yield based on starting C$_6$F$_5$IO and without loss of the catalyst [90] (Fig. 8).

Other oxidants containing a single oxygen atom donor such as tertiary amine oxides [91] (and references cited there in) or hypochlorites [78] associated with Fe- or Mn-porphyrins lead to results similar to those observed with PhIO, especially for alkene epoxidation. More recently, a biphasic system using

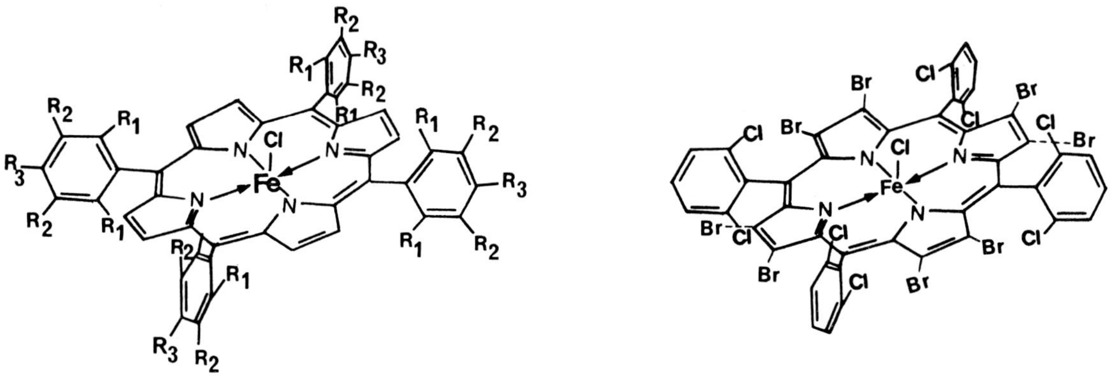

$R_1 = R_2 = R_3 = H$	Fe(TPP)(Cl)
$R_1 = Cl$; $R_2 = R_3 = H$	Fe(TDCPP)(Cl)
$R_1 = R_3 = CH_3$; $R_2 = H$	Fe(TMP)(Cl)
$R_1 = R_3 = C_6H_5$; $R_2 = H$	Fe(TTPPP)(Cl)
$R_3 = Cl$; $R_1 = R_2 = H$	Fe(TpClPP)(Cl)

Fe[tetra-(2,6-dichlorophenyl) octabromoporphyrin][Cl]

Fe(TDCPOBP) (Cl)

Fig. 8. *Formula of various meso-tetra-arylporphyrins used in model systems (including TDCPP and TDCPOBP which are very resistant towards oxidative degradation, and the very hindered TTPPP)*

the water-soluble oxidant $KHSO_5$ and Mn(III)-porphyrins in the presence of pyridines or imidazoles was found to be very efficient for alkane hydroxylation [92].

III-3. MODEL SYSTEMS USING ALKYLHYDROPEROXIDES OR H_2O_2

The use of oxidants like ROOH or H_2O_2 leads to supplementary problems linked to the two possible modes of cleavage of their O-O bond (Eqn 25). While a heterolytic cleavage will give the expected Fe(V) = O species, a homolytic cleavage will give undesired RO˙ radicals. In fact, simple Mn(porphyrin)(Cl) or Fe(porphyrin)(Cl) complexes fail to catalyze alkene epoxidation by alkylhydroperoxides. Moreover, they fail to catalyze alkane hydroxylation by alkylhydroperoxides by a cytochrome-*P*-450-like mechanism [93]. Actually, Mn(porphyrin)(Cl) complexes lead to a slow decomposition of alkylhydroperoxides [94, 95] and Fe(porphyrin)(Cl) complexes lead to a rapid decomposition of alkylhydroperoxides but without good epoxidation of alkenes [93, 96]. It is noting that alkanes are oxidized by cumylhydroperoxide in the presence of Fe-porphyrin catalysts to the corresponding alcohols and ketones, with good yields [93, 94]. However, in these systems, the active species is not linked to the metal and appears to be a cumylO˙ or cumylOO˙ radical derived from a homolytic cleavage of the O-O or OH bond of the starting hydroperoxide [79b, 93, 97].

$$Fe^{III} + ROOH \quad \xrightarrow{\text{homolytic}} Fe(IV)-OH + RO˙$$
$$\xrightarrow{\text{heterolytic}} Fe(V) = O + ROH \quad (25)$$

The ability of Fe- or Mn-porphyrins to catalyze alkene epoxidation by alkylhydroperoxides is considerably improved by the use of imidazole as cocatalyst. For instance, under conditions for which cumylOOH is unable to epoxidize 2-methyl-hept-2-ene in the presence of Fe(TPP)Cl, it leads to a 17% epoxidation yield upon addition of imidazole [96]. Addition of catalytic amounts of imidazole to the Mn(TPP)(Cl)–cumylhydroperoxide system makes it able to epoxidize cyclooctene, *cis*-stilbene and 2-methyl-hept-2-ene with yields between 20% and 50% [96]. Similarly, a 60% yield of epoxidation of the reactive alkene tetramethylethylene is obtained with Mn(TPP)(Cl) and *t*BuOOH in the presence of imidazole [95]. These results point to two important improvements of the Fe(porphyrin)(Cl)–ROOH or Mn(porphyrin)(Cl)–ROOH systems upon imidazole addition: (a) a large increase of the rate of ROOH reaction with Mn(III)-porphyrins, and (b) a very important increase of oxygen atom transfer to alkenes.

Among the various possible oxygen atom donors susceptible to transfer their oxygen atom to hydrocarbons, H_2O_2 is especially interesting since it is a readily available and cheap oxidant and since it gives only water as secondary product. However, simple Mn(III)- and Fe(III)-porphyrins were found unable to catalyze the transfer of an oxygen atom of H_2O_2 to hydrocarbons [98–100]. Also with this oxidant the addition of catalytic amounts of imidazole to these systems made them very efficient for alkene epoxidation and alkane hydroxylation. As shown in Fig. 9, stereospecific epoxidation of various alkenes are obtained with almost quantitative conversions and yields, and good rates (up to 100 turnovers/min) [98, 100]. The H_2O_2–Mn(TDCPP)(Cl)–imidazole system may be used for the conversion of alkanes by using a 2–6-fold excess of H_2O_2 [99]. Within 2 h at room temperature, 50–80% conversion of cyclohexane and other alkanes were obtained, the corresponding alcohols and ketones being formed with yields between 40% and 75% (Fig. 9). Under these conditions, adamantane is almost completely converted (95%) with preferential formation of 1-adamantanol (63%) and minor formation of 2-adamantanol and 2-adamantanone (19% and 3%).

Very recently, a comparison of the regioselectivities and stereospecificities of the Mn(porphyrin)–H_2O_2–imidazole and Mn(porphyrin)–PhIO–imidazole systems showed that

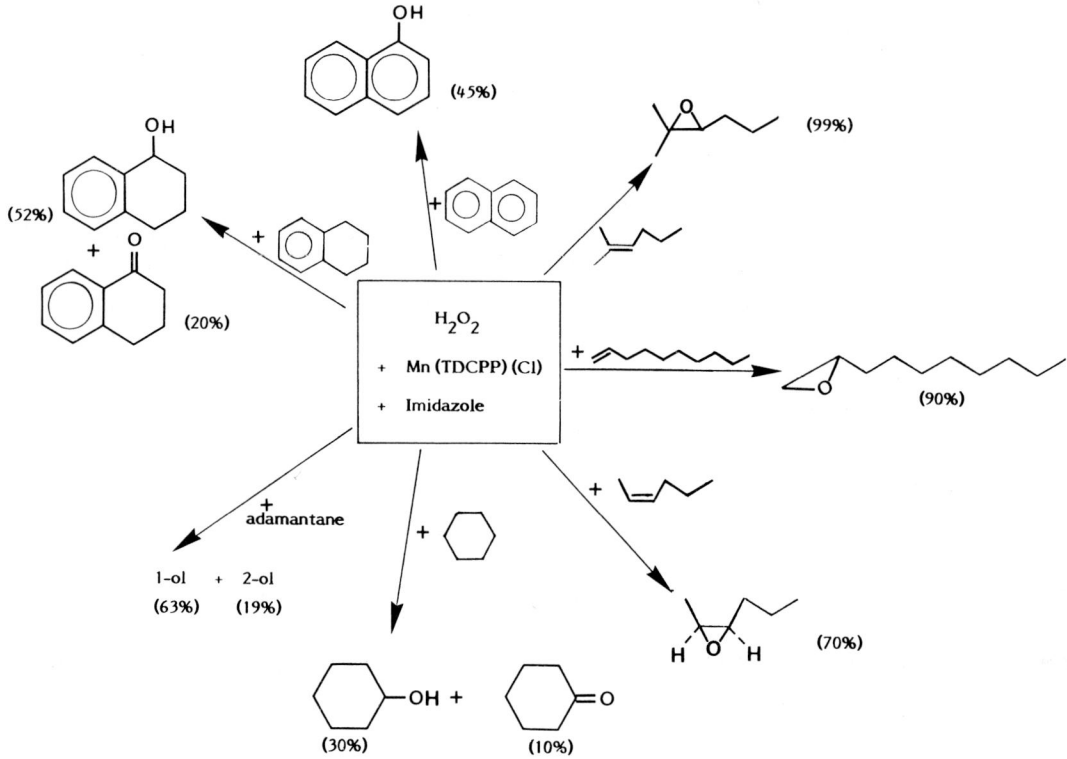

Fig. 9. *Examples of oxidations performed by H_2O_2 in the presence of catalytic amounts of Mn(TDCPP)(Cl) and imidazole (yields based on starting substrate)*

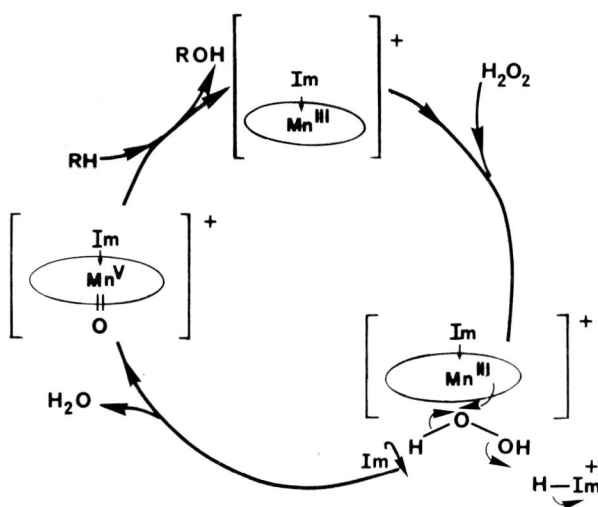

Fig. 10. *Catalytic cycle of substrate monooxygenation by the H_2O_2:Mn(TDCPP)(Cl)-imidazole system: the two roles of imidazole*

these two systems involved identical oxidizing species which should be high-valency imidazole–Mn(V) = O intermediates [100]. It was also shown [100, 101] that imidazole played at least two roles in these model H_2O_2-dependent monooxygenations; (a) as a ligand of Mn(III) presumably allowing enough electron density to be provided on the metal in order to cleave the O-O bond of H_2O_2 in a heterolytic manner, and (b) as an acid-base catalyst to facilitate H_2O_2 deprotonation and H_2O departure (Fig. 10).

With iron(III)-porphyrins as catalysts, the use of imidazole as a cocatalyst is also very important for transferring one oxygen atom from H_2O_2 or alkylhydroperoxides [100]. However, the yields of substrate monooxygenations catalyzed by iron(III)-porphyrins have so far remained lower than those of the same reactions catalyzed by Mn-porphyrins [100]. This could be due to the pronounced tendency of Fe(III)-porphyrins to catalyze the dismutation of H_2O_2 or ROOH [102].

III-4. MODEL SYSTEMS USING O_2 ITSELF AND A REDUCING AGENT

Several sytems using various reducing agents and catalytic amounts of Fe(III)- or Mn(III)-porphyrins have been found to transfer one oxygen atom from O_2 to hydrocarbons. A very recent review on this particular subject appeared last year [76]. Thus the purpose of this section is not to describe in detail all the systems reported so far but only to give an overview of the systems reported between 1979 and 1986 and to concentrate on the most recent results and the present tendency in this field.

Borohydrides were the first reducing agents used in such systems [103–106]. Sodium ascorbate was also used in a biphasic system, with the substrate and Mn(TPP)(Cl) in benzene and ascorbate in a buffer pH 8.5, in the presence of catalytic amounts of a phase-transfer agent [107, 108]. Hydrogen itself in the presence of catalytic amounts of colloidal platinum was also used in Mn- and Fe-porphyrin-dependent systems [109–111]. Another way employed to provide reducing equivalents necessary for dioxygen activation was electrochemistry in the presence of either Fe-porphyrins [112] or Mn-porphyrins [113, 114]. All these O_2-dependent systems reproduce, at least in a qualitative manner, the main reactions of cytochrome P-450. A crucial step in O_2 activation by these systems was found to be the heterolytic cleavage of the O-O bond of an Fe(III)-O-O$^-$ intermediate formed after O_2 binding to Fe(II) and one-electron reduction of the Fe(II)-O_2

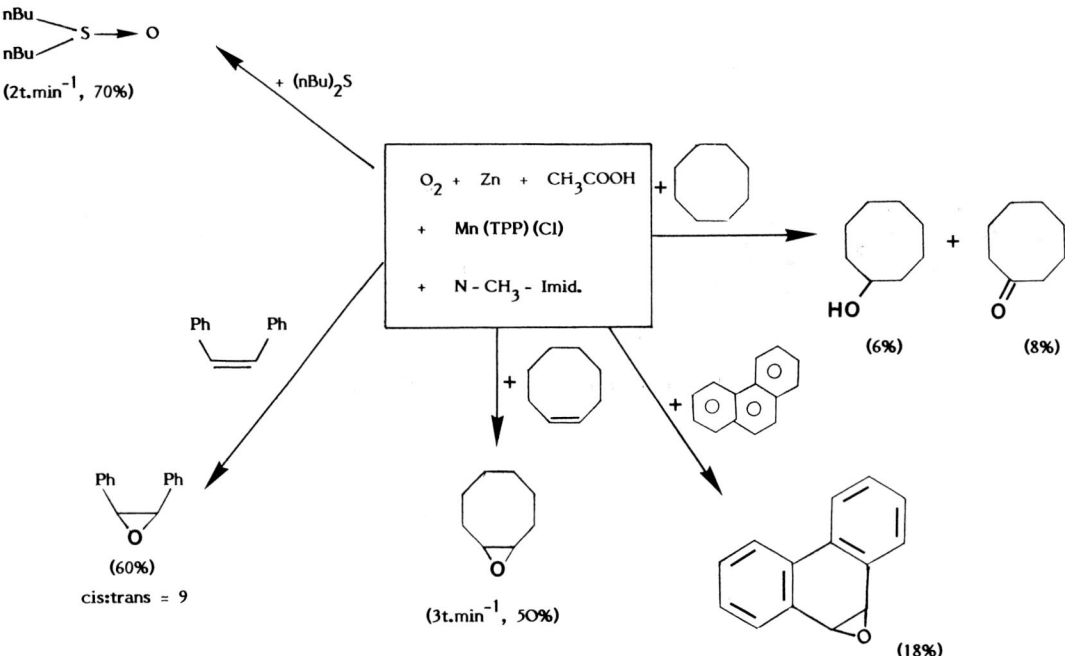

Fig. 11. *Examples of oxidations performed by the O_2-Zn-Mn(TPP)(Cl)-N-methyl-imidazole system*. Initial rates are given in turnovers/min Yields (%) are based on starting reducing agent

complex. This cleavage is greatly facilited by protonation or acylation of the Fe-O-O$^-$ intermediate. Accordingly, it was recently shown that the heterolytic cleavage of the O-O bond of porphyrin-Fe(III)-O-O-CO-R complexes involves a very low activation enthalpy of 17 kJ/mol and was acid-catalyzed [115]. The use of stoichiometric amounts of acid chlorides or acid anhydrides led to successful results in systems using O_2 or O_2^- and Fe-porphyrins [112] or Mn-porphyrins [113, 116] as catalysts.

A major problem with most of these systems was their relatively low catalytic activities and yields based on the reducing agent. The yields based on the reducing agent, which are between 0.1% and 5% for alkene epoxidation and between 0.01% and 0.5% for alkane hydroxylation by the borohydride-, ascorbate- and H_2-dependent systems [79c], are due to a competition between the hydrocarbon substrate and the reducing agent in excess for reaction with the high-valency metal-oxo intermediates. In contrast to the enzymatic system, in the model systems there is no separation between the active-oxygen species and the reducing agent in excess, so that reduction of the metal-oxo species by the reducing agent is faster than its reaction with the hydrocarbons [108, 111]. The best yield based on the reducing agent reported so far, 56%, was obtained for cyclooctene epoxidation by an electrochemical system using Mn(TPP)(Cl) as a catalyst and $(PhCO)_2O$ as a stoichiometric acylating agent. However, because of the slow arrival of the electrons in that system, the rate remained low [113].

Very recently, three systems were reported to give at the same time good yields based on the reducing agent (up to 50%) and rates (up to 9 turnovers/min) which are not too far from those observed with cytochromes P-450.

The first one employs a dihydropyridine as reducing agent in the presence of a flavin mononucleotide as an electron transfer catalyst and a water-soluble anionic Mn-porphyrin and N-methyl-imidazole. It epoxidizes nerol with a rate of 9 mol epoxide formed (mol catalyst)$^{-1}$ min^{-1} and yields based on the reducing agent of around 30% [117].

The second system uses Zn amalgam in the presence of methyl viologen as an electron-transfer catalyst, catalytic amounts of an iron-porphyrin and stoichiometric amounts of acetic anhydride as an acylating agent [118]. It hydroxylates cyclohexane with yields based on O_2 consumed up to 30% and rates up to 1.1 turnover/min. As expected for a monooxygenase-like system, $C_6H_{11}{}^{18}OH$ is formed when using $^{18}O_2$. Moreover, hydroxylation of cyclohexane occurs with a high kinetic isotope effect ($k_H/k_D = 7$), similar to those reported in the case of cytochrome P-450.

The third system is based on Zn powder as a very simple reducing agent, CH_3COOH as a proton source and Mn(TPP)(Cl) and N-methylimidazole as catalysts [119]. As shown in Fig. 11, this system is a good support for the main classes of reactions catalyzed by cytochromes P-450, i.e. the epoxidation of alkenes and of phenanthrene, the hydroxylation of alkanes and of aromatic rings, the S-oxidation of thioethers and the N-dealkylation of amines (P. Battioni, W. Lu, P. Gouvine and D. Mansuy, unpublished results). The rates and yields (based on the reducing agent) of the oxidations performed by this system are not far from those observed with cytochromes P-450. For instance, sulfoxidation of di-n-butylthioether, epoxidation of cyclooctene and hydroxylation of cyclooctane occur with respective rates of 2, 3 and 0.5 turnover/min and yields based on Zn of 70%, 50% and 15%. This system was also used under conditions of complete conversion of these substrates and led to a convenient and simple preparation of the corresponding oxidized products with high yields at the level of gram amounts.

III-5. FROM CHEMOSELECTIVE TO REGIOSELECTIVE SYSTEMS

The previous sections show that simple Fe- or Mn-porphyrin systems using various oxygen atom donors are able to reproduce the main reactions performed by the heme part of cytochromes P-450 since they involve high-valency metal-

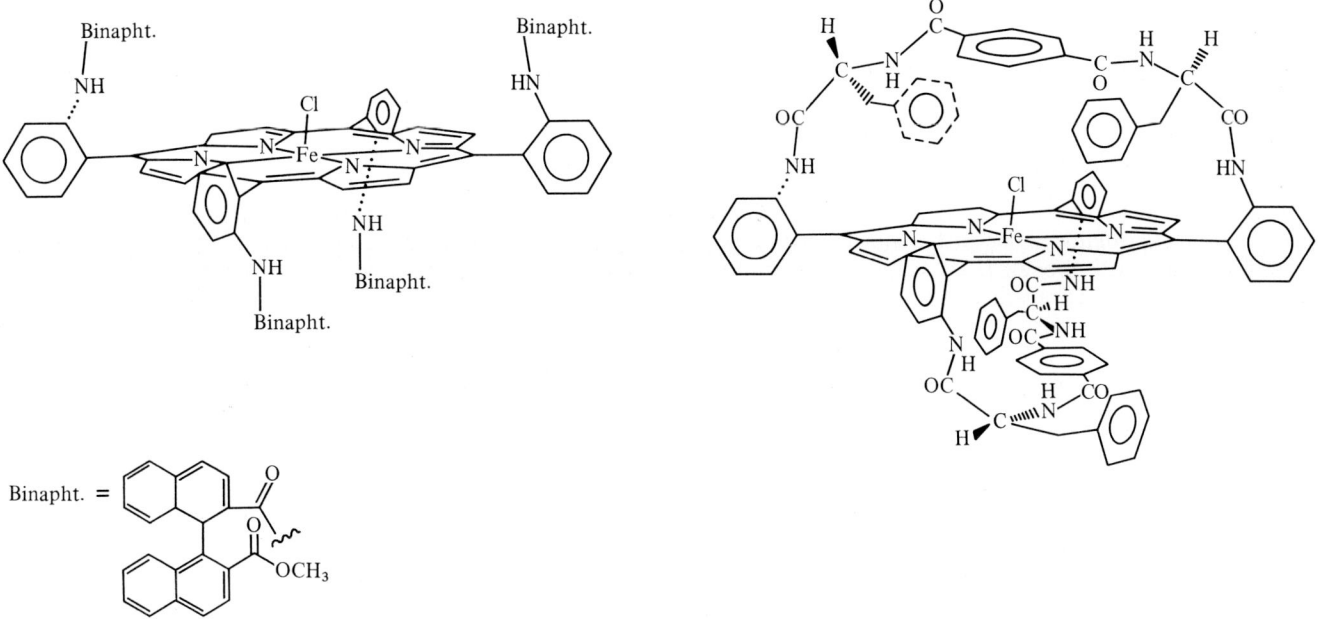

Fig. 12. *Chiral iron-porphyrins described as catalysts for the asymmetric epoxidation of alkenes (from [120, 121])*

oxo intermediates exhibiting a reactivity and chemioselectivity very similar the those of cytochromes *P*-450. However, in order to make these model systems regioselective and/or stereoselective like most cytochromes *P*-450, it is necessary to modify the environment of the metal in the porphyrin to control the access and the conformational mobility of the substrate. This is done in cytochrome *P*-450 by the protein part which is close to the heme.

III-5-1. Catalysts for asymmetric epoxidations

Iron-porphyrins bearing chiral centers or groups maintained in positions close to the iron have been prepared. They contain either pickets bearing chiral binaphthyl groups [120] or chiral "basket handles" including phenylalanine residues [121] on both sides of the porphyrin ring (Fig. 12). These catalysts were found to be able to catalyze epoxidation of alkenes with enantiomeric excess up to 50% in the case of styrene.

III-5-2. Catalysts for regioselective oxidations of alkanes or steroids

Hydroxylation of linear alkanes catalyzed by non-hindered Fe- or Mn-porphyrins such as Fe(TPP)(Cl) or Mn(TPP)(Cl) occurs almost exclusively on secondary CH_2 bonds and in an almost statistical manner. When the approach of the alkane toward the high-valency metal-oxo species is controlled by bulky substituents in the *ortho* position of the *meso*-aryl groups of the porphyrin, this regioselectivity changes in favor of the more accessible 1(ω) and 2 (ω-1)

positions of the alkane [93, 97, 122]. In particular, with the very hindered Mn(III)-tetra(2,4,6-triphenyl)phenyl-porphyrin catalyst (Fig. 8), hydroxylation of linear alkanes occurs mainly at the ω-1 but also in the ω position [97]. Interestingly, this steric control of the regioselectivity of the oxidation of linear alkanes may be obtained not only through the *meso*-aryl groups of the porphyrin but also through the ligand which is inserted into alkane C-H bonds. For instance, very recent results showed that the nitrene *N*-tosyl moiety of PhINTs, a nitrogen analogue of PhIO, is transferred to *n*-heptane in the presence of Mn(III)(TMP)(CF$_3$SO$_3$) as catalyst, with major formation of 1-tosylaminoheptane which is derived from the oxidation of CH_3 groups of *n*-heptane (about 50% of all tosylaminoheptanes formed in the reaction; Eqn 26) (J. P. Mahy, G. Bedi, P. Battioni, D. Mansuy, unpublished results).

Another interesting approach toward regioselective systems was to use iron-porphyrins bearing hydrophobic steroid pickets on both sides of the porphyrin ring included in lipid bilayer vesicles (Fig. 13). Such an organized system provides a channel for the access of substrates such as steroids with a hydrophobic area close to the iron centre. It was found to catalyze regioselective epoxidation of steroids [123].

III-6. APPLICATIONS OF MODEL SYSTEMS FOR THE OXIDATION OF SOME DRUGS OR XENOBIOTICS

Model systems using Fe(TPP)(Cl) or Mn(TPP)(Cl) as catalysts and iodosylbenzene or *t*-butylhydroperoxide as oxidants perform the oxidative dealkylation of *N*-nitrosoamines, a re-

$$\text{\textasciitilde}\text{\textasciitilde}\text{\textasciitilde} + \text{PhI} = \text{NTs} \xrightarrow{\text{Mn(III)(TMP)(CF}_3\text{SO}_3)} \text{\textasciitilde}\text{\textasciitilde}\text{\textasciitilde}\text{NHTs} \quad (48\%) \quad + \text{ secondary } N\text{-tosylamines } (52\%) \tag{26}$$

action which is involved in the toxicity and carcinogenicity of these compounds to experimental animals and which is, at least in part, dependent on cytochrome P-450. Evidence has been presented in favour of the abstraction of an α-hydrogen atom of the benzyl group by the active species [124] (Eqn 27).

Stereoselective epoxidations of steroid derivatives have been obtained with model systems. For instance, an estra-5(10), 9(11)-diene (Fig. 14) was oxidized by PhIO in the presence of iron(II)-phthalocyanine with major formation of the corresponding 5α,10α-epoxide [125]. Cholesterol acetate was epoxidized by air in the presence of a Ru(II)-porphyrin with the almost exclusive formation of the β-epoxide [126] (Fig. 14).

Very recently, the use of model systems has been applied to the study of drug oxidations. For instance two drugs containing a thiophene ring, tienilic acid and suprofen, were oxidized by C_6F_5IO in the presence of Mn(TDCPP)(Cl) with major formation of compounds hydroxylated at position 5 of the thiophene ring (Fig. 15) (E. Neau, P. Dansette, P. Battioni, D. Mansuy, unpublished results). These compounds are the major metabolites of these two drugs *in vivo* [127, 128], as well as *in vitro* in the presence of rat liver microsomes (E. Neau, P. Dansette, D. Mansuy, unpublished results). These are the first examples illustrating the potential interest of the use of such simple model systems for the prediction of drug metabolism and for the preparation of hydroxylated metabolites.

CONCLUSION

The use of iron-porphyrins to mimic cytochrome P-450 reactions has led to the complete characterization of many complexes containing various kinds of Fe-O, Fe-N and Fe-C bonds. This was important for our present understanding of the nature of P-450 – iron-metabolite complexes formed during the metabolism of many substrates. Moreover, this has made a large contribution to indicating the importance of the organometallic chemistry of cytochrome P-450 during its reactions with substrates. It has revealed the exceptional richness of the coordination chemistry of this cytochrome, and led to a better understanding of the detailed mechanisms of its reactions.

Major progress has been made towards efficient catalytically active model systems for cytochrome P-450 during this last decade. Very high catalytic activities for monooxygenase-type reactions have been obtained by using

$$(PhCH_2)_2NNO \xrightarrow[Fe(TPP)(Cl)]{PhIO} \underset{\underset{OH}{|}\underset{NO}{|}}{PhCH-N-CH_2Ph} \longrightarrow PhCH_2OH + PhCHO \qquad (27)$$

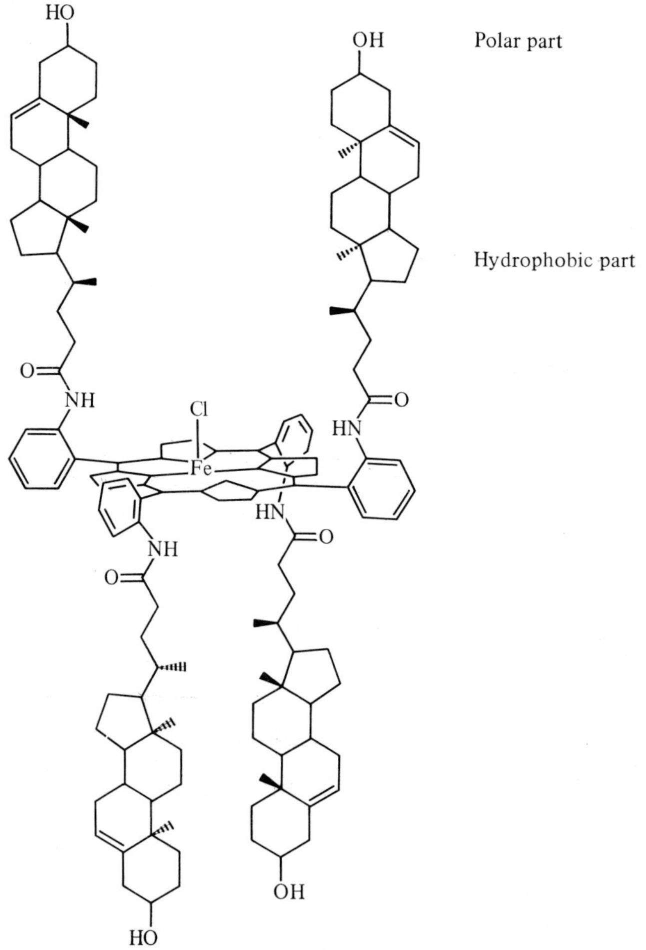

Fig. 13. *Iron-porphyrin, bearing steroid pickets and included in lipid bilayer vesicles, reported to catalyze regioselective oxidation of steroids (from [123])*

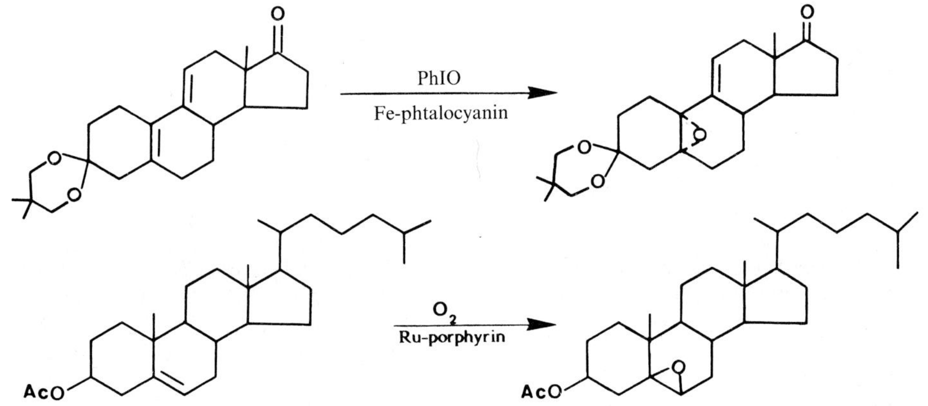

Fig. 14. *Regioselective epoxidations of steroids by model systems*

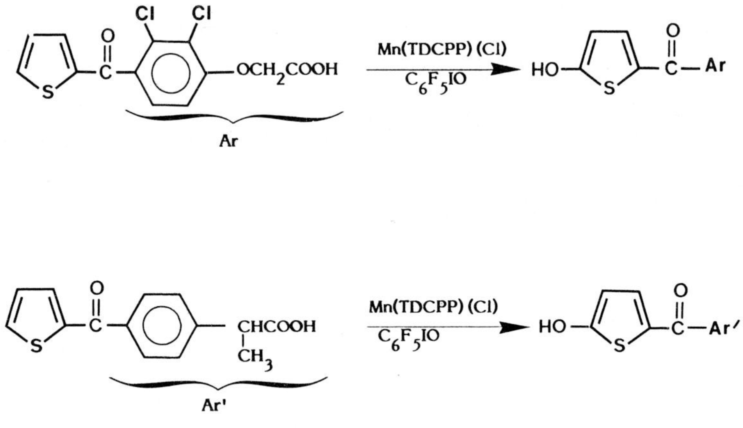

Fig. 15. *Oxidation of drugs containing a thiophene ring by model systems*

iodosylarenes and Fe- or Mn-porphyrins very resistant towards oxidative degradation. These catalysts are more potent and more resistant than cytochromes P-450 in the presence of the same oxidants. The use of pyridine, and especially imidazole, as cocatalysts in addition to Fe- and Mn-porphyrins has allowed more readily available oxidants such as ClO^-, alkylhydroperoxides, H_2O_2 or O_2 itself to be used in the presence of a reducing agent. These systems can now be employed for oxidations in fine chemistry as they lead to complete substrate conversions. They have already been used for drugs oxidation and appear to be good tools for the prediction of drug metabolism and the preparation of hydroxylated metabolites. Very much remains to be done to make these systems regioselective, although encouraging preliminary results have been obtained already for regioselective hydroxylation of linear alkanes and asymmetric epoxidation of alkenes.

REFERENCES

1. White, R. E. & Coon, M. J. (1980) *Annu. Rev. Biochem. 49*, 315−356.
2. Guengerich, F. P. & MacDonald, T. L. (1984) *Acc. Chem. Res. 17*, 9−16.
3. Ortiz de Montellano, P. R. (ed.) *Cytochrome P-450, structure, mechanism and biochemistry* (1986) Plenum Press, New York.
4. Ruckpaul K. & Rein, H. (eds) (1984) *Cytochrome P-450*, Akademie-Verlag, Berlin.
5. Nebert, D. W., Adesnik, M., Coon, M. J., Estabrook, R. W., Gonzalez, F. J., Guengerich, F. P., Gunsalus, I. C., Johnson, E. F., Kemper, B., Levin, W., Philipps, I. R., Sato, R. & Waterman, M. R. (1987) *DNA 6*, 1−11.
6. Collman, J. P. & Sorrell, T. N. (1977) *J. Am. Chem. Soc. Symp. Ser. 44*, 27−45.
7. Mansuy, D. (1981) in *Reviews in biochemical toxicology* (Hodgson, E., Bend, J. R. & Philpot, R. M., eds) pp. 283−320, Elsevier, New York.
8. Mansuy, D. (1983) in *The coordination chemistry of metalloenzymes in hydrolytic and oxidative process*, pp. 343−358 (Bertini, I., Drago, R. S. & Luchinat, C., eds) D. Reidel Publishing Company, Dordrecht.
9. Schappacher, M., Weiss, R., Montiel-Montoya, R., Trautwein, A. & Tabard, A. (1985) *J. Am. Chem. Soc. 107*, 3736−3738.
10. Poulos, T. L., Finzel, B. C., Gunsalus, I. C., Wagner, G. C. & Kraut, J. (1985) *J. Biol. Chem. 260*, 16122−16130.
11. Poulos, T. L., Finzel, B. C. & Howard, A. J. (1987) *J. Mol. Biol. 195*, 687−700.
12. Groves, J. T., Haushalter, R. C., Nakamura, M., Nemo, T. E. & Evans, B. J. (1981) *J. Am. Chem. Soc. 103*, 2884−2886.
13. Penner-Hahn J. E., McMurry J. T., Renner, M., Latos-Grazynski, L., Eble, K. S., Davis, I. M., Balch, A. L., Groves, J. T., Dawson, J. H. & Hodgson K. O. (1983) *J. Biol. Chem. 258*, 12761−12764.
14. Sugimoto, H., Tung, H. C. & Sawyer, D. T. (1988) *J. Am. Chem. Soc. 110*, 2465−2470.
15. Groves, J. T., McClusky, G. A., White, R. E. & Coon, M. J. (1978) *Biochem. Biophys. Res. Commun. 81*, 154−160.
16. Ortiz de Montellano, P. R. & Stearns, R. A. (1987) *J. Am. Chem. Soc. 109*, 3415−3420.

17. Ortiz de Montellano, P. R. (1984) *Annu. Rep. Med. Chem. 19*, 201–211.
18. Hines, R. N. & Prough R. A. (1980) *J. Pharmacol. Ther. 214*, 80–86.
19. Mansuy, D., Battioni, P. & Mahy, J. P. (1982) *J. Am. Chem. Soc. 104*, 4487–4489.
20. Mahy, J. P., Battioni, P., Mansuy, D., Fischer, J., Weiss, R., Mispelter, J., Morgenstern-Badarau, I. & Gans, P. (1984) *J. Am. Chem. Soc. 106*, 1699–1706.
21. Franklin, M. R. (1977) *Pharmacol. Ther. A. 2*, 227–245.
22. Mansuy, D. (1987) *Pharmacol. Ther. 23*, 41–45.
23. Pessayre, D., Larrey, D., Vitoux, J., Breil, P., Belghiti, J. & Benhamou, J. P. (1982) *Biochem. Pharmacol. 31*, 1699–1704.
24. Varoquaux, O., Advenier, C. & Renier, R. (1981) *Gaz. Med. Fr. 88*, 1625–1629.
25. Mansuy, D., Beaune, P., Chottard, J. C., Bartoli, J. F. & Gans, P. (1976) *Biochem. Pharmacol 25*, 609–612.
26. Jonsson, J. & Lindeke, B. (1976) *Acta Pharm. Suec. 13*, 313–320.
27. Mansuy, D., Gans, P., Chottard, J. C. & Bartoli, J. F. (1977) *Eur. J. Biochem. 76*, 607–615.
28. Mansuy, D., Battioni, P., Chottard, J. C. & Lange, M. (1977) *J. Am. Chem. Soc. 99*, 6441–6443.
29. Mansuy, D., Battioni, P., Chottard, J. C., Riche, C. & Chiaroni, A. (1983) *J. Am. Chem. Soc. 105*, 455–463.
30. Mansuy, D., Chottard, J. C. & Chottard, G. (1977) *Eur. J. Biochem. 76*, 617–624.
31. Wolf, C. R., Mansuy, D., Nastainczyk, W., Deutschmann, G. & Ullrich, V. (1977) *Mol. Pharmacol.*, 698–705.
32. Mansuy, D. & Fontecave, M. (1983) *Biochem. Pharmacol. 32*, 1871–1879.
33. Uehleke, H., Hellmer, K. H. & Tabarelli, S. (1973) *Naunyn-Schmiedeberg's Arch. Pharmacol. 279*, 39–52.
34. Mansuy, D., Fontecave, M. & Battioni, J. P. (1982) *J. Chem. Soc., Chem. Commun.*, 317–320.
35. Mansuy, D. & Battioni, J. P. (1982) *J. Chem. Soc., Chem. Commun.*, 638–639.
36a. Guilard, R. & Kadish, K. M. (1988) *Chem. Rev. 88*, 1121–1146.
36b. Brothers, P. J. & Collman, J. P. (1986) *Acc. Chem. Res. 19*, 209–215.
36c. Setsune, J. I. & Dolphin, D. (1987) *Can. J. Chem. 65*, 459–467.
37. Mansuy, D., Lange, M., Chottard, J. C., Guérin, P., Morière, P., Brault, D. & Rougée, M. (1977) *J. Chem. Soc., Chem. Commun.*, 648–649.
38. Mansuy, D., Lange, M., Chottard, J. C., Bartoli, J. F., Chevrier, B. & Weiss, R. (1978) *Angew. Chem., Int. Ed. Engl. 17*, 780–782.
39. Mansuy, D. (1980) *Pure Appl. Chem. 52*, 681–690.
40. Ruf. H. H., Ahr, H., Nastainczyk, W., Ullrich, V., Mansuy, D., Battioni, J. P., Montiel-Montoya, R. & Trautwein, A. (1984) *Biochemistry 23*, 5300–5306.
41. Ahr, H. J., King, L. J., Nastainczyk, W. & Ullrich, V. (1982) *Biochem. Pharmacol. 31*, 383–390.
42. Jonen, H. G., Werringloer, J., Prough, R. A. & Estabrook, R. W. (1982) *J. Biol. Chem. 257*, 4404–4411.
43. Battioni, P., Mahy, J. P., Delaforge, M. & Mansuy, D. (1983) *Eur. J Biochem. 134*, 241–248.
44. Delaforge, M., Battioni, P., Mahy, J. P. & Mansuy, D. (1986) *Chem-Biol. Interact. 60*, 101–114.
45. Battioni, P., Mahy, J. P., Gillet, G. & D Mansuy (1983) *J. Am. Chem. Soc. 105*, 1399–1401.
46. Saito, S. & Itano, H. A. (1981) *Proc. Natl Acad. Sci. USA 78*, 5508–5512.
47. Kunze, K. L. & Ortiz de Montellano, P. R. (1983) *J. Am. Chem. Soc. 103*, 1380–1382.
48. Mansuy, D., Battioni, P., Mahy, J. P. & Gillet, G. (1982) *Biochem. Biophys. Res. Commun. 106*, 30–36.
49. Ringe, D., Petsko, G. A., Kerr, D. E. & Ortiz de Montellano. P. R. (1984) *Biochemistry 23*, 2–4.
50. Ortiz de Montellano, P. R. & Kerr, D. E. (1983) *J. Biol. Chem. 258*, 10558–10563.
51. Ullrich, V. (1977) in *Biological reactive intermediates* (Jollow, D. J., Kocsis, J. J., Snyder, R. & Vaino, H., eds) pp. 65–80, Plenum Press, New York.
52. Mansuy, D., Battioni, J. P., Chottard, J. C. & Ullrich, V. (1979) *J. Am. Chem. Soc. 101*, 3971–3973.
53. Ortiz de Montellano, P. R. & Grab, L. A. (1986) *J. Am. Chem. Soc. 108*, 5584–5589.
54. Artaud, I., Grégoire, N., Battioni, J. P., Dupré, D. & Mansuy, D. (1989) *J. Am. Chem. Soc. 110*, in the press.
55. Mansuy, D., Chottard, J. C., Lange, M. & Battioni, J. P. (1980) *J. Mol. Catal. 7*, 215–226.
56. Groves, J. T., Avaria-Neisser, G. E., Fish, K. M., Imachi, M. & Kuczkowski, R. (1986) *J. Am. Chem. Soc. 108*, 3837–3838.
57. Collman, J. P., Kodadek, T., Raybuck, S. A., Brauman, J. I. & Papazian, L. M. (1985) *J. Am. Chem. Soc. 107*, 4343–4345.
58. Groves, J. T. & Watanabe, Y. (1986) *J. Am. Chem. Soc. 108*, 507–508.
59. Mansuy, D., Battioni, J. P., Dupré, D., Sartori, E. & Chottard, G. (1982) *J. Am. Chem. Soc. 104*, 6159–6161.
60. Lançon, D., Cocolios, P., Guilard, R. & Kadish, K. M. (1984) *J. Am. Chem. Soc. 106*, 4472–4478.
61. Balch, A. L. & Renner, M. W. (1986) *J. Am. Chem. Soc. 108*, 2603–2608.
62. Lavallee, D. K. (1987) *The chemistry and biochemistry of N-substituted porphyrins*, VCH Publishers, New York
63. Chevrier, B., Weiss, R., Lange, M., Chottard, J. C. & Mansuy, D. (1981) *J. Am. Chem. Soc. 103*, 2899–2901.
64. Latos-Grazynski, L., Cheng, R. J., La Mar, G. N. & Balch, A. L. (1981) *J. Am. Chem. Soc. 103*, 4270–4272.
65. Mahy, J. P., Battioni, P., Bedi, G., Mansuy, D., Fischer, J., Weiss, R. & Morgenstern-Badarau, I. (1988) *Inorg. Chem. 27*, 353–359.
66. Mansuy, D., Devocelle, L., Artaud, I. & Battioni, J. P. (1985) *Nouv. J. Chim. 9*, 711–716.
67. Mashiko, T., Dolphin, D., Nakano, T. & Traylor, T. G. (1985) *J. Am. Chem. Soc. 107*, 3735–3736.
68. Collman, J. P., Hampton, P. D. & Brauman, J. I. (1986) *J. Am. Chem. Soc. 108*, 7861–7862.
69. Traylor, T. G., Nakano, T., Miksztal, A. R. & Dunlap, D. P. (1987) *J. Am. Chem. Soc. 109*, 3625–3632.
70. Battioni, J. P., Artaud, I., Dupré, D., Leduc, P., Akhrem, I., Mansuy, D., Fischer, J., Weiss, R. & Morgenstern-Badarau, I. (1986) *J. Am. Chem. Soc. 109*, 5598–5607.
71. Artaud, I., Devocelle, L., Battioni, J. P., Girault, J. P. & Mansuy, D. (1987) *J. Am. Chem. Soc. 109*, 3782–3783.
72. Traylor, T. G. & Miksztal, A. R. (1987) *J. Am. Chem. Soc. 109*, 2770–2774.
73. Artaud, I., Devocelle, L., Grégoire, N., Battioni, J. P. & Mansuy, D. (1987) *Rec. Trav. Chim. Pays-Bas 106*, 336.
74. Gra, L. A., Swanson, B. A. & Ortiz de Montellano, P. R. (1988) *Biochemistry 27*, 4805–4814.
75. Komives, E. A., Tew, D., Olmstead, M. M. & Ortiz de Montellano, P. R. (1988) *Inorg. Chem. 27*, 3112–3117.
76. Tabushi, I. (1988) *Coord. Chem. Rev. 86*, 1–42.
77. Groves, J. T., Nemo, T. E. & Myers, R. S. (1979) *J. Am. Chem. Soc. 101*, 1032–1033.
78. Meunier, B. (1986) *Bull. Soc. Chim. Fr. 4*, 578–594.
79a. McMurry, T. J. & Groves, J. T. (1986) in *Cytochrome P-450, structure, mechanism and biochemistry* (Ortiz de Montellano, P. R., ed.) pp. 1–29, Plenum Press, New York.
79b. Bruice, T. C. (1986) *Ann. NY Acad. Sci. 471*, 83–98.
79c. Mansuy, D. (1987) *Pure Appl. Chem. 59*, 759–770.
79d. Mansuy, D. & Battioni, P. (1989) in *Frontiers of biotransformation* (Ruckpaul, K. & Rein, H., ed.) Akademie-Verlag, Berlin, in the press.
80. Groves, J. T. & Nemo, T. E. (1983) *J. Am. Chem. Soc. 105*, 6243–6248.
81. Lindsay-Smith, J. R. & Sleath, P. R. (1983) *J. Chem. Soc., Perkin Trans II*, 1165–1169.

82. Groves, J. T. & Nemo, T. E. (1983) *J. Am. Chem. Soc. 105*, 5786–5791.
83. Lindsay-Smith, J. R. & Sleath, P. R. (1982) *J. Chem. Soc., Perkin Trans II*, 1009–1015.
84. Groves, J. T. & Subramanian, D. V. (1984) *J. Am. Chem. Soc. 106*, 2177–2181.
85. Mansuy, D., Leclaire, J., Fontecave, M. & Momenteau, M. (1984) *Biochem. Biophys. Res. Commun. 119*, 319–325.
86. Collman, J. P., Kodadek, T. & Brauman, J. I. (1986) *J. Am. Chem. Soc. 108*, 2588–2594.
87. Chang, C. K. & Ebina, F. (1981) *J. Chem. Soc., Chem. Commun.*, 778–779.
88. Traylor, P. S., Dolphin, D. & Traylor, T. G. (1984) *J. Chem. Soc., Chem. Commun.*, 279–280.
89. Traylor, T. G., Marsters, J. C., Nakano, T. & Dunlap, B. E. (1985) *J. Am. Chem. Soc. 107*, 5537–5539.
90. Traylor, T. G. & Tsuchiya, S. (1987) *Inorg. Chem. 26*, 1338–1339.
91. Dicken, C. M., Lu, F. L., Nee, M. W. & Bruice, T. C. (1985) *J. Am. Chem. Soc. 107*, 5776–5789.
92. De Poorter, B. & Meunier, B. (1985) *Nouv. J. Chim. 9*, 393–394.
93. Mansuy, D., Bartoli, J. F. & Momenteau, M. (1982) *Tetrahedron Lett. 23*, 2781–2784.
94. Mansuy, D., Bartoli, J. F., Chottard, J. C. & Lange, M. (1980) *Angew. Chem. Int. Ed. Engl. 19*, 909–910.
95. Balasubramanian, P. N., Sinha, A. & Bruice, T. C. (1987) *J. Am. Chem. Soc. 109*, 1456–1462.
96. Mansuy, D., Battioni, P. & Renaud, J. P. (1984) *J. Chem. Soc., Chem. Commun.*, 1255–1257.
97. Cook, B. R., Reinert, T. J. & Suslick, K. S. (1986) *J. Am. Chem. Soc. 108*, 7281–7286.
98. Renaud, J. P., Battioni, P., Bartoli, J. F. & Mansuy, D. (1985) *J. Chem. Soc., Chem. Commun.*, 888–889.
99. Battioni, P., Renaud, J. P., Bartoli, J. F. & Mansuy, D. (1986) *J. Chem. Soc., Chem. Commun.*, 341–343.
100. Battioni, P., Renaud, J. P., Bartoli, J. F., Reina-Artiles, M., Fort, M. & Mansuy, D. (1988) *J. Am. Chem. Soc. 110*, 8462–8470.
101. Battioni, P., Renaud, J. P., Bartoli, J. F., Momenteau, M. & Mansuy, D. (1987) *Rec. Trav. Chim. Pays-Bas 106*, 332.
102. Zipplies, M. F., Lee, W. A. & Bruice, T. C. (1986) *J. Am. Chem. Soc. 108*, 4433–4445.
103. Tabushi, I. & Koga, N. (1979) *J. Am. Chem. Soc. 101*, 6456–6458.
104. Perrée-Fauvet, M. & Gaudemer, A. (1981) *J. Chem. Soc., Chem. Commun.* 874–875.
105. Santa T., Mori, T. & Hirobe, M. (1985) *Chem. Pharm. Bull. 33*, 2175–2178.
106. Mori, T., Santa, T. & Hirobe, M. (1985) *Tetrahedron Lett. 26*, 5555–5558.
107. Mansuy, D., Fontecave, M. & Bartoli, J. F. (1983) *J. Chem. Soc., Chem. Commun.*, 253–254.
108. Fontecave, M. & Mansuy, D. (1984) *Tetrahedron 40*, 4297–4311.
109. Tabushi, I. & Yazaki, A. (1981) *J. Am. Chem. Soc. 103*, 7371–7373.
110. Tabushi, I. & Morimitsu, K. (1984) *J. Am. Chem. Soc. 106*, 6871–6872.
111. Tabushi, I., Kodera, M. & Yokoyama, M. (1985) *J. Am. Chem. Soc. 107*, 4466–4473.
111. Tabushi, I., Kodera, M. & Yokoyama, M. (1985) *J. Am. Chem. Soc. 107*, 4466–4473.
112. Khenkin, A. M. & Shteinman, A. A. (1984) *J. Chem. Soc., Chem. Commun.* 1219–1220.
113. Creager, S. E., Raybuck, S. A. & Murray, R. W. (1986) *J. Am. Chem. Soc. 108*, 4225–4227.
114. Leduc, P., Battioni, P., Bartoli, J. F. & Mansuy, D. (1988) *Tetrahedron Lett. 29*, 205–208.
115. Groves, J. T. & Watanabe, Y. (1986) *J. Am. Chem. Soc. 108*, 7834–7836.
116. Groves, J. T., Watanabe, Y. & McMurry, T. J. (1983) *J. Am. Chem. Soc. 105*, 4489–4490.
117. Tabushi, I. & Kodera, M. (1986) *J. Am. Chem. Soc. 108*, 1101–1103.
118. Karasevich, E. I., Khenkin, A. M. & Shilov, A. E. (1987) *J. Chem. Soc., Chem. Commun.*, 731–732.
119. Battioni, P., Bartoli, J. F., Leduc, P., Fontecave, M. & Mansuy, D. (1987) *J. Chem. Soc., Chem. Commun.*, 791–792.
120. Groves, J. T. & Myers, R. S. (1983) *J. Am. Chem. Soc. 105*, 5791–5796.
121. Mansuy, D., Battioni, P., Renaud, J. P. & Guérin, P. (1985) *J. Chem. Soc., Chem. Commun.*, 155–156.
122. Khenkin, A., Koifman, O., Semeikin, A., Shilov, A. & Shteinman, A. (1985) *Tetrahedron Lett. 26*, 4247–4248.
123. Groves, J. T. & Neumann, R. (1987) *J. Am. Chem. Soc. 109*, 5045–5048.
124. Lindsay-Smith, J. R., Nee, M. W., Noar, J. B. & Bruice, T. C. (1984) *J. Chem. Soc. Perkin Trans II*, 255–260.
125. Rohde, R., Neef, G., Sauer, G. & Wiechert, R. (1985) *Tetrahedron Lett. 26*, 2069–2072.
126. Marchon, J. C. & Ramasseul, R. (1988) *J. Chem. Soc., Chem. Commun.*, 298–299.
127. Mansuy, D., Dansette, P., Jaouen, M., Moinet, G. & Bayer, N. (1984) *Biochem. Pharmacol. 33*, 1429–1435.
128. Mori, Y., Sakai, Y., Kuroda, N., Yokoya, F., Toyoshi, K., Horie, M. & Baba, S. (1984) *Drug. Metab. Dispos. 12*, 767–771.

Review

Growth factors as transforming proteins

Carl-Henrik HELDIN[1] and Bengt WESTERMARK[2]

[1] Ludwig Institute for Cancer Research, Biomedical Center, Uppsala
[2] Department of Pathology, University Hospital, Uppsala

(Received March 7, 1989) — EJB 89 0275

INTRODUCTION

The growth of cells in culture is controlled by stimulatory or inhibitory factors. More than 20 different polypeptide growth factors have been structurally and functionally characterized, including cloning of their cDNAs (Table 1). The total number of polypeptide growth factors is likely to be considerably higher, since additional factors, that have not yet been fully characterized, are known from their specific biological activities. There are also factors that specifically inhibit cell growth, e.g. transforming growth factor β (TGF-β) [1], interferons [2] and tumor necrosis factor [3]. Moreover, the same factor may, depending on cell type and culture conditions, act both as a growth factor and a growth inhibitor [4a].

The *in vivo* functions of growth regulatory factors are assumed to be to stimulate fetal and placental growth during development, to regulate growth and differentiation of continuously regenerating tissues, e.g. in hemopoiesis, and to stimulate tissue repair processes.

The purpose of this review is to summarize data indicating that unscheduled production of growth factors by cells carrying the corresponding receptor leads to an autocrine stimulation of cell growth, and to discuss the possibility that such autocrine mechanisms are involved in cell transformation and tumorigenesis.

Mechanism of action of growth factors

Growth factors exert their mitogenic effect by interaction with specific cell surface receptors on responsive cells. Many growth factors form families of structurally related molecules that bind to a common receptor (Table 1). The binding of a growth factor to its receptor elicits a cascade of events, including protein phosphorylation, inositol-lipid breakdown, ion fluxes and changes in gene expression. Although attempts are being made to integrate these events into a scheme of molecular mechanisms involved in growth stimulation, our understanding of even the early phase of mitogenesis is still poor.

Many growth factor receptors are equipped with protein tyrosine kinase activities that become activated after ligand binding (Table 1). Recent cDNA cloning of several such receptors, has revealed that they consist of an external ligand-binding domain that is connected with the internal effector domain by a single stretch of hydrophobic amino acids constituting the transmembrane part of the receptor (4b). Ligand binding induces activation of the tyrosine kinase leading to autophosphorylation of the receptor, as well as to phosphorylation of cytoplasmic substrates. In some cases, autophosphorylation has been found to serve a regulatory function for the kinase activity. The molecular mechanism whereby ligand binding leads to activation of the receptor kinase remains to be elucidated, but recent data suggest that receptor dimerization is involved in the activation both of the platelet-derived growth factor (PDGF) type B receptor [5] and the epidermal growth factor (EGF) receptor [6]. It is conceivable that ligand-induced dimerization makes possible an interaction between the tyrosine kinase domains of the cytoplasmic parts of two receptors, leading to their activation.

The fact that many growth-factor receptors are protein-tyrosine kinases indicates that phosphorylation of specific substrates on tyrosine residues is important in growth stimulation [7]. Although the knowledge of substrates for tyrosine kinases is still incomplete, recent observations indicate that some substrates are themselves kinases phosphorylating serine/threonine residues [8], as well as tyrosine residues [9, 10] in proteins. Thus, activation of growth-factor receptors appears to initiate a cascade of intracellular phosphorylation events.

Another early event in growth-factor-stimulated cells is stimulation of phosphatidylinositol turnover leading to the degradation of phosphatidylinositol bisphosphate and release of diacylglycerol and inositol trisphosphate, both of which are implicated in growth stimulation. Diacylglycerol stimulates the activity of protein kinase C, an enzyme that also is activated by certain tumor promoters [11]. Inositol trisphosphate mobilizes Ca^{2+} from internal stores, and thereby raises the cytoplasmic Ca^{2+} concentration [12]. The molecular mechanism whereby growth factors stimulate degradation of phosphatidylinositol bisphosphate remains to be elucidated, but there are some recent data to indicate that the enzyme involved, phospholipase C, may undergo phosphorylation on tyrosine residues [13]. The importance of the phosphatidylinositol cycle in regulation of cell growth remains to be established, however, since there is a poor correlation between the ability of different growth factors to stimulate phosphatidylinositol turnover and to stimulate cell growth. Moreover, PDGF also stimulates growth of cells that have been depleted of protein kinase C [14].

Correspondence to C.-H. Heldin, Ludwig Institute for Cancer Research, Box 595, Biomedical Center, S-751 23 Uppsala, Sweden

Abbreviations. CSF, colony-stimulating factor; EGF, epidermal growth factor; FGF, fibroblast growth factor; HTLV, human T-cell leukemia virus; IGF, insulin-like growth factor; IL, interleukin; MGSA, melanoma growth stimulatory activity; PDGF, platelet-derived growth factor; SSV, simian sarcoma virus; TGF, transforming growth factor.

Table 1. *Growth factors*

Factors are grouped in families according to their structural similarities, except in the cases of IL and CSF, where the designations refer to structurally unrelated factors that primarily affect lymphoid and hemopoietic cells, respectively. Factors that act primarily as growth inhibitors or differentiation factors, such as TGF-β, interferon and tumor necrosis factor, have not been included in the table. For references see the text and [205]. Molecular sizes are given as determined from migrations in SDS/gel electrophoresis. PD-ECGF, platelet-derived endothelial-cell growth factor; GRP, gastrin-related peptide

Family	Growth factor	Structure of mature product	Receptor	Target cells
PDGF	PDGF-AA PDGF-AB PDGF-BB	dimers of A chains (17 kDa) and B chains (c-*sis*) (16 kDa) in different combinations	type A — binds PDGF-AA, PDGF-AB and PDGF-BB, 170 kDa, protein tyrosine kinase; type B — binds PDGF-BB, and at lower affinity PDGF-AB, 180 kDa, protein tyrosine kinase	mesenchymal cells, glial cells
EGF	EGF TGF-α vaccinia-virus growth factor	monomers of 6—9 kDa displaying amino acid sequence similarity to each other, and formed by proteolytic cleavage of membrane-bound precursors	c-*erbB*, 175 kDa, protein tyrosine kinase	epithelial cells, mesenchymal cells, glial cells
FGF	acidic FGF basic FGF	monomers of $\approx$ 17 kDa displaying amino acid sequence similarity to each other and to the IL-1 family	130 kDa, protein tyrosine kinase	endothelial cells, mesenchymal cells, endocrine epithelial cells, neuronal cells
	hst/KS3 prod.	monomer of glycosylated 19-kDa peptide	?	endothelial cells, fibroblasts
	int-2 product	monomer of glycosylated 24-kDa peptide	?	?
	FGF-5	monomer of glycosylated 26-kDa peptide	?	fibroblasts
PD-ECGF		monomer of 45 kDa	?	endothelial cells
IGF	insulin	two chains of 5.7 kDa formed from proinsulin by proteolysis	(130 kDa, 90 kDa)$_2$, protein tyrosine kinase	various cells types
	IGF-I	monomer of 7 kDa homologous to proinsulin	(130 kDa, 90 kDa)$_2$, protein tyrosine kinase	various cell types
	IGF-II	monomer of 7 kDa homologous to proinsulin	250 kDa, identical to the mannose-6-phosphate receptor	various cell types
MGSA		monomer of 16 kDa	?	melanoma cells, fibroblasts
Bombesin	GRP neuromedin B neuromedin C litorin	monomers of short peptides with a homologous carboxy-terminal hepta-peptide	75—85 kDa	fibroblasts, bronchial epithelial cells
IL	IL-1α IL-1β	monomers of 15—30 kDa displaying amino acid sequence similarity to each other and to the FGF family	80 kDa, protein tyrosine kinase	various cell types
	IL-2	monomer of glycosylated 15-kDa peptide	55 kDa and 70 kDa	T cells, B cells
	IL-3 (multi CSF)	monomer of glycosylated 15-kDa peptide	140 kDa, protein tyrosine kinase	hemopoietic cells
	IL-4	monomer of glycosylated 15-kDa peptide	140 kDa and 70 kDa	B cells, T cells, mast cells
	IL-5	dimer of glycosylated 14-kDa peptides	?	B cells, eosinophilic granulocytes
	IL-6	monomer of 21 kDa	80 kDa	activated B cells, B cell hybridomas
	IL-7	monomer of glycosylated 15-kDa peptide	?	lymphoid progenitor cells
CSF	G-CSF	monomer of glycosylated 19-kDa peptide	150 kDa	granulocyte lineage of hemopoietic cells, endothelial cells
	M-CSF (CSF-1)	two variants due to differential splicing: homodimers of glycosylated 14-kDa or 21-kDa chains formed from membrane-bound precursors	c-*fms*, 150 kDa, protein tyrosine kinase	macrophage lineage of hemopoietic cells
	GM-CSF	monomer of glycosylated 14-kDa peptide	51 kDa	granulocyte/macrophage lineage of hemopoietic cells, endothelial cells
	erythropoietin	monomer of glycosylated 18-kDa peptide	?	erythroid lineage of hemopoietic cells

Specific, growth-regulated genes are induced 10 min to a few hours after the addition of growth factor [15—18]. The precise function of the corresponding products in the mitogenic pathway remains to be elucidated; some are secreted proteins with a potentially modulatory role on cell growth, others are, for example, transcription factors with a possible role in the regulation of the machinery that drives the cell cycle [18].

Subversion of the mitogenic pathway of growth factors in cell transformation

Recent studies on oncogenes have given several examples that the normal counterparts of oncogenes (proto-oncogenes) code for proteins that have a regulatory function in the mitogenic pathway of growth factors (reviewed in [19]). There are examples of transforming proteins that act as growth factors, others represent structurally altered versions of growth-factor receptors, and yet others are activated components of the intracellular messenger system. Available data support the hypothesis that oncogene products act by subverting the mitogenic pathway, thereby giving the cell a constitutive growth stimulus which leads to uncontrolled cell growth.

AUTOCRINE GROWTH FACTORS IN CELL TRANSFORMATION

In 1967, Howard Temin suggested that the growth of transformed cells may be driven by endogenously produced growth factors [20]. This idea was later developed into the autocrine concept of cell growth [21, 22]. A prerequisite for autocrine growth stimulation is that the same cell both produces a growth factor and expresses the corresponding receptor. There are several examples of growth-factor-producing cells lines. In some cases, evidence has been presented that such cell lines are dependent on the autocrine loop for their growth *in vitro*; conclusive evidence that autocrine stimulation has a role in tumor growth *in vivo* is, however, still lacking.

In theory, the aberrant expression of a growth-factor receptor in a cell that has as its normal function production of the corresponding growth factor, could also lead to autocrine growth stimulation. Anomalous growth-factor-receptor expression has been noticed in human tumor cell lines [23] but examples where such an expression is linked to a true autocrine response have not yet been found. This possibility should be kept in mind, however, since there are indications that different growth-factor receptors can use the same intracellular signaling system [24].

Platelet-derived growth factor

PDGF is a major growth factor for connective tissue cells (for review see [25, 26]). Structurally, it is a dimer of disulphide-bonded A and B polypeptide chains. All three dimeric combinations have been identified and purified from platelets and transformed cells [27—29]. The different isoforms were found to differ in functional activities. Thus, in contrast to PDGF-AB, PDGF-AA had only a low mitogenic activity and no effect on actin reorganization or chemotaxis in human foreskin fibroblasts [30].

PDGF-AA is, however, a potent mitogen for Swiss 3T3 cells [31] and oligodendrocyte progenitor cells of the developing rat optic nerve [32].

A molecular basis for the differences in functional activities between the isoforms of PDGF was elucidated by the demonstration that two different PDGF-receptor types exist, denoted type A and type B [33, 34]. The type-A receptor binds all three PDGF isoforms, whereas the type-B receptor binds only PDGF-BB with high affinity, PDGF-AB with lower affinity and PDGF-AA with very low or no affinity [33—35].

The transforming potential of the B chain of PDGF is illustrated by the fact that its gene has been transduced by two different retrovirus isolates, viz. simian sarcoma virus (SSV) [36] and Parodi-Irgens feline sarcoma virus [37]. Studies of SSV-transformed cells *in vitro* support the hypothesis that a factor similar to PDGF-BB exerts the transforming activity of SSV (reviewed in [38]). Firstly, SSV-transformed cells *in vitro* produce a growth factor that binds to PDGF receptor(s) and is recognized by anti-PDGF sera [39—44]. Secondly, only cell types that have PDGF receptors are transformed by SSV [45]. Thirdly, SSV transformation is inhibited by agents that interfere with the interaction between PDGF-BB and its receptor(s), such as anti-PDGF antibodies [43, 46] and suramin [47]. Moreover, the transforming activity of v-*sis* is lost when the hydrophobic *env*-derived leader sequence, which causes transfer of the v-*sis* product into the endoplasmic reticulum and allows it to interact with the ligand-binding domain of the receptors, is deleted [48, 49].

Genomic [50] or cDNA [51, 52] c-*sis* sequences under the control of viral promoters, have been shown to cause cell transformation in transfection experiments. Similar experiments with A-chain cDNA have revealed that it too has transforming potential, albeit less efficient compared to the B chain [53, 54].

The transforming potential of PDGF raises the question whether an unscheduled expression of PDGF-A or PDGF-B chain is involved in the genesis of spontaneous human tumors. The finding that PDGF-A and PDGF-B-chain mRNA is often expressed in cell lines of human tumors is in support of this possibility [25, 26]. For example, human glioma cell lines have been found to produce PDGF-like factors at high frequency; out of 23 glioma cell lines investigated, 23 and 17, were found to express PDGF-A- and PDGF-B-chain mRNA, respectively [55]. Analysis of the protein products of glioma cell lines revealed that all three dimeric combinations of PDGF were synthesized by one cell line expressing both A- and B-chain mRNA [56]. Since many of the cell lines also express PDGF-A- and PDGF-B-type receptors, autocrine stimulation of growth may indeed occur, although formal evidence is lacking. Investigation of mRNA expression in sections of human glioblastoma by *in situ* techniques revealed that PDGF-B-chain and B-type-receptor mRNA were expressed not only in the tumor cells but also in endothelial cells [57]. This suggests that autocrine stimulation involving PDGF-like factors may be of importance in the growth of normal as well as malignant cells in human glioblastoma.

There are also several examples of PDGF-receptor-negative cell types that produce PDGF, e.g. mammary carcinoma cell lines [58—60], lung carcinoma cell lines [61], melanoma cell lines [62] and leukemia cell lines [63, 64]. In these cases it is highly unlikely that the endogenously produced PDGF has any autocrine function; it is possible, however, that it, in some cases, could contribute to the formation of connective tissue stroma *in vivo*, via a paracrine mechanism.

Transforming growth factor-α

Transforming growth factor α (TGF-α) [65—67], epidermal growth factor [68, 69] and vaccinia virus growth factor

[70, 71] have about 40% amino acid similarity and bind to the EGF receptor (Table 1). All three molecules are synthesized as membrane-bound precursors; the active molecules are released from the cell membrane after specific proteolysis. TGF-α is synthesized by cells transformed by certain retroviruses; it is not the product of any oncogene *per se*, but its synthesis is regulated by a number of different oncogenes, e.g. *abl*, *mos*, *fes*, *fms* and *ras* [72–77]. TGF-α is also produced by certain cells transformed by DNA tumor viruses [74, 78] and by cell lines established from human tumors [55, 79–81]. In contrast, expression of EGF in tumor cell lines is less common; one example found, a human salivary gland adenocarcinoma cell line, may represent a normal function of this cell type [82].

The designation 'transforming' growth factor α represents somewhat of a misnomer; it has been given to the factor since it stimulates growth in soft agar of certain cells in synergy with TGF-β [83], an effect that TGF-α shares with other growth factors [84].

Temperature-sensitive mutants of Moloney and Kirsten sarcoma viruses were used to address the question whether TGF-α production serves an autocrine function in virus-induced cell transformation [73, 85]. At the permissive temperature, the cells produce TGF-α, have low serum requirement, downregulated EGF receptors and do not respond to exogenous EGF. At the non-permissive temperature the cells do not produce TGF-α, have EGF receptors and respond to EGF. These finding clearly indicate that oncogene expression regulates TGF-α production, but can not be taken as a proof that autocrine stimulation by TGF-α is important in the transformation process. Rather, data have been presented indicating that *ras* transformation occurs equally well in cells that are devoid of EGF receptors and thereby not responsive to TGF-α [86].

TGF-α [87–89] and EGF [90] gene sequences, placed under the control of viral promoters, were found to have transforming effects on cells *in vitro* and *in vivo*; however, they were considerably weaker than, for example, corresponding constructs with the *sis* oncogene/PDGF-B-chain gene. The frequent expression of TGF-α in EGF-receptor-containing cell lines [55, 81, 91] and the fact that factors interacting with the EGF receptor have been found in the urine of cancer patients [92], suggest that autocrine stimulation of cell growth involving TGF-α may operate in the development of human neoplasia.

Fibroblast growth factor and related factors

The fibroblast growth factor (FGF) family of growth factors currently consists of five members (Table 1). The prototype factors, acidic and basic FGFs, are endothelial cell mitogens with about 55% amino acid identity, that bind to the same receptor (reviewed in [93]). The two FGFs also share the property of having a strong affinity for heparin, which has simplified their identification, characterization and purification from various sources. FGFs have been shown to stimulate the formation of new blood vessels *in vivo*; factors in this family have therefore attracted considerable interest as potential tumor-derived angiogenic factors (reviewed in [94]).

Acidic and basic FGF were originally purified from brain and pituitary tissue, but have subsequently been found in a variety of normal [95–100] and transformed [96, 101–105] cells. Several of the malignant producer cell types also respond to FGF, e.g. glioma [103], melanoma [104] and rhabdomyosarcoma [102]; FGF may thus have the dual function of both stimulating the producer cells by an autocrine mechanism and inducing neovascularization of the tumor by a paracrine mechanism. Both these mechanisms, however, require FGF to be secreted from the producer cell; how this can occur remains to be elucidated, since cloning of cDNAs for acidic [106] and basic [107] FGF revealed no signal sequences. That the gene for basic FGF is indeed a potential oncogene was recently demonstrated by transfection of DNA constructs [108–110]. Interestingly, the transforming activity of basic FGF was dramatically increased when the FGF cDNA was linked to an immunoglobulin signal sequence [108]. Cells transfected with acidic-FGF cDNA had a low tumorigenic capacity [111]. Whether this indicates a difference in the intrinsic transforming efficiencies of acidic and basic FGF, or rather is the result of the use of different expression systems, remains to be determined.

The three additional members of the FGF family are the products of *int-2* [112, 113], *hst*/KS3 [114–117], and the FGF-5 [118, 119] genes. They have about 40% amino acid sequence similarity to each other and to FGFs. All three factors differ from acidic and basic FGFs in the sense that they have signal sequences, and thus are most likely efficiently secreted from their producer cells.

The *hst*/KS3 gene has been isolated from a human gastric carcinoma [114, 115] and from a Kaposi's sarcoma [116] using the transfection assay on NIH3T3 cells. Similarly, the FGF-5 gene was identified because of its ability to transform NIH3T3 cells in transfection experiments [118, 119]. The *int-2* gene is activated by proviral insertion and is implicated in murine breast cancer induced by mouse mammary tumor virus [113].

FGF-5 has affinity for heparin-Sepharose and has growth-factor activity [119]. The product of *hst*/KS3 has also been shown to have growth-promoting activity for NIH3T3 cells [116] as well as for endothelial cells [117]; the *int-2* protein may also be a growth factor. Thus, the mechanisms for the transforming activities of these oncogenes are most likely analogous to that of the *sis* oncogene.

Insulin-like growth factors

Insulin, and the insulin-like growth factors, IGF-I and IGF-II, are three homologous peptides that bind with different affinities to three distinct receptor types [120]. All three members in this family have been implicated in autocrine mechanisms. A teratoma-derived cell line, called 1246-3A, was found to produce a factor very similar to insulin [121]. Furthermore, the growth of this cell line was inhibited by monoclonal antibodies against insulin, indicating that the endogenously produced factor stimulated cell growth in an autocrine fashion. Analogously, a human lung carcinoma cell line, CALU-6, has been found to produce IGF-I and to be inhibited by monoclonal antibodies against IGF-I [122]. Autocrine stimulation of cell growth by IGF-I has also been suggested for human osteosarcoma and breast carcinoma cell lines [123, 124].

The finding that a Buffalo rat cell line produced a growth-promoting activity, denoted multiplication-stimulating activity and later shown to be very similar to IGF-II, led Temin to suggest the possibility that autocrine stimulation of growth occurred [20]. In this particular case, however, an autocrine effect of the endogenously produced growth factor is unlikely, since it was found that a clone of the cell line, that did not produce multiplication-stimulating activity, was equally capable of growing in serum-free medium [125]. An autocrine effect of IGF-II was, however, suggested by the frequent expression of IGF-II in hepatocellular carcinomas induced by woodchuck hepatitis virus [126]. It is also possible that IGF-II is involved in autocrine growth stimulation in Wilms' tumors [127, 128] and smooth-muscle-cell tumors [129], where IGF-II mRNA has been found in large amounts.

Melanoma-growth-stimulatory activity

Melanoma growth stimulatory activity (MGSA) was first identified in the culture medium of the Hs294T human malignant melanoma cell line [130], and has subsequently been found to be produced by about 70% of the primary cell cultures from human melanoma biopsies [131]. An autocrine role of MGSA is suggested by the fact that it stimulates the growth of Hs294T cells, and that the growth of these cells is inhibited by monoclonal antibodies against MGSA [132]. Cloning of MGSA cDNA [133] revealed that it is identical to the *gro* gene product [134] and related to connective-tissue-activating peptide III, a mitogenic polypeptide that is stored in platelets and released upon thrombin-induced platelet aggregation [135]. These proteins are part of a larger family of growth-regulated cytokines [136, 137].

Bombesin-like peptides

Several human peptides share an identical C-terminal heptapeptide with the amphibian tetradecapeptide bombesin; this family of short peptides has growth-promoting activity on certain cell types [138, 139]. A monoclonal antibody against the C-terminus of bombesin was found to inhibit growth of human small-cell lung carcinoma cells *in vivo*, suggesting an involvement of bombesin-like peptides as autocrine factors in this type of tumor [140]. However, the generality of this observation has recently been questioned [141], since it was found that the inhibitory effect of bombesin antagonists on small-cell lung cancer cells was not mediated via the bombesin receptor and that all small-cell lung cancer cells were not dependent on bombesin *in vitro*.

Colony-stimulating factors

Colony-stimulating factors (CSF) regulate normal hemopoiesis and have been identified by their ability to support growth of bone marrow cells in semisolid media [142–144]. At present five CSFs have been identified: M-CSF, G-CSF, GM-CSF, multi-CSF (also called IL-3) and erythropoietin (Table 1). G-CSF and M-CSF stimulate the proliferation of relatively late progenitors committed to the granulocyte and macrophage lineages, respectively. GM-CSF stimulates, in addition, more immature progenitor cells, and multi-CSF has the ability to support the growth of cells from relatively early pluripotent progenitors to mature cells of several lineages (see references in [142–144]. Erythropoietin stimulates growth and differentiation of cells in the erythrocyte lineage of hemopoietic cells [145].

Studies on myeloid leukemia cell lines in culture have shown that, in general, they are as dependent on the addition of exogenous CSFs as normal myeloid cells, suggesting that autocrine stimulation by CSFs is rarely involved in the pathogenesis of 'spontaneous' leukemia [142, 143]. There are, however, exceptions; GM-CSF has been found to induce an autocrine response in acute myeloblastic leukemia [146, 147].

Non-autocrine, non-tumorigenic myeloid cells in culture may be converted to an autocrine tumorigenic state by spontaneous mutation [148–150], or, experimentally, by introducing suitable CSF-cDNA expression vectors into immortalized cell lines [151, 152]. A systematic examination of factor-independent mutants arising from a factor-dependent myeloid precursor cell line revealed that 10 out of 11 released growth factors: in about two-thirds of the cases, GM-CSF; in about one third, multi-CSF, and in one case a possible new growth factor [149]. Only in one case was no growth factor released; this may represent an activation of the mitogenic pathway inside the cell. The development of a growth-factor-independent state probably occurs in two steps, the first through a true autocrine stimulation (which can be blocked by antibodies), and the second through secondary mutational events which cause abolition of growth-factor dependence by a non-autocrine mechanism [152–154].

Interleukins

Growth factors produced by and acting on cells in the lymphoid system have been designated interleukins (IL) [155]. The number of factors in this group is continuously increasing; currently seven have been cloned (IL-1 – IL-7; Table 1). Several of them have been implicated in autocrine mechanisms.

IL-2 stimulates the proliferation of T cells [156, 157] and B cells [158] via binding to a receptor consisting of two subunits, a 75-kDa α chain and a 55-kDa β chain. A transient co-expression of IL-2 and its receptor occurs after antigen stimulation of T cells. Transformation of T cells by human T-cell leukemia virus I (HTLV-I) is mediated by an aberrant activation of the IL-2 dependent pathway which is thought to be the initial step in the development of HTLV-I-related acute T-cell lymphoma. This concept is based on the finding that the viral gene product p40^x, which is a *trans*-acting transcriptional activator of the long terminal repeat, is known to induce IL-2 mRNA, as well as IL-2-receptor-mRNA expression in T cells [159–161]. In the development of the malignant tumor, HTLV-I infected cells may progress to total IL-2 independence, most probably as a function of additional genetic abnormalities. In support of this possibility, some IL-2-independent HTLV-I-transformed T-cell lines were found to lack IL-2 production [162]. Autocrine mechanisms were directly demonstrated in the T-cell-lymphoma cell line IARC 301 [163] and in T-cell leukemia cells in primary culture [164], since antibodies against IL-2 or its receptor retarded the growth of the cells. A reversible autocrine loop, involving production of IL-2 and constitutive expression of Il-2 receptors, also occurs in bovine lymphocytes infected with the parasite *Theilevia parva* [165].

IL-4, previously called B-cell-stimulatory factor 1 [166, 167], acts on B cells at several phases of their differentiation, but also on mast cells and T cells [168]. Antigen stimulation of certain T-helper-cell lines in the presence of adherent cells leads to release of IL-4 which apparently serves an autocrine function, since addition of a monoclonal antibody against IL-4 inhibited cell growth [169, 170]. In addition, a majority of transformed mast cell lines, and all IL-3-dependent non-transformed mast cells investigated, were found to express IL-4 mRNA [171], suggesting an autocrine mechanism of cell growth.

IL-6 is also known as interferon β2 [172–174], B-cell differentiation factor [175] and hybridoma growth factor [176]. It is thus a multifunctional molecule, and, for example, acts as a growth factor for hybridoma cells [176] and a growth inhibitor for fibroblasts [177]. Human myeloma cells in primary culture have been found to produce IL-6 and express the IL-6 receptor [178]. An autocrine effect of IL-6 in these cells was indicated by the growth inhibitory effect of anti-(IL-6) antibodies.

Other examples of autocrine growth factors

As indicated by the examples described above, known growth factors appear to be involved fairly frequently in autocrine mechanisms. There are, in addition, several reports that suggest the occurrence of autocrine mechanisms, but where the factors involved have not been identified. Examples include the production of B-cell-stimulatory factors by immortalized B lymphocytes [179] and a factor that is produced by and appears to act on Schwann cells [180].

MECHANISM OF ACTIVATION OF GROWTH FACTOR EXPRESSION IN CELL TRANSFORMATION

There are several examples of normal cells that produce autocrine factors, indicating that autocrine stimulation is not only connected with cell transformation. Presumably, autocrine loops in normal cells are tightly controlled and occur only transiently, as a response to external stimuli or in a specific phase of development. Examples include the production of basic FGF by endothelial cells [181], and of PDGF by vascular smooth muscle cells [182–184], placental cytotrophoblasts [185], endothelial cells in glioblastoma tumors [57] as well as by mitogen-stimulated human fibroblasts [186]. The finding of an autoinduction of PDGF [183] in fibroblasts is interesting and may suggest the presence of a positive autocrine feedback loop in growth stimulation. Analogously, TGF-α [187], TGF-β [188] and IL-1 [189] have recently been found to induce the synthesis of their own mRNA and protein.

It is possible that the establishment of an autocrine loop that drives cell proliferation is an important step in the development of the malignant phenotype. Given that autocrine loops also occur under certain conditions in normal cells, it is on the other hand possible that it reflects a normal feature of a cell at a certain developmental stage and thus is a result of transformation, rather than a causative event.

Only in a few cases has the mechanism behind the unscheduled growth-factor synthesis been elucidated. The PDGF-B-chain gene has been transduced by two retroviruses [36, 37] as the v-*sis* oncogene. In addition, there are some examples of nearby insertion of retroviral DNA: the constitutive expression of the IL-3 gene in WEHI-3B leukemia cells [190], the activation of the GM-CSF and multi-CFS genes in a myeloid precursor cell line [149], the activation of the M-CSF gene in a *myc*-retrovirus-induced murine monocyte tumor [150], the activation of the IL-2 gene in a T-cell-lymphoma cell line [191] and the activation of *int-2* in mammary carcinomas [113]. In all these cases, the normal regulatory functions are lost and growth-factor-gene expression is driven by *cis*-acting viral elements. *Trans*-acting mechanisms are also likely to occur, e.g. in the expression of TGF-α in cells transformed by *abl, mos, fes, fms* and *ras* (see above) and in the expression of myelomonocytic growth factor in cells transformed by v-*src* or v-*mil* [192–194]. For most growth factors, no examples of amplification of their genes have been found. The human *hst* and *int-2* genes, that are localized close to each other in the genome, have, however, been found to be amplified in several human malignancies [195, 196].

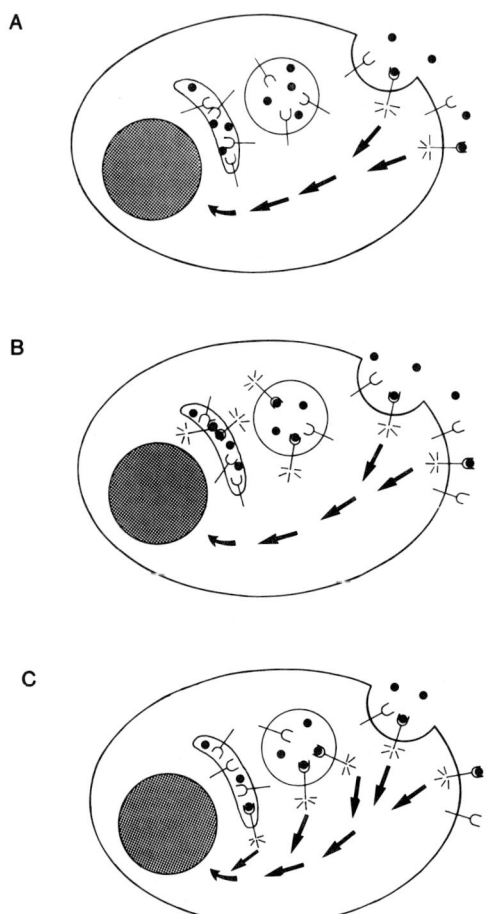

Fig. 1. *Schematic illustration of different possible mechanisms of autocrine growth stimulation.* (A) The growth factor is secreted and activates its receptor at the cell surface only. (B) The growth factor activates its receptor also in the Golgi apparatus or in the secretory vesicles, but the ligand-receptor complex must be transported to the cell membrane in order to transduce the mitogenic signal further. (C) The growth factor activates its receptor and initiates all signals associated with the mitogenic response already in the Golgi apparatus or in the secretory vesicles.

EXTRACELLULAR *VS* INTRACELLULAR ACTION OF AUTOCRINE GROWTH FACTORS

The findings that addition of a signal sequence to the FGF molecule increases its transforming properties [108], and that deletion of the signal sequence of the PDGF-B chain/*sis* product [48, 49], or of IL-3 [197], abolishes cell transformation, indicate that transport into the lumen of the endoplasmic reticulum of the growth factor is important in order for an autocrine effect to occur. In the lumen of the endoplasmic reticulum, and later during maturation in the Golgi apparatus, the growth factor will have the ability to interact with the extracellular domains of growth-factor receptors. The question then arises whether the growth factor can activate the corresponding receptor already inside the cell, or whether it first has to be secreted and thereafter bind to receptors at the cell surface.

There are several examples where antibodies to growth factor have been found to block autocrine growth (see above). Since antibodies enter the cells only very slowly, it is likely in these cases that the growth factor acts via receptors at the cell surface. In other cases antibodies have been found to be without effect [151, 152, 171, 198]. The lack of effect of antibodies may, however, not be taken as conclusive evidence for an intracellular autocrine loop; established cell lines that have been kept in culture for years and thereby selected for fast growth, most likely have accumulated several perturbations in the machinery that controls cell proliferation, and therefore may have escaped a dependence on a particular autocrine loop [198].

The complexity and technical difficulties involved in trying to answer the question whether a growth factor can activate its receptor already inside the producer cell, are exemplified by studies on SSV-transformed cells. Evidence has been presented that the transforming protein of SSV-transformed cells can activate the PDGF receptor kinase already inside the cell [199, 200]. On the other hand, anti-PDGF antibodies, that do not enter the cell, inhibited SSV transformation of normal human fibroblasts [46] and some SSV-transformed cell lines [43]. Furthermore, suramin, an agent that interferes with the interaction between PDGF and its receptors [42], completely reversed SSV transformation of human fibroblasts [47]. It can, however, not be excluded that suramin is to some extent taken up by cells. Alteration of cellular transport processes by monensin, that results in the accumulation of the v-*sis* product in the Golgi apparatus, led to the autophosphorylation of a 160-kDa precursor of the PDGF receptor and inhibited the induction of c-*fos* expression [201]. These data suggest that at least part of the autocrine effect in SSV-transformed cells is due to activation of the mitogenic pathway at the cell surface. Some of the apparent discrepancies between these observations may be due to differences between the various cell systems used. Most of the data obtained are, however, compatible with the hypothesis that activation of the receptor kinase occurs inside the cell, but that the ligand-receptor complex has to reach the cell surface to transduce all the signals involved in the mitogenic pathway; receptor-kinase substrates may thus be localized at the inner leaflet of the cell membrane. This mechanism is schematically illustrated in Fig. 1B and compared with the possibilities that the entire autocrine stimulation occurs inside the cell (Fig. 1C), or that the growth factor first has to be secreted and act exclusively via receptors at the cell surface (Fig. 1A). The possibility illustrated in Fig. 1B would be compatible with the finding that very high antibody concentrations were needed to inhibit SSV transformation [46]; higher antibody concentrations may be needed to displace a ligand from a ligand-receptor complex, than to bind and neutralize a secreted ligand. Moreover, a partial inhibitory effect of anti(IL-3) antibodies was observed in cells expressing low levels of IL-3 but no inhibition was seen in cells expressing high levels of the factor [197].

AUTOCRINE GROWTH STIMULATION
IN MULTISTEP CARCINOGENESIS

An unscheduled production of a growth factor by a cell that carries the corresponding receptor may not be expected to cause anything but a hyperplastic response of otherwise phenotypically normal cells (cf. [38]). Activation of the mitogenic pathway of a growth factor may be a necessary, but not sufficient, event in tumorigenesis. In terminally differentiating systems, such as in hemopoiesis, such an event must be complemented by a differentiation block in order to cause a malignancy and this is probably also true for the lymphoid system. It is notable, for instance, that transformation of erythroid cells by v-*erbB* requires the cooperation of v-*erbA*, the role of which probably is to block differentiation [202]. In other systems, an aberrant growth signal probably has to be complemented by an immortalizing event.

Autocrine mechanisms may be early as well as late phenomena in multistep tumorigenesis. In SSV-induced sarcomas and gliomas [203], and perhaps also in spontaneous tumors of the same kind, autocrine growth is the first event which is conceived to be complemented by secondary events in the proliferating population. An analogous situation may prevail in HTLV-associated acute T-cell lymphoma, via *trans*-activation of the IL-2 pathway [204]. It is conceivable that many of the cell lines that have been shown to coexpress a growth factor and its receptor, but which are not growth arrested by anti-(growth factor) antibodies, have evolved from autocrine parental cells. Autocrine growth stimulation may also occur as a late step in tumorigenesis, as inferred from experiments of introducing growth-factor-cDNA expression vectors into immortalized cells of hemopoietic or fibroblastic origin [87–90, 151].

We thank Lena Claesson-Welsh for valuable comments and Linda Baltell for skillful secretarial assistance.

REFERENCES

1. Sporn, M. B., Roberts, A. B., Wakefield, L. M. & de Crombrugghe (1987) *J. Cell. Biol. 105*, 1039–1045.
2. Romeo, G., Fiorucci, G. & Rossi, G. B. (1989) *Trends Genet. 5*, 19–24.
3. Old, L. (1988) *Sci. Am. 254*, 41–49.
4a. Sporn, M. B. & Roberts, A. B. (1985) *Nature 313*, 745–747.
4b. Yarden, Y. & Ullrich, A. (1988) *Annu. Rev. Biochem. 57*, 443–478.
5. Heldin, C.-H., Ernlund, A., Rorsman, C. & Rönnstrand, L. (1989) *J. Biol. Chem. 264*, 8905–8912.
6. Schlessinger, J. (1986) *J. Cell. Biol. 103*, 2067–2072.
7. Hunter, T. & Cooper, J. A. (1985) *Annu. Rev. Biochem. 54*, 897–930.
8. Morrison, D. K., Kaplan, D. R., Rapp, U. & Roberts, T. M. (1988) *Proc. Natl Acad. Sci. USA 85*, 8855–8859.
9. Ralston, R. & Bishop, J. M. (1985) *Proc. Natl Acad. Sci. USA 82*, 7845–7849.
10. Gould, K. L. & Hunter, T. (1988) *Mol. Cell. Biol. 8*, 3345–3356.
11. Nishizuka, Y. (1984) *Nature 308*, 693–698.
12. Berridge, M. J. & Irvine, R. F. (1984) *Nature 312*, 315–321.
13. Wahl, M. I., Daniel, T. O. & Carpenter, G. (1988) *Science 241*, 968–970.
14. Coughlin, S. R., Lee, W. M. F., Williams, P. W., Giels, G. M. & Williams, L. T. (1985) *Cell 43*, 243–251.
15. Cochran, B. H., Reffel, A. C. & Stiles, C. D. (1983) *Cell 33*, 939–947.
16. Linzer, D. F. H. & Nathans, D. (1983) *Proc. Natl Acad. Sci. USA 80*, 4271–4275.
17. Ferrari, S. & Baserga, R. (1987) *Bioessays 7*, 9–13.
18. Almendral, J. M., Sommer, D., MacDonald-Bravo, H., Bruckhardt, J., Perera, J. & Bravo, R. (1988) *Mol. Cell. Biol. 8*, 2140–2148.
19. Heldin, C.-H., Betsholtz, C., Claesson-Welsh, L. & Westermark, B. (1987) *Biochim. Biophys. Acta 907*, 219–244.
20. Temin, H. M. (1967) in *Growth regulating substances for animal cells in cultures* (Defendi, V. & Stoker, M., eds) vol. 7, pp. 103–116, The Wistar Symposium Monograph, Wistar Institute Press, Philadephia.
21. Sporn, M. B. & Todaro, G. J. (1980) *N. Eng. J. Med. 303*, 878–880.
22. Sporn, M. B. & Roberts, A. B. (1985) *Nature 313*, 745–747.
23. Heldin, N. E., Gustafsson, B., Claesson-Welsh, L., Hammacher, A., Heldin, C.-H. & Westermark, B. (1988) *Proc. Natl Acad. Sci. USA 85*, 9302–9306.

24. Pierce, J. H., Ruggiero, M., Fleming, T. P., Di Fiore, P. P., Greenberger, J. S., Varticovski, L., Schlessinger, J., Rovera, G. & Aaronson, S. A. (1988) *Science 239*, 628–631.
25. Heldin, C.-H., Wasteson, Å. & Westermark, B. (1985) *Mol. Cell. Endocrinol. 39*, 169–187.
26. Ross, R., Raines, E. W. & Bowen-Pope, D. F. (1987) *Cell 46*, 155–169.
27. Heldin, C.-H., Johnsson, A., Wennergren, S., Wernstedt, C., Betsholtz, C. & Westermark, B. (1986) *Nature 319*, 511–514.
28. Hammacher, A., Hellman, U., Johnsson, A., Östman, A., Gunnarsson, K., Westermark, B., Wasteson, Å. & Heldin, C.-H. (1988) *J. Biol. Chem. 263*, 16493–16498.
29. Stroobant, P. & Waterfield, M. D. (1984) *EMBO J. 3*, 2963–2967.
30. Nistér, M., Hammacher, A., Mellström, K., Siegbahn, A., Rönnstrand, L., Westermark, B. & Heldin, C.-H. (1988) *Cell 52*, 791–799.
31. Kazlauskas, A., Bowen-Pope, D. F., Seifert, R., Hart, C. E. & Cooper, J. A. (1988) *EMBO J. 7*, 3727–3735.
32. Pringle, N., Collarini, E. J., Mosley, M. J., Heldin, C.-H., Westermark, B. & Richardson, W. D. (1989) *EMBO J. 4*, 1049–1056.
33. Heldin, C.-H., Bäckström, G., Östman, A., Hammacher, A., Rönnstrand, L., Rubin, K., Nistér, M. & Westermark, B. (1988) *EMBO J. 7*, 1387–1392.
34. Hart, C. E., Forstrom, J. D., Kelly, R. A., Smith, E., Ross, R., Murray, M. J. & Bowen-Pope, D. F. (1988) *Science 240*, 1529–1532.
35. Escobedo, J. A., Navankasatussas, S., Cousens, L. S., Coughlin, S. R., Bell, G. I. & Williams, L. T. (1988) *Science, 240*, 1532–1534.
36. Devare, S. G., Reddy, E. P., Law, J. D., Robbins, K. C. & Aaronson S. A. (1983) *Proc. Natl Acad. Sci. USA 80*, 731–735.
37. Besmer, P., Snyder, H. W. Jr, Murphy, J. E., Hardy, W. D. Jr & Parodi, A. (1983) *J. Virol. 46*, 606–613.
38. Westermark, B., Betsholtz, C., Johnsson, A. & Heldin, C.-H. (1987) in *Viral carcinogens* (Kjeldgaard, N. O. & Forschhammer, J., eds) pp. 445–457, Munksgaard, Copenhagen.
39. Deuel, T. F., Huang, J. S., Huang, S. S., Stroobant, P. & Waterfield, M. (1983) *Science 221*, 1348–1350.
40. Bowen-Pope, D. F., Vogel, A. & Ross, R. (1984) *Proc. Natl Acad. Sci. USA 81*, 2396–2400.
41. Owen, A. J., Pantazis, P. & Antoniades, H. N. (1984) *Science 225*, 54–56.
42. Garrett, J. S., Coughlin, S. R., Niman, H. L., Tremble, P. M., Giels, G. M. & Williams, L. T. (1984) *Proc. Natl Acad. Sci. USA 81*, 7466–7470.
43. Huang, J. S., Huang, S. S. & Deuel, T. F. (1984) *Cell 39*, 79–87.
44. Johnsson, A., Betsholtz, C., von der Helm, K., Heldin, C.-H. & Westermark, B. (1985) *Proc. Natl Acad. Sci. USA 82*, 1721–1725.
45. Leal, F., Williams, L. T., Robbins, K. C. & Aaronson, S. A. (1985) *Science 230*, 327–330.
46. Johnsson, A., Betsholtz, C., Heldin, C.-H. & Westermark, B. (1985) *Nature 317*, 438–440.
47. Betsholtz, C., Johnsson, A., Heldin, C.-H. & Westermark, B. (1986) *Proc. Natl Acad. Sci. USA 83*, 6440–6444.
48. Hannink, M. & Donoghue, D. J. (1984) *Science 130*, 1197–1199.
49. King, C. R., Giese, N. A., Robbins, K. C. & Aaronson, S. A. (1985) *Proc. Natl Acad. Sci. USA 82*, 5295–5299.
50. Gazit, A., Igarashi, H., Chiu, I. M., Srinivasan, A., Yaniv, A., Tronick, S. R., Robbins, K. C. & Aaronson, S. A. (1984) *Cell 39*, 80–97.
51. Clarke, M. F., Westin, E., Schmidt, D., Josephs, S. F., Ratner, L., Wong-Staal, F., Gallo, R. C. & Reitz, M. S. (1984) *Nature 308*, 464–467.
52. Josephs, S. F., Ratner, L., Clarke, M. F., Westin, E. H., Reitz, M. S. & Wong-Staal, F. (1984) *Science 225*, 636–639.
53. Bywater, M., Rorsman, F., Bongcam-Rudloff, E., Mark, G., Hammacher, A., Heldin, C.-H., Westermark, B. & Betsholtz, C. (1988) *Mol. Cell. Biol. 8*, 2753–2762.
54. Beckmann, M. P., Betsholtz, C., Heldin, C.-H., Westermark, B., Di Marco, E., Di Fiore, P. P., Robbins, K. C. & Aaronson, S. A. (1988) *Science, 241*, 1346–1349.
55. Nistér, M., Libermann, T., Betsholtz, C., Pettersson, M., Claesson-Welsh, L., Heldin, C.-H., Schlessinger, J. & Westermark, B. (1988) *Cancer Res. 48*, 3910–3918.
56. Hammacher, A., Nistér, M., Westermark, B. & Heldin, C.-H. (1988) *Eur. J. Biochem. 176*, 179–186.
57. Hermansson, M., Nistér, M., Betsholtz, C., Heldin, C.-H., Westermark, B. & Funa, K. (1988) *Proc. Natl Acad. Sci. USA 85*, 7748–7752.
58. Rozengurt, E., Sinnett-Smith, J. & Taylor-Papadimitriou, J. (1985) *Int. J. Cancer 36*, 247–252.
59. Perez, R., Betsholtz, C., Westermark, B. & Heldin, C.-H. (1987) *Cancer Res. 47*, 3425–3429.
60. Bronzert, D. A., Pantazis, P., Antoniades, H. N., Kasid, A., Davidson, N., Dickson, R. B. & Lippman, M. E. (1987) *Proc. Natl Acad. Sci. USA 84*, 5763–5767.
61. Söderdahl, G., Betsholtz, C., Johansson, A., Nilsson, K. & Berg, J. (1988) *Int. J. Cancer. 41*, 636–641.
62. Westermark, B., Johnsson, A., Paulsson, Y., Betsholtz, C., Heldin, C.-H., Herlyn, M., Rodeck, U. & Koprowski, H. (1986) *Proc. Natl Acad. Sci. USA 83*, 7197–7200.
63. Alitalo, R., Andersson, L. C., Betsholtz, C., Nilsson, K., Westermark, B., Heldin, C.-H. & Alitalo, K. (1987) *EMBO J. 6*, 1213–1218.
64. Pantazis, P., Sariban, E., Kufe, D. & Antoniades, H. N. (1986) *Proc. Natl Acad. Sci. USA 83*, 6455–6459.
65. Bringman, T. S., Lindquist, P. B. & Derynck, R. (1987) *Cell 48*, 429–440.
66. Gentry, L. E., Twardzik, D. R., Lim, G. J., Ranchalis, J. E. & Lee, D. C. (1987) *Mol. Cell. Biol. 7*, 1585–1591.
67. Teixidó, J., Gilmore, R., Lee, D. C. & Massagué, J. (1987) *Nature 326*, 883–885.
68. Gray, A., Dull, T. J. & Ullrich, A. (1983) *Nature 303*, 722–725.
69. Scott, J., Urdea, M., Quiroga, M., Sanchez-Pescador, R., Fong, N., Selby, M., Rutter, W. J. & Bell, G. I. (1983) *Science 221*, 236–240.
70. Stroobant, P., Rice, A. P., Gullick, W. J., Chen, D. J., Kerr, I. M. & Waterfield, M. (1985) *Cell 42*, 383–393.
71. Twardzik, D. R., Brown, J. P., Ranchalis, J. E., Todaro, G. J. & Moss, B. (1985) *Proc. Natl Acad. Sci. USA 82*, 5300–5304.
72. De Larco, J. E. & Todaro, G. J. (1978) *Proc. Natl Acad. Sci. USA 75*, 4001–4005.
73. Todaro, G. J., De Larco, J. E., Marquardt, H., Bryant, M. L., Sherwin, S. A. & Sliski, A. H. (1979) in *Hormones and cell culture, book A* (Sato, G. H. & Ross, R., eds) vol. 6, pp. 113–127, Cold Spring Harbor, Conferences on Cell Proliferation, Cold Spring Harbor, New York.
74. Ozanne, B., Fulton, R. J. & Kaplan, P. L. (1980) *J. Cell. Physiol. 105*, 163–180.
75. Twardzik, D. R., Todaro, G. J., Marquardt, H., Reynolds, F. H. Jr & Stephenson, J. R. (1982) *Science 216*, 894–896.
76. Twardzik, D. R., Todaro, G. J., Reynolds, F. H. Jr & Stephenson, J. R. (1983) *Virology 124*, 201–207.
77. Marquardt, H., Hunkapiller, M. W., Hood, L. E., Twardzik, D. R., De Larco, J. E., Stephenson, J. R. & Todaro, G. J. (1983) *Proc. Natl Acad. Sci. USA 80*, 4684–4688.
78. Kaplan, P. L. & Ozanne, B. (1982) *Virology 123*, 372–380.
79. Todaro, G. J., Fryling, C. & De Larco, J. E. (1980) *Proc. Natl Acad. Sci. USA 77*, 5258–5262.
80. Derynck, R., Roberts, A. B., Winkler, M. E., Chen, E. Y. & Goeddel, D. V. (1984) *Cell 38*, 287–297.
81. Derynck, R., Goeddel, D. V., Ullrich, A., Gutterman, J. U., Williams, R. D., Bringman, T. S. & Berger, W. H. (1987) *Cancer Res. 47*, 707–712.
82. Sato, M., Yoshida, H., Hayashi, Y., Miyakami, K., Bando, T., Yanagawa, T., Yuna, Y., Azuma, M. & Ueno, A. (1985) *Cancer Res. 45*, 6160–6167.

83. Anzano, M. A., Roberts, A. B., Smith, J. M., Sporn, M. B. & DeLarco, J. E. (1983) *Proc. Natl Acad. Sci. USA 80*, 6264–6268.
84. van Zoelen, E. J. J, van Oostward, T. M. J. & de Laat, S. W. (1988) *J. Biol. Chem. 263*, 64–68.
85. Kaplan, P. L., Andersson, M. & Ozanne, B. (1982) *Proc. Natl Acad. Sci. USA 79*, 485–489.
86. McKay, I., Malone, P., Marshall, C. J. & Hall, A. (1986) *Mol. Cell. Biol. 6*, 3382–3387.
87. Rosenthal, A., Lindquist, P. B., Bringman, T. S., Goeddel, D. V. & Derynck, R. (1986) *Cell 46*, 301–309.
88. Finzi, E., Fleming, T., Segatto, O., Pennington, C. Y., Bringman, T. S., Derynck, R. & Aaronson, S. A. (1987) *Proc. Natl Acad. Sci USA. 84*, 3733–3737.
89. Watanabe, S., Lazar, E. & Sporn, M. B. (1987) *Proc. Natl Acad. Sci. USA 84*, 1258–1262.
90. Stern, D. F., Hare, D. L., Cecchini, M. A. & Weinberg, R. A. (1987) *Science 235*, 321–324.
91. Smith, J. J., Derynck, R. & Korc, M. (1987) *Proc. Natl Acad. Sci. USA 84*, 7567–7570.
92. Sherwin, S. A., Twardzik, D. R., Bohn, W. H., Cockly, K. D. & Todaro, G. J. (1983) *Cancer Res. 43*, 403–407.
93. Gospodarowicz, D., Neufeld, G. & Schweigerer, L. (1986) *Mol. Cell. Endocrinol. 46*, 187–204.
94. Folkman, J. & Klagsbrun, M. (1987) *Science 235*, 442–447.
95. Ferrara, N., Schweigerer, L., Neufeld, G., Mitchell, R. & Gospodarowicz, D. (1987) *Proc. Nat. Acad. Sci. USA 84*, 5773–5777.
96. Moscatelli, D., Presta, M., Joseph-Silverstein, J. & Rifkin, D. B. (1986) *J. Cell. Physiol. 129*, 273–276.
97. Vlodavsky, I., Folkman, J., Sullivan, R., Friedman, R., Ishai-Michaeli, R., Sasse, J. & Klagsbrun, M. (1987) *Proc. Natl Acad. Sci. USA 84*, 2292–2296.
98. Winkles, J. A., Friesel, R., Burgess, W. H., Mehlman, T., Weinstein, R. & Maciag, T. (1987) *Proc. Natl Acad. Sci. USA 84*, 7124–7128.
99. Sato, Y. & Rifkin, D. B. (1988) *J. Cell Biol. 107*, 1199–1205.
100. Sakaguchi, M., Kajio, T., Kawahara, K. & Kato, K. (1988) *FEBS Lett. 233*, 163–166.
101. Klagsbrun, M., Sasse, J., Sullivan, R. & Smith, J. A. (1986) *Proc. Natl Acad. Sci. USA 83*, 2448–2452.
102. Schweigerer, L., Neufeld, G., Mergia, A., Abraham, J. A., Fiddes, J. C. & Gospodarowicz, D. (1987) *Proc. Natl Acad. Sci. USA 84*, 842–846.
103. Libermann, T. A., Friesel, R., Jaye, M., Lyall, R. M., Westermark, B., Drohan, W., Schmidt, A., Maciag, T. & Schlessinger, J. (1987) *EMBO J. 6*, 1627–1632.
104. Halaban, R., Kwon, B. S, Gkos, P., Delli Bovi, P. & Baird, A. (1988) *Oncogene Res. 3*, 177–186.
105. Ensoli, B., Nakamura, S., Salahuddin, S. Z., Biberfeld, P., Larsson, L., Beaver, B., Wong-Staal, F. & Gallo, R. C. (1989) *Science 243*, 223–226.
106. Jaye, M., Howk, R., Burgess, W., Ricca, G. A., Chiu, I.-M., Ravera, M. W., O'Brian, S. J., Modi, W. S., Maciag, T. & Drohan, W. N. (1986) *Science 233*, 541–545.
107. Abraham, J. A., Whang, J. L., Tumolo, A., Mergia, A., Friedman, J., Gospodarowicz, D. & Fiddes, J. C. (1986) *EMBO J. 5*, 2523–2528.
108. Rogelj, S., Weinberg, R. A., Fanning, P. & Klagsbrun, M. (1988) *Nature 331*, 173–175.
109. Sasada, R., Kurokawa, T., Iwane, M. & Igarashi, K. (1988) *Mol. Cell. Biol. 8*, 588–594.
110. Neufeld, G., Mitchell, R., Ponte, P. & Gospodarowicz, D. (1988) *J. Cell. Biol. 106*, 1385–1394.
111. Jaye, M., Lyall, R. M., Mudd, R., Schlessinger, J. & Sarver, N. (1988) *EMBO J. 7*, 963–969.
112. Moore, R., Casey, G., Brookes, S., Dixon, M., Peters, G. & Dickson, C. (1986) *EMBO J. 5*, 919–924.
113. Nusse, R. (1988) *Trends Genet 4*, 291–295.
114. Sakamoto, H., Mori, M., Taira, M., Yoshida, T., Matsukawa, S., Shimizu, K., Sekiguchi, M., Terada, M. & Sugimura, T. (1986) *Proc. Natl Acad. Sci. USA 83*, 3997–4001.
115. Taira, M., Yoshida, T., Miyagawa, K., Sakamoto, H., Terada, M. & Sugimura, T. (1987) *Proc. Natl Acad. Sci. USA 84*, 2980–2984.
116. Delli-Bovi, P., Curatola, A. M., Kern, F. G., Greco, A., Ittman, M. & Basilico, C. (1987) *Cell 50*, 729–737.
117. Delli-Bovi, P., Curatola, A. M., Newman, K. M., Sato, Y., Moscatelli, D., Hewick, R. M., Rifkin, D. & Basilico, C. (1988) *Mol. Cell. Biol. 8*, 2933–2941.
118. Zhan, X., Culpepper, A., Reddy, M., Loveless, J. & Goldfarb, M. (1987) *Oncogene 1*, 369–376.
119. Zhan, X., Bates, B., Hu, X. & Goldfarb, M. (1988) *Mol. Cell. Biol. 8*, 3487–3495.
120. Froesch, E. R., Schmid, C., Schwander, J. & Zapf, J. (1985) *Annu. Rev. Physiol. 47*, 443–467.
121. Yamada, Y. & Serrero, G. (1988) *Proc. Natl Acad. Sci. USA 85*, 5936–5940.
122. Minuto, F., Del Monte, P., Barreca, A., Alama, A., Cariola, G. & Giordano, G. (1988) *Cancer Res. 48*, 3716–3719.
123. Blatt, J., White, C., Dienes, S., Friedman, H. & Foley, T. (1984) *Biochem. Biophys. Res. Commun. 123*, 373–376.
124. Huff, K. K., Kaufman, D., Gabbay, K. H., Spencer, E. M., Lippman, M. E. & Dickson, R. B. (1986) *Cancer Res. 46*, 4613–4619.
125. Nissley, S. P., Short, P. A., Rechler, M. M., Podskalny, J. M. & Coon, H. (1978) *Cell 11*, 441–446.
126. Fu, X.-X., Su, C. Y., Lee, Y., Hintz, R., Biempica, L., Snyder, R. & Rogler, C. E. (1988) *J. Virol. 62*, 3422–3430.
127. Reeve, A. E., Eccles, M. R., Wilkins, R. J., Bell, G. I. & Millow, L. J. (1985) *Nature 317*, 258–260.
128. Scott, J., Cowell, J., Robertson, M. E., Priestly, L. M., Wadey, R., Hopkins, B., Pritchard, J., Bell, G. I., Rall, L. N., Graham, C. F. & Knott, T. J. (1985) *Nature 317*, 260–262.
129. Höppener, J. W. M., Mosselman, S., Roholl, P. J. M., Lambrechts, C., Slebos, R. J. C., de Pagter-Holthuizen, P., Lips, C. J. M., Jansz, H. S. & Sussenbach, J. S. (1988) *EMBO J. 7*, 1379–1385.
130. Richmond, A., Lawson, D. H., Nixon, D. W., Stevens, J. S. & Chawla, R. K. (1982) *Cancer Res. 42*, 3175–3180.
131. Richmond, A., Fine, R., Murray, D., Lawson, D. H., & Priest, L. (1986) *J. Invest. Dermatol. 86*, 295–302.
132. Lawson, D. H., Thomas, H. G., Roy, R. G. B., Gordon, D. J., Chawla, R. K., Nixon, D. W. & Richmond, A. (1987) *J. Cell. Biochem. 34*, 169–185.
133. Richmond, A., Balentien, E., Thomas, H. G., Flaggs, G., Barton, D. E., Spiess, J., Brodini, R., Francke, U. & Derynck, R. (1988) *EMBO J. 7*, 2025–2033.
134. Anisowicz, A., Bardwell, L. & Sager, R. (1987) *Proc. Natl Acad. Sci. USA 84*, 7188–7192.
135. Castor, C. W., Miller, J. W. & Walz, D. A. (1983) *Proc. Natl Acad. Sci. USA 80*, 765–769.
136. Oquendo, P., Alberta, J., Wen, D., Graycar, J., Derynck, R. & Stiles C. D. (1989) *J. Biol. Chem. 264*, 4133–4137.
137. Kawahara, R. S. & Deuel, T. F. (1989) *J. Biol. Chem. 264*, 679–682.
138. Zacchary, I. & Rozengurt, E. (1985) *Proc. Natl Acad. Sci. USA 82*, 7616–7620.
139. Carney, D. N., Cuttitta, F., Moody, T. W. & Minna, J. D. (1987) *Cancer Res. 47*, 821–825.
140. Cuttitta, F., Carney, D. N., Mulshine, J., Moody, T. W., Fedorko, J., Fischler, A. & Minna, J. N. D. (1985) *Nature 316*, 823–826.
141. Layton, J. E., Scanlon, D. B., Soveny, C. & Morstyn, G. (1988) *Cancer Res. 48*, 4783–4789.
142. Metcalf, D. (1985) *Science 229*, 16–22.
143. Sachs, L. (1985) *Science 238*, 1374–1379.
144. Clark, S. C. & Kamen, R. (1987) *Science 236*, 1229–1237.
145. Jacobs, K., Shoemaker, C., Rudersdorf, R., Neill, S. D., Kaufman, R. J., Mufson, A., Seehra, J., Jones, S. S., Hewick, R., Fritsch, E. F., Kawakita, M., Shimizu, T. & Miyake, T. (1985) *Nature 313*, 806–809.
146. Young, D. & Griffin, J. D. (1986) *Blood 68*, 1178–1181.
147. Young, D., Wagner, K. & Griffin, J. D. (1987) *J. Clin. Invest. 79*, 100–106.

148. Schrader, J. W. & Crapper, R. M. (1983) *Proc. Natl Acad. Sci. USA. 80*, 6892–6896.
149. Stocking, C., Löliger, C., Kawai, M., Sucin, S., Gough, N. & Ostertag, W. (1988) *Cell 53*, 869–879.
150. Baumbach, W. R., Colstone, E. M. & Cole, M. D. (1988) *J. Virol. 9*, 3151–3155.
151. Lang, R. A., Metcalf, D., Gough, N. M., Dunn, A. R. & Gonda, T. J. (1985) *Cell 43*, 531–542.
152. Laker, C., Stocking, C., Bergholz, U., Hess, N., De Lamarter, J. F. & Ostertag, W. (1987) *Proc. Natl Acad. Sci. USA 84*, 8458–8462.
153. Pierce, J. H., Di Fiore, P. P., Aaronson, S. A., Potter, M., Pumphrey, J., Scott, A. & Ihle, J. N. (1985) *Cell 41*, 685–693.
154. Cook, W. D., Metcalf, D., Nicola, N. A., Burgess, A. W. & Walker, F. (1985) *Cell 41*, 677–683.
155. O'Garra, A., Umland, S., De France, T. & Christiansen, J. (1988) *Immunol. Today 9*, 45–54.
156. Smith, K. A. (1988) *Science 240*, 1169–1176.
157. Taniguchi, T., Matsui, H., Fujita, Y., Hatakeymana, M., Kashima, N., Fuse, A., Hamuro, J., Hishi-Takaoka, C. & Yamada, G. (1986) *Immunol. Rev. 92*, 121–133.
158. Mingari, M. C., Gerosa, F., Carra, G., Accolla, R. S., Moretta, A., Zubler, R. H., Waldmann, T. A. & Moretta, L. (1984) *Nature 312*, 641–643.
159. Inoue, J., Seiki, M., Taniguchi, T., Tsuru, S. & Yoshida, M. (1986) *EMBO J. 5*, 2883–2888.
160. Maruyama, M., Shibuya, H., Harada, H., Hatakeyama, M., Seiki, M., Fujita, T., Inoue, J., Yoshida, M. & Taniguchi, T. (1987) *Cell 48*, 343–350.
161. Siekevitz, M., Feinberg, M. B., Holbrook, N., Wong-Staal, F. & Greene, W. C. (1987) *Proc. Natl Acad. Sci. USA 84*, 5389–5393.
162. Arya, S. L., Wong-Staal, F. & Gallo, R. C. (1984) *Science 223*, 1086–1087.
163. Duprez, V., Lenoir, G. & Dautry-Varsat, A. (1985) *Proc. Natl Acad. Sci. USA 82*, 6932–6936.
164. Arima, N., Daitoku, Y., Oghaki, S., Fukumori, J., Tanaka, H., Yamamoto, Y., Fujimoto, K. & Onoue, K. (1986) *Blood 68*, 779–782.
165. Dobbelaere, D. A. E., Coquerelle, T. M., Roditi, I. J., Eichhorn, M. & Williams, R. O. (1988) *Proc. Natl Acad. Sci. USA 85*, 4730–4734.
166. Lee, F., Yokota, T., Otsuka, T., Meyerson, P., Villaret, D., Coffman, R., Mosmann, T., Rennick, D., Roehm, N., Smith, C., Zlotnik, A. & Arai, K. (1986) *Proc. Natl Acad. Sci. USA 83*, 2061–2065.
167. Noma, Y., Sideras, P., Naito, T., Bergstedt-Lindquist, S., Azuma, C., Severinson, E., Tanabe, T., Kinashi, T., Matsuda, F., Yaoita, Y. & Honjo, T. (1986) *Nature 319*, 640–646.
168. Ohara, J. & Paul, W. E. (1987) *Nature 325*, 537–540.
169. Fernandez-Botran, R., Sanders, V. M., Olivier, K. G., Chen, Y. M., Krammer, P. H., Uhr, J. W. & Vitetta, E. S. (1986) *Proc. Natl Acad. Sci. USA 83*, 9689–9693.
170. Lichtman, A. H., Kurt-Jones, E. A. & Abbas, A. K. (1987) *Proc. Natl Acad. Sci. USA 84*, 824–827.
171. Brown, M. A., Pierce, J. H., Watson, C. J., Falco, J., Ihle, J. N. & Paul, W. E. (1987) *Cell 50*, 809–818.
172. Haegeman, G., Content, J., Volckaert, G., Derynck, R., Tavernier, J. & Fiers, W. (1986) *Eur. J. Biochem. 159*, 625–632.
173. May, L. T., Helgott, D. C. & Sehgal, P. B. (1986) *Proc. Natl Acad. Sci. USA 83*, 8957–8961.
174. Zilberstein, A., Ruggieri, R., Korn, J. H. & Revel, M. (1986) *EMBO J. 5*, 2529–2537.
175. Van Snick, J., Cayphas, S., Vink, A., Uyttenhove, C., Coulie, P. G., Rubira, M. R. & Simpson, R. J. (1986) *Proc. Natl Acad. Sci. USA 83*, 9679–9683.
176. Hirano, T., Yasukawa, K., Harada, H., Taga, T., Watanabe, Y., Matsuda, T., Kashiwamura, S., Nakajima, K., Koyama, K., Iwamatsu, A., Tsunasawa, S., Sakiyama, F., Matsui, H., Takahara, Y., Taniguchi, T. & Kishimoto, T. (1986) *Nature 324*, 73–76.
177. Kohase, M., Henriksen-DeStefano, D., May, L. T., Vilcek, J. & Sehgal, P. B. (1986) *Cell 45*, 659–666.
178. Kawano, M., Hirano, T., Matsuda, T., Taga, T., Horii, Y., Iwato, K., Asaoku, H., Tang, B., Tanabe, O., Tanaka, H., Kuramoto, A. & Kishimoto, T. (1988) *Nature 332*, 83–85.
179. Gordon, J., Ley, S. C., Melamed, M. D., English, L. S. & Hughes-Jones, N. C. (1984) *Nature 310*, 145–147.
180. Porter, S., Glaser, L. & Bunge, R. P. (1987) *Proc. Natl Acad. Sci. USA 84*, 7768–7772.
181. Sato, Y. & Rifkin, D. B. (1988) *J. Cell Biol. 107*, 1199–1205.
182. Seifert, R. A., Schwartz, S. M. & Bowen-Pope, D. F. (1984) *Nature 311*, 669–671.
183. Nilsson, J., Sjölund, M., Palmberg, L., Thyberg, J. & Heldin, C.-H. (1985) *Proc. Natl Acad. Sci. USA 82*, 4418–4422.
184. Sejersen, T., Betsholtz, C., Sjölund, M., Heldin, C.-H., Westermark, B. & Thyberg, J. (1986) *Proc. Natl Acad. Sci. USA 83*, 6844–6848.
185. Goustin, A. S., Betsholtz, C., Pfeifer-Ohlsson, S., Persson, H., Rydnert, J., Bywater, M., Holmgren, G., Heldin, C.-H., Westermark, B. & Ohlsson, R. (1985) *Cell 41*, 301–312.
186. Paulsson, Y., Hammacher, A., Heldin, C.-H. & Westermark, B. (1987) *Nature 318*, 715–717.
187. Coffey, R. J. Jr, Derynck, R., Wilcox, J. N., Bringman, T. S., Goustin, A. S., Moses, H. L. & Pittelkow, M. R. (1987) *Nature 328*, 817–820.
188. van Obbergen-Schilling, E., Roche, N. S., Flanders, K. C., Sporn, M. B. & Roberts, A. B. (1988) *J. Biol. Chem. 263*, 7741–7746.
189. Dinarello, C. A., Ikejima, T., Warner, S. J. C., Oreucole, S. F., Lonnemann, G., Cannon, J. G., & Libby, P. (1987) *J. Immunol. 139*, 1902–1910.
190. Ymer, S., Tucker, W. Q. J., Sanderson, C. J., Hapel, A. J., Campbell, H. D. & Young, I. G. (1985) *Nature 317*, 255–258.
191. Chen, S. J., Holbrook, N. J., Mitchell, K. F., Vallone, C. A., Freegard, J. S., Crabtree, G. R. & Lin, Y. (1985) *Proc. Natl Acad. Sci. USA 82*, 7284–7288.
192. Adkins, B., Lautz, A. & Graf, T. (1984) *Cell 39*, 439–445.
193. Graf, T., von Weizsäcker, F., Grieser, S., Coll, J., Stehelin, D., Patschinsky, T., Bister, K., Bechade, C., Calothy, G. & Leutz, A. (1986) *Cell 45*, 357–364.
194. von Weizsäcker, F., Beug, H. & Graf, T. (1986) *EMBO J. 5*, 1521–1528.
195. Yoshida, M. C., Wada, M., Satoh, H., Yoshida, T., Sakamoto, H., Miyagawa, K., Yokota, J., Koda, T., Kakinuma, M., Sugimura, T. & Terada, M. (1988) *Proc. Natl Acad. Sci. USA 85*, 4861–4864.
196. Ali, I. U., Merlo, G., Callahan, R. & Liderau, R. (1989) *Oncogene*, 4, 89–92.
197. Browder, T. M., Abrams, J. S., Wong, P. M. C. & Nienhuis, A. W. (1989) *Mol. Cell. Biol. 9*, 204–213.
198. Betsholtz, C., Westermark, B., Ek, B. & Heldin, C.-H. (1984) *Cell 39*, 447–457.
199. Keating, M. T. & Williams, L. T. (1988) *Science 239*, 914–916.
200. Huang, S. S. & Huang, J. S. (1988) *J. Biol. Chem. 263*, 12608–12618.
201. Hannink, M. & Donoghue, D. J. (1988) *J. Cell Biol. 107*, 287–298.
202. Kahn, P., Frykberg, L., Brady, C., Stanly, I., Beug, H., Vennström, B. & Graf, T. (1986) *Cell 45*, 349–356.
203. Deinhardt, F. (1980) in *Viral oncology* (Klein, G., ed.) pp. 359–398, Raven Press, New York.
204. Greene, W., Leonard, W., Wano, Y., Svetlik, P., Pfeffer, N., Sodorski, J., Rosen, C. A., Goh, W. C. & Haseltine, W. A. (1986) *Science 232*, 877–880.
205. Sporn, M. B. & Roberts, A. B., eds (1989) *Peptide growth factors and their receptors, Handbook of Experimental Pharmacology*, Springer Verlag, Heidelberg, in the press.

Review

A chromosomal basis of lymphoid malignancy in man

Thomas BOEHM and Terence H. RABBITTS
Medical Research Council Laboratory of Molecular Biology, Cambridge

(Received May 3/June 2, 1989) — EJB 89 0560

Antigen receptors, antibodies and T cell receptors (TCR), are proteins made by B and T cells respectively whose function is to recognise diverse antigenic determinants. Accordingly, variability is required in that part of the molecule which recognises and binds to antigen, whereas constant functions (effector functions) are necessary as well. This dual requirement is met by the genes which encode the various receptor proteins: each of the loci, in inherited (germ-line) DNA, possesses separate chromosomal segments which can rearrange and join to each other in the differentiation of B and T cells. Such rearrangements are unique to individual lymphoid cells and serve to mark that cell with particular mature receptor genes. This process of DNA rearrangement also allows errors to occur, which include aberrant intra-gene rearrangements and in addition the creation of chromosomal abnormalities which may involve proto-oncogene activation resulting in appearance of lymphoid tumours.

We discuss such chromosomal abnormalities in this review article with reference to mechanism of the formation of these abnormalities and their consequences for tumour pathogenesis. Firstly, we have outlined some key features of Ig and TCR gene organisation and discuss some characteristics of the expression of these genes which might be important in the interpretation of the effects of chromosome abnormalities.

Structure and rearrangement of human antigen receptor genes

Immunoglobulin gene organisation

Variable genes

There are three separate immunoglobulin (Ig) loci; the heavy (H) chain locus (Igh) and two light (L) chain loci, κ and λ. Each of the three loci are encoded by separate autosomal chromosomes (Table 1). A summary of the organisation of these loci appears in Figs 1 and 2. The heavy chain locus (Fig. 1) is the most complex, consisting of a set of variable (V) region, diversity (D) region segments and joining (J) region segments which somatically rearrange and join to create the active H-chain V gene. All variable region gene segments have a similar organisation (illustrated in Fig. 1 for the heavy-chain

Correspondence to T. H. Rabbitts, MRC Laboratory of Molecular Biology, Hills Road, Cambridge, CB2 2QH, England

Abbreviations. TCR, T-cell receptor; CLL, chronic lymphocytic leukaemia; BL, Burkitt's lymphoma; PLL, prolymphocytic leukaemia; ALL, acute lymphocytic leukaemia; C, constant region; D, J and V, diversity, joining and variable segments; EBV, Epstein-Barr virus; KDE, kappa-deleting element; CDR, complementarity determining region; AT, Ataxia telangiectasia; LCL, lymphoblastoid cell line; LTR, long-terminal repeat.

Table 1. *Chromosomal localisation of human antigen receptor genes*

(A) Immunoglobulin		(B) T-cell receptor	
Gene	Chromosome band	Gene	Chromosome band
heavy chain	14q32.3	α	14q11
κ light chain	2p12	δ	14q11
λ light chain	22q11	β	7q35
		γ	7p15

V gene). The bulk of the V segment is encoded in a single large exon and an intron (of about 100—150 bp) occurs near the end of the leader sequence. This leader sequence encodes a hydrophobic peptide which plays a role in directing the synthesised immunoglobulin peptides to the outside of the cell. The 3' end of the V segment is flanked by recombinase signals (see below).

Initial estimates of the size of the V_H germline repertoire in man suggested about 100 genes, based on the existence of four approximately equal sized subgroups [1]. More recently, two new subgroups have been defined and, using probes from each subgroup, it is estimated that 100—200 V_H genes lie in a region of 2500 kb on human chromosome 14 [2]. The first functional V_H (a member of the V_H VI family) is located 96 kb upstream of the $C\mu$ gene [3, 4] and the major cluster of D_H segments lie between this V_H and the J_H segments [5, 6], this cluster of D_H segments being only 36 kb from $C\mu$. A single D_H segment (DHQ52) [7] is located just upstream of six JH segments. At least one other cluster of D segments occurs at 14q32 [5], and this seems to be located at the most distal part of the V_H cluster (i.e. farthest from the C_H genes). The V_H genes are interspersed as the various members of different subgroups [8] and a large number of pseudo-V_H genes have been identified in the locus. Typically V_H segments are found to be only about 5—15 kb apart [9], and although the transcriptional orientation of the V_H segments in the locus as a whole is unclear, cloned examples tend to be in the same orientation with respect to each other.

The two light-chain loci only have V and J segments (neither has D segments). Thus, only V-J joining occurs to create the active loci. In the human κ chain locus, there are five J segments located about 3 kb upstream of the single $C\kappa$ gene. The first $V\kappa$ segment (the single $V\kappa IV$ gene [10]) is located about 25 kb upstream of $J\kappa$ and in reverse orientation (i.e. the $V\kappa IV$ and $J\kappa$-$C\kappa$ genes are head-to-head) (Fig. 2). The next $V\kappa$ gene, $V\kappa B2$ is only about 12 kb away from $V\kappa IV$ and

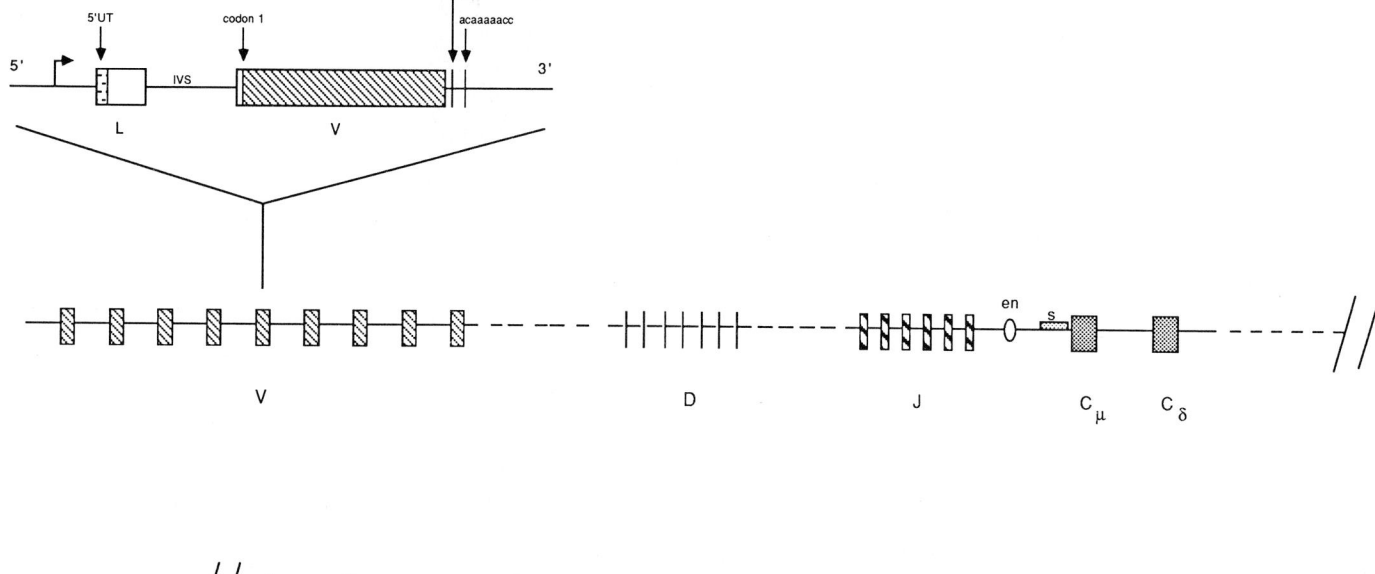
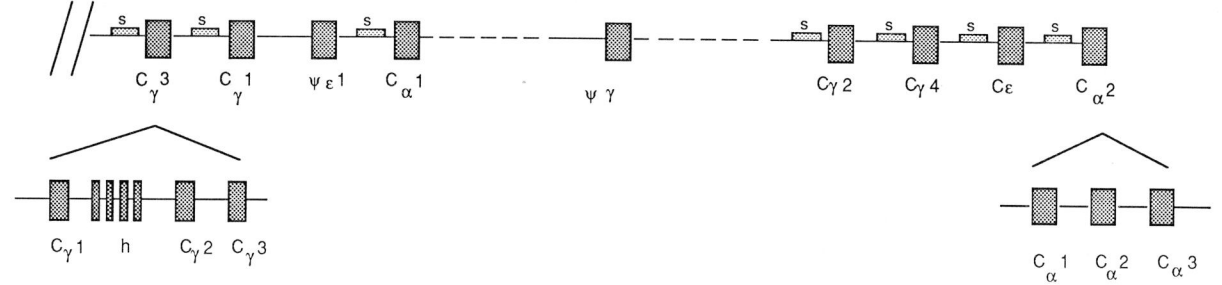

Fig. 1. *Organisation of the human immunoglobulin heavy-chain locus.* The schematic picture depicts the clusters of V, D, J and C segments. There are of the order of 100–200 V segments; the detailed organisation of an individual V gene is shown above. A cluster of D_H elements exists upstream of J_H but other D_H elements (not shown) occur within the body of the V_H cluster. The major Igh transcription enhancer (en) is located between $C\mu$ and J_H. Switch (s) region elements are indicated near each C_H gene which undergoes the class switch. The domain structure of exons (boxes) and introns (lines) of the C_H genes are illustrated in detail for $C\gamma 3$ (h = hinge exons) and for $C\alpha 2$. The linkage of $C\delta$ and $C\gamma 3$ has not yet been achieved nor has the precise position of the $\psi C\gamma$ segment

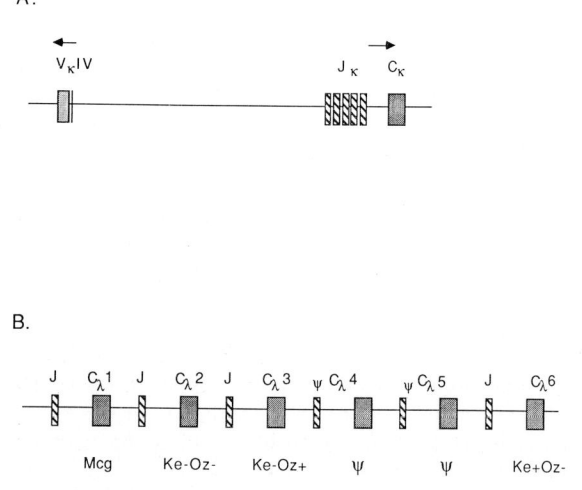

Fig. 2. *The human light-chain constant-region genes.* (A) The single human $C\kappa$ gene has five associated J elements. No D elements exist for the light chains. The $V\kappa$ gene nearest to the $C\kappa$ is $V\kappa$IV and this V gene is organised in the opposite transcriptional orientation to the $C\kappa$ gene. The five $J\kappa$ segments are shown as striped boxes. (B) The $C\lambda$ gene cluster. Six $C\lambda$ genes, each with associated J segments, are shown (not to scale). Four of the $C\lambda$ segments are non-allelic genes and two are pseudo (ψ) genes

is in the same orientation as the $V\kappa$IV gene. Rearrangement of such inverted $V\kappa$ segments takes place via inversion, rather than deletional joining, leaving behind 'relics' of the join called F fragments. Although not yet orientated with reference to $C\kappa$, the major part of the $V\kappa$ locus has a uniform transcriptional polarity. There are around a total of 80 $V\kappa$ genes [11, 12], including a large duplicated region encompassing 28 $V\kappa$ genes. The latter germline regions, designated Aa and Ab, themselves span 250 kb [12]. Other regions have been characterised; 190 kb containing 15 $V\kappa$ genes, 210 kb containing 18 $V\kappa$ genes and another 190 kb containing 10 $V\kappa$ genes. More than half of these $V\kappa$ segments may be pseudo-genes.

Little is presently known about the size of the λ L-chain V-gene pool. Two different $V\lambda$ probes have been isolated and used in genomic hybridisation experiments to show the presence of about 10 cross-hybridising restriction fragments in each case [13, 14]. It is possible that a rather limited number of $V\lambda$ genes exist in humans. Studies of chromosomal translocations have shown that the V genes occur on the telomeric side of the $C\lambda$ cluster on the long arm of chromosome 22.

Constant regions

The three immunoglobulin loci have varying complexity of constant-region gene segments. The most complex is the heavy-chain locus. The human C_H locus contains 11 constant-region genes contained within about 200 kb. As indicated above, the first C_H segment is $C\mu$ followed closely by $C\delta$ (Fig. 1); then comes two clusters of C_H elements, firstly $C\gamma3$, $C\gamma1$, $\psi\epsilon1$ and $C\alpha1$ and secondly $C\gamma2$, $C\gamma4$, $C\epsilon$ and $C\alpha2$ [15]. A pseudo-γ gene also exists in humans, probably located between the two clusters. The $\psi\epsilon$ gene found in the first cluster of C_H elements is a pseudogene because it is truncated at its 5' end. (A second $\psi\epsilon$ gene, $\psi\epsilon2$, also exists in human DNA: this is a processed gene and is not located on chromosome 14.) All but one C_H gene has immediately upstream a switch or S sequence (see below) responsible for the H-chain class switch; the $C\delta$ segment is the exception. Clinically normal individuals have been studied who have large deletions, sometimes homozygous, within their C_H loci [16]; the lack of immunodeficiency problems in such individuals is surprising but argues for the ability of the immune system to compensate (via cell selection) for this lack of genetic material.

The C_H gene segments, unlike C_L, have domains represented by exons which are separated by introns (Fig. 1). Apart from the membrane exons (which are discussed below), the $C\mu$ and $C\epsilon$ genes have four domains, $C\delta$ has three (including a hinge), the γ genes have four domains (one of these being a hinge which is quadruplicated in $\gamma3$) and the α genes have three domains, with the hinge as part of the second domain.

There is a single $C\kappa$ gene about 3 kb from the set of $J\kappa$ segments [17] (Fig. 2A). The $C\kappa$ element exists as a single exon with a splice acceptor site on its upstream side to which the rearranged $J\kappa$ segment joins by RNA splicing. The λ light-chain locus has a more complex constant-region cluster than the κ locus. The constant region of λ light chains are the products of at least four non-allelic (isotypic) genes and differ by small amino-acid substitutions defined, by serology, as Kern, Oz and mcg isotypes. Studies of the genomic genes show that there are six, non-allelic $C\lambda$ genes [18], two of which are pseudogenes [19] (Fig. 2B). Each of the $C\lambda$ genes apparently has one (or more) associated $J\lambda$ segments.

Immunoglobulin gene rearrangement programme

The immunoglobulin genes exhibit allelic exclusion. This means that normally only one of the two alleles at each locus is expressed (in addition they show L-chain isotype exclusion, see below). Allelic exclusion seems to be the result of imprecise joining at V-D-J junctions (H chain) or V-J junctions (L chains) such that many rearrangements are non-productive (for example, the protein translation frame of the V and J regions after joining may not coincide) and therefore only one allele can usually make a product. In the H chain, abortive D-J joining (not involving V_H) can also occur (as for example D_H-J_H joining in T cells) which clearly cannot result in Ig chain production.

A hierachy of rearrangements appears to exist for the immunoglobulin genes in B cell maturation [20]. It seems certain that the Igh locus is the first to rearrange bringing a V-D-J join to allow expression of μ H chains. In pre-B-cells, there is first an expression of H chains as cytoplasmic μ chains in the absence of L chains. At this stage, L chain rearrangement may not have occurred. The κ L chains undergo rearrangement next, resulting in the production of L-chain proteins and therefore of mature Ig as IgM and IgD (reviewed [21]); these two types of immunoglobulin are produced therefore, by individual cells using the same joined V_H segment in association with $C\mu$ and $C\delta$ and the same κ L chain. This co-expression of a single V_H with the two C_H segments is achieved by alternative RNA splicing of a common precursor molecule (reviewed [22]). The functionally rearranged V_H segment can also be expressed with one of the downstream C_H genes (γ, ϵ or α giving rise to IgG, IgE and IgA antibodies respectively). This occurs after the so-called H-chain class switch (see below) which rearranges the functional V_H gene to, for example, $C\gamma3$, with associated deletion of $C\mu$ and $C\delta$ segments (reviewed [23]).

The κ genes are usually deleted in λ-chain-producing cells: this takes place via a sequence found on the downstream of the $C\kappa$ gene, called the kappa-deleting element (KDE) [24]. Apparently, after deletion of the κ genes, λ gene rearrangement takes place and IgM with λ L chains is produced.

A further type of rearrangement is known to take place. This is the replacement of a functionally joined V_H by previously unrearranged upstream V_H. In mice, the most proximal V_H elements tend to be the first to be rearranged to produce V-D-J joins [25]. This V_H replacement phenomenon involves a secondary V_H rearrangement in which a germ-line V_H can join to the rearranged J_H, deleting the previously joined V_H [26, 27]. This may be a mechanism for expanding the Ig repertoire and for ensuring full usage of the germ-line V_H genes, even in otherwise null (heavy-chain-negative) pre-B cells.

Sequence requirements for gene rearrangement

Two forms of gene rearrangement (chromosomal DNA rearrangement) take place in the generation of the mature Ig genes and one in TCR genes. The first is V-D-J joining which is common to Ig and to TCR genes; the second, Igh class switching, is restricted to the Ig-H chain genes.

V-D-J joining signals

V-D, D-J or V-J joining is mediated by the recombinase enzyme. This enzyme has not yet been isolated but sequence requirements have been deduced from comparative sequences of Ig and TCR gene segments and, to a more limited extent, from functional studies on transfected genes [28]. Joining signals flanking the downstream side of an element have the consensus CACAGTG-spacer-ACAAAAACC while those on the upstream side have the consensus GGTTTTTGT-spacer-CACTGTG. Fig. 3 depicts this for all known rearranging genes of the T cell receptor and immunoglobulins. The spacer sequence (for which there is no consensus definable, although there may be conservation within families) must be 23 base pairs (2 helix turns) and 12 base pairs (1 helix turn) on opposite sides of the join.

The process of V_H gene replacement, discussed above, in which a rearranged V_H can be replaced by an upstream germ-line V_H, is apparently mediated only by heptamer and not nonamer homology [26, 27]. A conserved 'cryptic' heptamer is found seven bases upstream of the 3' end of V_H genes; therefore, the V_H of joined V_HDJ_H can use this heptamer and a new join can occur via the 3' joining signal of an unrearranged V_H.

Gene rearrangement seems to require transcriptional activity of the loci to be rearranged; this is the essence of the accessibility model of recombinase activity [25]. The 'opening' of chromatin thus appears to be a crucial pre-requisite for gene rearrangement, as judged by a number of transfection experiments. Transcription from either unrearranged V segments or from around D and/or J clusters (sterile transcripts) precedes gene rearrangement. Two phenomena, however, cannot be explained by the accessibility hypothesis. Firstly, although V_H and D_H elements have been shown to be acces-

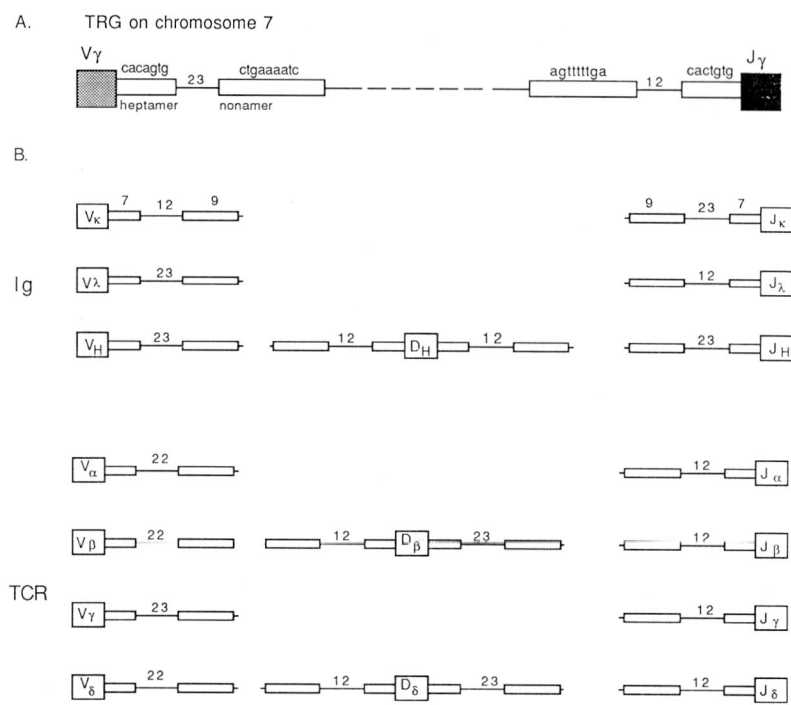

Fig. 3. *Organisation of recombinase signal sequences within the various human antigen receptor loci.* (A) The downstream recombinase signals are shown with typical sequences of heptamers and nonamers for the TCR γ locus of man. (B) Each of the rearranging genes within immunoglobulin and T cell receptor loci are depicted, with D elements where applicable, with appropriate spacing of the signal sequences (12 or 23 base pairs)

sible at the same time, it seems that DJ_H joins generally precede VD_H joins. Secondly, in mice Vκ genes join more frequently to Jκ1 or Jκ2 even though it seems very likely that all four Jκ elements are in a region of 'open' chromatin and therefore presumably equally accessible.

Immunoglobulin heavy-chain class switch

The Igh class switch is the switching of V_HDJ_H join from expression with the Cμ constant region to one of the other C_H genes, e.g. Cα. This process results from a recombination event between so-called switch or S sequences which are found upstream of all C_H genes except Cδ (see Fig. 1). The switch recombination process results in gene deletion of all the DNA between recombining S sequences. For instance, if a switch occurs between Sμ and Sγ1, the segments for Cμ, Cδ and Cγ3 are deleted along with flanking sequences, and the result is a new 'joined' S region (Sμ/Sγ1); the V_HDJ_H join is therefore effectively relocated from one C_H to another, 'carrying' with it the H chain gene transcriptional enhancer. The switch regions seem to be of variable length (2–8 kb) and composed of simple repeat sequences organised in a tandem array (typical sequences are $G_4T(GAGCT)_3$, reviewed in [23]), but divergence is evident when the various S regions are compared [29].

The κ-deleting element

In the hierachy of B cell gene rearrangement, κ L-chain rearrangement precedes λ rearrangement (see above) and the κ genes are frequently deleted in a cell rearranged for λ. A κ-deleting element (KDE) is responsible for this type of deletion. The KDE is located downstream of Cκ and the recombination site in the germ-line has nonamer–23-spacer–heptamer on the 5′ side [24]. Vκ genes can join directly to the KDE (obeying the 12/23 spacing rule of joining), thereby deleting the Cκ gene.

Human immunoglobulin gene variability

The variability afforded by inherited germ-line V_H genes has been discussed above. This V-region repertoire is enhanced in several ways by somatic processes.

Somatic mutation

Studies of rearranged L and H chain genes have shown that base changes occur within the V segments by somatic mutation [30–32]. These alterations have a tendency to cluster within or close to the complementarity determining regions (CDRs) rather than frame work regions (FRs) [33]. Furthermore, mutations can be found in regions adjacent to the rearranged V while rarely within the C region. There is evidence for the existence of a specific enzymatic hypermutation process acting on rearranged V gene, apparently independent of whether it is a functional gene. This evidence is strengthed by the observation of frequently occurring, repeated silent mutations indicative of enzymatically sought hypermutable regions of V genes [32].

Combinatorial diversity

The extent of theoretical diversity which can result from the DNA rearrangement of antibody genes is the product of the number of V, D (if present) and J segments. (This calculation could, for example, lead to a minimum of about 6×10^3 different human V_H segments; i.e. 10^2 V × 10 D × 6 J.) These calculations must, however, be tempered by the fact that some combinations of V, D and J segments may not make functional combinations.

Junctional diversity

Receptor diversity can be significantly amplified by two forms of junctional diversity viz. imprecision of V-J, V-D and D-J joining and N-region diversity. In κ L chains, the amino acid position 96 is a highly variable residue even though it is normally encoded by the first codon of the J [34, 35]. This is because V-J joining can occur in such a way that only the second or third residue of the triplet comes from J and the first or second, respectively, from the the incoming V [28]. In addition, codon 96 can be deleted entirely or out-of-frame joins compensated by base deletions upstream of the join. Analogous imprecisions in joining have been noted in V-D and D-J assembly of the H chain locus.

N-region diversity was originally observed in H chain V_H-D and D-J joins [36]. This was postulated to have arisen from the activity of terminal transferase [36] which was used to explain the random addition of nucleotides (i.e. not encoded by the genome) at the junctions of the fused segments.

Immunoglobulin gene expression

Ig promoters

The 5' end of germline V genes carry the transcriptional promoter sequences which become active after joining to the J segments. All known V genes carry TATA consensus homologies thought to be responsible for transcription activation by RNA polymerase II. An octanucleotide (ATGCAAAT) is found in V_H promoters [37], its complement is found in V_L promoters [38] and in the Igh enhancer. This element is needed for V_H and $V\kappa$ promoter activity.

Enhancers in H and L chains

An enhancer element is located within the intron between J_H and $C\mu$ genes (but upstream of the S sequence, see Fig. 1 and below) which stimulates the transcription of the V_H segment promoter joined to the J_H [39–42]. However, although the rearranged V_H is predominantly activated, some transcription of V_H genes immediately upstream of the productively rearranged V_H does occur [43].

An enhancer element has been identified in the mouse κ locus [44, 45] and its sequence contains an element related to one within the IgH chain enhancer. No enhancer has been identified within the λ locus as yet.

Genomic organisation of TCR gene loci

Unlike immunoglobulins, the clonotypic antigen receptors on T cells are always membrane-bound and non-covalently associated with a complex of invariant glycoproteins, collectively referred to as CD3. This complex is thought to act as an amplifier that signals the binding of antigen to the T-cell receptor to the second messenger system (for a recent review see [46]). To date, two types of TCR are known to be expressed by T cells. Bone-marrow-derived T-cell precursors enter the thymus and become functional T lymphocytes carrying either an α-β heterodimer (the 'classical' T-cell receptor [47]) or a γ-δ heterodimer [48, 49] although other routes are possible. The various genes for the TCR polypeptides are on separate autosomes as indicated in Table 1.

Variable region gene segments

TCR α and γ chains use variable (V) and joining (J) elements to form the variable-region gene, TCR β and δ chains use variable, diversity (D) and joining elements to form the variable regions. V genes of TCR polypeptides encode (in two exons) a hydrophobic leader sequence of about 20 amino acids and about a further 90–100 representing the mature protein.

The organisation of the human TCR loci appears in Fig. 4. An estimated 100 Vα genes can be grouped into 12 families [50] operationally defined as having more than 50% sequence homology on the protein level. About 40 Vβ genes are divided among 14 families [51, 52]. The eight known functional Vγ genes can be subdivided into at least four families, with three families having only one member [53]. The Vδ gene pool in man is not yet defined, since only a few Vδ genes have yet been described [54, 55]; the Vδ1 gene does not cross-hybridise with other V genes [56]. It is conceivable that Vα and Vδ gene pools are not entirely separate, since a highly homologous V gene has already been found in α and δ cDNA clones [57]. The regions occupied by V genes have been analysed by cosmid cloning and pulse-field gel electrophoresis; only the Vγ cluster has yet been linked on clones to its J regions [58].

The Vα/δ locus comprises about 1 Mb and all Vα genes identified so far are located upstream of Dδ1 [59]. The same is true for Vδ genes [60]; however, one Vδ element (termed Vδ2) has been located at the 3' end of Cδ [55] (very much like the situation in the mouse [61]). Members of Vα families are found both adjacent or widely apart from each other and interspersed among members of other Vα families; this may suggest partial duplication processes in that locus; although not yet physically linked, Vα and Dδ1 are at most 60 kb apart [59].

The physical map of the human Vβ locus which spans at least 0.6 Mb shows that all known Vβ genes map to the 5' side of the two DJC clusters and members of individual Vβ families show a similar pattern of clustered and interspersed arrangement [62], again suggesting internal duplication in the Vβ locus. The DJC clusters are within 80 kb of a Vβ11 family gene segment, the 3'-most Vβ element so far known [62].

The Vγ genes are located within 16 kb of the 5'-most Jγ element and span a region of about 100 kb [53, 58, 63].

Where known, the transcriptional orientation of V genes is similar to that of their corresponding D, J, and C regions, except for V genes of 3' of C regions which show opposite transcriptional orientation. In the mouse, V genes have been identified 3' of Cδ [61] Cβ [64] and genes (for mouse V_γ/C_γ organisation) [65], whereas in human, V genes have only been found 3' of Cδ [55]. V gene rearrangement thus occurs by both deletional [66–68] and inversional [55, 64] mechanisms, although evidence for sister chromatid exchange has also been obtained in one instance at the TCR β locus [69].

The human D elements of the TCR β locus are very similar to each other, both in sequence and in length [70, 71]. The D elements in the TCR δ locus are shorter and very dissimilar [55]. A comparison of J elements from the T-cell receptor loci [50, 70, 72, 73] reveals that they are slightly longer than Ig J elements but all very homologous to each other.

Constant region genes

The constant-region genes of TCR polypeptides are associated with their corresponding D (in TCRβ and TCRδ) and J elements as shown in Fig. 4 [55, 70, 73–80]. The constant

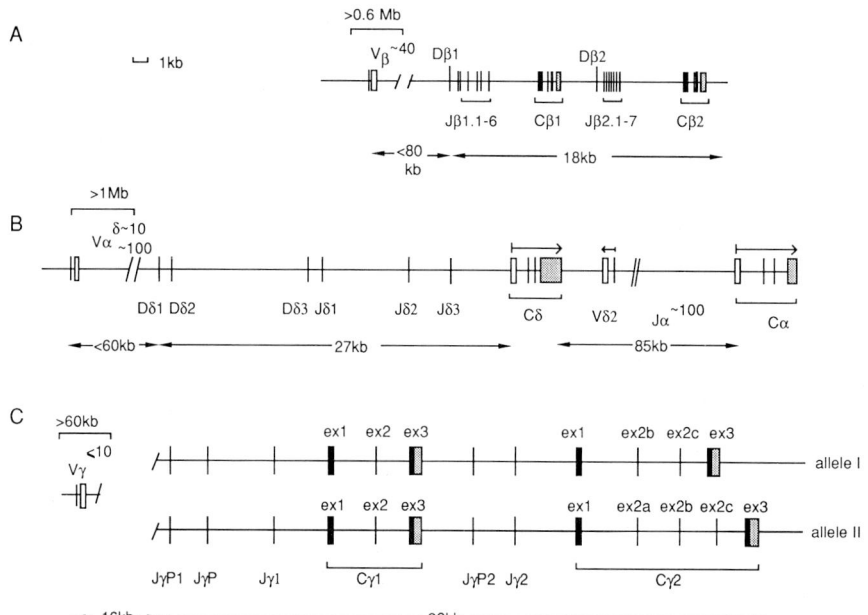

Fig. 4. *Genomic organisation of TCR D, J and C complexes in man (drawn to scale)*. (A) T-cell receptor β-chain locus. Each of the homologous Cβ genes (encoded in four exons) are associated with a single D element and a number of closely linked J elements. An estimated 40 Vβ genes are located within 0.6 Mb, with the most 3′ member less than 80 kb away from Dβ1. All Vβ genes so far analysed lie upstream of the DJC complexes. The transcriptional orientation of Cβ genes is from left to right, that of Vβ genes is presumed to be the same. (B) T-cell receptor δ- and α-chain locus. Two related C complexes are located in tandem with similar transcriptional orientation (arrow above exons). A single Vδ element is known to be located 3′ of Cδ and shows opposite transcriptional orientation. The Jα region is about 80 kb in length and contains an estimated 100 Jα segments. Three Dδ and three Jδ elements have so far been described. An estimated 100 Vα and 10 Vδ genes are located in a region of about 1 Mb upstream of Dδ1; the distance between the first Vα/δ and Dδ1 is thought to be less than 60 kb. The transcriptional orientation of V elements is presumed to be similar to that of C regions. (C) T-cell receptor γ-chain locus. The two haplotypes thus far described in humans are depicted. The Cγ1 gene has three associated Jγ elements, the Cγ2 gene is preceded by two Jγ elements. In allele I, the Cγ2 gene occurs as a gene with a duplicated exon 2, whereas the Cγ2 gene in allele II carries a triplicated second exon. The Vγ gene cluster is located less than 20 kb upstream of JγP1

regions generally contain three cysteine residues (for exceptions see below), two of which reside in the first exon and again are believed to be involved in creating an immunoglobulin-like fold of the C domain. The third cysteine residue, encoded in a separate exon, is thought to be involved in interchain disulphide bonds. In the TCR γ locus, however, the Cγ2 gene exists in two allelic variants, containing duplicated or triplicated second exons, that do not contain the third cysteine residue [78, 79, 81]. Thus, depending on the usage of the Cγ gene, the TCR γ polypeptide is capable of forming disulphide-linked and non-covalently associated heterodimers with the TCR δ chain [82, 83]. The TCR β chain contains one extra cysteine residue located within the first exon. The significance of this cysteine for alternative disulphide bonding is not clear at present, although alignment of protein sequences for secondary structure predictions indicate that this extra cysteine residue is not used. A hydrophobic stretch of about 20 amino acids in the carboxy-terminal part of the TCR polypeptides is believed to be the membrane-spanning region. It contains approximately in the centre at least one positively charged residue (lysine) [TCR α and δ polypeptides contain the motif $R(X)_5K$ and may thus actually have two positive charges in the transmembrane region]. Positively charged residue(s) may form a salt-bridge with CD3 molecules that contain negatively charged residues (Asp or Glu) in their transmembrane regions (for review see [46]). The cytoplasmic tail of TCR polypeptides is rather short (about 5 residues) in α and δ chains, and longer (about 15 residues) in β and γ chains, the latter being highly charged.

Diversity of TCR polypeptides

TCR α and γ variable region genes are formed by V-J rearrangements, very much like that of IgL chains. Examination of the sequence of V-J junctions, however, reveals two features not seen in IgL chains, i.e. the apparent flexibility in the 3′ position of joining of Vα [84], Vδ [56] and Vγ [85] and presence of prominent N regions [56, 63, 84, 86, 87]. Thus, for example, although only very few V and J regions are used in the TCR γ locus, the potential exists to produce more than 10^5 different variable-region genes. Diversity of TCR α variable regions is effected by a relatively large number of Vα elements, but also by an enormous number of Jα elements (an estimated 100 Jα are scattered over about 80 kb between Cα and Cδ, see Fig. 4). The above-mentioned flexibility in 3′ joining and the introduction of N regions result in an estimated 10^{10} different variable-region genes. TCR β rearrangements proceed via D-J intermediates [88] to form a complete V-D-J variable-region gene. About 10^{10} different variable region genes can be made from 40 Vβ, 2 Dβ, and 13 Jβ elements. TCR δ rearrangements in adult thymocytes are characterised by a different pattern of rearrangements to form a variable-region gene. Evidence for V-D, D-D, D-D-J intermediates has been obtained in mouse [89]; in human, the complete δ chain variable-region gene is composed of V-D-D-D-J joints [55]. The relatively small number of V elements in TCR β, γ and δ genes (Fig. 4) limits the combinatorial diversity in heterodimers based on V elements alone, very different from the situation in IgH. In contrast to immunoglobulins, diversity of TCR heterodimers is greatly enhanced by the marked junctional heterogeneity. The TCR δ chain is the prime example for this phenomenon: only about 10 variable, 3 diversity and 3 joining elements suffice to generate about 10^{15} different δ chains because of N-region diversification (which has been observed in all TCR polypeptides, but only in the heavy-chain gene of the Ig loci), usage of D elements in all three reading frames and flexibility in the 3′ joining position of V elements. In addition, the δ chain is unique in the usage of all three D elements in the formation of the variable-region gene; this also results in the presence of four N regions. Hypermutation in variable-region genes appears to be essentially absent in the T-cell-mediated immune response [56, 57, 60, 63, 84, 87, 90].

Patterns of rearrangement and expression in TCR loci

The sequence requirements for TCR gene rearrangements appear similar to those of Ig genes. The heptamer/nonamer signal sequences in TCR genes are indistinguisable from those found in Ig genes; in addition, the 12/23-spacer rule for the assembly of variable region genes seems to be valid for TCR genes as well. This strongly suggests that the recombinase machinery acts on both Ig and TCR genes. The 'accessibility' model for rearrangements [25] that links an open chromatin configuration (i.e. via transcription) to the process of recombination probably also holds for TCR genes, although sterile (i.e. germ-line) transcripts from TCR genes have not yet been described. The fact that phenotypically mature T cells sometimes carry abortive Igh (i.e. D-J) rearrangements [91, 92] suggests that the control over the accessibility of Ig and TCR genes is not absolutely tissue-specific.

TCR β genes assemble their variable-region gene from Vβ, Dβ and Jβ segments. Although D-D rearrangements are formally possible according to the 12/23 role of recombination signal utilisation [28] (see Fig. 3), no evidence for such recombination (such as in the TCR δ locus) have been found. An intermediate step in the formation of the variable region gene is the D-J recombination (most extensively studied in the mouse, see [93, 94]), analogous to the situation in the Igh locus; such DJ-C transcripts are frequently observed [88].

Rearrangements in the TCR δ locus appear to be the earliest events in T-cell ontogeny [95, 96]. cDNA clones obtained from mouse fetal thymocytes have shown that TCR δ chains are predominantly composed of a VDJ structure in the variable region, with very small, if any, N regions [89]. In contrast, TCR δ chains from adult thymocytes in human have a variable region predominantly composed of VDDDJ joints with very large N regions as discussed earlier, very similar to the situation in mouse [97]. Due to the unique geometry of TCR δ/α loci, rearrangement of a V element to a Jα element invariably deletes the TCR δ chain D, J, and C genes on that allele. Therefore, VJα rearrangements can be seen as a means to eliminate the possibility of δ-chain expression. However, a specific deletion process might also be involved. In analogy to the κ-deleting element, a δ-chain deletion element has also been invoked [98].

The T-cell repertoire is elaborated in the thymus by mechanisms still largely unknown. Thymic precursor cells arising in the bone marrow migrate into the thymus, and it is there that they differentiate and become selected for both self-tolerance and MHC restriction in antigen recognition. The current hypothesis is that clonal deletion is a major principle in achieving the final T-cell repertoire (for recent review see [99]). In the mouse, some δ-chain rearrangements occur, in thymic development, before any other TCR rearrangements, and some γ and δ gene transcription precedes the appearance of full-length [i.e. V(D)J] transcripts for β and α chains; it appears, therefore, that the rearrangement and expression of TCRs occurs in a sequential and ordered fashion [95, 100, 101].

A recent study in the mouse system suggests the appearance of at least three waves of T-cells in the developing thymus [102]; two waves of γ/δ-bearing T-cells are followed by thymocytes carrying α/β receptors.

Chromosomal abnormalities in lymphoid tumours

Many lymphoid tumours carry chromosomal abnormalities. These are principally chromosomal translocations, inversion or less commonly, deletions. Chromosome translocation occurs between chromosomes (either allelic or non-allelic pairs) and generally only involves two chromosomes. Translocation may be reciprocal (balanced) exchange of chromosomal material with no loss or gain of material or non-reciprocal (unbalanced) exchange where loss of material occurs at one or both chromosomal junctions. Most frequently, translocation results in one longer and one shorter chromosome; designations follow the convention that the shorter chromosome is called p^- or q^- (depending on whether short or long arms are involved respectively). Chromosome inversion occurs between two breakpoints on the same chromosome, inverting all material between the two breaks with respect to the original centromere and telomere.

Rearranging genes and chromosome abnormalities

Since B and T cell tumours have such a large number of well characterised chromosomal abnormalities which involve the rearranging genes (immunoglobulin genes for B cells and T cell receptor genes for T cells) it is generally agreed that the process of chromosome DNA rearrangement, discussed above, is incorrectly used, resulting in the formation of a chromosomal abnormality, rather than the normal V-D-J joining or class switching. A refinement of this idea has been postulated in which heptamer/nonamer recombinase signal-like sequences are involved on the chromosome not bearing the rearranging gene [103–105]. This has certainly been validated in many cases but is not true of a number of others [106]. Nonetheless, the general picture is that recombinase or the class-switch recombination enzyme (in some cases of Igh involvement) is involved at the rearranging gene locus providing the stimulus for directed (i.e. via signal-like sequences) or random cutting at the second location selected into the tumour phenotype.

Putative oncogene activation via chromosome translocation or inversion

The study of chromosome abnormalities associated with the immunoglobulin genes is useful since these are mainly tumour-associated chromosomal abnormalities and are thought to play a role in tumour pathogenesis of these cases carrying the abnormal chromosome.

The idea that chromosomal abnormalities are involved in the generation of the tumour phenotype has been frequently proposed using paradigms of translocation in Burkitt's lymphoma (see below) and in chronic myelogeneous leukaemia (Philadelphia chromosome) and is thought to involve

Table 2. *Summary of chromosomal abnormalities in B-cell and T-cell tumours*
For detailed references, see text. An asterisk indicates that these tumours and cells are from patients with ataxia telangiectasia

Translocation	Tumour	Rearranging Gene	Proto-oncogene	Other Gene
t(8;14)(q24;q32)	Burkitt's Lymphoma	Igh	c-myc	
t(2;8)(p12;q24)		Igκ	c-myc	
t(8;22)(q24;q11)		Igλ	c-myc	
t(14;18)(q32;q21)	Follicular lymphoma	Igh		bcl-2
t(11;14)(q13;q32)	B-PLL	Igκ		bcl-1/11q13
	B-CLL			
	Myeloma			
inv(14)(q11;q32)	T-lymphoma (SUP T1)	TCRα		V_H
inv(14)(q11;q32)	B-ALL	TCRα		V_H
inv(14)(q11;q32)	T-CLL	TCRα		14q32.1
	T-CLL(AT5BI)*	TCRα		14q32.1
t(14;14)(q11;q32)	T-PLL	TCRα		14q32.1
	T-CLL(AT)*	TCRα		14q32.1
	non-leukaemic*	TCRα		14q32.1
t(8;14)(q24;q11)	T-CLL(SKW-3)	TCRα	c-myc	
t(8;14)(q24;q11)	T-ALL	TCRα	c-myc	
t(8;14)(q24;q11)	T-ALL(DeF)	TCRδ	c-myc	
t(10;14)(q23;q11)	T-ALL	TCRα or δ		10q23
t(11;14)(p15;q11)	T-ALL (RPMI 8402)	TCRδ		11p15
t(11;14)(p13;q11)	T-ALL	TCRδ		T-ALL^{bcr}
t(7;9)(q34;q34.2)	T-ALL	TCRβ		9q34.3
t(7;14)(q35;q32)	T-ALL [AT]*	TCRβ		14q32.1
t(7;10)(q35;q24)	T-ALL	TCRβ		10q24

oncogene activation. The model is as follows. Chromosomal rearrangement at one or the other of the rearranging genes (e.g. immunoglobulin genes in the lymphoid progenitors) erroneously involves a second chromosome. At or very near to the junction of the breakage on the second chromosome is a proto-oncogene which, as a result of translocation, is activated to become an oncogene. Thus we define here an oncogene as a gene which can be involved in the unrestricted growth phenotype of tumour cells.

Chromosome abnormalities in B cell tumours

B-cell-tumour-associated chromosomal abnormalities tend to involve the Igh locus at 14q32 rather than the two L-chain loci (Table 2). The reason for this is not clear, but if V-D-J recombinase is involved in the mechanism of creation of the abnormality it may reflect the time of Igh rearrangement and the requirement for multiple events to occur in the tumour aetiology. The range of B-cell tumour types which carry translocations is very wide but nonetheless the number of different abnormalities is relatively small, suggesting that these lesions are important for B-cell neoplasia in general and not necessarily restricted to a tumour sub-type.

Burkitt's lymphoma

Chromosomal translocation t(8;14) and the c-myc oncogene

Burkitt-type lymphoma is a B-cell tumour which is found in the malarial belt of Africa (endemic BL, with associated infection by Epstein-Barr virus) and in so-called sporadic forms elsewhere. This tumour always carries a chromosome translocation involving the chromosome band 8q24 at the site of the c-myc proto-oncogene [107]; there are three forms of this translocation: t(8;14)(q24;q32), t(2;8)(p12;q24)

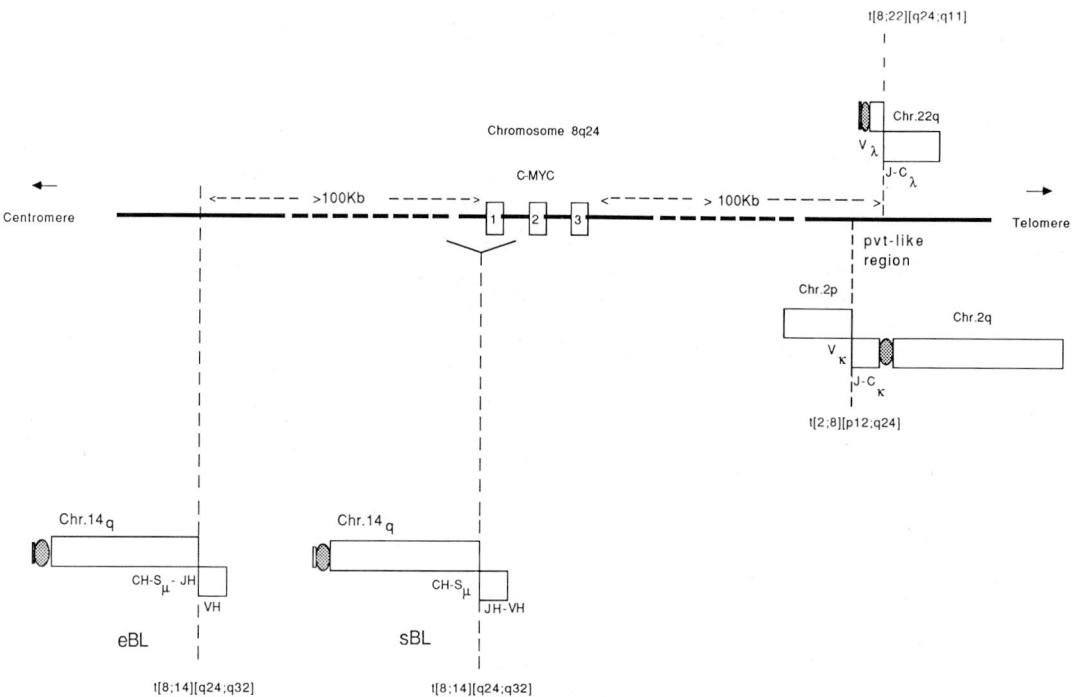

Fig. 5. *Diagrammatic representation of chromosomal translocations found in Burkitt's lymphoma*. The central line represents the position of the c-myc proto-oncogene (consisting of three exons) at chromosome 8 band q24. The direction of telomere and centromere are indicated. The region where the various chromosome translocation breakpoints occur are indicated by vertical dotted lines with the respective translocations within the immunoglobulin loci shown. Endemic BL (eBL) and sporadic BL (sBL) generally have different breakpoints at 14q32 and 8q24 (see diagram and text). The variant Burkitt's lymphomas, involving the L-chain loci, break downstream (telomeric) of c-myc and generally not immediately adjacent to c-myc but rather in a region, designated the pvt-like region, at least 100000 bp from the c-myc gene

and t(8;22)(q24;q11). These involve c-myc (8q24) and immunoglobulin H-chain, κL-chain and λL-chain genes respectively. The t(8;14) form of the translocation most frequently occurs (about 90% of cases). Various studies have established the organisation and reorganisation of the c-myc and immunoglobulin genes. The c-myc is orientated in 8q24 with the 5' end of the gene nearest to the centromere (Fig. 5). In t(8;14) the chromosomal break on 8q24 occurs upstream of c-myc (i.e. between centromere and the 5' end of the gene), thereby causing c-myc to be translocated to the chromosome 14q. The Ig H chain locus is orientated at 14q32 telomere – V_H-D_H-J_H-C_H-centromere and the translocation is generally found within or near J_H or S regions. Thus, after translocation, the 14q$^+$ chromosome has the complex structure telomere-8q24-3'c-myc5'-(breakpoint)-5'J_H-C_H3'-centromere 14 (Fig. 5), with c-myc and Igh in opposite transcriptional orientations (head-to-head).

Detailed molecular biology has revealed a great deal of information on the normal and abnormal activity of the c-myc gene. The gene has three exons, and protein-coding reading frame begins with an AUG codon at the beginning of exon 2 [108]. A second, non-AUG codon has also been identified near the 3' end of the otherwise non-coding exon 1 which gives rise to identical protein except for the very N-terminal residues [109]. The normal c-myc protein is a nuclear protein with a rather short half-life whose function is as yet unknown.

t(8;14) breakpoints at c-myc and Igh genes

The positions of translocation breakpoints relative to the c-myc gene and relative to immunoglobulin genes have received much attention. Examples have been described in BL t(8;14) of breaks within the gene between exon 1/2, just 5' to the gene and a large distance upstream of the gene (Fig. 5). The majority of endemic BL samples have chromosome 8q24 breakpoints which cannot be mapped close to the c-myc gene [110–114]. A noteable exception to this is the Raji cell line where the breakpoint occurs 2 kb upstream of the promoter [112, 114]. On the other hand, most sporadic BL show breaks close to the 5' end of c-myc or within the gene itself (within exon 1 or between exons 1 or 2) [113]. These varying breakpoints, however, have consistent mutational consequences for the translocated c-myc gene as will be discussed below.

A similar disparity of breakpoints at chromosome 14q32 have been described for endemic and sporadic BL with t(8;14). In endemic BL, the predominant site of breakage is within the J_H region [115] sometimes with associated D_H-J_H joins [5]. These breakpoints presumably reflect translocations as errors of V-D-J_H rearrangement (Fig. 5). Again, the Raji cell line represents an exception in that this endemic BL has a breakpoint within the S region [111, 112, 114]. Switch region breakpoints are most frequently observed in sporadic BL [116–118]. The correlation of Igh breakage position and endemic or sporadic BL may reflect the fact that the afflicted B cells in these two forms of disease came from different stages of differentiation as suggested by several studies of cytomorphic and immunological properties of these cells. Endemic BL is associated with EBV (which can immortalise infected cells carrying EBV receptors), and occurs in cells which do not secrete IgM. Presumably, translocation occurs here during the process of V-D-J_H joining. Sporadic BL is found in cells which are capable of IgM secretion. These are more mature cells undergoing the Ig class switch, which occurs after V-D-J$_H$ joining, during which translocation presumably takes place.

Translocations in variant BL

The two so-called variant BL (vBL) account for about 10% of BL translocations and carry t(2;8)(p12;q24) or (8;22)(q24;q11) translocations. In the variant translocations, breakpoints occur on the 3' side of the c-myc (Fig. 5) and, most frequently, at a considerable distance (at least greater than 45 kb) from c-myc [119]. There are exceptions of both t(2;8) and t(8;22) whose breakpoints have been mapped close to the 3' end of c-myc [120]. A cluster of variant breakpoints has been analysed containing t(2;8) breakpoints and this region was designated the pvt-like locus due to the similarities with the mouse plasmacytoma variant breakpoint region [121, 122]. In this pvt-like region, a segment homologous to mis-1 was found [122] which corresponds to a rat sequence characterised by retrovirus insertions in T cell lymphomas. In addition, an interstitial deletion occurs in the pvt-like region within the DNA of a cell line derived from an HTLV-I$^+$, adult T-cell leukaemia [122]. These findings imply that breakage within the pvt region is crucial to development of different types of malignancy.

The translocation breakpoints on chromosome 2 or 22 split the Ig κ or λ L-chain loci respectively. Normally, the break seems to occur at Jκ or Jλ segments [123–125] so that the V segments remain on the original chromosomes (Fig. 5). In one studied case of t(2;8), the breakage seems to be within the Vκ locus. However, in all known cases the Ig C genes move behind c-myc in the variant translocations.

Evidence that chromosomal translocation activates the c-myc proto-oncogene

It is assumed from the consistently occurring translocations in Burkitt-type lymphoma that the c-myc proto-oncogene is activated by its proximity to one of the immunoglobulin loci after translocation. Direct and indirect evidence in support of this view is derived from both transfection studies and transgenic mouse experiments.

Lymphoblastoid cells can be established in culture by infecting B-cell cultures with EBV. Such lines are non-tumorigenic in nude mice, except after prolonged cultures. It has been demonstrated that such lines can be converted to tumorigenic phenotype by transfection with various types of plasmids containing the c-myc gene controlled by strong enhancers, including a truncated c-myc gene, from the BL line Manca, which included the Ig transcription enhancer [126]. Thus, it seems that EBV infection and c-myc activation are sufficient for tumorigenic conversion of B cells, as apparently occurs in pathogenesis of Burkitt's lymphoma.

Strong support for a role in B-cell tumour pathogenesis for c-myc has come from insertion of various forms of c-myc genes in the germ-line of mice (transgene) and subsequent tumour induction [127–129]. Particularly effective in generation of transgenic mouse strains in which individuals develop B-cell tumours are c-myc gene constructs with strong enhancers. Tissue-specific tumour induction of transgenic strains carrying mammary tumour virus LTR–c-myc fusion genes demonstrates that inappropriate expression of c-myc can result in tumour formation [130], even though synergistic effects of the activity of other oncogenes exacerbates the tumorigenic effect [130]. The case for the pathogenic effect of the translocated 'activated' c-myc is overwhelming and thus BL translocations represent a paradigm for the study of lymphoid-specific translocations.

How does translocation activate the c-myc oncogene?

The data on the variety of chromosomal breakpoints around the c-myc gene (upstream and downstream) present a paradox. Can a unifying hypothesis be formulated to explain the consequences of translocation on the c-myc gene? The key seems to be transcriptional deregulation rather than the increase in the absolute levels of mRNA production in BL compared with, for example, LCLs. A relatively consistent feature of BL-translocated c-myc genes is the presence of somatic mutations (whether the translocation occurs upstream or downstream of the c-myc gene). Such somatic mutations where first shown in a t(8;14) case (the Raji cell line) [112] where mutations including point mutations, deletions and small duplications, extended through exon-1–intron-1–exon-2 of the translocated c-myc. This observation was later extended to cases bearing only exon 1 mutations [111, 123] and to cases (relatively rare compared with exon 1 only changes) with mutations in exon 1 and exon 2 [111, 112, 131]. An extensive study [113], carried out in the light of the early reports of somatic mutations within translocated c-myc genes, showed the frequent occurrence of mutations within a region near the 3' end of exon 1. This hot-spot site may be significant in one of two ways. As described above, c-myc protein can apparently originate from two distinct initiation codons giving rise to two proteins with different N-termini [109]. In BL, the loss of exon 1 by truncation in translocation or by somatic mutations at the end of exon 1 may be involved in the loss of synthesis of the larger of the two myc proteins; this may contribute to the oncogenic activation of c-myc. A different, but not necessarily exclusive, mechanism for activation is that the exon 1 mutations correlate with altered transcription control by removal of a transcriptional elongation block [132]. Since somatic mutations occur in both t(8;14) and variant BL translocations [t(2;8) and t(8;22)], acquisition of somatic mutations which accumulate as a consequence of translocation could be sufficient to be a general activator of translocated c-myc.

One intriguing feature of BL is that the normal (non-translocated) allele of c-myc is generally transcriptionally silent (reviewed [105]). A similar effect was observed in EBV$^+$ LCLs transfected by activated c-myc constructs [126]. This feature is reflected in changes in the pattern of DNase-I-hypersensitivity sites upstream of and within c-myc. There are five major sites [133] mapping within 2 kb of the 5' end of the gene. Studies of the pattern of such sites on the normal allele and various translocated alleles have shown that only site I (the most 5' of the DNase-hypersensitivity sites) remains on the silent normal c-myc allele [133, 134]. Such a situation is also seen in HL60 cells which have been induced to differentiate and which cease to express c-myc cytoplasmic mRNA [135]. The pattern of hypersensitive sites does not vary between, for example, that seen in an actively growing, non-tumorigenic LCL and that in a translocated c-myc gene (except, of course, where translocation breakpoints occur near the gene, Fig. 5). No difference is seen between patterns of sites where breakage is far upstream or far downstream of c-myc.

t(11;14)(q13;q32) in B cell tumours

A specific translocation of the immunoglobulin heavy-chain locus at 14q32.3 has been described [136] which involves

11q13. This abnormality has been observed in chronic lymphocytic leukaemia, prolymphocytic leukaemia, diffuse B cell lymphoma and multiple myeloma. DNA cloning has established that the J_H region of the immunoglobulin heavy chain gene is broken by the translocation [137, 138] and in all cases so far published this involves breakage to the end of J_H4 [137–139]. The breakage of the precise position where normal B cell development involves V-D-J joining strongly implicates the normal V-D-J recombinase in the process of chromosomal translocation [103], a feature which has also been highlighted for translocations and inversions involving T cell receptor genes. This idea has been further supported by the sequence analysis of the 11q13 region where it seems that heptamer-like and nonamer-like similarities exist in some cases [106]. However, such similarities are not present in all cases (see later section for discussion). The involved region of 11q13 is fairly clustered in chronic lymphocytic leukaemia, prolymphocytic leukaemia and diffuse B-cell lymphomas and a region of 11q13 which occurs abut 36 kb away from the cluster has also been reported in a multiple myeloma sample. One tumour breakpoint at chromosome 11q13 occurs within a middle repetitive element.

The involvement of this 11q13 breakpoint region in the pathogenesis of those tumours carrying abnormalities of that region, is strongly implied by fairly consistent occurrence of the abnormality in various B-cell tumours. Further support is derived from the relatively short stretch of DNA at 11q13 which contains all cloned breakpoints. Nonetheless, as yet, no gene has been identified in this region as far as 40 kb from a known t(11;14) breakpoint [139]. It should, of course, be borne in mind that, for example, the variant BL breakpoints occur about 300 kb away from the c-myc oncogene [122] and, therefore, that the inability to detect a gene near a breakpoint is not necessarily a negative indication. The expectation would thus be that a transcribed region will be located, revealing a gene showing deregulation similar to the c-myc gene as a consequence of its juxtaposition to the J_H segments.

B-cell follicular lymphoma with translocation t(14;18)

B-cell follicular lymphoma is a common B-cell neoplasm in man. A cytogenetic hallmark of this disease is the presence of a translocation t(14;18)(q32;q21) [136]. As many as 60–80% of these tumours possess this translocation. Some cases of follicular lymphoma progress to diffuse histiocytic or large cell lymphoma and some examples of the latter have been reported which also carry t(14;18).

Molecular junctions of translocation t(14;18)

The cytogenetic breakpoint at 14q32 was investigated via Igh H-chain probes because the 14q32 breakpoint corresponds cytogenetically to this locus. Indeed, the Igh locus is involved in this translocation with cloned breakpoints again occurring within the J_H cluster [140–146], generally precisely at the 5′ end of J_H segments. In these tumours, unlike t(11;14)(q13;q32) where only one J_H has been found to be involved so far, at least two different J_H segments have been shown at translocation junctions. Thus FL translocations arise from breakages within the immunoglobulin H chain and at positions adjacent to the normal site of DNA rearrangement associated with V-D-J joining. This is one reason for believing that the V-D-J recombinase is aberrantly involved in the process of translocation.

The bcl-2 gene at the chromosome 18 junction of t(14;18)

A gene has been identified close to the chromosome 18 junction of the follicular lymphoma t(14;18) [140, 142, 144] and this gene, bcl-2, is thereby implicated in the pathogenesis of those follicular lymphomas carrying the translocation. Detailed analysis of the bcl-2 gene has been carried out, revealing a gene with three exons. There is a long 5′ untranslated first exon, a small 220-bp intron between exons I and II and a very large intron (370 kb) between exons II and III [141, 144–146]. The promoter elements of bcl-2 contain TATA and CCAAT motifs as well as multiple potential Sp1 binding sites. Somatic mutations resulting in coding-region changes, like those seen in translocated c-myc, have been found in translocated bcl-2 [147].

Two breakpoint regions have been described for the bcl-2 gene on chromosome 18q21. The majority of these occur within the 3′ non-coding region of the third exon. However, a second region of breakage near the 5′ end of bcl-2 has been found to occur at a reasonable frequency [148]. In one such case, an unexpected order of the involved genes was observed [141], being 3′ Cγ Sγ/Sμ J_H-5′: 5′ bcl-2 3′. For this tumour translocation to have happened, inversion must have occurred to derive the new gene order; this inversion could be either of bcl-2 or of the Igh locus.

Mechanism of t(14;18) translocation

The precise mechanism of this chromosomal translocation remains controversial. The fact that translocation breakpoints have so far all been found within the J_H cluster obviously argues for errors of V-D-J recombinase activity in the creation of the translocation. However, there are two distinct view points. One set of data argues for the existence of heptamer-like and nonamer-like recombinase signal sequences at the junction of the translocations on chromosome 18 [149]. These sequences are suggested as the target for V-D-J recombinase resulting in an inter-chromosomal rearrangement (i.e. translocation) instead of the normal intra-chromosomal V-D-J joining. In contrast to this, other data [143] argue that chromosome 18 cuts do not occur at recombinase signals. A model was proposed in which illegitimate pairing occurs between D_H and J_H ends from chromosome 14 and staggered double-stranded breaks from chromosome 18, resulting in a translocation. Clearly, these two models differ significantly in the role of recombinase.

Effect of translocation of bcl-2 to the immunoglobulin H-chain locus

The various translocation t(14;18) junctions which have been studied so far do not appear to truncate in any way the coding region of the bcl-2 and the only alterations reported are some point somatic mutations [147]. The best hypothesis at the moment appears to be deregulation via transcription perturbation. The bcl-2 mRNA has a fairly short half-life [147, 150] and it displays transcriptionally regulated expression in T and B cells in response to mitogens [150]. The translocated bcl-2 transcription product displays a similar half-life to normal bcl-2 mRNA and, therefore, the higher level of mRNA in cells with translocations must reflect higher levels of transcription or post-transcriptional processes. Interestingly, it seems that the untranslocated bcl-2 allele in FL is transcriptionally silent [147] again suggesting, as in the case of the c-myc oncogene, that tight controls are exerted on the gene.

The precise nature of the transcriptional deregulation remains uncertain. Many translocation breakpoints do occur in the 3′ untranslated region of bcl-2 which might remove control elements. On the Igh side, all the breakpoints of the t(14;18) have been found in J_H segments which leaves the H-chain transcription enhancer close to the translocation junction and therefore potentially active in bcl-2 transcription. However, since most breakpoints occur at the 3′ end of bcl-2 and since the bcl-2 gene is about 370 kb long, the enhancer would need to operate over this large span if it is involved in bcl-2 activation.

The protein product of the bcl-2 gene

Analysis of genomic and cDNA clones has identified two possible bcl-2 open reading frames which could yield two proteins of 239 amino acids (designated α) and one of 205 amino acid (designated β). The putative α and β chains derived from overlapping reading frames would be identical except at the carboxy terminus [141, 146, 151]. No similarities with any known protein have been found. Antibodies were raised against the protein of bcl-2 (made in a bacterial expression vector system) and subcellular fractionation indicates that the protein is associated with the cell membrane [152] but the absence of a transmembrane domain makes it unlikely to span the membrane and behave as a receptor-like protein.

Obviously, these studies are preliminary and much work needs to be done to determine the function of the bcl-2 protein and the role of this protein in tumorigenesis. One recent study bears directly on the latter problem [153]. A retrovirus vector was constructed containing the bcl-2 sequence and this was introduced into normal mouse bone marrow cells or in bone marrow derived from a strain of transgenic mice carrying the c-myc gene (as a transgene) controlled by the Igh transcriptional enhancer. The latter cells developed a low number of colonies in soft agar (6.7×10^{-3}% of cells plated) whereas normal cells infected with bcl-2 construct gave no colonies. However, none of the colonies proliferated in culture. Further experiments on growth factor dependence in the presence of the bcl-2 construct were taken to mean that bcl-2 provides a survival signal to the cells and thus might contribute to neoplasia by allowing a clone to persist in an individual [after the translocation t(14;18)] until other oncogenes become activated by other means; e.g. c-myc. This conclusion complements the observed co-existence of t(14;18) and t(8;14) in acute B cell leukaemias affecting bcl-2 and c-myc respectively [153a].

Other translocations involving the Igh locus

A large number of B cell tumours have been described in which cytogenetically the chromosome band 14q32 is involved. One of these cases, translocation t(14;19)(q32;q13) [154], has recently been the subject of cloning studies [155]. A translocation breakpoint was cloned from a t(14;19) (which was part of a three-way translocation) in a B-CLL and the junction appears to have occurred on chromosome 14 within the C-region cluster of the Igh locus and probably within the switch region sequence associated with the $C\alpha 2$ gene [155]. Chromosome 19 sequences were joined to this region of the $C\alpha$ gene but as yet no analysis of this area has been carried out. The presumption is, again by analogy to the BL paradigm, that a new proto-oncogene will be identified at or near the translocation junction of chromosome 19 and that this will have been activated by the translocation. Thus, the putative oncogene would be involved in pathogenesis of tumours carrying t(14;19) or possibly any other analogous abnormality of this chromosome 19 region.

Chromosomal abnormalities in T cells

A large number of different chromosomal abnormalities have been described in T-cell tumours. These are summarised in Table 2 and discussed in detail below. The TCR δ and α loci are the main targets for these aberrations and the diversity of breakpoints in these genes is illustrated in Fig. 6. A number of abnormalities associated with TCR β have been described (see below) but, as yet, none has been shown to involve the TCR γ in humans.

Inv(14)(q11;q32) and t(14;14)(q11;q32)

The involvement of the TCR α-chain locus in the generation of the inv(14)(q11;q32) and t(14;14)(q11;q32) is now well established. In general, Jα elements are fused to sequences from the 14q32 region; the latter segment in fact comprises two different loci, the Igh locus at 14q32.3 and another locus at 14q32.1. The former is involved in a chromosomal inversion observed in the SUP-T1 cell line [156–158]. Detailed molecular analysis of sequences at both centromeric and telomeric breakpoints have shown that two rearranging genes from opposite ends of the chromosome (i.e. TCR α and Igh) were fused together. A scheme whereby this fusion occurred is shown in Fig. 7. After an initial Vα-Jα rearrangement, the actual inversion occurred between a V_H and a Jα segment. This event obeyed the 12/23 rule of recombination, since V_H sequences carry recombination signals with a 23-nucleotide spacer, whereas Jα elements have a 12-nucleotide spacer. Further analysis, however, has shown that the initial V_H-Jα fusion was superseded by another V_H-Jα rearrangement to result in a V_H-Jα-Cα (or IgT) fusion gene. A similar IgT fusion gene has been described in an acute lymphoid leukaemia of B-cell phenotype [159]. These results indicate that recombinase can recognise both TCR and Ig loci and fuse them into transcriptionally active units. Their pathogenic significance, however, is doubtful and IgT fusions may simply represent rare events in normal TCR α gene assembly; if this were the case, then the actual tumorigenic event must have occurred elsewhere; indeed, other chromosomal aberrations in the SUP-T1 cell line have been identified and are discussed below.

There is evidence for paracentric inversions of chromosome 14 in a variety of non-malignant conditions, most notably in ataxia telangiectasia. In the latter condition, chromosomal translocations and inversions involving chromosomes 7q35 (the site of the TCRβ locus), 7p15 (the site of the TCRγ locus), 14q11 (the site of the TCRδ/α locus) and 14q32 (the site of the Igh locus) have been observed [160]. Similar observations have been made in normal lymphocytes but occurring at very low frequencies [161]. Nevertheless, these findings argue that such quasi-normal rearrangements not only appear as rare aberrations in normal lymphocytes, but should occasionally also be present in their malignant counterparts. Thus it was important to analyse whether a chromosomal heterogeneity in 14q32 breakpoints exists. Indeed, evidence has been obtained that the 14q32.1 locus, frequently involved in t(14;14)(q11;q32) and inv(14)(q11;q32) abnormalities [162–164], is approximately 15–20 Mb centromeric to the Ig locus at band 14q32.1 [165]. Furthermore, it has been shown that

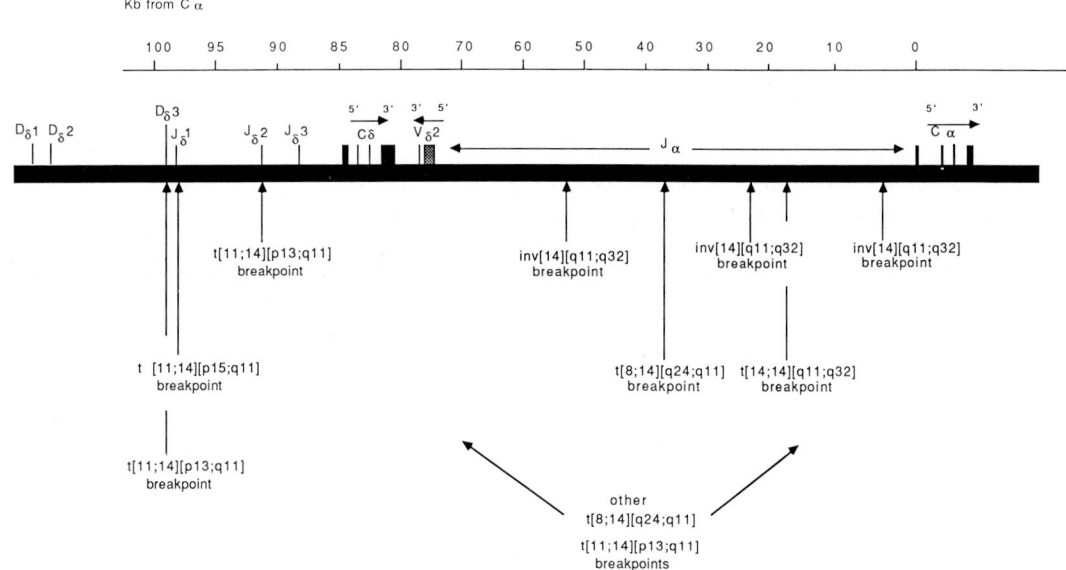

Fig. 6. *The position of breakpoints of chromosomal aberrations involving the TCR δ/α locus.* The organisation of TCR δ/α genes is as indicated (Fig. 4). The region between Vδ2 and Cα has many Jα segments

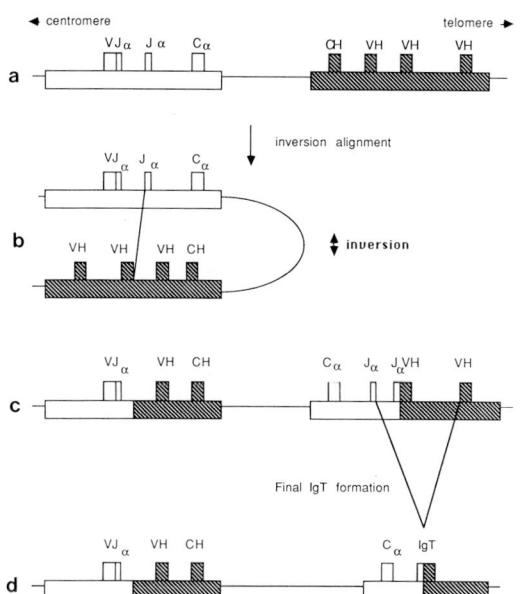

Fig. 7. *The mechanism of chromosome 14 inversion involving the TCR α locus and Igh.* (a) Schematic diagram of the Igh locus at 14q32 (hatched boxes) and the TCR α locus at 14q11 (open boxes) in the cell line SUP-T1. (b) Alignment of the two loci by inversion places Igh and the TCR α locus in the correct orientation with respect to each other to allow V-J joining (heptamer and nonamer signals are separated by 23 base pairs downstream of V_H and 12 base pairs upstream of Jα, facilitating this otherwise illegitimate rearrangement). This joining causes chromosome inversion to occur. (c) The IgT fusion gene created at the telomere of the aberrant chromosome was not the final IgT gene; therefore, a subsequent rearrangement at 14q32 must have occurred. This rearrangement would then have unusually involved V_H and Jα which form part of the same locus in the inverted chromosome. (d) The final configuration of the inverted chromosome of SUP-T1 showing the IgT fusion gene

the breakpoints in T-cell tumours from AT patients and non-AT patients fall into the same region at 14q32.1 [165]. A recent report described a translocation t(7;14)(q35;q32) between the

TCR β locus and a region proximal to the Igh locus on chromosome 14q32 [166]. It is possible, therefore, that the 14q32.1 locus was also involved in that translocation.

In summary, it seems that translocations and inversions to chromosome 14q32 in fact represent two classes. The first class is probably the reflection of rare but quasi-normal recombinations between Igh and TCRα loci. Therefore, their pathogenic significance is at best doubtful. The second class may, in fact, involve a proto-oncogene, located about 15–20 Mb proximal to the Igh locus at chromosome 14q32.1 which becomes activated by fusion to an antigen-receptor gene.

The progressive neurological disease ataxia telangiectasia (AT) is associated with an increased frequency of leukaemia, often associated with specific translocations and inversions of chromosome 14; typically inv(14)(q11;q32) and t(14;14)(q11;q32) [167–169]. In addition to this, there is a great deal of evidence that clonally associated abnormalities of the type discussed above also occur in circulating T cells of AT patients who do not have leukaemia symptoms [170–173]. Although these aberrations do occur in T cells of normal people, the latter are probably different from the abnormalities observed in the T-cell tumours and AT clonal proliferations [174–176]. The chromosomal breakpoint of an AT-associated clonal translocation (14;14) was investigated by *in situ* hybridisation [177] and the breakpoint at 14q32 was found to be in the same region (i.e. 14q32.1) as the T-cell-leukaemia-associated breaks of both AT and non-AT cancer patients. This observation suggests that progression from clonal non-leukaemia to overt leukaemia occurs via secondary changes occurring in an immortalised, but not yet malignant, cell resulting from the inversion or translocation of chromosome 14. Thus the AT-associated preponderance to T-cell leukaemia with chromosome abnormalities provides an interesting model for preleukaemia and for progression from this state to overt cancer.

t(8;14)(q24;q11)

Translocations involving chromosome 8q24 are also seen in T-cell tumours. Two examples of a translocation involving breakpoints to the 3' side of the c-myc gene have been described [178–180]. In one of these tumours, the translocation occurred probably by use of a cryptic heptamer located at the 3' end of a Vα gene and a site located at an undetermined distance 3' of the c-myc gene [180]. Although the authors claim the existence of a heptamer/nonamer-like signal sequence at the breakpoint on chromosome 8q24, there is only a poor match to consensus signal sequences. Nevertheless, this indicates that a sequence-specific cut by the recombinase on at least one chromosome participated in the formation of the translocation. In another case of T-ALL [179], it was later reported that the breakpoint on 14q11 was located in the Jδ region [181], although molecular cloning of the rearranged allele was not undertaken. A recent report [182] describes the analysis of a t(2;8)(q34;q24) translocation which involves the 3' end of the c-myc proto-oncogene. Although the sequences on chromosome 2q34 were not defined further, it was observed that, in addition to the normal 2.4-kb c-myc transcript, two additional transcripts of 3.8 kb and 6.8 kb hybridised to both c-myc and chromosome 2q34 probes. Since the breakpoint at the 3' end of c-myc is located 3' of the translation termination codon, the coding potential of these putative fusion transcripts is unclear.

t(11;14)(p13;q11)

This translocation was orginally identified in four paediatric cases of $CD2^+$ T-ALL by Williams et al. [183] and subsequently shown to split the TCR α-chain locus between Vα and Cα [184, 185] by analysis of somatic cell hybrids made from the tumours. Molecular cloning of the breakpoints of these tumours indicated the involvement of the TCR δ-chain locus in the majority of cases [186]. The structure of the $11p^+$ chromosomes suggested that the translocation occurred during an attempt to rearrange V segments to either J, DD or DDJ elements and thus reflects the type of rearrangement normally seen in the TCR δ locus. Interestingly, the breakpoints of all tumours studied on chromosome 11p13 cluster within a very short region; in fact, five tumours have translocation breaks within 2 kb [186, 187]. This suggests that the disruption of the T-ALL^{bcr} locus is a critical event in the aetiology of tumours carrying the t(11;14)(p13;q11) translocation. The relationship of the T-ALL^{bcr} locus to the Wilms' predisposition gene was investigated by using somatic cell hybrid lines carrying chromosomes 11 with varying degrees of 11p13 deletions [187]. The results indicated that the order of genes at 11p13 is: centromere-CAT-T-ALL^{bcr}-WT-AN2-FSHB-telomere. (CAT = catalase; WT = Wilms' predisposition locus as defined by tumour-associated deletions; AN2 = aniridia locus; FSHB = follicle-stimulating hormone β subunit.) This suggested that the T-ALL^{bcr} locus lies very close to but may be distinct from the Wilms' predisposition locus at 11p13.

Sequence analysis of chromosomal junctions of $11p^+$ and $14q^-$ chromosomes revealed insights into the mechanisms by which these translocations occurred. Two breakpoints were analysed [186] and only one of them showed a heptamer sequence at the breakpoint. This suggested that sequence-specific recognition by the recombinase machinery is not generally involved, especially since the two breakpoints are only about 600 bp apart. This finding also suggests that the mechanism by which the T-ALL^{bcr} locus becomes disrupted is irrelevant with respect to the phenotypic consequences of this lesion. The selective advantage of cells aquiring this mutation is probably reflected in the out-growth of the tumour.

No gene has yet been identified at the T-ALL^{bcr} locus. Although it is difficult to rule out the existence of the gene near the vicinity of the T-ALL^{bcr}, the failure to find a gene could mean that, again by analogy to the BL-variant tumours, the breaks may cluster at a considerable distance from the actual gene involved.

t(11;14)(p15;q11)

This translocation has been identified in the T-ALL cell line RPMI 8402 [188] and analysed by molecular cloning [80]. It represents a rare (sporadic) kind of abnormality that lacks the consistency of breakpoints as seen in t(11;14)(p13;q11). Therefore, this abnormality poses an important question: does the formation of the t(11;14)(p13;q11) have pathogenic consequences for the RPMI 8402 cell line? To address this question, it is important to know by which mechanism the translocation arose. Detailed molecular analysis has shown the presence of a heptamer-like signal sequence exactly at the breakpoint [80], again suggesting that two-site recognition by recombinase played a role. However, since no other abnormality has been identified in that region so far, it is possible to argue that the translocation was the mere result of recombinase error and its formation innocuous to the cell. This represents an interesting challenge of the central dogma of cancer cytogenetics that chromosomal abnormalities are involved in tumour pathogenesis. In contrast to the situation of the T-ALL^{bcr} locus, a gene has been found at the chromosomal breakpoint on chromosome 11p15. This provides the opportunity to study its role, if any, in malignant transformation of the RPMI 8402 cell line.

Other translocations in T-cell tumours

The diverse translocation types observed in T-ALL tumours are summarized in Table 2 and the most intensively studied are discussed above. A number of other translocations have been studied to a lesser extent. A t(10;14)(q24;q11) translocation has been described in a case of T-ALL [189]. Analysis of somatic cell hybrids indicated that the break occurred between Cα and Vα and was later reported to be within Jδ [181]. No information on the sequences involved on chromosome 10q24 are available. Recently, a translocation t(7;10)(q35;q24) has been described in which the TCR B has translocated with a region on 10q24 [190]. Molecular cloning of the junction, in this case again, illustrates the occurrence of heptamer-like similarities at the chromosome junctions. No mRNA was detectable from any region near the translocation point on 10q24 [190].

A t(7;9)(q34;q34) translocation involving the T-cell receptor β locus in the SUP-T1 cell line was shown to be due to recombination on the 5' side of Dβ2.1 (fused to Jβ2.2) and a locus on the long arm of chromosome 9, which becomes transcriptionally activated [191]. No information on this gene is yet available. Interestingly, a region on chromosome 9q32 also appears to be translocated to the TCR β locus in another tumour, SUP-T3 [191]; both regions are apparently rearranged in other T-lymphoblastic tumours. The SUP-T1 cell line thus carries at least two potentially pathogenic abnormalities: (a) an IgT fusion gene, as discussed above, and (b) a seemingly activated locus on chromosome 9q34. However, yet another abnormality involving a region 3' of the c-myc gene has been described [192]. This abnormality is most likely due to a net deletion event and may parallel changes seen in

Burkitt's lymphoma breakpoints. It is clear, therefore, that only functional assays can determine which of the described chromosomal aberrations was involved in malignant transformation.

REFERENCES

1. Rabbitts, T. H., Matthyssens, G. & Hamlyn, P. H. (1980) *Nature* 284, 238–243.
2. Berman, J. E., Mellis, S. J., Pollock, R., Smith, C. L., Suh, H., Heinke, B., Kowal, C., Surti, U., Chess, L., Cantor, C. R. & Alt, F. W. (1988) *EMBO J.* 7, 727–738.
3. Buluwela, L. & Rabbitts, T. H. (1988) *Eur. J. Immunol.* 18, 1843–1845.
4. Schroeder, H. W., Walter, M. A., Hofker, M. H., Ebens, A., van Dijk, K. W., Liao, L. C., Cox, D. W., Milner, E. & Perlmutter, R. M. (1988) *Proc. Natl. Acad. Sci. USA* 85, 8196–8200.
5. Buluwela, L., Albertson, D., Sherrington, P., Rabbitts, P. H., Spurr, N. & Rabbitts, T. H. (1988) *EMBO J.* 7, 2003–2010.
6. Matsuda, F., Lee, K. L., Nakai, A., Sato, T., Kodaira, M., Zong, S. Q., Ohno, H., Fukuhara, S. & Honjo, T. (1988) *EMBO J.* 7, 1047–1051.
7. Siebenlist, U., Ravetch, J. V., Korsmeyer, S., Waldmann, T. & Leder, P. (1981) *Nature* 294, 631–635.
8. Kodaira, M., Kinashi, T., Umemura, I., Matsuda, F., Noma, T., Ono, Y. & Honjo, T. (1986) *J. Mol. Biol.* 190, 529–541.
9. Matthyssens, G. & Rabbitts, T. H. (1980) *Proc. Natl. Acad. Sci. USA* 77, 6561–6565.
10. Klobeck, H.-G., Bornkamm, G. W., Combriato, G., Mocikat, R., Pohlenz, H.-D. & Zachau, H. G. (1985) *Nucleic Acids Res.* 13, 6514–6526.
11. Bentley, D. B. & Rabbitts, T. H. (1981) *Cell* 24, 613–623.
12. Straubinger, B., Huber, E., Lorenz, W., Osterholzer, E., Pargent, W., Pech, M., Pohlenz, H.-D., Zimmer, F.-J. & Zachau, H. G. (1988) *J. Mol. Biol.* 199, 23–34.
13. Anderson, M. L. M., Szajnert, M. F., Kaplan, J. C., McColl, L. & Young, B. D. (1984) *Nucleic Acids Res.* 12, 6647–6661.
14. Tsujimoto, Y. & Croce, C. M. (1984) *Nucleic Acids Res.* 12, 8407–8413.
15. Flanagan, J. G. & Rabbitts, T. H. (1982) *EMBO J.* 1, 655–660.
16. Lefranc, M.-P., Lefranc, G. & Rabbitts, T. H. (1982) *Nature* 300, 760–762.
17. Hieter, P. A., Max, E. E., Seidman, J. G., Maizel, J. V. & Leder, P. (1980) *Cell* 22, 197–207.
18. Hieter, P. A., Hollis, G. F., Korsmeyer, S. J., Waldmann, T. A. & Leder, P. (1981) *Nature* 294, 536–540.
19. Dariavach, P., Lefranc, G. & Lefranc, M.-P. (1987) *Proc. Natl. Acad. Sci. USA* 84, 9074–9078.
20. Hieter, P. A., Korsmeyer, S. J., Waldmann, T. A. & Leder, P. (1981) *Nature* 290, 368–372.
21. Wall, R. & Kuehl, J. M. (1983) *Annu. Rev. Immunol.* 1, 393–422.
22. Blattner, F. R. & Tucker, P. W. (1984) *Nature* 307, 417–422.
23. Honjo, T., Shimizu, A. & Yaoita, Y. (1989) in *Immunoglobulin genes* (Alt, F., Honjo, T. & Rabbitts, T. H., eds) pp. 123–149, Academic Press, New York.
24. Siminovitch, K. A., Moore, M. W., Durdik, J. & Selsing, E. (1987) *Nucleic Acids Res.* 6, 2699–2705.
25. Yancopoulos, G. D. & Alt, F. W. (1985) *Cell* 40, 271–281.
26. Reth, M., Gehrmann, P., Petrac, E. & Wiese, P. (1986) *Nature* 322, 840–842.
27. Kleinfield, R., Hardy, R. R., Tarlinton, D., Dangl, J. M., Herzenberg, L. A. & Weigert, M. (1986) *Nature* 322, 843–846.
28. Tonegawa, S. (1983) *Nature* 302, 575–581.
29. Flanagan, J. G. & Rabbitts, T. H. (1982) *Nature* 300, 709–713.
30. Weigert, W. G., Cesari, I. M., Yonkovich, S. J. & Cohn, M. (1970) *Nature* 228, 1045–1048.
31. Bernard, O., Hozumi, N. & Tonegawa, S. (1978) *Cell* 15, 1133–1140.
32. Berek, C. & Milstein, C. (1987) *Immunol. Rev.* 96, 23–41.
33. Griffiths, G. M., Berek, C., Kaartinen, M. & Milstein, C. (1984) *Nature* 312, 271–274.
34. Sakano, H., Huppi, K., Heinrich, G. & Tonegawa, S. (1979) *Nature* 280, 288–294.
35. Max, E. E., Seidman, J. G. & Leder, P. (1979) *Proc. Natl. Acad. Sci. USA* 76, 3450–3454.
36. Alt, F. W. & Baltimore, D. (1982) *Proc. Natl. Acad. Sci. USA* 79, 4118–4122.
37. Parslow, T. G., Blair, D. L., Murphy, W. J. & Granner, D. K. (1984) *Proc. Natl. Acad. Sci. USA* 81, 2650–2654.
38. Falkner, F. G. & Zachau, H. G. (1984) *Nature* 310, 71–74.
39. Banerji, J., Olson, L. & Schaffner, W. (1983) *Cell* 33, 729–740.
40. Gillies, S. D., Morrison, S. L., Oi, V. T. & Tonegawa, S. (1983) *Cell* 33, 717–728.
41. Neuberger, M. S. (1983) *EMBO J.* 2, 1373–1378.
42. Rabbitts, T. H., Forster, A., Baer, R. & Hamlyn, P. H. (1983) *Nature* 306, 806–809.
43. Wang, X.-F. & Calame, K. (1985) *Cell* 43, 659–665.
44. Picard, O. & Schaffner, W. (1984) *Nature* 302, 80–82.
45. Stafford, J. & Queen, C. (1984) *Mol. Cell. Biol.* 4, 1042–1049.
46. Clevers, H., Alarcon, B., Wileman, T. & Terhorst, C. (1988) *Annu. Rev. Immunol.* 6, 629–662.
47. Meuer, S. C., Acuto, O., Hussey, R. E., Hodgdon, J. C., Fitzgerald, K. A., Schlossman, S. F. & Reinherz, E. L. (1983) *Nature* 303, 808–810.
48. Brenner, M. B., McLean, J., Dialynas, D. P., Strominger, J. L., Smith, J. A., Owen, F. L., Seidman, J. G., Ip, S., Rosen, F. & Krangel, M. S. (1986) *Nature* 322, 145–149.
49. Bank, I., DePinho, R. A., Brenner, M. B., Cassimeris, J., Alt, F. W. & Chess, L. (1986) *Nature* 322, 179–181.
50. Klein, M. H., Concannon, P., Everett, M., Kim, L. D. H., Hunkapiller, T. & Hood, L. (1987) *Proc. Natl. Acad. Sci. USA* 84, 6884–6888.
51. Concannon, P., Pickering, L. A., Kung, P. & Hood, L. (1986) *Proc. Natl. Acad. Sci. USA* 83, 6598–6602.
52. Kimura, N., Toyonaga, B., Yoshikai, Y., Triebel, F., Debre, P., Minden, M. D. & Mak, T. W. (1986) *J. Exp. Med.* 164, 739–750.
53. Forster, A., Huck, S., Ghanem, N., Lefranc, M.-P. & Rabbitts, T. H. (1987) *EMBO J.* 6, 1945–1950.
54. Hata, S., Brenner, M. B. & Krangel, M. S. (1987) *Science* 238, 678–682.
55. Loh, E. Y., Cwirla, S., Serafini, A. T., Phillips, J. H. & Lanier, L. L. (1988) *Proc. Natl. Acad. Sci. USA* 85, 9714–9718.
56. Hata, S., Satyanarayana, K., Devlin, P., Band, H., McLean, J., Strominger, J. L., Brenner, M. B. & Krangel, M. S. (1988) *Science* 240, 1541–1544.
57. Guglielmi, P., Davi, F., D'Auriol, L., Bories, J.-C., Dausset, J. & Bensussan, A. (1988) *Proc. Natl. Acad. Sci. USA* 85, 5634–5638.
58. Lefranc, M.-P. & Rabbitts, T. H. (1989) *Trends Biochem. Sci.* 14, 214–218.
59. Griesser, H., Champagne, E., Tkachuk, D., Takihara, Y., Lalande, M., Baillie, E., Minden, M. & Mak, T. W. (1988) *Eur. J. Immunol.* 18, 641–644.
60. Satyanarayana, K., Hata, S., Devlin, P., Roncarolo, M. G., de Vries, J. E., Spits, H., Strominger, J. L. & Krangel, M. S. (1988) *Proc. Natl. Acad. Sci. USA* 85, 8166–8170.
61. Iwashima, M., Green, A., Davis, M. M. & Chien, Y.-H. (1988) *Proc. Natl. Acad. Sci. USA* 85, 8161–8165.
62. Lai, E., Concannon, P. & Hood, L. (1988) *Nature* 331, 543–546.
63. Lefranc, M. P., Forster, A., Baer, R., Stinson, M. A. & Rabbitts, T. H. (1986) *Cell* 45, 237–246.
64. Malissen, M., McCoy, C., Blanc, D., Trucy, J., Devaux, C., Schmitt-Verhulst, A.-M., Fitch, F., Hood, L. & Malissen, B. (1986) *Nature* 319, 28–33.
65. Woolf, T., Lai, E., Kronenberg, M. & Hood, L. (1988) *Nucleic Acid. Res.* 16, 3863–3875.
66. Fujimoto, S. & Yamagishi, H. (1987) *Nature* 327, 242–243.
67. Okazaki, K., Davis, D. D. & Sakano, H. (1987) *Cell* 49, 477–485.

68. Toda, M., Fujimoto, S., Iwasato, T., Takeshita, S., Tezuka, K., Ohbayashi, T. & Yamagishi, H. (1988) *J. Mol. Biol. 202*, 219–231.
69. Duby, A. D., Klein, K. A., Murre, C., Seidman, J. G. (1985) *Science 228*, 1204–1206.
70. Toyonaga, B., Yoshikai, Y., Vadasz, V., Chin, B. & Mak, T. W. (1985) *Proc. Natl Acad. Sci. USA 82*, 8624–8628.
71. Tunnacliffe, A. & Rabbitts, T. H. (1985) *Nucleic Acid. Res. 13*, 6651–6661.
72. Huck, S. & Lefranc, M.-P. (1987) *FEBS Lett. 224*, 291–296.
73. Takihara, Y., Tkachuk, D., Michalopoulos, E., Champagne, E., Reimann, J., Minden, M. & Mak, T. W. (1988) *Proc. Natl Acad. Sci. USA 85*, 6097–6101.
74. Yoshikai, Y., Clark, S. P., Taylor, S., Sohn, U., Wilson, B. I., Minden, M. D. & Mak, T. W. (1985) *Nature 316*, 837–840.
75. Baer, R., Lefranc, M.-P., Minowada, J., Forster, A., Stinson, M. A. & Rabbitts, T. H. (1986) *Mol. Biol. Med. 3*, 265–277.
76. Baer, R., Boehm, T., Yssel, H., Spits, H. & Rabbitts, T. H. (1988) *EMBO J. 7*, 1661–1668.
77. Lefranc, M.-P. & Rabbitts, T. H. (1985) *Nature 316*, 464–466.
78. Lefranc, M.-P., Forster, A. & Rabbitts, T. H. (1986) *Proc. Natl Acad. Sci. USA 83*, 9596–9600.
79. Pelicci, P. G., Subar, M., Weiss, A., Dalla-Favera, R. & Littman, D. R. (1987) *Science 237*, 1051–1055.
80. Boehm, T., Baer, R., Lavenir, I., Forster, A., Waters, J. J., Nacheva, E. & Rabbitts, T. H. (1988) *EMBO J. 7*, 385–394.
81. Krangel, M. S., Band, H., Hata, S., McLean, J. & Brenner, M. B. (1987) *Science 237*, 64–67.
82. Brenner, M. B., McLean, J., Scheft, H., Riberdy, J., Ang, S.-L., Seidman, J. G., Devlin, P. & Krangel, M. S. (1987) *Nature 325*, 689–694.
83. Hochstenbach, F., Parker, C., McLean, J., Gieselmann, V., Band, H., Bank, I., Chess, L., Spits, H., Strominger, J. L., Seidman, J. G. & Brenner, M. B. (1988) *J. Exp. Med. 168*, 761–776.
84. Luria, S., Gross, G., Horowitz, M. & Givol, D. (1987) *EMBO J. 6*, 3307–3312.
85. Huck, S., Dariavach, P. & Lefranc, M.-P. (1988) *EMBO J. 7*, 719–726.
86. Quertermous, T., Strauss, W., Murre, C., Dialynas, D. P., Strominger, J. L. & Seidman, J. G. (1986) *Nature 322*, 184–187.
87. Leiden, J. M. & Strominger, J. L. (1986) *Proc. Natl Acad. Sci. USA 83*, 4456–4460.
88. Clark, S. P., Yoshikai, Y., Taylor, S., Siu, G., Hood, L. & Mak, T. W. (1984) *Nature 311*, 387–389.
89. Chien, Y.-h., Iwashima, M., Wettstein, D. A., Kaplan, K. B., Elliott, J. F., Born, W. & Davis, M. M. (1987) *Nature 330*, 722–727.
90. Ikuta, K., Ogura, T., Shimizu, A. & Honjo, T. (1985) *Proc. Natl Acad. Sci. USA 82*, 7701–7705.
91. Forster, A., Hobart, M., Hengartner, H. & Rabbitts, T. H. (1980) *Nature 286*, 897–899.
92. Kurosawa, Y., van Boehmer, H., Haas, W., Sakano, H., Trauneker, A. & Tonegawa, S. (1981) *Nature 290*, 565–570.
93. Born, W., Yague, J., Palmer, E., Kappler, J. & Marrack, P. (1985) *Proc. Natl Acad. Sci. USA 82*, 2925–2929.
94. Kronenberg, M., Goverman, J., Haars, R., Malissen, M., Kraig, E., Phillips, L., Delovitch, T., Suciu-Foca, N. & Hood, L. (1985) *Nature 313*, 647–653.
95. Chien, Y.-h., Iwashima, M., Kaplan, K. B., Elliott, J. F. & Davis, M. M. (1987) *Nature 327*, 677–682.
96. Lindsten, T., Fowlkes, B. J., Samelson, L. E., Davis, M. M. & Chien, Y.-h. (1987) *J. Exp. Med. 166*, 761–775.
97. Elliott, J. F., Rock, E. P., Patten, P. A., Davis, M. M. & Chien, Y.-h. (1988) *Nature 331*, 627–631.
98. de Villartay, J.-P., Hockett, R. D., Coran, D., Korsmeyer, S. J. & Cohen, D. I. (1988) *Nature 335*, 170–174.
99. Marrack, P. & Kappler, J. (1988) *Immunol. Today 9*, 308–315.
100. Raulet, D. H., Garman, R. D., Saito, H. & Tonegawa, S. (1985) *Nature 314*, 103–107.
101. Snodgrass, H. R., Dembic, Z., Steinmetz, M. & von Boehmer, H. (1985) *Nature 315*, 232–233.
102. Havran, W. L. & Allison, J. P. (1988) *Nature 335*, 443–445.
103. Finger, L. R., Harvey, R. C., Moore, R. C. A., Showe, L. C. & Croce, C. M. (1986) *Science 234*, 982–985.
104. Haluska, F., Tsujimoto, Y. & Croce, C. M. (1987) *Trends Genet. 3*, 11–15.
105. Showe, L. C. & Croce, C. M. (1987) *Annu. Rev. Immunol. 5*, 253–277.
106. Rabbitts, T. H., Boehm, T. & Mengle-Gaw, L. (1988) *Trends Genet. 4*, 300–304.
107. Leder, P., Battey, J., Lenoir, G., Moulding, C., Murphy, W., Potter, H., Steward, T. & Taub, R. (1983) *Science 222*, 765–771.
108. Colby, W. W., Chen, E. Y., Smith, D. H. & Levinson, A. D. (1983) *Nature 301*, 722–725.
109. Hann, S. R., King, M. W., Bentley, D. L., Anderson, C. W. & Eisenman, R. N. (1988) *Cell 52*, 185–195.
110. Bernard, O., Cory, S., Gerondakis, S., Webb, E. & Adams, J. M. (1983) *EMBO J. 2*, 2375–2380.
111. Rabbits, T. H., Forster, A., Hamlyn, P. H. & Baer, R. (1984) *Nature 309*, 592–597.
112. Rabbits, T. H., Hamlyn, P. H. & Baer, R. (1983) *Nature 306*, 760–765.
113. Pelicci, P. G., Knowles, D. M., Magrath, I. T. & Dalla-Favera, R. (1986) *Proc. Natl Acad. Sci. USA 83*, 2984–2987.
114. Hamlyn, P. H. & Rabbitts, T. H. (1983) *Nature 304*, 135–139.
115. Haluska, F. G., Finger, S., Tsujimoto, Y. & Croce, C. M. (1986) *Nature 324*, 158–161.
116. Gelmann, E. P., Psallidopoulos, C. M., Papas, T. S. & Dalla-Favera, R. D. (1983) *Nature 306*, 799–801.
117. Wiman, K. G., Clarkson, B., Hayday, A. C., Saito, H., Tonegawa, S. & Hayward, W. S. (1984) *Proc. Natl Acad. Sci. USA 81*, 6798–6802.
118. Showe, L. C., Ballantine, M., Nishikura, K., Erikson, J., Kaji, H. & Croce, C. M. (1985) *Mol. Cell. Biol. 5*, 501–507.
119. Davis, M., Malcolm, S. & Rabbitts, T. H. (1984) *Nature 308*, 286–288.
120. Sun, L. K., Showe, L. C. & Croce, C. M. (1986) *Nucleic Acids Res. 14*, 4037–4042.
121. Graham, M. & Adams, J. M. (1986) *EMBO J. 5*, 2845–2850.
122. Mengle-Gaw, L. & Rabbitts, T. H. (1987) *EMBO J. 6*, 1959–1965.
123. Taub, R., Kelly, K., Battey, J., Latt, S., Lenoir, G. M., Tantravahi, V., Tu, Z. & Leder, P. (1984) *Cell 37*, 511–520.
124. Croce, C. M., Thierfelder, W., Erikson, J., Nishikura, K., Finan, J., Lenoir, G. M. & Nowell, P. C. (1983) *Proc. Natl Acad. Sci. USA 80*, 6922–6926.
125. Rappold, G. A., Hameister, H., Cremer, T., Adolph, S., Henglein, B., Freese, U. K., Lenoir, G. M. & Bornkamm, G. W. (1984) *EMBO J. 3*, 2951–2965.
126. Lombardi, L., Newcomb, E. W. & Dalla-Favera, R. (1987) *Cell 49*, 161–170.
127. Adams, J. M., Harris, A. W., Pinkert, C. A., Corcoran, L. M., Alexander, W. S., Cory, S., Palmiter, R. D. & Brinster, R. L. (1985) *Nature 318*, 533–538.
128. Leder, A., Pattengale, P. K., Kuo, A., Stewart, T. A. & Leder, P. (1986) *Cell 45*, 485–495.
129. Harris, A. W., Pinkert, C. A., Crawford, M., Langdon, W. Y., Brinster, R. L. & Adams, J. M. (1988) *J. Exp. Med. 167*, 353–371.
130. Sinn, E., Muller, W., Pattengale, P., Tepler, I., Wallace, R. & Leder, P. (1987) *Cell 49*, 465–475.
131. Murphy, W., Sarid, J., Taub, R., Vasicek, T., Battcy, J., Lenoir, G. & Leder, P. (1986) *Proc. Natl Acad. Sci. USA 83*, 2939–2943.
132. Cesarman, E., Dalla-Favera, R., Bentley, D. & Groudine, M. (1987) *Science 238*, 1272–1275.
133. Siebenlist, U., Henninghausen, L., Battey, L. & Leder, P. (1984) *Cell 37*, 381–391.
134. Dyson, P. J. & Rabbitts, T. H. (1985) *Proc. Natl Acad. Sci. USA 82*, 1984–1988.
135. Dyson, P. J., Littlewood, T. D., Forster, A. & Rabbitts, T. H. (1985) *EMBO J. 4*, 2885–2891.

136. Yunis, J. J. (1983) *Science 221*, 227–235.
137. Tsujimoto, Y., Yunis, J., Onorato-Showe, L., Erikson, J., Nowell, P. C. & Croce, C. M. (1984) *Science 224*, 1403–1406.
138. Tsujimoto, Y., Jaffe, E., Cossman, J., Gorham, J., Nowell, P. C. & Croce, C. M. (1985) *Nature 315*, 340–343.
139. Rabbitts, P. H., Douglas, J., Fischer, P., Nacheva, E., Karpas, A., Catovsky, D., Melo, J. V., Stinson, M. A. & Rabbitts, T. H. (1988) *Oncogene 3*, 99–103.
140. Tsujimoto, Y., Finger, L. R., Yunis, J., Nowell, P. C. & Croce, C. M. (1984) *Science 226*, 1097–1099.
141. Tsujimoto, Y., Bashir, M. M., Givol, I., Cossman, J., Jaffe, E. & Croce, C. M. (1987) *Proc. Natl Acad. Sci. USA 84*, 1329–1331.
142. Bakhshi, A., Jensen, J. P., Goldman, P. G., Wright, J. J., McBride, O. W., Epstein, A. L. & Korsmeyer, S. J. (1985) *Cell 41*, 889–906.
143. Bakhshi, A., Wright, J. J., Graninger, W., Seto, M., Owens, J., Cossman, J., Jensen, J. P., Goldman, P. & Korsmeyer, S. J. (1987) *Proc. Natl Acad. Sci. USA 84*, 2396–2400.
144. Cleary, M. L. & Sklar, J. (1985) *Proc. Natl Acad. Sci. USA 82*, 7439–7443.
145. Cleary, M. L., Meeker, T. C., Levy, S., Lee, E., Trela, M., Sklar, J. & Levy, R. (1986) *Cell 44*, 97–106.
146. Cleary, M. L., Smith, S. D. & Sklar, J. (1986) *Cell 47*, 19–28.
147. Seto, M., Jaeger, U., Hockett, R. D., Graninger, W., Bennett, S., Goldman, P. & Korsmeyer, S. J. (1988) *EMBO J. 7*, 123–131.
148. Cleary, M. L., Galili, N. & Sklar, J. (1986) *J. Exp. Med. 164*, 315–320.
149. Tsujimoto, Y., Gorham, J., Cossman, J., Jaffe, E. & Croce, C. M. (1985) *Science 229*, 1390–1393.
150. Reed, J. C., Tsujimoto, Y., Alpers, J. D., Croce, C. M. & Nowell, P. C. (1987) *Science 236*, 1295–1299.
151. Tsujimoto, Y. & Croce, C. M. (1986) *Proc. Natl Acad. Sci. USA 83*, 5214–5218.
152. Tsujimoto, Y., Ikegki, N. & Croce, C. M. (1987) *Oncogene 2*, 3–7.
153. Vaux, D. L., Cory, S. & Adams, J. M. (1988) *Nature 335*, 440–442.
153a. Pegoraro, L., Palumbo, A., Erikson, J., Falda, M., Giovanazzo, B., Emanuel, B. S., Rovera, G., Nowell, P. C. & Croce, C. M. (1984) *Proc. Natl Acad. Sci. USA 81*, 7166–7170.
154. Ueshima, Y., Rowley, J. D., Variakojis, D., Winter, J. & Gordon, L. (1984) *Blood 63*, 1028–1038.
155. McKeithan, T. W., Shima, E. A., LeBeau, M. M., Minowada, J., Rowley, J. D. & Diaz, M. D. (1986) *Proc. Natl Acad. Sci. USA 83*, 6636–6640.
156. Baer, R., Chen, K.-C., Smith, S. D. & Rabbitts, T. H. (1985) *Cell 43*, 705–713.
157. Baer, R., Forster, A. & Rabbitts, T. H. (1987) *Cell 50*, 97–105.
158. Denny, C. T., Yoshikai, Y., Mak, T. W., Smith, S. D., Hollis, G. F. & Kirsch, I. R. (1986) *Nature 320*, 549–551.
159. Denny, C. T., Hollis, G. F., Hecht, F., Morgan, R., Link, M. P., Smith, S. D. & Kirsch, I. R. (1986) *Science 234*, 197–200.
160. Aurias, A., Dutrillaux, B., Buriot, D. & Lejeune, J. (1980) *Mutat. Res. 69*, 369–374.
161. Hecht, F., Hecht, B. K. & Kirsch, I. R. (1987) *Cancer Genet. Cytogenet. 26*, 95–104.
162. Mengle-Gaw, L., Willard, H. F., Smith, C. I. E., Hammarstrom, L., Fischer, P., Sherrington, P., Lucas, G., Thompson, P. W., Baer, R. & Rabbitts, T. H. (1987) *EMBO J. 6*, 2273–2280.
163. Baer, R., Heppell, A., Taylor, A. M. R., Rabbitts, P. H., Boullier, B. & Rabbitts, T. H. (1987) *Proc. Natl Acad. Sci. USA 84*, 9069–9073.
164. Davey, M. P., Bertness, V., Nakahara, K., Johnson, J. P., McBride, O. W., Waldmann, T. A. & Kirsch, I. R. (1988) *Proc. Natl Acad. Sci. USA 85*, 9287–9291.
165. Mengle-Gaw, L., Albertson, D. G., Sherrington, P. D. & Rabbitts, T. H. (1988) *Proc. Natl Acad. Sci. USA 85*, 9171–9175.
166. Russo, G., Isobe, M., Pegoraro, L., Finan, J., Nowell, P. C. & Croce, C. M. (1988) *Cell 53*, 137–144.
167. Bridges, B. A. & Harnden, D. G. (1981) *Nature 289*, 222–223.
168. Oxford, J. M., Harnden, D. G., Parrington, J. M. & Delhanty, J. D. A. (1975) *J. Med. Genet. 12*, 251–262.
169. Taylor, A. M. R. & Butterworth, S. V. (1986) *Int. J. Cancer 37*, 511–516.
170. Aurias, A., Dutrillaux, B. & Griscelli, C. (1983) *Hum. Genet. 63*, 320–322.
171. Aurias, A. & Dutrillaux, B. (1986) *Hum. Genet. 72*, 210–214.
172. Aurias, A., Croquette, M. F., Nuyts, J. P., Griscelli, C. & Dutrillaux, B. (1986) *Hum. Genet. 72*, 22–24.
173. Taylor, A. M. R., Oxford, J. M. & Metcalf, J. A. (1981) *Int. J. Cancer 27*, 311–319.
174. Beatty-DeSana, J. W., Haggard, M. J. & Cooledge, J. W. (1975) *Nature 255*, 242–243.
175. Hecht, F., McCaw, B. K., Peakman, D. & Robinson, A. (1975) *Nature 255*, 243–244.
176. Welch, J. P. & Lee, C. L. Y. (1975) *Nature 255*, 241–242.
177. Kennaugh, A. A., Butterworth, S. V., Hollis, R., Baer, R., Rabbitts, T. H. & Taylor, A. M. R. (1986) *Hum. Genet. 73*, 254–259.
178. Shima, E. A., LeBeau, M. M., McKeithan, T. W., Minowada, J., Showe, L. C., Mak, T. W., Minden, M. D., Rowley, J. D. & Diaz, M. O. (1986) *Proc. Natl Acad. Sci. USA 83*, 3439–3443.
179. Erikson, J., Finger, L., Sun, L., Ar-Rushdi, A., Nishikura, K., Minowada, J., Finan, J., Emanuel, B. S., Nowell, P. C. & Croce, C. M. (1986) *Science 232*, 884–886.
180. Bernard, O., Larsen, C.-J., Hampe, A., Mauchauffe, M., Berger, R. & Mathieu-Mahul, D. (1988) *Oncogene 2*, 195–200.
181. Isobe, M., Russo, G., Haluska, F. G. & Croce, C. M. (1988) *Proc. Natl Acad. Sci. USA 85*, 3933–3937.
182. Finger, L. R., Huebner, K., Cannizzaro, L. A., McLeod, K., Nowell, P. C. & Croce, C. M. (1988) *Proc. Natl Acad. Sci. USA 85*, 9158–9162.
183. Williams, D. L., Look, A. T., Melvin, S. L., Roberson, P. K., Dahl, G., Flake, T. & Stass, S. (1984) *Cell 36*, 101–109.
184. Erikson, J., Williams, D. L., Finan, J., Nowell, P. C. & Croce, C. M. (1985) *Science 229*, 784–786.
185. Lewis, W. H., Michalopoulos, E. E., Williams, D. L., Minden, M. D. & Mak, T. W. (1985) *Nature 317*, 544–546.
186. Boehm, T., Buluwela, L., Williams, D., White, L. & Rabbitts, T. H. (1988) *EMBO J. 7*, 2011–2017.
187. Boehm, T., Lavenir, I., Forster, A., Wadey, R. B., Cowell, J. K., Harbott, J., Lampert, F., Waters, J., Sherrington, P., Couillin, P., Azoulay, M., Junien, C., van Heyningen, V., Porteous, D. J., Hastie, N. D. & Rabbitts, T. H. (1988) *Oncogene 3*, 691–695.
188. LeBeau, M. M., McKeithan, T. W., Shima, E. A., Goldman-Leikin, E., Chan, S. J., Bell, G. I., Rowley, J. D. & Diaz, M. O. (1986) *Proc. Natl Acad. Sci. USA 83*, 9744–9748.
189. Kagan, J., Finan, J., Letofsky, J., Besa, E. C., Nowell, P. C. & Croce, C. M. (1987) *Proc. Natl Acad. Sci. USA 84*, 4543–4546.
190. Boehm, T., Mengle-Gaw, L., Kees, U., Spurr, N., Lavenir, I., Forster, A. & Rabbitts, T. H. (1989) *EMBO J. 8*, 2621–2631.
191. Reynolds, T. C., Smith, S. D. & Sklar, J. (1987) *Cell 50*, 107–117.
192. Rabbitts, T. H., Baer, R., Buluwela, L., Mengle-Gaw, L., Taylor, A. M. & Rabbitts, P. H. (1986) *Cold Spring Harbor Symp. Quant. Biol. 51*, 923–930.

Review

The nucleoskeleton and the topology of transcription

Peter R. COOK

Sir William Dunn School of Pathology, University of Oxford, England

(Received March 29, 1989) — EJB 89 0379

Transcription is conventionally believed to occur by passage of a mobile polymerase along a fixed template. Evidence for this model is derived almost entirely from material prepared using hypotonic salt concentrations. Studies on subnuclear structures isolated using hypertonic conditions, and more recently using conditions closer to the physiological, suggest an alternative. Transcription occurs as the template moves past a polymerase attached to a nucleoskeleton; this skeleton is the active site of transcription. Evidence for the two models is summarised. Much of it is consistent with the polymerase being attached and not freely diffusible. Some consequences of such a model are discussed.

The accepted model for transcription contains three essential participants — the template, polymerase and nascent RNA [1, 2]. Transcription is initiated by a diffusible polymerase binding to a promoter; the polymerase then processes along the DNA (Fig. 1A). As knowledge has increased, other occasional participants have been added (e.g. transcription factors, topoisomerases) but not any other central players. Evidence for this model comes almost entirely from studies using hypotonic salt concentrations, largely because chromatin aggregates in physiological concentrations and so becomes difficult to use. In contrast, studies on preparations isolated using hyper- or iso-tonic salt concentrations suggest that active RNA polymerase is associated with a nucleoskeleton and is not freely diffusible. This nucleoskeleton is seen as the active site of RNA synthesis and transcription occurs by movement of DNA past the attached polymerase (Fig. 1B).

Ultimately models will be distinguished by reconstructing efficient transcription *in vitro* from purified constituents. Proof of the model involving an attached polymerase will require measurement of its association constant for a skeleton. In the meantime, the two models can be distinguished operationally by determining whether the polymerase is freely diffusible or attached to a larger structure. Much evidence for the 'text-book' model is consistent with the alternative; this alternative has a number of important consequences and these are examined. Discussion centres on RNA polymerase II, the enzyme that transcribes most eukaryotic genes, but applies equally to other polymerases.

ARTEFACTS

Structure within isolated nuclei cannot be discussed sensibly without some consideration of problems caused by artefacts. They arise because nuclei and chromatin aggregate in physiological salt concentrations [3]; therefore unphysiological conditions are almost invariably used. Controversy centres on whether structures seen *in vitro* are any more than isolation artefacts. RNA, DNA and protein, each concentrated in the nucleus at about 0.1 g/ml, might be expected to aggregate as soon as ion concentrations are altered. (For reviews of the controversy, see [4—6].) Nuclei are usually isolated by lysing cells in about 1/10 the physiological salt concentration [7]. Intuition suggests that such low concentrations are 'mild', but they destroy the 30-nm chromatin fibre, decondense heterochromatin, extract a quarter of nuclear protein (including active polymerases [8, 9]) and aggregate ribonucleoprotein particles [10]. The 'stabilizing' cations that are usually added also activate nucleases, so supercoiling is lost and polymerases can initiate aberrantly at resulting nicks.

These, then, are just the initial conditions that are traditionally used. Subsequently these hypotonically extracted nuclei are themselves re-extracted to prepare various subnuclear structures. For example, 'matrices' are prepared by extraction in 2 M NaCl (for a review see [11]) and 'scaffolds' by extraction with lithium diiodosalicylate following a mandatory heat-shock [12]. What relationship such structures bear to those *in vivo* is open to argument. Ultimately the criticism that artefacts have been generated are best countered using physiological conditions, so discussion will concentrate on the few studies which use them.

EVIDENCE FOR THE 'TEXT-BOOK' MODEL

Despite almost complete acceptance of the 'text-book' model, there seem to be only two kinds of evidence to support it.

'Miller' spreads

'Miller' spreads apparently provide the best direct evidence [13]. They are prepared by dropping nuclei isolated using conditions described above in a solution that is little more than distilled water (sometimes containing the detergent 'Joy'). Removing counterions charges chromatin, which expands and bursts the nucleus; individual chromatin fibres and

Correspondence to P. R. Cook, Sir William Dunn School of Pathology, University of Oxford, South Parks Road, Oxford OX1 3RE, UK

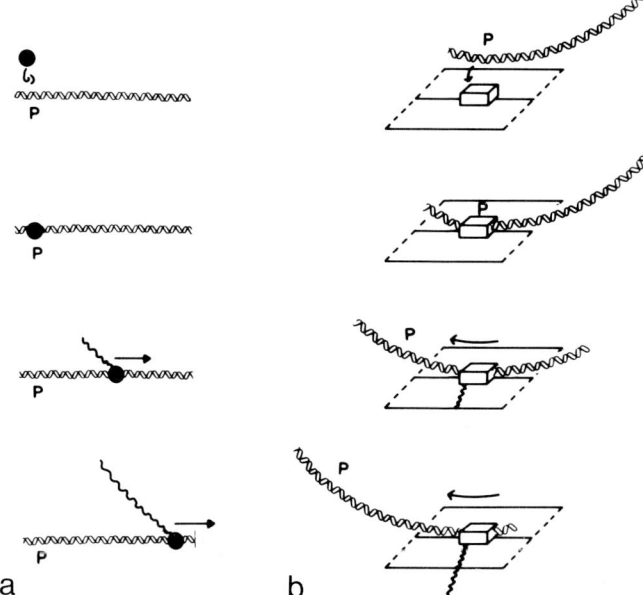

Fig. 1. *Two models for transcription.* Four different stages in the process are shown (top to bottom): formation of the polymerase–template complex, initiation of RNA synthesis at P and elongation of RNA. (A) The 'text-book' model. The polymerase (circle) moves along a fixed template to generate the transcript (wavy line). (B) The attachment model. The template moves past a polymerase (rectangular box) attached to a nucleoskeleton (grid)

'text-book' transcription complexes can then be seen at the edge of the spread chromatin. The most striking of these resemble Christmas trees and contain highly transcribed genes like the silk fibroin gene (Fig. 2a) or ribosomal DNA, a polymerase I unit. No skeleton can be seen.

A priori, it would seem dangerous to draw general conclusions about structures *in vivo* using such a disruptive procedure and based on visualization of a minority of transcription complexes.

Soluble polymerases

Modern biochemistry has proved very successful at dissecting details of transcription [14–16] and this very success lends support to the generally accepted model; if soluble polymerases work, then the model must be correct.

It is the general experience of those that purify RNA polymerase II that most enzyme is insoluble [17, 18]. A large pool of soluble enzyme is found in frog's eggs [19] but this is inactive, awaiting activation later in development. It is rarely appreciated how inefficient such soluble polymerases are. Rates of transcription *in vitro* cannot be expressed relative to rates *in vivo* as the latter are not known, since rates of RNA turnover cannot be assessed accurately. Additionally, extremely low absolute synthetic rates can be measured because ^{32}P-labelled precursors of high specific activities are available. Overall efficiencies are then easily overlooked. Pure polymerases transcribe 'activated' templates at > 50 nucleotides/s, initiating incorrectly at nicks. In contrast, they are almost completely inactive on intact chromatin. Systems that do initiate correctly are impure, involving long preincubations in crude extracts; one of the most efficient mammalian systems, a 'Manley' extract, polymerises correctly initiated transcripts at < 10 nucleotides/h, or perhaps 0.01% of the rate *in vivo* [20]. In a yeast system, $< 10^{-3}$ transcripts are accurately initiated per added template molecule during a typical incubation [21]. Chromatin templates are transcribed even less efficiently [22, 23] but appropriate preincubations improve rates slightly [24]. Quite possibly, some templates assemble into larger structures during preincubations and prime what little specific transcription there is.

Until a soluble system is developed that initiates correctly at rates approaching those found *in vivo*, this kind of evidence cannot provide definitive proof for a skeleton-free model. Indeed, observed activities seem to be partial reactions, lesser and non-specific activities resulting from disruption of some larger and more active complex. It is also as well to remember that the largest stimulatory effects seen to date *in vitro* are small when compared to effects that occur during development. Thus, sequences like enhancers and upstream activating sequences stimulate transcription *in vitro* by factors of a few hundred or less; they have similar effects in 'transient expression' assays. These are to be compared with the 10^8-fold differences in rate of transcription of the growth hormone gene that probably occur during development [25].

LOOPS AND SUPERCOILS IN NUCLEAR DNA

The organization of the chromatin fibre into loops or domains is an essential part of the alternative model. It is also a recurrent motif in many chromosome models. Again, despite wide acceptance, evidence for looping is inconclusive and controversial, derived almost entirely from studies using unphysiological conditions.

Early evidence

The best evidence remains the observation of meiotic lampbrush chromosomes in living amphibian cells, where lateral loops of native chromatin are specifically attached to a central core [26]. However, transcription here is very odd as globin genes [27] and both strands of some repeated sequences are transcribed promiscuously [28, 29].

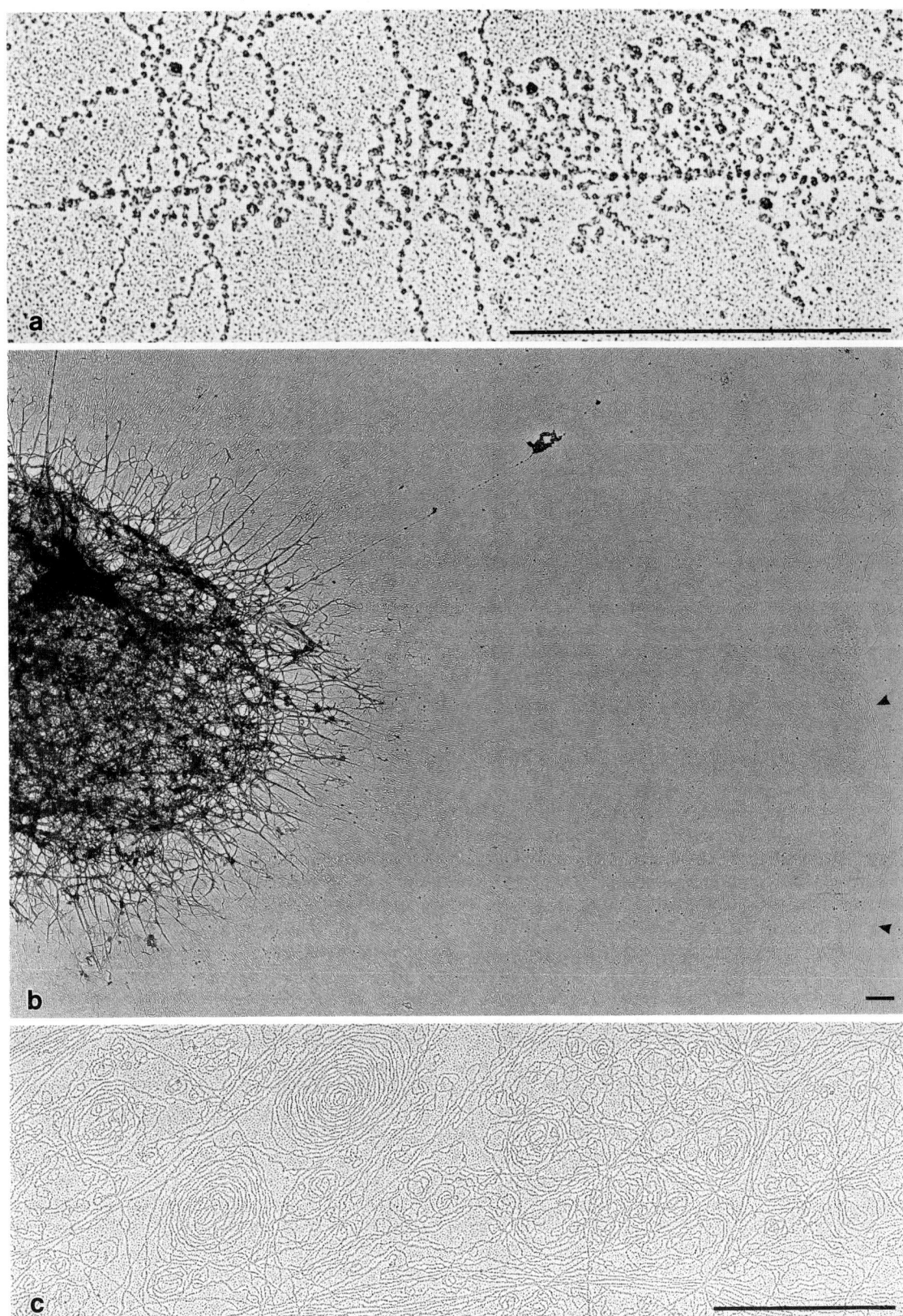

Fig. 2. *Electron microscopy of nuclei prepared under hypotonic (a) or hypertonic (b, c) conditions gives very different views of how transcription might occur.* (a) Part of a 'Miller' spread showing the silk fibroin transcription unit. Chromatin is dispersed by diluting counterions in buffered distilled water. A typical 'Christmas tree' structure is visible, but no 'nucleoskeleton'. From McKnight et al. [236] with permission. (b) Low-power view of a 'nucleoid' prepared by lysing cells in 2 M NaCl and then spreading DNA using Kleinschmidt's procedure. A tangled network of superhelical fibres extend from the prominent central skeleton (left) to the edge of the field (arrowheads). See [47]. (c) High-power view of the edge of a spread like that in (b). No nascent RNA can be seen amongst the DNA, all remains associated with the central skeleton in (b). From McCready et al. [47] with permission. Bars are 1 μm

Loops are more generally seen in fixed mitotic or polytene chromosomes [30, 31] or after stripping histones from DNA [32, 33] but, again, these might be preparative artefacts. The kinetics of nuclease digestion of conventionally prepared rat liver nuclei are also consistent with a looped structure; progressive detachment of DNA leaves a resistant fraction, presumably the base of the loop, attached to a pelletable structure [34].

Supercoiling also provides evidence for looping. Supercoils are maintained in DNA circles but not in broken or linear molecules. In 1973 it was suggested that linear eukaryotic DNA was supercoiled by organisation into loops and that differential supercoiling underlay differential gene expression [35]. Supercoiled DNA has distinctive properties [36]; these are shared by 'nucleoids' isolated by stripping off histones with 2 M NaCl [37]. Both sediment biphasically in gradients containing intercalators [37–42] and they bind ethidium [43], denature [43] and change shape [44, 45] similarly. Electron microscopy of spread nucleoids reveals supercoiled fibres attached to a collapsed nuclear 'cage' (Fig. 2 b, c) [46, 47], suggesting that this linear DNA is looped and attached. Supercoiling in the loop stabilises any right-handed interwound superhelix at its base [43]. Supercoils arise because dissociation of nucleosome cores in 2 M NaCl leaves their imprints in DNA if it is looped [48, 49]. The DNA of nuclei isolated using hypotonic conditions also melts at the temperature characteristic of circular DNA [50, 51].

Loop size

Loop sizes have been estimated at 10–200 kb using these rather unsatisfactory preparations [12, 32, 34, 37–39, 43]; they change little during the cell cycle [52]. Recent measurements using nuclease digestion of HeLa cells extracted using physiological conditions give an average of 86 kb (Jackson, D. A., Dickinson, P. and Cook, P. R., unpublished results), suggesting that there might be one transcription unit per loop.

Specific attachments

Sequences attached to the skeleton can be mapped by progressively detaching DNA with a nuclease; sequences close to attachment sites should resist detachment and so be enriched in a pelleted fraction: those lying further away should be depleted. If attachments are generated artefactually, there should be no specificity in the aggregate and any given sequence will neither be enriched nor depleted. In fact, α globin sequences in HeLa nucleoids can be enriched eightfold whereas β and γ genes are depleted; α globin must lie closer to the attachment site than β or γ [53]. This 'detachment mapping' has been extended to many different preparations (e.g. matrices and scaffolds), but with variable results [5].

Attachment sequences can also be selected by incubating nuclei with DNA fragments to see which bind specifically. Little specificity is seen unless nuclei are first extracted with 2 M NaCl or heat-shocked and treated with lithium diiodosalicylate; then fragments containing enhancers or topoisomerase II sites bind specifically [54–60]. It is difficult to know whether such complexes are analogous to those *in vivo* or artefacts due to aggregated topoisomerase trapping its consensus sequence.

It seems prudent to leave this evidence until controversies are resolved, especially bearing in mind the unphysiological conditions used. It is hardly decisive evidence in favour of looping *in vivo*.

Loops and supercoils in native chromatin

Whether unrestrained supercoils (and, by inference, loops) exist in native chromatin is important because different loops might contain different degrees of supercoiling and these differences might affect transcription. Bases in DNA are buried and can only be read by unwinding the duplex. Supercoiling around the nucleosome is of opposite sense to that of the double helix so its torsional energy is potentially available for unwinding. Can this energy assist eukaryotic polymerases *in vivo* and do enzymes that alter supercoiling (i.e. topoisomerases) influence transcription? Such questions provoked the suggestion that nuclear DNA might be supercoiled [61] and lie behind the continuing search for torsionally strained or 'dynamic' chromatin. Early studies showed conclusively that supercoiling stimulates transcription *in vitro* [62, 63]; whether it does *in vivo*, and whether some chromatin is torsionally strained, are now the issues.

Chromatin containing free energy of supercoiling should bind more of an intercalator like psoralen than the sample when relaxed; extra psoralen binding was found in living bacteria but not in human and *Drosophila* cells [64, 65]. However, unrestrained supercoils in a eukaryotic chromatin fraction, perhaps the functional fraction, might have gone undetected by this insensitive method.

Various workers have claimed to have discovered such a minor fraction of 'dynamic' chromatin [66–74], but the evidence is at best only correlative, sometimes irreproducible [75–78] and, at worst, flawed. Some depends on the isolation of a minor fraction [66, 67], but accidental histone loss will inevitably introduce free supercoils. Other evidence involves novobiocin, which is assumed to inhibit topoisomerase II specifically, but it also inhibits RNA polymerases I, II and III directly [79], affects RNA attachments [80] and is used at such high concentrations that histones precipitate [81, 82] and oxidative phosphorylation is inhibited [83, 84]. In other experiments [85], topoisomerase I was injected into oocytes at concentrations equal to the enormous pool of free histone [86]. Such evidence is justly treated sceptically [76], but does not disprove the existence of a minor fraction containing free supercoils.

Recently, better but only circumstantial evidence has come from yeast cells with mutant topoisomerases and an appreciation that supercoiling is the inevitable consequence of transcription.

Topoisomerase mutants

Transcription of closed loops or circles poses topological problems that apply equally to models involving mobile polymerases or mobile templates [87–90]. Consider a 1-kb plasmid, transcribed by the polymerase in *Escherichia coli*. If the 'text-book' model were applied strictly, the polymerase (radius 7.5 nm), plus nascent transcript, attached ribosomes (each with radius 15 nm) and nascent protein would all track along a helical path, passing through the centre of the circular template (radius 9 nm if condensed by supercoiling) on transcription of each helical turn (i.e. every 10 bp). If this miracle were possible, the resulting transcript would be intertwined around the template once for every helical turn transcribed and could only be untwined by passing one of its ends through the circle, again once for every turn transcribed. Of the formal solutions, the likeliest involves no net rotation of polymerase and transcript about the helical axis, simply because they are too bulky. Then, the template must rotate, becoming positively supercoiled ahead of the polymerase and negatively supercoiled behind [89]. This compensatory coiling will quickly limit transcription unless removed by topoisomerases.

Various results show that topoisomerases do indeed play a central role. Topoisomerase I is closely associated with genes transcribed by polymerase I and II [91–105]. When the locus for topoisomerase I is inactivated, yeast cells remain viable [106–108], presumably because topoisomerase II rescues them. Topoisomerase II is indispensable as mutations in its gene are lethal [106–108]; although its main role in eukaryotes may be in replication and chromosome segregation [109, 110], it nevertheless forms 'open' transcription complexes [111].

Recently, Liu and Wang [89] suggested that transcription changes the superhelical density of bacterial DNA if positive and negative supercoils are not removed equally. As negative supercoils are relaxed only by topoisomerase I and positive supercoils only by topoisomerase II, supercoils of appropriate sense should accumulate if either enzyme is inactivated using inhibitors or temperature-sensitive mutants. In fact, positive supercoils accumulate on transcription of a test plasmid when topoisomerase II is inactivated [112]. In eukaryotes, both topoisomerases I and II relax positive and negative supercoils so their hypothesis is more difficult to test. Nevertheless, plasmids isolated from yeast cells lacking topoisomerase I are more negatively supercoiled than their less-transcribed counterparts [113–115]. Furthermore, plasmids become positively supercoiled when transcribed in specially engineered yeast expressing bacterial topoisomerase I but not mutated yeast topoisomerases I and II [115]. Positive supercoils also accumulate on transcription of circular DNA *in vitro* in the presence of topoisomerase I [116].

Interpretation of these experiments depends on complete inactivation of topoisomerases and on there being no other activities. This is not an academic reservation as bacterial topoisomerase mutants usually have compensatory mutations elsewhere [117–119]. Note also that they involve 'minichromosomes' whose behaviour may not be representative of chromosomal DNA. Furthermore, the existence of supercoils or torsionally strained chromatin does not necessarily prove the existence of loops. Supercoils might persist locally because their rate of diffusion to the place where they could be lost, chromosome ends, might be too slow. Slow diffusion might be expected down a long chromatin fibre, non-specifically snagged with other fibres in the dense tangle that is chromatin. So, again, the existence of loops *in vivo* still awaits formal proof.

EVIDENCE FOR ATTACHED TRANSCRIPTION COMPLEXES

Studies using 2 M NaCl

The first hints that transcription complexes might be attached came from studies on nucleoids prepared using 2 M NaCl [120] and so are rightly treated cautiously. However the central conclusions have now been confirmed using isotonic conditions (see below). Nucleoids have two advantages for this type of study: unlike matrices, they are prepared only by exposure to hypertonic salt concentrations (and not sequential treatments with both hypo- and hyper-tonic conditions) and their DNA remains supercoiled and unbroken. Therefore they have not accumulated artefacts due to exposure to hypotonic solutions, nor from binding nicked DNA [37].

If the text-books are correct, 2 M NaCl treatment, which dissociates pure polymerase from the template, should extract all pulse-labelled RNA from nucleoids: quite the opposite is found [44]. When the analogous experiment to Miller's is performed with nucleoids (i.e. DNA is spread) no nascent RNA is seen at the edges of the spread (Fig. 2c); all of it remains associated with a central skeleton (Fig. 2b) [47, 121]. Digestion with ribonuclease removed internal parts of nascent RNA but not 5' caps, suggesting they were attached; they should be detached if the text-books were correct [121]. Another powerful control made non-specific aggregation unlikely: transcripts of an infecting nuclear virus, influenza, were also attached but those of a cytoplasmic rhabdovirus were not [122].

Transcribed genes were also attached [121]. In one extensive analysis the site of integration of a single integrated avian sarcoma virus or polyoma virus was 'detachment' mapped in various transformed clones [123]. In every case where the integrated virus was expressed, proviral sequences, particularly enhancers, resisted detachment. In some cases the transforming virus integrated into a site remote from an attachment point; then adjacent cellular DNA became attached, presumably through the provirus, but sequences on the other unaffected chromosome remained unattached. Some transformants spontaneously revert and lose the transformed phenotype; subsequently retransformed clones can be reselected following treatment with azacytidine. Integrated proviral sequences in the revertants, now non-transcribed, lost their close attachment, but regained it when reexpressed in the retransformants. These correlations with powerful internal controls provide strong evidence for an association of transcribed sequences with this skeletal structure.

As discussed earlier, other workers using superficially similar material obtained variable results so the interpretation of all remains controversial [5]. For example, transcribed regions of genes are rarely, if ever, attached to scaffolds isolated using lithium diiodosalicylate [12, 55, 124] (but see also [125]). But there is no reason to believe these results any more than the others, especially as they involve a hypotonic treatment and a mandatory heat-shock, which is known to induce aggregation [126–128].

Studies using 'physiological' conditions

It is difficult to know which buffer to choose when trying to reproduce physiological conditions because the precise ionic milieu *in vivo* remains unknown. There are also practical problems: the major cytoplasmic counterion is protein, an expensive additive to buffers, so Cl^- is commonly used, but this may reduce transcription rates to one tenth [21]. Notwithstanding such difficulties, it has at least now become possible to use isotonic salt concentrations, if not truly physiological ones, during isolation. Aggregation is avoided by first encapsulating cells in agarose microbeads of about 50 μm diameter [8, 129]. Agarose is permeable to small molecules, so cells can be regrown or extracted in a 'physiological' buffer containing Triton [130]; then most cytoplasmic proteins and RNA diffuse out to leave encapsulated chromatin surrounded by the cytoskeleton [8, 129, 130]. These fragile cell remnants are protected by the agarose coat, but accessible to probes like antibodies and enzymes.

To what extent is nuclear structure and function preserved? Heterochromatin, which decondenses readily when ion concentrations are altered, provides a marker for gross structural preservation: supercoiling, generated by subsequent histone removal, provides another for molecular structure as a nick anywhere in a loop relaxes it. Both indicate that structure is preserved [130]. Function is also preserved. At least two different DNA polymerase activities can be demonstrated by lysing encapsulated cells. One is an aberrant soluble activity that uses nicked templates and is found irrespective of cell-cycle stage: it is the activity generally purified by biochemists. Another pellets with the microbeads and syntheses DNA *in vitro* at a rate equivalent to that *in vivo*: it is only found in S-phase cells and uses the native chromatin template [9, 131, 132]. Transcriptional rates are also well preserved. As rates *in vivo* are unknown, relative efficiencies cannot be determined but this preparation transcribes twice as efficiently as nuclei prepared conventionally and, as before, the activity resists extraction and uses the native chromatin template [130, 133, 134]. By these functional criteria, then, this preparation is as proficient in the vital nuclear activities as the living cell.

The two models for transcription can be distinguished by fragmenting chromatin with an endonuclease and removing any unattached material electrophoretically [133]. If the transcription complex is attached it should remain in beads: if unattached, it should electroelute from beads with most chromatin. Isotonic buffers can be used from cell lysis, through nuclease treatment and electroelution to subsequent polymerase assay. Removing 75% of the chromatin in this way hardly reduced transcriptional activity. Combined treatment with RNase and *Eco*RI, followed by electrophoresis, removed > 95% of nascent RNA (and so RNP) and 73% of DNA (and chromatin) but only 30% of the activity. A slight reduction in activity might be expected as the template has been cut into 10-kb pieces, so some genes are inevitably truncated. Clearly little, if any, activity escapes with chromatin, degraded RNA and associated RNP. Nascent RNA and the transcribed template also resisted electroelution [133].

These results are simply explained by an attached polymerase; nevertheless other possibilities should be excluded. The transcription complex cannot fortuitously have no net charge and so be unable to electroelute as the same result is obtained at a different pH [130]. If the complex is unattached, it must be so large that the polymerase is effectively attached. Even if it has a structure like a 'Christmas tree', RNase removes all 'branches' (i.e. RNP) and the 'trunk' has been cut into pieces very much smaller than the 150-kb fragments that can electroelute as chromatin [133]. In addition, transcribing 'minichromosomes' only a few thousand residues long also resist electroelution (unpublished work).

NUCLEOSKELETONS

Molecular nature

Unfortunately the molecular nature of any nucleoskeleton remains controversial. Structures isolated using 2 M NaCl (e.g. matrices, scaffolds, nucleoid cages) are probably derived from it but which parts are true constituents and which are artefactual additions is unclear [5]. To cite but one example: hypotonic conditions used to isolate nuclei convert pure RNP particles into fibres that cannot be redissolved in 2 M NaCl [10] so it is hardly surprising that RNP skeletons can then be seen.

Studies using the 'physiological' buffer provide a strong candidate for a nucleoskeleton that might exist *in vivo* [135]. When encapsulated cells are lysed using Triton, treated with *Hae*III and most chromatin electroeluted, electron microscopy of thick resinless sections then reveals a diffuse skeleton which ramifies throughout the nucleus (Fig. 3). Individual elements are about 10 nm wide with the axial repeat characteristic of intermediate filaments [136]. In view of the history of artefacts, images such as those in Fig. 3 must necessarily be treated cautiously, however appealing their structure. Nevertheless, if such a skeleton is an artefact formed prior to fixation, its creation cannot interfere with replication and transcription which continue at, or close to, *in vivo* rates. If an artefact is created on fixation, it is difficult to see why a diffuse network and not an aggregate is formed. Obviously it is important to demonstrate whether active polymerases are associated with this skeleton and whether other cytoskeletal elements like actin and tubulin, which cosediment with various subnuclear structures [7], also extend into nuclei and are associated with different functions.

'Miller' spreads

If a nucleoskeleton is composed of intermediate filaments, it is easy to explain why no skeleton is seen in 'Miller' spreads; some intermediate filaments are soluble in hypotonic solutions [137] so the skeleton might dissolve, leaving the 'Christmas tree'. Alternatively, spreading might strip the complex from the skeleton. Why, then, are skeletons not seen in conventional sections, especially those from highly active material isolated prior to fixation in 'physiological' conditions [31]? Perhaps they are too thin to allow visualisation of diffuse skeletons.

Relationship with other skeletons

Ribosomes (diameter 30 nm) pass through densely packed chromatin to the cytoplasm which has diffusional pores only 100 nm wide [138]. This makes it likely that they travel along tracks to their destination. Intermediate filaments could well provide such tracks as they form a nucleoskeleton, lamins [139–141] and the more familiar cytoskeleton, from nuclear pore to cell membrane [142, 143]. Then mRNA would remain attached to members of this one family at all stages of its life-cycle [121], from synthesis to translation [144] (but see [145]).

Targeting

Connecting genes physically with specific cytoplasmic destinations allows mRNA transfer along the connecting filaments to specific places; messages and the proteins they specify are indeed localised in the cytoplasm [146–149]. As nuclei rotate *in vivo* [150], interactions between cytoplasmic and nuclear filaments must be dynamic. Perhaps all nucleofilaments lead ribosomes to the nuclear periphery; then, rotation allows cytofilaments to be scanned so specific ones can be selected for ribosome transhipment.

Such connections also allow information flow from membrane to gene. Contacts between cell membranes generated during differentiation might stabilise underlying cytoskeletons, and hence nucleoskeletons, which in turn could influence gene expression. In this case it is the structure, rather than a second messenger, that transmits the information.

Duplication of the skeleton

If skeleton and attached DNA are duplicated simultaneously rather than separately, the replication site is also a nucleoskeleton-assembly site. This raises the possibilities that

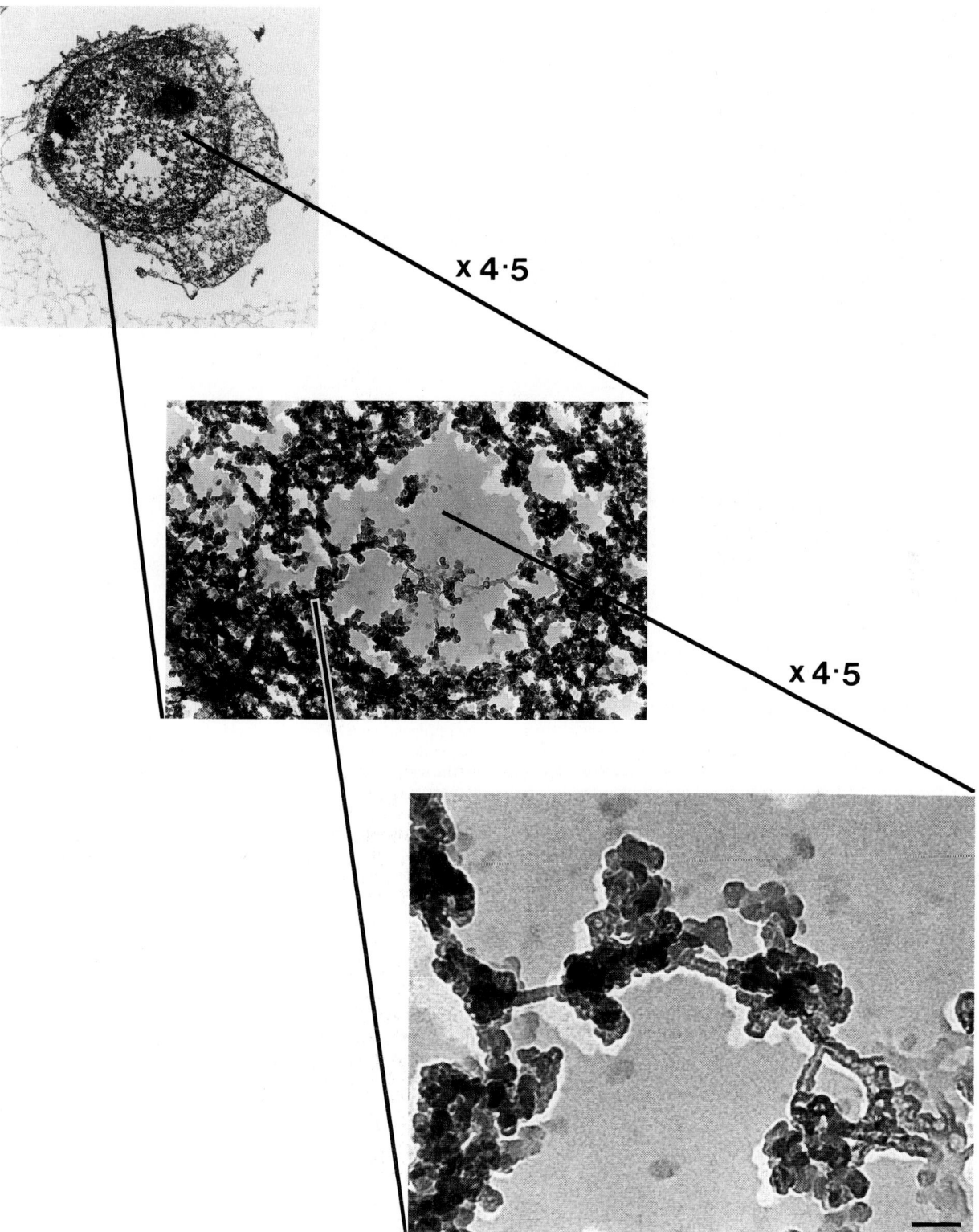

Fig. 3. *A candidate nucleoskeleton*. Electron micrographs at 4.5× higher magnifications of a thick resinless section of a HeLa cell derivative. Cells were encapsulated, lysed, obscuring chromatin cut with *Hae*III and then removed by electroelution. All procedures up to fixation took place in a 'physiological' buffer. At the lowest magnification, the cell remnant can be seen surrounded by agarose and at the highest, residual chromatin clumps attached to a skeleton. The bar is 100 nm. See [135]

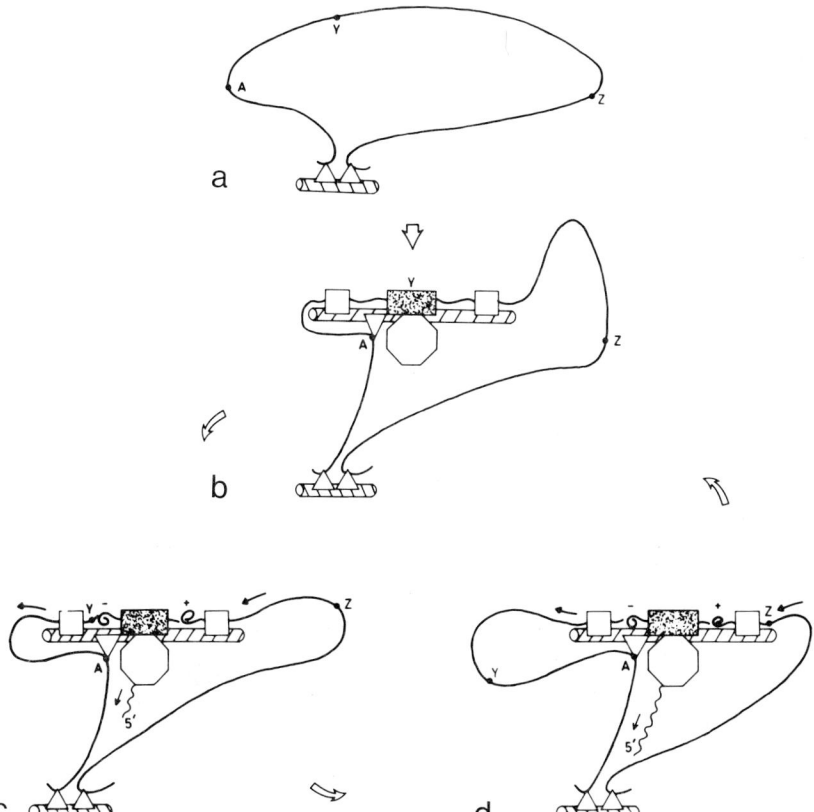

Fig. 4. *A schematic model for transcription (not to scale)*. (a) A loop of DNA is shown attached to the skeleton (rod) at two sites (△). These attachments probably persist whether or not the loop is transcribed or replicated. The gene (Y–Z) out in the loop cannot be transcribed as it is remote from any attached polymerase. A marks an upstream activating sequence. (b) During development, the gene is activated by binding to the skeleton and assembly into an attached transcription complex containing polymerase (stippled rectangle), upstream binding sites (▽), topoisomerases (small squares) and RNA processing site (octagon). For the sake of simplicity, the complex is assembled on an additional skeletal element; transcription factors and another loop formed by an enhancer are also excluded. The upstream binding site now permanently tethers the gene to the complex and abuts the polymerase so that they can inter-communicate through physical contact or indirectly through variations in supercoiling of the connecting loop. (c, d) After initiation, DNA moves (arrows) through the complex and RNA (wavy line) is synthesised and processed. Probably the 5′ RNA end is attached [121] and a loop of RNA is extruded, rather than as shown. Positive and negative supercoils appear transiently as shown but are removed by topoisomerases. After the transcript is completed, A remains attached so the gene can easily return to its position in (b) and reinitiate synthesis

specific attachments, and so specific functions, might be inherited by the two progeny structures and that both might be replicated semi-conservatively.

Motors

Whichever model for transcription proves to be correct, it seems likely that additional motors drive the contortions of template and transcript. Actin is an obvious candidate; it copurifies with ribonucleoprotein complexes [151] and polymerase II [152], it is a known transcription factor [153] and injection of anti-actin antibodies into living newt cells inhibits transcription of lampbrush chromosomes [154].

A MODEL FOR AN ATTACHED POLYMERASE

A nucleoskeleton, a still ill-defined structure, is the structure to which the polymerase and associated activities are attached. These include transcription factors (e.g. those bound to upstream and downstream sites like enhancers), topoisomerases and processing enzymes (e.g. those involved in capping, splicing, methylation and polyadenylation). Some are tightly bound, for example >95% of RNA polymerase II pellets with nuclear fragments [18] and little is displaced even by 600 mM $(NH_4)_2SO_4$ [8]; others (e.g. TFIIA) are less tightly bound [155]. The whole complex must be very large, dwarfing the template and associated proteins. Transcription occurs as the template passes through the polymerisation site; the resulting transcript remains attached during subsequent processing and transfer to the cytoplasm.

A specific model is illustrated in Fig. 4. A gene in a loop of DNA is initially remote from the polymerase and cannot be transcribed. On activation, it attaches to the skeleton and is assembled into a transcription complex. The DNA at the polymerisation site can be imagined as being connected to adjacent sequences through two ball-bearing races, topoisomerases, that allow it to rotate. These topoisomerases are drawn spatially separated from the polymerase, but may well abut it and consequently few supercoils will normally accumulate.

As the DNA moves past the polymerisation site it rotates so that the transcribed base on the helical template strand maintains the same topological relationship to the skeleton (Fig. 5). Template movements are analogous to those of a bolt driven through a fixed nut using a ratchet screwdriver (Fig. 5, below). The nut 'sees' the whole length of the thread as it passes through; the fixed polymerase 'sees' the transcribed

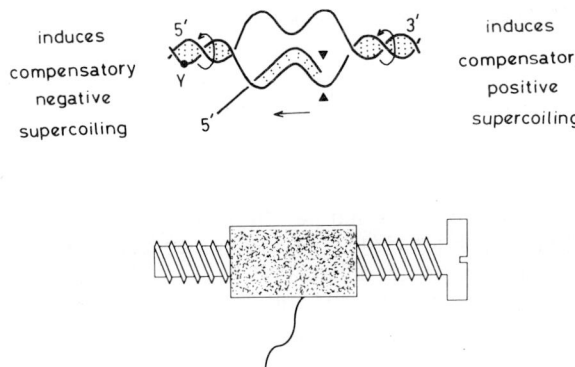

Fig. 5. *DNA movements at the polymerisation site.* DNA moves through the fixed polymerisation site (above) like a screw through a fixed nut (below). Above: Y is the first base to be copied. DNA moves to the left (arrow) and spins (arrows) so the transcribed base between the triangles always retains the same stereochemical relationship to the page (i.e. the skeleton). RNA is synthesised and extruded downwards to the left. Rotation induces compensatory supercoils to accumulate. Below: the bolt (DNA) rotates and passes leftwards through the fixed nut (polymerase). The wavy line shows the analogous position of the transcript

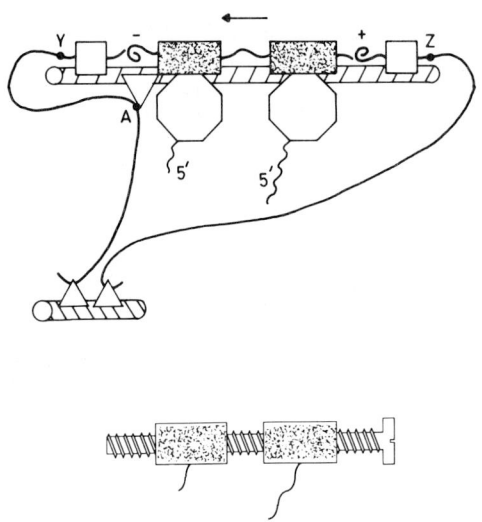

Fig. 6. *A transcription unit containing two polymerisation sites with the analogous bolt and two fixed nuts below.* Extra sites can be added to the right. Supercoils do not arise between sites if DNA moves through them at the same speed. Removal of the skeleton, perhaps during the preparation of a 'Miller' spread, yields the conventional view of a transcription unit. Symbols as for Figs 4 and 5

strand in the same way. As a right-handed twist drives the bolt, a right-handed twist accompanies DNA translocation and just as spinning the ratchet relieves wrist-strain, so a topoisomerase spins the DNA to release accumulated supercoils. The template is truly dynamic.

Highly active transcription units would contain additional polymerases attached to the right-hand side of the complex (Fig. 6). The template necessarily passes through them at the same speed so its axial rotation at each is identical and no supercoils build up between. The analogy here is that of one bolt being driven through two fixed nuts (Fig. 6, below); no topological problems arise within the bolt, only at its ends, so topoisomerases would be needed there. Intriguingly, topoisomerase I cuts are concentrated at the ends of the ribosomal locus [93, 94, 100, 105].

AN ATTACHMENT HYPOTHESIS FOR GENE ACTIVATION

In higher eukaryotes in which only a fraction of DNA is transcribed, most genes will be remote from the skeleton and so remote from polymerases. Sequences out in the loop will only be transcribed if they first attach (Fig. 7). Genes are switched on and off during development by attachment or detachment and cells in different tissues possess different arrays of attachments.

Some evidence supports this. Of different cell types (e.g. fibroblasts, lymphocytes, hepatocytes, teratocarcinoma cells from man, mouse, bird and insect) the only ones that fail to yield superhelical DNA and a nucleoid cage on lysis in Triton and 2 M NaCl are those that are transcriptionally inactive, i.e. mature hen erythrocytes and human sperm [38] (and unpublished work). Furthermore, when hen erythrocytes are fused with growing cells, they begin transcribing again as a matrix reforms [156]. And as described earlier, a sequential inactivation and reactivation of integrated pro-viral genes correlates with their detachment and reattachment [123].

What triggers specific attachments of target sequences during development? It could involve selective changes in chemical constitution (e.g. by methylation [157]), conformation (e.g. coiling or supercoiling in a different sense or degree [61, 158]) or binding of specific activators or repressors [159]. As all sequences associate transiently with the skeleton during replication (see later), this might be a prerequisite for transcriptional attachment [160–163]. Consider the α and β globin clusters which are probably each in one loop [53, 164]; genes in both clusters are arranged along the chromosome in order of their expression during development [165]. As the

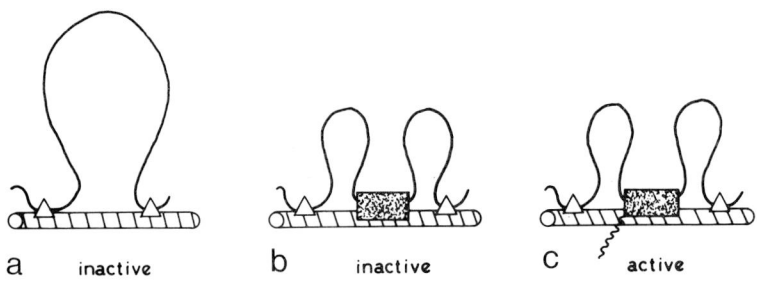

Fig. 7. *The attachment hypothesis.* A gene in a loop can only be transcribed when it attaches to the polymerase (stippled rectangle) at the skeleton. For example, the globin loop is shown inactive in a fibroblast (a). It attaches and becomes potentially active (but not expressed) in an erythroid stem cell (b), and expressed after addition of the appropriate transcription factors in an erythroblast (c). Symbols as for Fig. 4

first gene in the complex is replicated it becomes attached and so expressed. This attachment, and consequent expression, is retained subsequently and interferes with the transcriptional attachment, and so expression, of adjacent genes (see below). On replication later in development it detaches and the next gene along the chromosome attaches and it in turn becomes expressed. In this way gene order determines the sequence of expression during development.

This hypothesis requires that transcriptional attachments are sufficiently stable to survive when the template is replicated. The remarkable patterns seen in 'Miller' spreads of transcription units on sister chromatids supports this; the two patterns are similar, sometimes with the same transcript-free gaps in their middles [166]. This is explained with difficulty by the conventional model, by assuming two sets of diffusible polymerases somehow initiate together on the two units. However, such patterns are the inevitable consequence of transcription by attached complexes which are duplicated along with the DNA.

The inheritance of specific attachments might be allied to the inheritance of specific structures within loops [35, 61]. Consider the globin sequence in two different cells, with the same attachments but differing degrees of supercoiling. If during replication, detachments and net rotation of DNA are prevented, it is an inevitable consequence of semi-conservative replication that daughter loops inherit the superhelical density of parent loops. Then, structures of DNA (as supercoiling) or its attachments would contain all the heritable epigenetic information needed to trigger differentiation.

CONSEQUENCES OF THE ATTACHMENT MODEL

Loops, supercoils and topoisomerases

Ptashne [167] has recently explained various *cis* effects by a template looping that brings together upstream sites and the promoter. Such a looping has been incorporated into the specific model presented in Fig. 4. Indeed, the two models converge when the complex bridging the two sites becomes sufficiently large.

Loops also enable differing superhelical densities to be maintained locally and these differences could impinge upon transcription at a number of different points. In bacteria, supercoiling, whilst having little effect on transcriptional elongation [14, 63], has complicated effects on initiation [168]; the free energy of supercoiling often aids promoter 'opening' or unwinding. There is every chance that it also affects initiation in eukaryotes [169–173]. However, it is not yet clear how nucleosomes influence transcription; although they inhibit initiation *in vitro* [22–24, 174], they might do so by steric blocking or by competing for the free energy [175]. Their dissociation would release free energy of supercoiling to assist opening but whether they do so is controversial [176–178]. It is worth noting that the supercoils that accumulate ahead of the transcription site have the opposite sense to those in the nucleosome and so might destabilise it.

Recent models for transcription assume that topoisomerases will be soluble like the polymerase, diffusing to their site of action and removing supercoils as they arise [89]. However, if the polymerase is attached, then it seems likely that topoisomerases are too (Fig. 4). Topoisomerase II is closely associated with nuclear matrices and scaffolds [179–182] and activity is not extracted from encapsulated cells by Triton and isotonic salt concentrations (unpublished work). In addition, if topoisomerase were part of a larger complex, it would probably act at sites determined by three-dimensional structure; if freely diffusible, it would act randomly. In fact, topoisomerase I cuts are sometimes localised to only one strand [101].

Stable transcription complexes

The model requires that the template is stably attached to the skeleton and associated polymerases. Indeed, DNA forms stable complexes with polymerase I [183–185], II [24, 155, 186–189] and III [190, 191] and their stability does not depend on transcription *per se* [155, 188]. For example, when the U2 snRNA gene is injected into oocytes, it sequesters transcription factors so templates added subsequently cannot be transcribed. Remarkably, this complex is stable even after the polymerase has moved away from the initiation site and when transcription is inhibited with α-amanitin [192]. It is difficult to imagine how a diffusible polymerase could return to compete efficiently at the same initiation site but this is inevitable if the gene is attached. In Fig. 4, sequence A tethers the gene so that it remains permanently associated with a particular polymerase.

Capping, methylation, splicing and polyadenylation

Nascent RNA is capped [193, 194], methylated [195], spliced [196] and polyadenylated [197] before the polymerase completes synthesis. Therefore a model involving a mobile polymerase requires that all associated processing activities are dragged along the template. This seems unlikely as they are so bulky; it seems inevitable that the smaller template moves relative to them. For example, 'spliceosomes' are 40–60 nm in diameter and contain > 50 different polypeptides [198] and approximately every 500 bases of heterogeneous nuclear RNA is complexed with > 36 polypeptides [145]. In addition, heterogeneous nuclear RNA is associated with nuclear matrices [199] but the status of this association is controversial, especially bearing in mind that pure RNP particles aggregate in the hypotonic conditions used to prepare them [10].

The fact that *in vitro* systems splice [200, 201] and methylate [202] added RNA might be taken as evidence that no larger structure is involved. However, these involve crude extracts which, like *in vitro* transcription systems, always require preincubations when large complexes may form on added RNA. Conversely, there is no temporal lag if some skeletal structure is maintained. Globin pre-messenger is associated with matrices isolated from HeLa cells transfected with plasmids containing the rabbit β-globin gene. When such matrices are incubated with a splicing extract and ATP, the amount of matrix-bound pre-messenger falls and free intron lariat increases, without any lag [203]. Thus it seems likely that these post-transcriptional processes are all associated with the skeleton.

Stereochemical consequences

A symmetrical molecule like DNA can, in principle, bind to the polymerase in one of two orientations; some asymmetry must tell the polymerase which way to transcribe. A larger asymmetric structure would orient a symmetric polymerase correctly if it bound DNA at three or more sites [204]. Perhaps this explains why transcription units have at least three sites essential for initiation (e.g. the TAATA box, UAS and initiation site).

A number of general stereochemical consequences stem from the inflexibility of chromatin and the inability of the polymerase to diffuse freely; whilst certain attachments remain, this ensures that some regions of the template can never approach the polymerisation site. A promoter in any loop will be sufficiently close to perhaps as few as one attached transcription complex; every site is restricted to transcribing only those genes within range. The dedication of polymerases to transcribe particular genes is the inevitable consequence of the model and makes it easy to imagine how stable patterns of expression might be established during development. Specific stereochemical consequences will now be discussed.

Position effects, enhancers and transvection

Position effects were initially discovered in *Drosophila* and result in gene activity being suppressed by an adjacent (or *cis*) chromosomal rearrangement involving heterochromatin [205]. Suppression can affect any translocated gene. It occurs early in development and does not necessarily affect all cells carrying the rearrangement, but the closer the gene is to the breakpoint in heterochromatin, the more likely it is to be affected. Once initiated, the suppression is inherited by progeny cells within the fly, so that gene expression in different clones of cells within one tissue may produce a variegated phenotype. Such position effects are now commonly found in 'transgenic' animals; whether the transgene is expressed depends on its integration site [206–211].

The inactivation of genes by position effects is explained with difficulty by models involving diffusible molecules but is naturally explained by structural ones [61]. Translocation of a hitherto active gene from its usual environment, which presumably contains the appropriate attachment sites and associated polymerases, into a heterochromatic region devoid of them, would inevitably inactivate it.

'Enhancers' provide examples of position effects at the molecular level. They increase expression of adjacent genes, but not those on other chromosomes; they may lie 5′ or 3′ to the initiation site and act over many thousands of base pairs [212, 213]. Their effects are usually explained in terms of 'entry' sites for diffusible polymerases, but if so the polymerase must then 'scan' thousands of base pairs both upstream and downstream for the initiation site. The attachment model sees enhancers as sequences that bind to the nucleoskeleton, bringing adjacent genes into close proximity to bound polymerases. Indeed, enhancers are the sequences most closely associated with nucleoid cages [123].

Recently, particularly powerful enhancers or 'dominant control regions' have been uncovered at the extreme ends of the β-globin locus [164]. These sequences, selected because of their hypersensitivity to nucleases, were used to construct a 'mini-locus' containing the β-globin gene; on introduction into mice, the transgene was expressed in a tissue-specific manner, independently of position. These sequences also bind to scaffolds [60] and presumably represent attachment sequences, isolating the β-globin gene from the effects of adjacent DNA in the chromosome by forming a loop.

If a fully active transcription complex is assembled at the nucleoskeleton from the promoter and distant *cis*-acting sequences by looping out intervening DNA, a promoter from one chromosome might occasionally be incorporated into a complex with a *cis*-acting sequence from another. Such an event would go undetected unless functional sequences on one chromosome complemented deficiencies on another and unless the complementing chromosomes were together. Just such an effect may underlie 'transvection' in the bithorax complex in *Drosophila* which depends on chromosome pairing [214, 215]. It has been explained in terms of nuclear messages with limited diffusional ranges or trans-splicing; recent experiments make such explanations unlikely [216]. The expression of *Ubx*, a gene in the complex, is regulated by *cis*-acting elements lying 40–60 kbp on either side. Some of these can regulate a second copy of *Ubx* on another chromosome, but only if paired with it. Perhaps a skeleton brings *cis*-acting enhancers from one chromosome together with the promoter from another.

Transcriptional interference

The phenomenon of 'transcriptional interference' is commonly found when two functional genes carried by retroviral vectors are inserted into a chromosome. Assay of the population shows both genes to be transcribed, but assay of individual cells or clones shows only one of the two promoters to be active [217–221]. Inactivation of one promoter improves expression of the other [222]. Such effects are also found with rearranged cellular genes [223] or transfected minichromosomes packed with transcription units [224, 225]. Transcriptional interference is usually interpreted in terms of mobile polymerases running from one transcription unit into another, but then it is difficult to see how transcription of a downstream gene might inhibit one upstream or why transcription of the upstream gene does not stimulate transcription of the downstream one. (Note that interference cannot be seen in most 'transient' and stable transfectants because cells generally contain > 1 plasmid.)

Interference is simply explained if stereochemical constraints determine how closely adjacent attached complexes can be spaced; only one of two adjacent promoters can attach at any one time and be active. Both would be activated by increasing the interstitial DNA above a critical minimum which must be > 5.2 kb [220]. Perhaps such interference normally controls expression of adjacent genes during development so that only one of them could be active at any one time [224]. Examples might include switching between the early and late promoters during viral growth [163] and between the *Adh* promoters during development in *Drosophila* [221].

Analogous stereochemical constraints should affect how closely processing sites can be spaced: sites on the transcript are usually further apart than this minimum spacing. Such interference might explain why the transcriptional machinery ignores a polyadenylation signal in the 5′ long terminal repeat of retroviruses and then uses the identical sequence further downstream in the 3′ repeat [226]. This has been explained by transcript looping [227], but could equally result from steric hindrance of any polyadenylation at sites too close to the polymerisation site or an attached 5′ cap. Again, increasing the separation between the cap site and the 5′ signal should allow polyadenylation and give some estimate of how far apart the two sites are in space.

Role of introns during transcription

Genes with introns are transcribed in transgenic mice at least tenfold more efficiently than their counterparts without introns [228]. As no sequence-specific signals have been detected in some of the introns tested, it seems that introns must play some general structural role. Just such general effects might be expected if the gene was looped and attached to a number of polymerisation sites; thus, in Fig. 6, intron loss might make it impossible for a shorter template to loop back and attach to two polymerases simultaneously. Here, interference within a transcription unit, rather than between units, reduces the number of attachments and so the transcription rate.

Replicational interference

A replication fork would interfere with a transcription unit in an analogous way. Most transcription units in *E. coli* happen to be aligned on the chromosome so that the direction of replication and transcription are similar; it has been suggested that this results from evolutionary pressures to prevent polymerases colliding [229]. But if active DNA and RNA polymerases are both attached (see below) they cannot collide. Instead, passage of DNA during replication and transcription through the two complexes in the same direction would minimize interference; movement in opposite directions would be impossible so one process, presumably transcription, must stop.

Other polymerases and functions of DNA

This discussion has concentrated on the role of the nucleoskeleton with respect to transcription. However, the attachment model can equally be applied to other functions of DNA (i.e. replication, repair and recombination). Indeed, very similar kinds of evidence to that reviewed above, especially that derived using isotonic conditions, show that nascent DNA and the relevant DNA polymerases (α and β), if active, are associated with the nucleoskeleton [131, 132]. Activation is again seen as a binding of sequences to polymerases associated with a skeleton [230]. Structural models [231, 232] (but see [233]) would seem essential to explain how damage induced by ultraviolet light could be removed selectively from the transcribed, but not the non-transcribed strand, of the DHFR gene [234] and how adjacent replicons might initiate coordinately [235].

CONCLUSIONS

The 'text-book' model for transcription sees the polymerase and transcript moving along DNA unattached to any skeleton. Like many received ideas, this one seems to have little decisive evidence for it. Accurate transcription at *in vivo* rates by a soluble system would provide strong evidence, but existing systems are still very inefficient. Something that influences rates by a few orders of magnitude is clearly lacking. Therefore it seems worthwhile to examine an alternative in which the template moves past a polymerase attached to a nucleoskeleton; this nucleoskeleton is the active site. The best evidence for this alternative is circumstantial and stems from studies on cells fractionated using only one kind of procedure (lysis in a 'physiological' buffer, followed by nuclease digestion and electrophoretic removal of detached chromatin) and so must be corroborated. This crude system transcribes very efficiently. Obivously, analysis of such an insoluble polymerase – nucleoskeleton complex poses a difficult challenge to biochemists and formal proof of the alternative model will be difficult.

Our perception of whether template or polymerase moves is determined by our perception of their relative sizes. We now know that the polymerase and associated activities (including transcription factors, topoisomerases and associated splicing, capping, methylation and polyadenylation complexes) must dwarf the template; they bind to each other and to a number of sites on the DNA, forming it into loops. When that complex becomes sufficiently large, the two models inevitably converge.

If active polymerases are indeed stably attached, our DNA-centred universe becomes a skeleton-centred one. Enhancers bind to the skeleton, not the polymerase. Transcription factors bind to the skeleton as well as to DNA. Most importantly, as biochemists we look in the pellet rather than in the supernatant.

I thank the Cancer Research Campaign for their continued support and my colleagues Drs J. Lang, S. J. McCready and D. A. Jackson for their help. I am especially indebted to Dr O. L. Miller for supplying Fig. 2a and allowing me to use it freely.

REFERENCES

1. Alberts, B., Bray, D., Lewis, J., Raff, M., Roberts, K. & Watson, J. D. (1983) *Molecular biology of the cell*, Garland, New York.
2. Darnell, J., Lodish, H. & Baltimore, D. (1986) *Molecular cell biology*, Scientific American Books, New York.
3. Fredericq, E. (1971) in *Histones and nucleohistones* (Phillips, D. M. P., ed.) pp. 135–186, Plenum Press, London.
4. Razin, S. V., Yarovaya, O. V. & Georgiev, G. P. (1985) *Nucleic Acids Res. 13*, 7427–7444.
5. Cook, P. R. (1988) *J. Cell Sci. 90*, 1–6.
6. Roberge, M., Dahmus, M. E. & Bradbury, E. M. (1988) *J. Mol. Biol. 201*, 545–555.
7. MacGillivray, A. J. & Birnie, G. D. (1986) *Nuclear structures: isolation and characterization*, Butterworths, London.
8. Jackson, D. A. & Cook, P. R. (1985) *EMBO J. 4*, 913–918.
9. Jackson, D. A. & Cook, P. R. (1986) *EMBO J. 5*, 1403–1410.
10. Lothstein, L., Arenstorf, H. P., Chung, S.-Y., Walker, B. W., Wooley, J. C. & LeStourgeon, W. M. (1985) *J. Cell Biol. 100*, 1570–1581.
11. Verheijen, R., Van Venrooij, W. & Ramaekers, F. (1988) *J. Cell Sci. 90*, 11–36.
12. Mirkovitch, J., Mirault, M.-E. & Laemmli, U. K. (1984) *Cell 39*, 223–232.
13. Miller, O. L. (1984) *J. Cell Sci. Suppl. 1*, 81–93.
14. McClure, W. R. (1985) *Annu. Rev. Biochem. 54*, 171–204.
15. Sollner-Webb, B. & Tower, J. (1986) *Annu. Rev. Biochem. 55*, 801–830.
16. Geiduschek, E. P. & Tocchini-Valentini, G. P. (1988) *Annu. Rev. Biochem. 57*, 873–914.
17. Beebee, T. J. C. (1979) *Biochem. J. 183*, 43–54.
18. Weil, P. A., Luse, D. S., Segall, J. & Roeder, R. G. (1979) *Cell 18*, 469–484.
19. Engelke, D. R., Shastry, B. S. & Roeder, R. G. (1983) *J. Biol. Chem. 258*, 1921–1931.
20. Manley, J. L., Fire, A., Cano, A., Sharp, P. A. & Gefter, M. L. (1980) *Proc. Natl Acad. Sci. USA 77*, 3855–3859.
21. Lue, N. F. & Kornberg, R. D. (1987) *Proc. Natl Acad. Sci. USA 84*, 8839–8843.
22. Knezetic, J. A. & Luse, D. S. (1986) *Cell 45*, 95–104.
23. Lorch, Y., LaPointe, J. W. & Kornberg, R. D. (1987) *Cell 49*, 203–210.
24. Workman, J. L. & Roeder, R. G. (1987) *Cell 51*, 613–622.
25. Ivarie, R. D., Schacter, B. S. & O'Farrell, P. H. (1983) *Mol. Cell Biol. 3*, 1460–1467.
26. Callan, H. G. (1977) *Proc. R. Soc. Lond. Ser. B. 214*, 417–448.
27. Perlman, S. M., Ford, P. J. & Rosbash, M. M. (1977) *Proc. Natl Acad. Sci. USA 74*, 3835–3839.
28. Diaz, M. O., Barsacchi-Pilone, G., Mahon, K. A. & Gall, J. G. (1981) *Cell 24*, 649–659.
29. Jamrich, M., Warrior, R., Steele, R. & Gall, J. G. (1983) *Proc. Natl Acad. Sci. USA 80*, 3364–3367.
30. DuPraw, E. J. (1965) *Nature 206*, 338–343.

31. Bjorkroth, B., Ericsson, C., Lamb, M. M. & Daneholt, B. (1988) *Chromosoma (Berl.)* 96, 333–340.
32. Paulson, J. R. & Laemmli, U. K. (1977) *Cell* 12, 817–828.
33. Marsden, M. P. F. & Laemmli, U. K. (1979) *Cell* 17, 849–858.
34. Igo-Kemenes, T. & Zachau, H. G. (1977) *Cold Spring Harbor Symp. Quant. Biol.* 42, 109–118.
35. Cook, P. R. (1973) *Nature* 245, 23–25.
36. Bauer, W. & Vinograd, J. (1974) *Basic principles in nucleic acid chemistry* (T'so, P. O. P., ed.) pp. 265–303, Academic Press, London.
37. Cook, P. R. & Brazell, I. A. (1975) *J. Cell Sci.* 19, 261–279.
38. Cook, P. R. & Brazell, I. A. (1976) *J. Cell Sci.* 22, 287–302.
39. Benyajati, C. & Worcel, A. (1976) *Cell* 9, 393–407.
40. Pinon, R. & Salts, Y. (1977) *Proc. Natl Acad. Sci. USA* 74, 2850–2854.
41. Hartwig, M. (1978) *Acta Biol. Med. Ger.* 37, 421–432.
42. Stoilov, L. M., Zlatanova, J. S., Vassileva, A. P., Ivanchenko, M. G., Krachmarov, C. P. & Genchev, D. D. (1988) *J. Cell Sci.* 89, 243–252.
43. Cook, P. R. & Brazell, I. A. (1978) *Eur. J. Biochem.* 84, 465–477.
44. Cook, P. R., Brazell, I. A. & Jost, E. (1976) *J. Cell Sci.* 22, 303–324.
45. Vogelstein, B., Pardoll, D. M. & Coffey, D. S. (1981) *Cell* 22, 79–85.
46. Mullinger, A. M. & Johnson, R. T. (1979) *J. Cell Sci.* 38, 369–389.
47. McCready, S. J., Akrigg, A. & Cook, P. R. (1979) *J. Cell Sci.* 39, 53–62.
48. Kornberg, R. D. (1977) *Annu. Rev. Biochem.* 46, 931–954.
49. Morse, R. H. & Simpson, R. T. (1988) *Cell* 54, 285–287.
50. Touchette, N. A. & Cole, R. D. (1985) *Proc. Natl Acad. Sci. USA* 82, 2642–2646.
51. Touchette, N. A., Anton, E. & Cole, R. D. (1986) *J. Biol. Chem.* 261, 2185–2188.
52. Warren, A. C. & Cook, P. R. (1978) *J. Cell Sci.* 30, 211–226.
53. Cook, P. R. & Brazell, I. A. (1980) *Nucleic Acids Res.* 8, 2895–2906.
54. Cockerill, P. N. & Garrard, W. T. (1986) *Cell* 44, 273–282.
55. Gasser, S. M. & Laemmli, U. K. (1986) *Cell* 46, 521–530.
56. Cockerill, P. N., Yuen, M.-H. Garrard, W. T. (1987) *J. Biol. Chem.* 262, 5394–5397.
57. Phi-Van, L. & Stratling, W. H. (1988) *EMBO J.* 7, 655–664.
58. Mirkovitch, J., Gasser, S. M. & Laemmli, U. K. (1988) *J. Mol. Biol.* 200, 101–109.
59. Izaurralde, E., Mirkovitch, J. & Laemmli, U. K. (1988) *J. Mol. Biol.* 200, 111–125.
60. Jarman, A. P. & Higgs, D. R. (1988) *EMBO J.* 7, 3337–3344.
61. Cook, P. R. (1974) *Biol. Rev.* 49, 51–84.
62. Chamberlin, M. J. (1974) *Annu. Rev. Biochem.* 43, 721–775.
63. Wang, J. C. (1985) *Annu. Rev. Biochem.* 54, 665–697.
64. Sinden, R. R., Carlson, J. O. & Pettijohn, D. E. (1980) *Cell* 21, 773–783.
65. Esposito, F. & Sinden, R. R. (1987) *Nucleic Acids Res.* 15, 5105–5124.
66. Luchnik, A. N., Bakayev, V. V., Zbarsky, I. B. & Georgiev, G. P. (1982) *EMBO J.* 1, 1353–1358.
67. Luchnik, A. N., Bakayev, V. A., Yugai, A. A., Zbarsky, I. B. & Georgiev, G. P. (1985) *Nucleic Acids Res.* 13, 1135–1149.
68. Luchnik, A. N., Hisamutdinov, T. A. & Georgiev, G. P. (1988) *Nucleic Acids Res.* 16, 5175–5190.
69. Glikin, G. C., Ruberti, I. & Worcel, A. (1984) *Cell* 37, 33–41.
70. Villeponteau, B., Lundell, M. & Martinson, H. (1984) *Cell* 39, 469–478.
71. Borsoum, J. & Berg, P. (1985) *Mol. Cell Biol.* 5, 3048–3057.
72. Kmiec, E. B. & Worcel, A. (1985) *Cell* 41, 945–953.
73. Kmiec, E. B., Razvi, F. & Worcel, A. (1986) *Cell* 45, 209–218.
74. Kmiec, E. B., Ryoji, M. & Worcel, A. (1986) *Proc. Natl Acad. Sci. USA* 83, 1305–1309.
75. Petryniak, B. & Lutter, L. C. (1987) *Cell* 48, 289–295.
76. Wolffe, A. P., Andrews, M. T., Crawford, E., Losa, R. & Brown, D. D. (1987) *Cell* 49, 301–303.
77. Worcel, A. (1988) *Cell* 54, 919.
78. Kmiec, E. B. (1988) *Cell* 54, 919–920.
79. Webb, M. L. & Jacob, S. T. (1988) *J. Biol. Chem.* 263, 4745–4748.
80. Schroder, H. C., Trolltsch, D., Friese, U., Bachmann, M. & Muller, W. E. G. (1987) *J. Biol. Chem.* 262, 8917–8925.
81. Cotten, M., Bresnahan, D., Thompson, S., Sealy, L. & Chalkley, R. (1986) *Nucleic Acid Res.* 14, 3671–3686.
82. Sealy, L., Cotten, M. & Chalkley, R. (1986) *EMBO J.* 5, 3305–3311.
83. Downes, C. S., Ord, M. J., Mullinger, A. M., Collins, A. R. S. & Johnson, R. T. (1985) *Carcinogenesis* 6, 1343–1352.
84. Gallagher, M., Weinberg, R. & Simpson, M. V. (1986) *J. Biol. Chem.* 261, 8604–8607.
85. Ryoji, M. & Worcel, A. (1984) *Cell* 37, 21–32.
86. Woodland, H. R. & Adamson, E. D. (1977) *Dev. Biol.* 57, 118–135.
87. Maaloe, O. & Kjeldgaard, N. O. (1966) *Control of macromolecular synthesis*, Benjamin, New York.
88. Gamper, H. B. & Hearst, J. E. (1982) *Cell* 29, 81–90.
89. Liu, L. F. & Wang, J. C. (1987) *Proc. Natl Acad. Sci. USA* 84, 7024–7027.
90. Jackson, D. A. & Cook, P. R. (1988) in *Chromosomes and chromatin*, vol. III (Adolph, K. W., ed.) pp. 97–118, CRC Press, Boca Raton, FL.
91. Higashimakagawa, T., Wahn, H. & Reeder, R. H. (1977) *Dev. Biol.* 55, 375–386.
92. Bauer, W. R., Ressner, E. C., Kates, J. & Patzke, J. V. (1977) *Proc. Natl Acad. Sci. USA* 74, 1841–1845.
93. Gocke, E., Leer, J. C., Nielsen, O. F. & Westergaard, O. (1983) *Nucleic Acids Res.* 11, 7661–7678.
94. Fleischmann, G., Pflugfelder, G., Steiner, E. K., Javaherian, K., Howard, G. C., Wang, J. C. & Elgin, S. (1984) *Proc. Natl Acad. Sci. USA* 81, 6958–6962.
95. Bonven, B. J., Gocke, E. & Westergaard, O. (1985) *Cell* 41, 541–551.
96. Muller, M. T., Pfund, W. P., Mehta, V. B. & Trask, D. F. (1985) *EMBO J.* 4, 1237–1243.
97. Gilmour, D. S., Pflugfelder, G., Wang, J. C. & Lis, J. T. (1986) *Cell* 44, 401–407.
98. Uemura, T. & Yanagida, M. (1986) *EMBO J.* 5, 1003–1010.
99. Brill, S. J., DiNardo, S., Voelkel-Meiman, K. & Sternglanz, R. (1987) *Nature* 326, 414–416.
100. Gilmour, D. S. & Elgin, S. C. (1987) *Mol. Cell Biol.* 7, 141–148.
101. Stewart, A. F. & Schutz, G. (1987) *Cell* 50, 1109–1117.
102. Egyhazi, E. & Durban, E. (1987) *Mol. Cell Genet.* 7, 4308–4316.
103. Garg, L. C., DiAngelo, S. & Jacob, S. T. (1987) *Proc. Natl Acad. Sci.* 84, 3185–3188.
104. Zhang, H., Wang, J. C. & Liu, L. F. (1988) *Proc. Natl Acad. Sci.* 85, 1060–1064.
105. Ness, P. J., Koller, T. & Thoma, F. (1988) *J. Mol. Biol.* 200, 127–139.
106. Uemura, T. & Yanagida, M. (1985) *EMBO J.* 3, 1737–1744.
107. DiNardo, S., Voelkel, K. A. & Sternglanz, R. (1984) *Proc. Natl Acad. Sci. USA* 81, 2616–2620.
108. Goto, T. & Wang, J. C. (1985) *Proc. Natl Acad. Sci. USA* 82, 7178–7182.
109. Heck, M. M. S. & Earnshaw, W. C. (1986) *J. Cell Biol.* 103, 2569–2581.
110. Heck, M. M. S., Hittelman, W. N. & Earnshaw, W. C. (1988) *Proc. Natl Acad. Sci. USA* 85, 1086–1090.
111. Tabuchi, H. & Hirose, S. (1988) *J. Biol. Chem.* 263, 15282–15287.
112. Wu, H.-Y., Shyy, S., Wang, J. C. & Liu, L. F. (1988) *Cell* 53, 433–440.
113. Osborne, B. I. & Guarente, L. (1988) *Genes Dev.* 2, 766–772.
114. Brill, S. J. & Sternglanz, R. (1988) *Cell* 54, 403–411.
115. Giaever, G. N. & Wang, J. C. (1988) *Cell* 55, 849–856.
116. Tsao, Y.-P., Wu, H.-Y. & Liu, L. F. (1989) *Cell* 56, 111–118.

117. DiNardo, S., Voelkel, K. A., Sternglanz, R., Reynolds, A. E. & Wright, A. (1982) *Cell 31*, 43−51.
118. Pruss, G. J., Manes, S. H. & Drlica, K. (1982) *Cell 31*, 35−42.
119. Raji, A., Zabel, D. J., Laufer, C. S. & Depew, R. E. (1985) *J. Bacteriol. 162*, 1173−1179.
120. Jackson, D. A., McCready, S. J. & Cook, P. R. (1984) *J. Cell Sci. Suppl. 1*, 59−79.
121. Jackson, D. A., McCready, S. J. & Cook, P. R. (1981) *Nature 292*, 552−555.
122. Jackson, D. A., Caton, A. J., McCready, S. J. & Cook, P. R. (1982) *Nature 296*, 366−368.
123. Cook, P. R., Lang, J., Hayday, A., Lania, L., Fried, M., Chiswell, D. J. & Wyke, J. A. (1982) *EMBO J. 1*, 447−452.
124. Amati, B. B. & Gasser, S. M. (1988) *Cell 54*, 967−978.
125. Razin, S. V. & Yarovaya, O. V. (1985) *Exp. Cell Res. 158*, 273−275.
126. Littlewood, T. D., Hancock, D. C. & Evan, G. I. (1987) *J. Cell Sci. 88*, 65−72.
127. McConnell, M., Whalen, A. M., Smith, D. E. & Fisher, P. A. (1987) *J. Cell Biol. 105*, 1087−1098.
128. Berrios, M. & Fisher, P. A. (1988) *Mol. Cell Biol. 8*, 4573−4575.
129. Cook, P. R. (1984) *EMBO J. 3*, 1837−1842.
130. Jackson, D. A., Yuan, J. & Cook, P. R. (1988) *J. Cell Sci. 90*, 365−378.
131. Jackson, D. A. & Cook, P. R. (1986) *EMBO J. 5*, 1403−1410.
132. Jackson, D. A. & Cook, P. R. (1986) *J. Mol. Biol. 192*, 65−76.
133. Jackson, D. A. & Cook, P. R. (1985) *EMBO J. 4*, 919−925.
134. Thorburn, A., Moore, R. & Knowland, J. (1988) *Nucleic Acids Res. 16*, 7183.
135. Jackson, D. A. & Cook, P. R. (1988) *EMBO J. 7*, 3667−3677.
136. Steinart, P. M. & Roop, D. R. (1988) *Annu. Rev. Biochem. 57*, 593−626.
137. Zackroff, R. V. & Goldman, R. D. (1979) *Proc. Natl Acad. Sci. USA 76*, 6226−6230.
138. Gershon, N. D., Porter, K. R. & Trus, B. L. (1985) *Proc. Natl Acad. Sci. USA 82*, 5030−5034.
139. McKeon, F. D., Kirschner, M. W. & Caput, D. (1986) *Nature 319*, 463−468.
140. Aebi, U., Cohn, J., Buhle, L. & Gerace, L. (1986) *Nature 323*, 560−564.
141. Fisher, D. Z., Chaudhary, N. & Blobel, G. (1986) *Proc. Natl Acad. Sci. USA 83*, 6450−6454.
142. Georgatos, S. D. & Blobel, G. (1987) *J. Cell Biol. 105*, 105−115.
143. Georgatos, S. D. & Blobel, G. (1987) *J. Cell Biol. 105*, 117−125.
144. Cervera, M., Dreyfuss, G. & Penman, S. (1981) *Cell 23*, 113−120.
145. Dreyfuss, G. (1986) *Annu. Rev. Cell Biol. 2*, 459−498.
146. Blobel, G. (1985) *Proc. Natl Acad. Sci. USA 82*, 8527−8529.
147. Lawrence, J. B. & Singer, R. H. (1986) *Cell 45*, 407−415.
148. Garner, C. C., Tucker, R. P. & Matus, A. (1988) *Nature 336*, 674−677.
149. Pavlath, G. K., Rich, K., Webster, S. G. & Blau, H. M. (1989) *Nature 337*, 570−573.
150. Paddock, S. W. & Albrecht-Buehler, G. (1988) *Exp. Cell Res. 175*, 409−413.
151. Nakayasu, H. & Ueda, K. (1984) *Cell Struct. Funct. 9*, 317−326.
152. Smith, S. S., Kelly, K. H. & Jockusch, B. M. (1979) *Biochem. Biophys. Res. Commun. 86*, 161−166.
153. Egly, J. M., Miyamoto, N. G., Moncollin, V. & Chambon, P. (1984) *EMBO J. 3*, 2363−2371.
154. Scheer, U., Hinssen, H., Franke, W. W. & Jockusch, B. M. (1984) *Cell 39*, 111−122.
155. Reinberg, D., Horikoshi, M. & Roeder, R. G. (1987) *J. Biol. Chem. 262*, 3322−3330.
156. Woodcock, C. L. F. & Woodcock, H. (1986) *J. Cell Sci. 84*, 105−127.
157. Holliday, R. (1987) *Science 238*, 163−170.
158. Cantor, C. R. (1981) *Cell 25*, 293−295.
159. Dynan, W. S. & Tjian, R. (1985) *Nature 316*, 774−778.
160. Smithies, O. J. (1982) *J. Cell Physiol. Suppl. 1*, 137−143.
161. Bendig, M. M. & Williams, J. G. (1984) *Mol. Cell Biol. 4*, 2109−2119.
162. Enver, T., Brewer, A. C. & Patient, R. K. (1988) *Mol. Cell Biol. 8*, 1301−1308.
163. Vales, L. D. & Darnell, J. E. (1989) *Genes Dev. 3*, 49−59.
164. Grosveld, F., van Assendelft, G. B., Greaves, D. R. & Kollias, G. (1987) *Cell 51*, 975−985.
165. Maniatis, T., Fritsch, E. F., Lauer, J. & Lawn, R. M. (1980) *Annu. Rev. Genet. 14*, 145−178.
166. McKnight, S. L. & Miller, O. L. (1979) *Cell 17*, 551−563.
167. Ptashne, M. (1988) *Nature 335*, 683−689.
168. Pruss, G. J. & Drlica, K. (1989) *Cell 56*, 521−523.
169. Lescure, B., Chestier, A. & Yaniv, M. (1978) *J. Mol. Biol. 124*, 73−85.
170. Akrigg, A. & Cook, P. R. (1980) *Nucleic Acids Res. 8*, 845−854.
171. Weintraub, H., Cheng, P. F. & Conrad, K. (1986) *Cell 46*, 115−122.
172. DiMauro, E., Caserta, M., Negri, R. & Carnevali, F. (1985) *J. Biol. Chem. 260*, 152−159.
173. Hirose, S. & Suzuki, Y. (1988) *Proc. Natl Acad. Sci. USA 85*, 718−722.
174. Matsui, T. (1987) *Mol. Cell Biol. 7*, 1401−1408.
175. Levin, J. M. & Cook, P. R. (1981) *J. Cell Sci. 50*, 199−208.
176. Lorch, Y., LaPointe, J. W. & Kornberg, R. D. (1988) *Cell 55*, 743−744.
177. Solomon, M. J., Larsen, P. L. & Varshavsky, A. (1988) *Cell 53*, 937−947.
178. Han, M. & Grunstein, M. (1988) *Cell 55*, 1137−1145.
179. Berrios, M., Osheroff, N. & Fisher, P. (1985) *Proc. Natl Acad. Sci. USA 82*, 4142−4146.
180. Earnshaw, W. C., Halligan, B., Cooke, C. A., Heck, M. M. S. & Liu, L. F. (1985) *J. Cell Biol. 100*, 1706−1715.
181. Earnshaw, W. C. & Heck, M. M. S. (1985) *J. Cell Biol. 100*, 1716−1725.
182. Gasser, S. M., Laroche, T., Falquet, J., Boy de la Tour, E. & Laemmli, U. K. (1986) *J. Mol. Biol. 188*, 613−629.
183. Henderson, S. & Sollner-Webb, B. (1986) *Cell 47*, 891−900.
184. Grummt, E. I., Kuhn, A., Bartsch, I. & Rosenbauer, H. (1986) *Cell 47*, 901−911.
185. McStay, B. & Reeder, R. H. (1986) *Cell 47*, 913−920.
186. Davison, B. L., Egly, J. M., Mulvihill, E. R. & Chambon, P. (1983) *Nature 301*, 680−685.
187. Fire, A., Samuels, M. & Sharp, P. A. (1984) *J. Biol. Chem. 259*, 2509−2516.
188. Cai, H. & Luse, D. S. (1987) *J. Biol. Chem. 262*, 298−304.
189. Robertson, M. (1988) *Nature 336*, 522−524.
190. Brown, D. D. (1984) *Cell 37*, 359−365.
191. Wolffe, A. P. & Brown, D. D. (1988) *Science 241*, 1626−1632.
192. Tebb, G. & Mattaj, I. W. (1988) *EMBO J. 7*, 3785−3792.
193. Furuichi, Y. (1978) *Proc. Natl Acad. Sci. USA 75*, 1086−1090.
194. Salditt-Georgieff, M., Harpold, M., Chen-Kiang, S. & Darnell, J. E. (1980) *Cell 19*, 68−79.
195. Chen-Kiang, S., Nevins, J. R. & Darnell, J. E. (1979) *J. Mol. Biol. 135*, 733−752.
196. Osheim, Y. N., Miller, O. L. & Beyer, A. L. (1985) *Cell 43*, 143−151.
197. Birnsteil, M. L., Busslinger, M. & Strub, K. (1985) *Cell 41*, 349−359.
198. Reed, R., Griffith, J. & Maniatis, T. (1988) *Cell 53*, 949−961.
199. Verheijen, R., Van Venrooij, W. & Ramaekers, F. (1988) *J. Cell Sci. 90*, 11−36.
200. Sharp, P. A. (1987) *Science 235*, 766−771.
201. Maniatis, T. & Reed, R. (1987) *Nature 325*, 673−678.
202. Narayan, P. & Rottman, F. M. (1988) *Science 242*, 1159−1162.
203. Zeitlin, S., Parent, A., Silverstein, S. & Efstratiadis, A. (1987) *Mol. Cell Biol. 7*, 111−120.
204. Ogston, A. G. (1958) *Nature 181*, 1462.
205. Baker, W. K. (1968) *Adv. Genet. 14*, 133−169.
206. Grosschedl, R., Weaver, D., Baltimore, D. & Constantini, F. (1984) *Cell 38*, 647−688.

207. Magram, J., Chada, K. & Constantini, F. (1985) *Nature 315*, 338–340.
208. Townes, T. M., Lingrel, J. B., Chen, H. Y., Brinster, R. L. & Palmiter, R. D. (1988) *EMBO J. 4*, 1715–1723.
209. Kollias, G., Wrighton, N., Hurst, J. & Grosveld, F. (1986) *Cell 46*, 89–94.
210. Hammer, K., Krumlauf, R., Campen, S., Brinster, R. & Tilghman, S. (1987) *Science 235*, 53–58.
211. Allen, N. D., Cran, D. G., Barton, S. C., Hettle, S., Reik, W. & Surani, M. A. (1988) *Nature 333*, 852–855.
212. Guarente, L. (1988) *Cell 52*, 303–305.
213. Muller, M. M., Gerster, T. & Schaffner, W. (1988) *Eur. J. Biochem. 176*, 485–495.
214. Lewis, E. B. (1985) *Cold Spring Harbor Symp. Quant. Biol. 50*, 155–164.
215. Judd, B. H. (1988) *Cell 53*, 841–843.
216. Biggin, M. D., Bickel, S., Benson, M., Pirrotta, V. & Tjian, R. (1988) *Cell 53*, 713–722.
217. Cullen, B. R., Lomedico, T. & Ju, G. (1984) *Nature 307*, 241–245.
218. Emerman, M. & Temin, H. M. (1984) *Cell 39*, 459–467.
219. Emerman, M. & Temin, H. M. (1986) *Nucleic Acids Res. 14*, 9381–9396.
220. Emerman, M. & Temin, H. M. (1986) *Mol. Cell Biol. 6*, 792–800.
221. Corbin, V. & Maniatis, T. (1989) *Nature 337*, 279–282.
222. Hawley, R. G., Covarrubias, L., Hawley, T. & Mintz, B. (1987) *Proc. Natl Acad. Sci. USA 84*, 2406–2410.
223. Bingham, P. M. & Zachar, Z. (1985) *Cell 40*, 819–825.
224. Proudfoot, N. J. (1986) *Nature 322*, 562–568.
225. Weintraub, H. (1988) *Proc. Natl Acad. Sci. USA 85*, 5819–5823.
226. Temin, H. M. (1982) *Cell 28*, 3–5.
227. Benz, E. W., Wydro, R. M., Nadal-Ginard, B. & Dina, D. (1980) *Nature 288*, 665–669.
228. Brinster, R. L., Allen, J. M., Behringer, R. R., Gelinas, R. E. & Palmiter, R. D. (1988) *Proc. Natl Acad. Sci. USA 85*, 836–840.
229. Brewer, B. J. (1988) *Cell 53*, 679–686.
230. McCready, S. J., Godwin, J., Mason, D. W., Brazell, I. A. & Cook, P. R. (1980) *J. Cell Sci. 46*, 365–386.
231. McCready, S. J. & Cook, P. R. (1984) *J. Cell Sci. 70*, 189–196.
232. Harless, J. & Hewitt, R. R. (1987) *Mutat. Res. 183*, 177–184.
233. Mullenders, L. H. F., van Leeuwen, A. C. van K, van Zeeland, A. A. & Natarajan, A. T. (1988) *Nucleic Acids Res. 16*, 10607–10622.
234. Mellon, I., Spivak, G. & Hanawalt, P. C. (1987) *Cell 51*, 241–249.
235. Edenberg, H. J. & Huberman, J. A. (1975) *Annu. Rev. Genet. 9*, 245–284.
236. McKnight, S. L., Sullivan, N. L. & Miller, O. L. (1976) *Prog. Nucleic Acid Res. Mol. Biol. 19*, 313–318.

19th Sir Hans Krebs Lecture

Engineering of protein bound iron-sulfur clusters
A tool for the study of protein and cluster chemistry and mechanism of iron-sulfur enzymes

Helmut BEINERT[1,2] and Mary Claire KENNEDY[1]

[1] Department of Biochemistry and [2] National Biomedical ESR Center, Medical College of Wisconsin, Milwaukee, Wisconsin, USA

(Delivered at the 19th FEBS Meeting in Rome, July 2, 1989) — EJB 89 1060

An increasing number of iron-sulfur (Fe-S) proteins are found in which the Fe-S cluster is not involved in net electron transfer, as it is in the majority of Fe-S proteins. Most of the former are (de)hydratases, of which the most extensively studied is aconitase. Approaches are described and discussed by which the Fe-S cluster of this enzyme could be brought into states of different structure, ligation, oxidation and isotope composition. The species, so obtained, provided the basis for spectroscopic and chemical investigations. Results from studies by protein chemistry, EPR, Mössbauer, ^{1}H, ^{2}H and ^{57}Fe electron-nuclear double resonance spectroscopy are described. Conclusions, which bear on the electronic structure of the Fe-S cluster, enzyme-substrate interaction and the enzymatic mechanism, were derived from a synopsis of the recent work described here and of previous contributions from several laboratories. These conclusions are discussed and summarized in a final section.

It seems almost more than a coincidence that in this lecture, in honor of Sir Hans Krebs, our main subject will be one of the enzymes of the tricarboxylic acid cycle, as Sir Hans used to call it. The senior author had the privilege of knowing Sir Hans personally. An incident comes back to mind that occurred during a visit of his to Madison some 20 years ago. It was during the days of unrest on campuses, when students frequently challenged the wisdom of the established older generation. In the question period following Sir Hans' lecture a student asked him what he thought was the most important thing to do today. Somewhat surprised, Sir Hans looked through his heavy glasses and countered with a slight smile: "If you ask me: of course, it is doing biochemistry. We hope we can convey some of this spirit in this lecture."

In today's biochemical research we see much emphasis on and interest in protein engineering, an expression unknown just a decade ago. We would like here to draw attention to an activity we might call 'metal-cluster engineering' of protein-bound clusters, which has occupied us for the past few years.

It probably has become fairly common knowledge that iron-sulfur (Fe-S) proteins may not only function as electron carriers, as e.g. the well known ferredoxins, but may be directly involved in catalytic processes through their Fe-S clusters. Examples of such proteins are hydrogenases [1] and (de)hydratases [2–8] of the type of aconitase [9]. However, details of the work leading to the concept of 'Fe-S enzymes', as we propose to call these proteins, are less well known. Since these aspects are equally interesting from the standpoint of Fe-S cluster chemistry as well as protein and enzyme chemistry, we would like to present here a survey of them.

The protein that has been most useful to us in gathering such information is citrate(isocitrate) hydro-lyase, or more commonly referred to as aconitase (Fig. 1). The enzyme, discovered by Martius in 1937 [10], now has a half-century-old history. However, the fact that it is an Fe-S protein has only been known since 1972 [11] and was not fully appreciated until the 1980s [9, 12]. In the following we will briefly describe some of the approaches, which we have called cluster engineering above, as well as experiments, mostly spectroscopic, which these approaches have allowed us to perform. Finally, we will present the conclusions which we think we can draw from this work.

CLUSTER FORMS AND NOMENCLATURE

For purposes of orientation, we introduce at this point the species of Fe-S clusters that can be bound to aconitase and the nomenclature used below. The enzyme has so far only been observed to bind a single Fe-S cluster. When the enzyme is purified aerobically beyond 80% purity this cluster is in the [3Fe-4S]$^+$ state and the enzyme is inactive [13, 14]. The inactive enzyme can be converted to the active [4Fe-4S]$^{2+}$ form (activation) by addition of Fe under reducing conditions or simply by reduction, in which case the Fe is mobilized from decaying clusters [13] (Fig. 2). The reduced 3Fe form [3Fe-4S]^0, is therefore very prone to conversion to the active [4Fe-4S]$^{2+}$ state. This active 4Fe form can be reduced by one electron to the [4Fe-4S]$^+$ state, which has $\approx 30\%$ of the activity of the $2+$ form [15] and binds substrate and analogs tightly, as does the $2+$ form. This is in contrast to the [3Fe-4S]$^+$ form which has no activity and binds substrate at best weakly, as it also binds other divalent and trivalent anions. A scheme illustrating these interconversions is shown in Fig. 3. Note that only species

---

Correspondence to H. Beinert, Department of Biochemistry, Medical College of Wisconsin, 8701 Watertown Plank Road, Milwaukee, Wisconsin 53226, USA

Abbreviation. ENDOR, electron-nuclear double resonance.

Enzymes. Citrate(isocitrate) hydro-lyase, aconitase (EC 4.2.1.3).

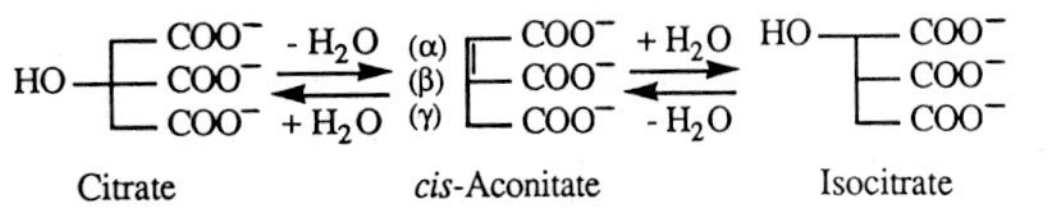

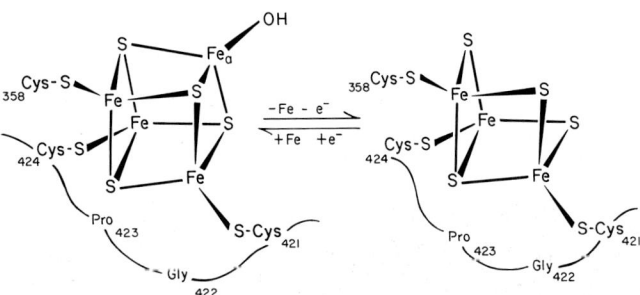

Fig. 1. *Scheme describing the aconitase reaction.* Note that aconitase recognizes the portion of substrate derived from oxaloacetate (α and β carbons) and acetyl-CoA (γ carbon) and that any one substrate, in the presence of active enzyme, is converted to an equilibrium mixture of the three substrates

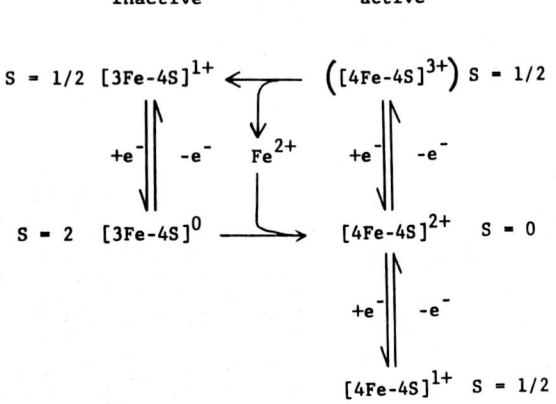

Fig. 2. *Schematic description of the interconversion of the 4 Fe and 3 Fe cluster of aconitase.* Fe_a is the labile iron. The numbers identifying the Cys residues are not those used by Robbins and Stout [16, 17], but are lower by one unit according L. Zheng, P. C. Andrews, M. A. Hermodson, J. E. Dixon and H. Zalkin (*J. Biol. Chem.*) who have now identified the N-terminus of the peptide chain

Fig. 3. *Scheme showing the relationship between various cluster forms of aconitase excluding enzyme-substrate complexes and the linear form.* For each cluster type the oxidation state (i.e. the charge balance of the cluster core) and the spin state are indicated. For $[3Fe-4S]^+$ the EPR signal is centered at $g = 2.01$ and for $[4Fe-4S]^+$ (substrate-free) the g values are 2.06, 1.93, 1.86; $[3Fe-4S]^0$ has a broad signal between 0 and 50 mT at 9.2 GHz similar to that found for other substances in the $S = 2$ state. The existence of a $[4Fe-4S]^{3+}$ cluster has not been demonstrated but there is evidence that oxidation precedes release of Fe [13, 15]

with spin $S > 0$ can have EPR signals. The spin states and g values are also given in Fig. 3. On exposure of the enzyme in the $[3Fe-4S]^+$ form to pH 9–10.5 the 'linear' cluster is formed nearly quantitatively [18]. It is also seen on denaturation with urea [18, 19] or guanidine hydrochloride [20]. The conversions that are possible involving the linear cluster are shown in Fig. 4.

In contrast to the $[3Fe-4S]^{+,0}$ species, which we call inactive or the 3Fe enzyme, we refer in this paper to apoenzyme as a species devoid of cluster iron, although the other cluster constituents may still be present.

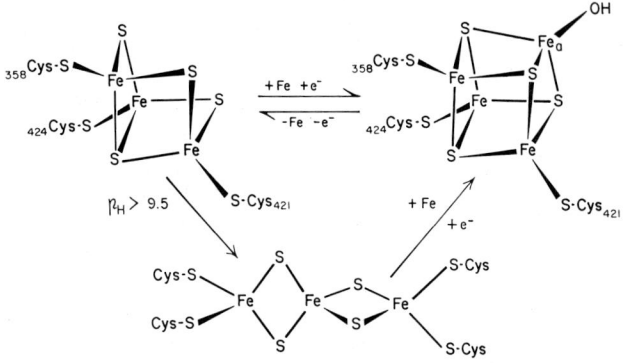

Fig. 4. *Schematic description of the interconversions involving the 'linear' 3Fe cluster* [18]. Concerning the numbers identifying the Cys residues see legend to Fig. 2 and the text. The Cys residues bound to the linear cluster are those at positions 250, 257, 421 and 424. Their precise disposition at the cluster is not known (cf. [20])

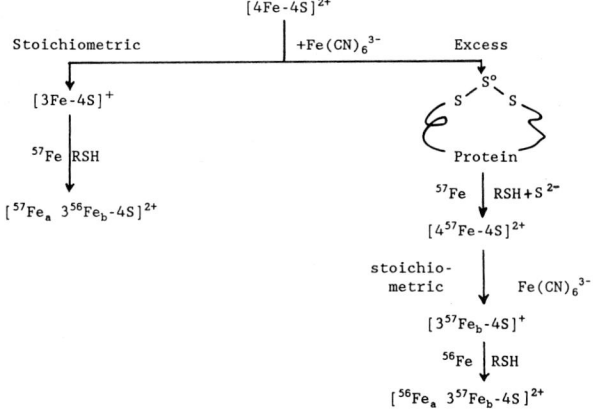

Fig. 5. *Scheme describing preparation of specifically ^{57}Fe-labeled Fe-S cluster of aconitase*

SUBSTITUTIONS OF CLUSTER IRON AND SULFIDE. APOENZYME

While labeling through replacement by direct exchange is possible for both iron [13] and sulfide [21] it is not very effective when a stable isotope is to be highly enriched for certain spectroscopies. It is however, convenient for the incorporation of a radiolabel where partial substitution is satisfactory.

One of the most obvious routes to specifically metal-labeled clusters is offered by the fact that only one specific Fe (called Fe_a) of the originally present $[4Fe-4S]^{2+}$ cluster is lost during the routine preparation of the enzyme [13]. Thus, it becomes possible, by insertion of ^{57}Fe into the vacant fourth position, to observe exclusively this Fe_a by Mössbauer spectroscopy which detects only ^{57}Fe (Fig. 5). Spectroscopic data show that, when aconitase is activated at pH 7–8.5 with a slight excess of Fe, no significant scrambling of cluster irons occurs [22].

To achieve a different distribution of label, more complicated routes have to be pursued, with the next most obvious labeling being $4^{57}Fe$ or $^{56}Fe_a\,3^{57}Fe_b$. In attempts to accomplish this, the elaboration, in good yield, of a stable apoenzyme was a crucial step. Experiments that followed the protocols found useful for smaller and more sturdy proteins such as ferredoxins, involving acidification or mercurial treatment [23], led to ill-defined products which contained relatively little viable apoprotein. Since the 3Fe protein is readily produced from the native 4Fe form by addition of stoichiometric amounts of hexacyanoferrate (III), whereas on overtitration the clusters are lost altogether, we became curious whether this might not be a useful route to the preparation of an apoenzyme. Oxidation by an excess of hexacyanoferrate (III), in the presence of excess EDTA, indeed leads to an air-stable apoenzyme in minutes. Analysis indicates that the SH ligands of the cluster are now tied up in disulfide, trisulfide or tetrasulfide structures and are thus preserved together with part of the original labile sulfide of the cluster, which has been converted to S^0 [24]. In this way a protein structure suitable for reconstitution seems to be maintained so that in the presence of Fe, S^{2-} and dithiothreitol the apoprotein is readily converted to the active form $[4Fe-4S]^{2+}$ in $\approx 70\%$ yield, as measured by its enzymatic activity. Interestingly, it is important during reconstitution to add Fe with or before dithiothreitol [24]; otherwise once the disulfides, trisulfides or tetrasulfides are reduced (and if Fe is not present for immediate cluster formation) the protein may unfold in an uncontrolled fashion such that sulfhydryls and sulfides become engaged in other interactions. This procedure allows us to introduce Fe isotopes and perhaps in the future other metals of our choice into all four positions of the cluster. By oxidative removal of Fe_a and activation to the 4Fe form with ^{56}Fe we can readily produce a form of cluster, Mössbauer or electron-nuclear-double-resonance (ENDOR)-silent in the Fe_a position, but with detectable Fe_b sites (Fig. 5). Scrambling of cluster irons has never been observed under these conditions.

Labeling of sulfide, e.g. with ^{34}S or ^{33}S, is also readily accomplished starting with the apoenzyme prepared by oxidative cluster destruction. However, for efficient labeling an excess of the labeled component is required to swamp out the sulfur (S^0) bound in the apoprotein. Excess sulfide, however, destroys clusters by extraction of iron, unless iron is added stoichiometric with the sulfide. If this is carried out, iron sulfide species are formed which are hard or impossible to separate from the protein. Plans to eventually produce a selenoprotein also called for a more efficient method of reconstitution. It should be mentioned here that Petering et al. [25] had observed 15 years ago that on oxidation of spinach ferredoxin, the labile sulfide of the cluster was incorporated in a trisulfide linkage in the protein. The iron was partly bound to non-cluster sites on the protein from which it could be removed by chelators. Under reducing conditions, the ferredoxin could be reconstituted. This procedure has apparently never been used for preparative purposes. A suggestion to attempt this is found in a recent review by Thomson [26].

THE CYANOLYZED APOPROTEIN

As mentioned above, the apoenzyme prepared by oxidative cluster destruction has most ($\approx 75\%$) of its original S^{2-} preserved in the form of S^0. Even if an excess of the desired isotope was used in reconstitution, attempts to substitute Se^{2-} for S^{2-} might well be thwarted in case the enzyme showed preference for S^{2-}, as do other Fe-S proteins [27, 28], so that S and Se hybrids would be formed. Likewise, for labeling with sulfur isotopes in high yield the $^{32}S^0$ present in the apoprotein is a disadvantage. A known procedure for removal of S^0 is cyanolysis [29]. It is indeed possible to cyanolyze the apoprotein without serious losses so that $\geq 90\%$ of its S^0 is removed with preservation of the disulfides in which

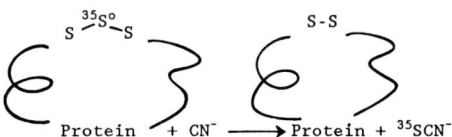

Fig. 6. *Schematic description of removal of S^0 from apoaconitase*

the S^0 had been incorporated during oxidative degradation (Fig. 6). With this cyanolyzed apoprotein incorporation of $^{34}S^{2-}$, $^{33}S^{2-}$ and Se^{2-} is much more expedient, since only a minimal excess of the isotope is required and thus formation of large quantities of undesirable sulfides or selenides can be avoided. In this fashion, the 4Fe selenoprotein could be readily obtained and, by its oxidation with hexacyanoferrate (III), the 3Fe selenoprotein. One of the great surprises was that the 4Fe selenoprotein was twice as active in our routine assay as the Fe-S protein. Detailed studies of this effect have yet to be made.

LIGANDS TO THE CLUSTER

In the context of the work discussed above it was of great interest which of the 11 or 12 Cys residues [30, 31], or other amino acids of the protein, furnished the cluster ligands. According to Plank et al. and Plank and Howard [20, 31] Cys421, Cys424 (peptide 7) and Cys358 (peptide 3) are the cluster ligands in agreement with the data derived from the structure as determined by X-ray diffraction on the $[3Fe-4S]^+$ form [16]. Thus there appear to be only three Cys ligands. The $[4Fe-4S]^{2+}$ form, prepared by soaking a crystal of the 3Fe form with a solution containing iron and dithionite, also shows only three Cys ligands [17]. The question, therefore, arises, which Cys residues form the two disulfide bonds which we observed in the apoenzyme, namely those that apparently stabilized the apoenzyme in a conformation suitable for reconstitution? Plank et al. and Plank and Howard [20, 31] find Cys383 (peptide 2) involved in a disulfide bond in addition to those listed above. This would require a substantial rearrangement of the protein structure determined by Robbins and Stout [16, 17].

PURPLE ACONITASE: THE 'LINEAR' CLUSTER

Related to the considerations just presented are observations made on the so-called purple aconitase, a form of the enzyme that is obtained on exposure of the 3Fe enzyme to pH 9 – 10.5 or on addition of urea [18] before complete cluster destruction sets in. Purple aconitase was shown to be a $[3Fe-4S]^+$ form of a structure different from the cubane-type 3Fe form. The cubane-type 3Fe cluster is derived from the cubic 4Fe form by removal of one Fe atom (Fig. 2). Purple aconitase has a linear arrangement of Fe atoms, somewhat like an extended [2Fe-2S] cluster (Fig. 4). From the striking similarity of the Mössbauer, EPR, magnetic circular dichroism, resonance Raman and electronic spectra of purple aconitase to those of the model $[3Fe-4S]^+$ linear cluster described by Hagen et al. [32], it seems almost certain that purple aconitase has four Cys ligands as does the synthetic model [32]. This has recently been supported by protein-chemical approaches [20]. The question then again arises: which Cys residues are the ligands of the linear cluster? According to the work in [20] these are Cys421 and Cys424 (peptide 7) of the cubane-type cluster, as expected, and in addition the two Cys residues at

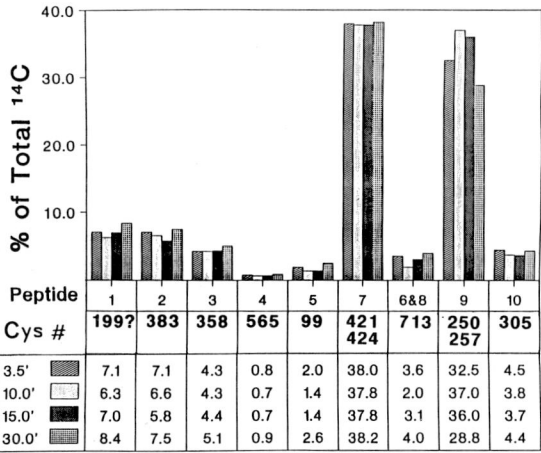

Fig. 7. *Labeling pattern of Cys residues of purple aconitase (i.e. containing the linear 3Fe cluster) first exposed in 6M guanidine (pH 8) to unlabeled iodoacetate for the times indicated and thereafter reduced by dithiothreitol and carboxymethylated with radiolabeled iodoacetic acid.* The cysteinyl-tryptic peptides were mapped and analyzed for radioactivity. It has been demonstrated previously that the cysteine-containing peptides 6 and 8 represent the same cysteine [31]. Therefore, the labeling of these peptides is expressed as one value; 6 and 8. The exposure times to the label are indicated by shading of the bars from left to right (3.5 min, 10 min, 15 min and 30 min, respectively) and the numbers directly under the bars indicate the peptides in which the label was found. The numbers in the next lower row indicate the positions of cysteines in the amino acid sequence (see legend to Fig. 2); from Plank et al. [20]

positions 250 and 257 of peptide 9 (Figs 4 and 7) which are located six residues apart on the same side of a helix which is in close proximity in domain 2 [16, 17]. Thus, obviously, on formation of the linear cluster a rather different mode of ligation of the cluster comes into play. According to simulations by computer graphics the necessary rearrangement of the protein is entirely feasible (Stout, C. D., personal communication). What we find remarkable here is not that under denaturing conditions sulfhydryl exchange occurs at the Fe-S cluster (this has been observed previously [33]), but that a unique and stable new cluster type should be formed almost quantitatively. It is also of interest in this context that Martin et al. [34] have shown, by replacement of one of the Cys ligands to the [4Fe-4S] cluster of ferredoxin I from *Azotobacter vinelandii* by Ala, that the Fe-S cluster is nevertheless formed utilizing another, originally more distant Cys residue as the fourth ligand.

The purple enzyme has no aconitase activity, but it is surprising that $\approx 60\%$ of the activity can be regained on anaerobic reduction of the purple form in the presence of iron, indicating that a large fraction of the enzyme rebounds to its native structure and mode of cluster ligation. It is also worth mentioning here that the linear cluster of aconitase seems to be the most stable 3Fe cluster, although the enzyme is inactive and the protein is more prone to denaturation than that bearing the cubane 3Fe cluster. The remarkable stability of the linear cluster is in keeping with the experience in R. H. Holm's laboratory [35], namely that a cubane type [3Fe-4S] cluster with RS^- ligands could not be synthesized, whereas the linear form was readily obtained in such attempts.

It becomes obvious from these observations that only under the influence of the protein is the arrangement of the cluster directed in favor of the cubane-type cluster. Our previous report that there are several more reactive Cys residues in purple aconitase than in the native enzyme [18] is also in keeping with the observation that the protein containing the linear cluster is less stable and probably partly unfolded. All Cys residues that are not cluster ligands are now exposed, with Cys99 (peptide 5), Cys305 (peptide 10), Cys358 (the original cluster ligand in peptide 3) and Cys713 (peptide 6 or 8) [20] becoming most reactive. Native aconitase (3Fe or 4Fe) has only a single reactive Cys (Cys565 in peptide 4). Interestingly, despite its demonstrated vicinity to the cluster [16, 17, 36, 37], this particular Cys residue never seems to become a cluster ligand, as we have shown by covalently blocking this group. Blocking this Cys residue does not prevent insertion of the fourth Fe into the [3Fe-4S] cluster during the [3Fe-4S] to [4Fe-4S]$^{2+}$ conversion, nor does it interfere with the formation of the [3Fe-4S]$^+$ (linear) from the [3Fe-4S]$^+$ (cubane) cluster [37]. We would also like to recall here our previous experience [18] that, so far, we have not been able to prepare a stable reduced [3Fe-4S]^0 (linear) form. On reduction, as mentioned above, the [4Fe-4S]$^{2+}$ enzyme is formed concomitant with breakdown products containing what appear to be [2Fe-2S]$^+$ fragments [18].

SPECTROSCOPY

We have described in the foregoing sections possibilities and obstacles in preparing various free and ligated forms of aconitase. We shall now briefly summarize the most significant observations that the available forms and their specific labeling have allowed us to make. We restrict ourselves below to observations by Mössbauer, EPR and ENDOR spectroscopies.

Mössbauer spectroscopy

By far the most informative technique as far as the cluster structure is concerned has been Mössbauer spectroscopy, since it looks directly at the Fe nuclei and allows us to discriminate between individual Fe atoms unless they are in an identical state; but even then the number of equivalent atoms can usually be estimated. Mössbauer spectroscopy has been instrumental in deciphering the complexity of the behavior of aconitase, as it is schematically depicted in Figs 2−4 [18, 22, 38], a complexity unrecognized until eight years ago. For details, the original reports will have to be consulted. In addition to this contribution of Mössbauer spectroscopy three observations stand out.

(1) While the Fe atoms of the cluster remain spin-coupled [22, 38], states with significantly different valence localization at individual iron atoms arise during the following reactions. (a) Reduction: [3Fe-4S] $\xrightarrow{e^-}$ [3Fe-4S]^0, one Fe$_b$ remains unchanged ferric, the two others share the electron (see Fig. 8) [38]. (b) Reduction: [4Fe-4S]$^{2+}$ $\xrightarrow{e^-}$ [4Fe-4S]$^+$, Fe$_{b2}$ and Fe$_{b3}$ form a delocalized Fe(II)−Fe(III) pair, Fe$_a$ and Fe$_{b1}$ show ferrous character [38]. (c) Substrate addition: [4Fe-4S]$^{2+,+}$ + substrate → [4Fe-4S]$^{2+,+}$ · substrate. For 2+, Fe$_a$ shows increased ferrous character; Fe$_{b2,3}$ remain unchanged (see Fig. 9), for 1+, Fe$_a$ assumes strong ferrous character, Fe$_{b2,3}$ changes little and Fe$_{b1}$ is intermediate (see Table 1) [22, 38].

The isomer shift (δ), a parameter measured in Mössbauer spectroscopy (in analogy to the chemical shift in NMR) has typical ranges of values for divalent and trivalent iron and for tetrahedral or octahedral coordination. Such values are assembled in Tables 1 of [39] and [40]. Typically, for [4Fe-4S]$^{2+}$ clusters values between 0.41 mm/s and 0.47 mm/s are found at 4.2 K. They increase by 0.10−0.12 mm/s on reduction to the 1+ state. For the 2+ state of [2Fe-2S] clusters

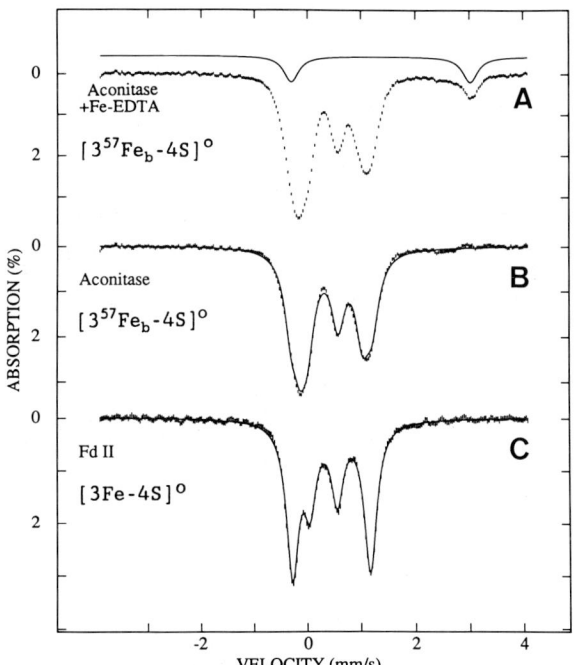

Fig. 8. *Mössbauer spectra of the [3Fe-4S]^0 cluster of aconitase and ferredoxin II of Desulfovibrio gigas.* EDTA has to be added to prevent conversion to the [4Fe-4S]$^{2+}$ form, which results in the formation of some ferrous EDTA (A). The contribution of ferrous EDTA is removed in (B). There are two quadrupole doublets in a 2:1 ratio, indicating the presence of one Fe of ferric character and two Fe of more ferrous character that share the extra electron. The low-velocity limbs of the doublets overlap for aconitase but are clearly separated in the analogous spectrum of ferredoxin II (C), which is shown in order to explain the interpretation of the aconitase spectra

Table 1. *Mössbauer isomer shifts of Fe in Fe-S clusters*
C. pasteurianum, Clostridium pasteurianum; Ch. vinosum, Chromatium vinosum; HiPIP, high-potential iron protein; Fd, ferredoxin; sub., substrate

[4Fe-4S] cluster charge	Protein	Isomer shift
		mm/s
2+	C. pasteurianum Fd	0.44
2+	Ch. vinosum HiPIP	0.44
2+	aconitase	0.45 (Fe$_a$, Fe$_{b1-3}$)
2+	aconitase + sub.	0.84–0.89 (Fe$_a$)
		0.45 (Fe$_{b1-3}$)
+	C. pasteurianum Fd	0.58
+	aconitase	0.65 (Fe$_a$)
+	aconitase + sub.	1.00 (Fe$_a$)
		0.64 (Fe$_{b1}$)
		0.49 (Fe$_{b2,3}$)

or ferric rubredoxin the δ values lie around 0.25 mm/s. For comparison with these values, we list in Table 1 δ values observed with aconitase under various conditions. Values of the magnitude measured for the substrate complex of the [4Fe-4S]$^{2+}$ and [4Fe-4S]$^+$ clusters of aconitase have never been reported before for 4Fe clusters. They fall into the range of isomer shifts found for high-spin ferrous complexes [22, 40, 41] for which δ lies between 0.7 mm/s and 1.4 mm/s. It is therefore clear that the cluster sites for which these values are measured, have acquired distinct ferrous character. Table 1 also shows that the isomer shifts of the four iron sites differ greatly from each other, indicating the formation of localized or 'trapped' valence [42, 43]. It is important to note that in spite of localization of valence the three or four iron atoms of the three- or four-iron clusters, respectively, remain spin-coupled as shown by Mössbauer spectroscopy in a magnetic field. This pronounced localization of valence in structures such as Fe-S clusters, which we generally think of as quite symmetrical (see, however [44]), can only arise under the influence of the protein, exerted by the particular ligand environment that it furnishes and through conformational constraint. The influence of added ligands such as substrates or inhibitors must also be considered; but this again is directed by the protein environment.

(2) As pointed out under (1), on addition of substrate to the [4Fe-4S] forms the most dramatic change occurs at Fe$_a$, the iron that is labile toward oxidation. The Mössbauer spectra for this site are no longer compatible with a tetrahedral environment of iron. It was therefore concluded on the basis of observation of model compounds with pentacoordinated iron [45, 46] and other iron proteins of higher coordination number [22, 40] that an expansion of the coordination sphere of Fe$_a$ takes place to pentacoordination or hexacoordination [22]. It is also important to note that according to the almost twofold increase in the quadrupole splitting and isomer shift of Fe$_a$, this Fe atom has now attracted the highest d-electron density of the cluster irons. From the standpoint of the catalytic mechanism of the enzyme it is of interest that this electron density may in part have been acquired from the bound substrate, inasmuch as the other three cluster Fe atoms (Fe$_b$) show little change (Fig. 9) [22, 38]. We must remember, however, that the sulfur atoms of the cluster also represent a significant reservoir of electron density [47] which we cannot observe with Mössbauer spectroscopy.

(3) On addition of any one of the three substrates (cf. Fig. 1) to [4Fe-4S]$^{2+}$ not only do the events described under (1) and (2) above occur, but at least two new species arise [22], a phenomenon most likely related to the fact that we are dealing with three substrates that are readily equilibrated by the enzyme. The exact distribution of bound species is not known. On rapid mixing at 0 °C with the different substrates, freeze quenching and observation by Mössbauer spectroscopy, different mixtures of Fe$_a$ species are seen (Fig. 10) [38].

It must be kept in mind, however, in considering these results as well as those described for EPR and ENDOR below, that the observations have been made at low temperature ≥ 2 K) and may therefore not precisely describe states of the enzyme at room temperature in liquid solution [48]. However, they do show capabilities of the system at the low temperature limit, which are not likely to be entirely unrelated to those applying in liquid solution. Unfortunately, at this time, we have no choice but using extreme conditions if we want to use some of the most penetrating techniques at all, such as Mössbauer, ENDOR or magnetic circular dichroism spectroscopy.

EPR spectroscopy

EPR has been, as in most studies of this nature, an invaluable tool in trailblazing ahead of the more informative but also more demanding and laborious techniques, and also in confirming or refuting hints arising from those techniques. In contrast to what we can see by Mössbauer spectroscopy, only the various 3Fe states and the [4Fe-4S]$^+$ state [9, 15, 22] are

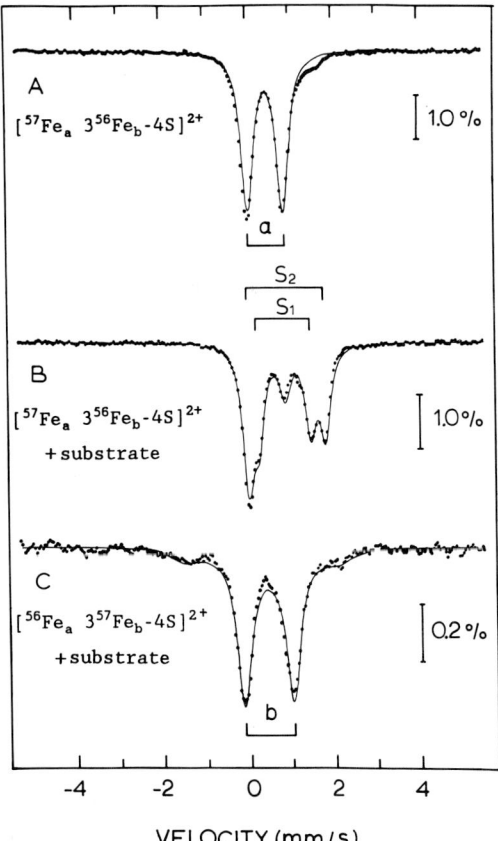

Fig. 9. *Mössbauer spectra of the [4Fe-4S]$^{2+}$ cluster of aconitase without (A) and with (B and C) substrate.* For (A) and (B), Fe$_a$ was labeled with ^{57}Fe, for (C) the Fe$_b$ sites were labeled with ^{57}Fe and Fe$_a$ contained pure ^{56}Fe. After substrate addition (B) two new doublets appear with increased isomer shift (see also Table 1) and quadrupole splitting. A minor fraction of Fe$_a$ remains unchanged (note the central small peak). These features may be related to the fact that the enzyme may be frozen in with anyone of the three substrates bound or in a state, when substrate is being exchanged and temporarily no substrate happens to be bound. The b sites (C) show no change on addition of substrate (from [38])

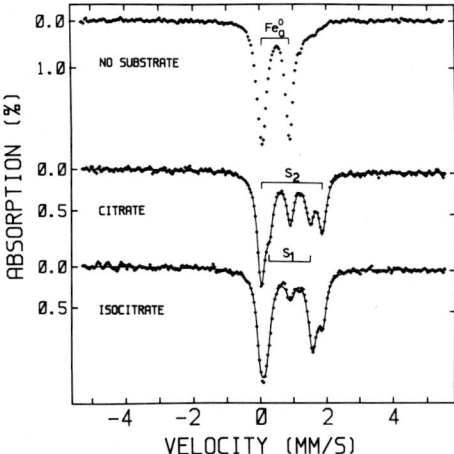

Fig. 10. *Mössbauer spectra of the [4Fe-4S]$^{2+}$ cluster of aconitase after rapid freezing 35 ms after mixing with 2.5 mM citrate or isocitrate at 0°C.* Only Fe$_a$ was labeled with ^{57}Fe. The features are qualitatively the same as in Fig. 9, but it can be seen that the intensities of the newly appearing doublets (S$_1$ and S$_2$) initially have different proportions depending on the particular substrate added (from [38])

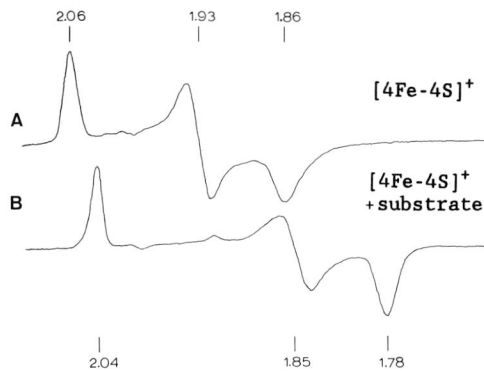

Fig. 11. *EPR spectra of the [4Fe-4S]$^+$ cluster of aconitase in the absence (A) and the presence (B) of substrate.* Prominent features of the spectra are marked of a g-value scale. Prior to freezing, the samples were photoreduced anaerobically in the presence of deazaflavin and oxalate. The spectra were recorded at a frequency of 9.24 GHz, 1 mW power and 13 K

detectable. With these limitations, the conversions between different cluster forms can be readily followed by EPR [15]. The effect of substrate binding was first discovered by EPR with the [4Fe-4S]$^+$ form (Fig. 11) [15]. Similarly, the strongly bound nitro analogs of citrate and isocitrate [49], and fluorocitrate change the EPR spectrum of [4Fe-4S]$^+$ significantly and even the less potent inhibitor *trans*-aconitate produces its own specific signature in EPR. EPR, however, as a technique tracing electron spins provided a tool to glean information not only on the cluster *per se* but also on the nature of the substrate-derived ligands, which may share spin density with the cluster components. While such work had to rely on observation of line broadening by hyperfine interactions [22], since hyperfine structure remained unresolved, it formed the basis for proceeding to the much more penetrating approach by ENDOR spectroscopy.

ENDOR spectroscopy

The superior resolution of hyperfine splitting provided by ENDOR has then made it possible to inquire into the nature of the non-protein ligands bound to Fe$_a$ and of the actual sites of binding on the substrate molecules. In ENDOR spectroscopy involving a nucleus of nuclear spin $I = 1/2$, the simplest case, two lines (a doublet) are expected. Their position in the spectrum is determined by the resonant frequency v_0 of the nucleus and the electron-nuclear coupling constant, A. The fact that v_0 for the various magnetic nuclei is known, greatly facilitates the assignment of the observed lines. In many cases the intensities of the two lines are not equal so that occasionally only one of the lines can be resolved. With nuclei of $I > 1/2$ there is additional (quadrupolar) splitting of each of these lines into $2 \times I$ lines, as it is seen for ^{17}O ($I = 5/2$) [50, 51].

^{1}H and ^{2}H ENDOR

With ^{1}H and/or ^{2}H ENDOR, it is in principle possible to obtain direct information not only about protons on cluster ligands from the protein but also on other nearby groups of the protein and on substrates or analogs. Furthermore, the exchangeability of protons in these groups or molecules and the ligation of solvent to the paramagnetic center can be probed. In addition to varying the medium between ^{1}H$_2$O and

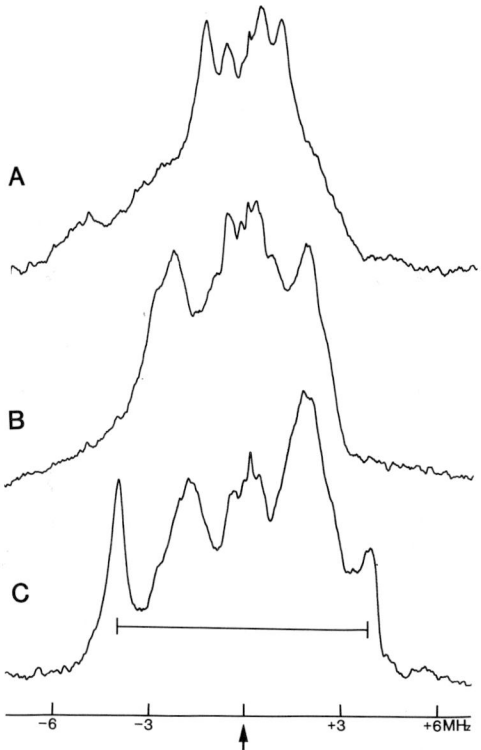

Fig. 12. ^{1}H-ENDOR spectra of the $[4Fe-4S]^+$ cluster of aconitase at g_{min} without substrate (A), with perdeuterated cis-aconitate (B) and with ordinary citrate (C). For (A) and (B) the solvent was ^2H_2O; for (C) it was ^1H_2O. Aconitase was photoreduced as for Fig. 11. The conditions of spectroscopy were as follows: microwave and modulation frequencies, 35.4 GHz and 100 KHz, respectively; modulation amplitude, 0.16 mT; r.f. scanning rate, 0.5 MHz/s; time constant, 0.128 s; H_0 for (A), (B) and (C) were 1.359 T, 1.4188 T and 1.4180 T, respectively. Temperature, 2 K

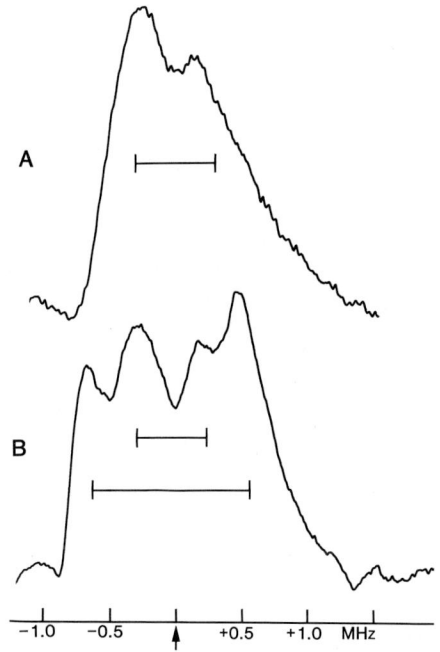

Fig. 13. ^{2}H-ENDOR spectra of the $[4Fe-4S]^+$ cluster of aconitase in the absence (A) and presence (B) of substrate. The other conditions were as for Fig. 12A with the following exceptions: plain substrate was added to (A); the modulation amplitude was 0.063 mT; r.f. scanning rate, 3 MHz/s; time constant, 0.03 s; H_0 for (A) and (B) were 0.1356 T and 1.4188 T, respectively

^2H_2O we used in these experiments cis- and trans-aconitate perdeuterated in the carbon-bound hydrogen positions. With ^{1}H ENDOR of enzyme in ^2H_2O and in absence of substrate, four different protons or sets of equivalent protons could be distinguished with widely varying coupling strengths (A = 1.2 MHz, 2.3 MHz, 5.0 MHz and 8.0 MHz). Some of the respective signals probably originate from protons of the Cys ligands; they could, however, also be due to protons of His101, His147 and His167 which are located in the active site. ^{1}H ENDOR also showed that there are significant changes in the interaction of the protons on the protein with the $[4Fe-4S]^+$ cluster of aconitase when substrate is bound (Fig. 12). Corresponding changes on substrate addition were found by Mössbauer spectroscopy [22] and ^{57}Fe ENDOR (see below). With deuterated cis-aconitate and enzyme in ^2H_2O, there was no change in the ENDOR spectrum as compared to that recorded with normal cis-aconitrate and enzyme in ^2H_2O, indicating that non-exchangeable protons of the substrate are not represented in the ^{1}H-ENDOR spectrum of enzyme plus substrate. There is, however, a considerable difference between the ^{1}H-ENDOR signal of the enzyme with bound protonated substrate in ^1H_2O (Fig. 12C) and that in ^2H_2O (Fig. 12B). The principal difference is the presence of a signal from a strongly coupled proton (or set of equivalent protons) with A = 7.8 MHz. Because the hyperfine coupling is large, we assign this proton signal to a cluster-bound H_xO (x = 1, 2) species that we detected in previous ^{17}O-ENDOR studies [50].

We have used the combination of ^{1}H and ^{2}H ENDOR to characterize this solvent species. The ^{2}H-ENDOR signal obtained from ENDOR measurements of the enzyme-substrate complex in ^2H_2O shows two doublet signals corresponding to $A_1(^2H)$ = 1.2 MHz and $A_2(^2H)$ = 0.5 MHz (Fig. 13B). The coupling $A_1(^2H)$ = 1.2 MHz multiplied by the ratio of the nuclear g values $[g_n(^1H)/g_n(^2H)]$ predicts a hyperfine coupling for the exchangeable proton of $A_1(^1H)$ = 7.8 MHz, precisely that observed by ^{1}H ENDOR. The second coupling of $A_2(^2H)$ = 0.5 MHz corresponds to an exchangeable proton with a coupling $A_2(^1H)$ = 3.25 MHz; this signal is not readily detected in the ^{1}H ENDOR because it would strongly overlap with the non-exchangeable ^{1}H resonances. Because the solvent species that is interacting with the enzyme-substrate complex exhibits two distinct protons, detectable as deuterons in ^{2}H ENDOR, we assign it to a water molecule, H_2O.

Recently, ^{17}O-ENDOR measurements also have disclosed that the substrate-free enzyme binds a solvent species, H_xO. We have used ^{1}H and ^{2}H ENDOR to characterize this species also. ^{1}H ENDOR of substrate-free enzyme in ^2H_2O as compared to the corresponding sample in ^1H_2O shows loss of a doublet from an exchangeable proton with $A(^1H)$ = 4 MHz. The ^{2}H ENDOR shows a doublet with $A(^2H)$ = 0.6 MHz from the replacing deuteron (Fig. 13A). However, unlike the ^{2}H-ENDOR spectrum of the enzyme-substrate complex in ^2H_2O (Fig. 13B) the spectrum of the substrate-free enzyme is lacking a second ^{2}H doublet. This suggests that the solvent species in the substrate-free enzyme is a hydroxyl ion.

We consider this conclusion that solvent is already bound to the free enzyme the most significant positive result of the ^{1}H- and ^{2}H-ENDOR experiments. Robbins and Stout also suggest from the crystal structure that the fourth cluster ligand is a water molecule or hydroxyl group [16].

^{17}O and ^{13}C ENDOR

With the use of substrates or analogs labeled with ^{17}O or ^{13}C, we were able to identify the groups of the substrate bound to Fe_a [50, 51]. Specific labeling of various sites on the substrate with these isotopes can be accomplished on a small scale by the use of enzymatic reactions, when central metabolites such as citrate or isocitrate are involved, that are accessible by various metabolic routes. The conclusions from the ^{17}O- and ^{13}C-ENDOR experiments can be summarized as follows. At equilibrium, after addition of substrate and binding of a carboxyl group to the active $[4Fe-4S]^+$ enzyme, the OH^- ligand at Fe_a is protonated (see above) and the hydroxyl of the citrate or isocitrate is bound to Fe_a. We have not been able, though, to directly observe this bound hydroxyl after addition of substrate; however, we did see it with the nitro analog of isocitrate (1-hydroxy-2-nitro-1,3-propanedicarboxylate, referred to here as nitroisocitrate), in which the central (β) carboxyl is replaced by a nitro group [50, 51]. The hydroxyl of this nitro analog does not exchange with solvent. When this compound was added to the enzyme, we could observe ENDOR signals from bound $H_2^{17}O$ and $R^{17}OH$ simultaneously [50]. As we did not observe binding of the hydroxyl from substrate, we concluded that at equilibrium cis-aconitate is the prevailing form of substrate bound to the $[4Fe-4S]^+$ cluster so that ENDOR signals from the bound minor substrate species were not observable. The studies with specifically carboxyl-labeled substrates showed that only the central (β) carboxyl is bound to Fe_a (the α- and β-carboxyls are derived from oxaloacetate, the γ-carboxyl from acetyl-CoA) [51]. This now extends the picture just described to specify also the mode of binding of the substrate. It has been known for over two decades that the hydrogen which is removed from substrate and replaced by hydroxyl in the conversions between citrate and isocitrate does not readily exchange with the medium, whereas the hydroxyl does [52]. This hydrogen must therefore be held by the enzyme during a single turnover. It is also known that the addition of H^+ and OH^- to cis-aconitate is trans [53]. In order for both of these conditions to be fulfilled the enzyme must bind cis-aconitate in two discrete modes, the citrate and isocitrate modes, with different carbons (β or α) of the substrate attached to Fe_a via the hydroxyl bridge. Our finding that at equilibrium the central carboxyl (β) is bound, specifies that the bound cis-aconitate (see above) is predominantly held in the citrate mode. We could, however, show [51], again with the nitro analog of isocitrate, that the enzyme will bind the analog, and by inference also isocitrate or cis-aconitate, in the isocitrate mode, viz. at the α-carboxyl, if there is no choice of carboxyl; the nitro analog has no central (β) carboxyl but a nitro group at the β-carbon, the carbon that normally bears the carboxyl of citrate which binds to the cluster iron. With respect to the foregoing picture, there is again the limitation imposed by the inability of EPR and consequently of ENDOR to detect the active species $[4Fe-4S]^{2+}$ which has $S = 0$, so that the results strictly only apply to the $[4Fe-4S]^+$ form which exhibits about 30% of the activity of the 2+ form [15]. It seems, however, unlikely that this form of the enzyme, which binds substrate tightly, should follow a radically different pathway in catalysis. The equilibrium of bound species, however, may well be different for the $[4Fe-4S]^{2+}$ form.

^{57}Fe ENDOR

In addition to ENDOR spectroscopy involving solvent or substrates labeled with isotopic nuclei of $I > 0$ we have also

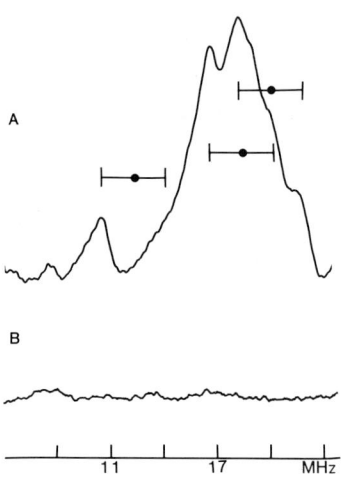

Fig. 14. ^{57}Fe-ENDOR spectra at $g_{min} = 1.85$ of the $[4Fe-4S]^+$ cluster of aconitase labeled with ^{57}Fe in the b sites (A) and with Fe of natural abundance in all sites (B), both in the presence of substrate. The conditions of reduction and spectroscopy were as in Fig. 12 with the following exceptions: microwave frequency, 34.75 GHz; modulation amplitude, 0.6 mT; r.f. scanning rate, 6 MHz/s; time constant, 0.032 s; H_0, 1.34 T

Table 2. *Hyperfine coupling constants for cluster irons in the $[4Fe-4S]^+$ cluster in the presence of substrate*

MB, Mössbauer. The Mössbauer values have been taken from [38]. Note that Mössbauer spectroscopy on polycrystalline samples of fairly isotropic compounds does not determine the relationship between the A and g tensor. The labeling of A values as x, y, z in [38] is therefore arbitrary. ENDOR, as carried out in the work reported, does however relate the A and g tensor. The designations used in the present paper are therefore preferred

Cluster iron	Hyperfine coupling constant from					
	A_1		A_2		A_3	
	MB	ENDOR	MB	ENDOR	MB	ENDOR
	MHz					
Fe_a	+26	23	+34	32	+34	32
Fe_{b1}	+12	–	+22	25	+12	23
Fe_{b2}	–32	35	–39	42	–39	43
Fe_{b3}	–32	31	–39	38	–39	39

sought to obtain new information by labeling the cluster subsites with ^{57}Fe, ^{33}S or ^{77}Se, respectively. In these instances, the ENDOR spectra had to be recorded at a higher EPR frequency (35 GHz instead of 9 GHz) in order to resolve the signals from the predominating proton ENDOR signals (cf. [44]). Data from ^{57}Fe ENDOR would be expected to conform to the results from Mössbauer spectroscopy, which also derives information via the ^{57}Fe nuclei. Indeed ENDOR spectroscopy clearly showed that in the $[4Fe-4S]^+$ form all four Fe atoms are inequivalent (Fig. 14) while with Mössbauer spectroscopy only three different types could be distinguished [38]. Mössbauer spectroscopy has the advantage that it can give us information about the 2+ and 1+ forms whereas only the paramagnetic 1+ form is suitable for ENDOR. However, the requirements for sample concentration are less demanding for ENDOR so that we were able to prepare the substrate-free $[4Fe-4S]^+$ form in sufficient purity, whereas we have not yet succeeded in preparing this particular state for Mössbauer

Table 3. *Hyperfine coupling constants and orientation of* A *with respect to* g *tensor for* Fe_a *in the absence and presence of substrate*

The A tensor is coaxial with the g tensor in the absence of substrate; it is rotated by $\Phi = 25°$ around the g_3 axis in the presence of substrate

Orientation of $\tilde{A}$	Hyperfine coupling constant of Fe_a	
	without substrate	with bound substrate
	MHz	
A_1	33	23
A_2	41	32
A_3	42	32

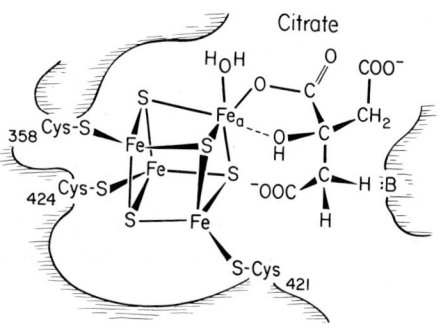

Fig. 15. *Scheme summarizing information obtained from the work described or mentioned above (cf. particularly Gahan et al. [59]) on the active site of aconitase and the binding of substrate to the Fe-S cluster. The figure shows citrate bound. When isocitrate is bound, the molecule is flipped by 180° so that the* $-CH_2-COO^-$ *group points downward and the carboxyl at the bottom in the figure now becomes bound to* Fe_a

spectroscopy. Therefore, we can strictly compare only the Mössbauer and ENDOR data for the [4Fe-4S]$^+$ form of the enzyme with bound substrate (Table 2). Note that the accuracy in the determination of hyperfine coupling constants is higher for ENDOR, while with Mössbauer spectroscopy in a magnetic field it is also possible to determine the sign of the coupling constants (see Table 2). For systems with a fairly isotropic g tensor, Mössbauer spectroscopy is insensitive to the spatial relationship of the axis systems of the g and A tensors. This can, however, be ascertained by ENDOR with the additional possibility of comparing the properties of the enzyme with and without substrate. The axis system of the A tensor of Fe_a is coaxial with the g tensor in the absence of substrate. However, upon addition of substrate the A tensor is related to the g frame by a rotation of about 25° around the g_3 axis when substrate is added. At the same time each of the principal values of $\tilde{A}$ is decreased by 20 – 30% (Table 3). These are unusually strong effects of a ligand, such as the substrate in this case, on an Fe-S cluster, confirming the conclusions drawn from Mössbauer spectroscopy [38]. The decrease in the A values indicates that there is a significant change in the orbital makeup and, perhaps, in the spin coupling of Fe_a when substrate is added. There is also a pronounced change in spin relaxation in Fe_a, recognized by the change of the relative intensity of the two components of the ENDOR signal from Fe_a. All the major changes of addition of substrate are limited to Fe_a, in agreement with the Mössbauer results [22, 38]. It is in agreement with the Mössbauer data that two of the Fe_b components are very similar, although not equivalent. On the other hand, Fe_{b1} differs substantially from $Fe_{b2,3}$ [38]. According to present concepts [54 – 56] of the substructure of Fe-S clusters we can conclude from the signs [38] and magnitudes of the A values of the cluster irons that Fe_{b2} and Fe_{b3} (A < 0) represent the mixed valence pair [Fe(III)/Fe(II)] subcomponent of the cluster, whereas Fe_a and Fe_{b1} (A > 0) are the ferrous pair.

^{33}S and ^{14}N ENDOR

We observed strong ENDOR signals from ^{33}S substituted into the labile sulfide positions of the cluster; however, because of the quadrupole splittings associated with the ^{33}S (I = 3/2) isotope, there is uncertainty about the number of nonequivalent sulfides represented in the signals; the signals observed cannot be interpreted in terms of a single type of S^{2-}. For the 4Fe-cluster ^{14}N-ENDOR signals were observed in the low frequency range, with A values of 18 MHz and 9 MHz, respectively. These resonances are almost identical in the presence and absence of substrate. Moreover, with the 3 Fe cluster nitrogen signals with significantly higher A values, viz. 32 MHz and 23 MHz were found. The strong couplings suggest covalent binding of these nitrogens to the cluster. The significance of these findings in view of the X-ray diffraction data [16] and the attempts of protein-chemical determination of the cluster ligands [20] is not clear at this time.

The significant results of ENDOR spectroscopy using various paramagnetic nuclei are summarized as follows. The unique contribution of ENDOR has been the identification by ^{17}O and ^{13}C ENDOR of the groups on the substrate which interact directly with the [4Fe-4S] cluster, namely the α or β carboxyl and the α or β hydroxyl, and the detection of cluster-bound OH^- and H_2O from the solvent. The knowledge that neither exchangeable nor non-exchangeable protons of the substrate interact with the Fe-S cluster is also uniquely due to ENDOR. Non-exchangeable protons of the protein were, however, observed. A combination of ^1H and ^2H ENDOR clearly showed a H_xO ligand at Fe_a with x apparently being 1 in the absence and becoming 2 in the presence of substrate. ^{57}Fe ENDOR confirmed and extended results obtained by Mössbauer spectroscopy in discriminating between all four Fe atoms of the cluster and ^{33}S ENDOR demonstrated that significant unpaired spin density resides on the labile sulfides, as expected from EPR spectroscopy on Fe-S proteins in general [57] and from theoretical calculations [47]. The spectra indicate that there are at least two inequivalent ^{33}S sites. Most interestingly ^{14}N ENDOR showed strong interaction of two nitrogen species with the cluster (Fe_b sites), which is as yet unexplained.

CONCLUSIONS

From the work reported above, on the background of previous work by other investigators [52, 53, 58 – 60] and in our own laboratory [9, 12 – 15, 18, 21, 22, 24, 30, 36 – 38, 50, 51], cf. [61], we draw the following conclusions. Inactive aconitase, as obtained on purification, has a cubane type [3Fe-4S]$^+$ cluster, ligated to the protein through three Cys residues (positions 358, 421 and 424; see, however, ^{14}N ENDOR). The active enzyme is thought to have the same ligands and in addition, as a fourth ligand on Fe_a, a hydroxyl group (Fig. 2), which is protonated on addition of substrate. The substrate is bound to Fe_a, the same iron that is lost on oxidative damage. The groups from the hydrated substrates that bind to Fe_a are the α (for isocitrate) or β (for citrate) carboxyl and the hydroxyl. A five-membered ring structure is formed as shown in Fig. 15 [59]. The dehydrated intermediate, *cis*-aconitate, is bound in the isocitrate (α carboxyl) or citrate (β carboxyl) mode via a bridging and exchangeable hydroxyl at the α or β carbon. The *cis*-aconitate intermediate must rotate by 180° between these modes during catalysis. At pH > 9 the inactive [3Fe-4S]$^+$ form is converted to the linear cluster (Fig. 4). During this process, Cys358 becomes detached and Cys250 and Cys257 become ligands. When a reconstitutable apoenzyme is prepared by oxidative destruction of the cubane-type 3Fe cluster, two disulfide bridges are formed involving Cys421, Cys424, Cys358 and possibly Cys383. The major portion of the labile sulfide is incorporated into these disulfides as S^0 in the form of trisulfide or tetrasulfide. S^0 can be removed from this apoenzyme by cyanolysis without major loss of viable protein. According to Mössbauer and ENDOR spectroscopy, the four Fe atoms in the [4Fe-4S]$^+$ form are not equivalent and there is a major change in the electron distribution when a substrate or analog is added to the 2+ or 1+ form. The [3Fe-4S]^0 form, the [4Fe-4S]$^{2+}$ form with substrate, and the [4Fe-4S]$^+$ form with and without substrate show distinct features of trapped valence in the iron atoms of the cluster. One of the most dramatic shifts of e$^-$ density is observed on adding substrate to the [4Fe-4S]$^{2+}$ form, with Fe$_a$ changing from Fe$^{\approx 2.5+}$ to Fe$^{\approx 2+}$. These localizations and shifts of charge, obviously inducible by the protein and/or substrate, in a spin-coupled metal-ligand system, are thought to form the basis for the catalytic usefulness of Fe-S clusters as active centers of enzymes.

The foregoing is an account of experimental work and intellectual contributions from several laboratories and colleagues without whose interest and efforts this work could not have been done. We express our appreciation to M. H. Emptage, B. M. Hoffman, J. B. Howard, T. A. Kent, H. Merkle, E. Münck, D. W. Plank, A. H. Robbins, C. D. Stout, J. Telser and M. Werst. The unpublished ENDOR data are part of the Ph. D. thesis of M. Werst. We thank Dr J. V. Schloss for providing us with the nitro analogs of citrate and isocitrate, Dr E. Kun for the inhibitory isomer of fluorocitrate, Dr J. D. Otvos for determining isotope enrichments by NMR, Dr H. Zalkin for communicating to us the amino acid sequence of pig heart aconitase as deduced from the DNA nucleotide sequence and allowing us to use this information in advance of publication, Mrs E. Ruzicka for preparation of the enzyme and the Institute for Enzyme Research, University of Wisconsin, Madison, for letting us use their facilities. The work carried out in the author's laboratory was supported by research grant R01 GM34812 from the Institute of General Medical Sciences of the National Institutes of Health, USPHS and by the Medical College of Wisconsin.

REFERENCES

1. Adams, M. W. W. & Mortenson, L. E. (1984) *J. Biol. Chem. 259*, 7045 – 7055.
2. Kuchta, R. D., Hanson, G. R., Holmquist, B. & Abeles, R. H. (1986) *Biochemistry 25*, 7301 – 7307.
3. Scopes, R. K. & Griffiths-Smith, K. (1984) *Anal. Biochem. 136*, 530 – 534.
4. Kelly, J. M. & Scopes, R. K. (1986) *FEBS Lett. 202*, 274 – 276.
5. Dreyer, J.-L. (1985) *Eur. J. Biochem. 150*, 145 – 154.
6. Schweiger, G., Dutscho, R. & Buckel, W. (1987) *Eur. J. Biochem. 169*, 441 – 448.
7. Flint, D. H. & Emptage, M. H. (1988) *J. Biol. Chem. 263*, 3558 – 3564.
8. Emptage, M. H. (1988) *Biochemistry 27*, 3104.
9. Kent, T. A., Dreyer, J.-L., Kennedy, M. C., Huynh, B. H., Emptage, M. H., Beinert, H. & Münck, E. (1982) *Proc. Natl Acad. Sci. USA 79*, 1096 – 1100.
10. Martius C. (1937) *Z. Physiol. Chem. 247*, 104 – 110.
11. Kennedy, C., Rauner, R. & Gawron, O. (1972) *Biochem. Biophys. Res. Commun. 47*, 740 – 745.
12. Ruzicka, F. J. & Beinert, H. (1978) *J. Biol. Chem. 253*, 2514 – 2517.
13. Kennedy, M. C., Emptage, M. H., Dreyer, J.-L. & Beinert, H. (1983) *J. Biol. Chem. 258*, 11098 – 11105.
14. Beinert, H., Emptage, M. H., Dreyer, J.-L., Scott, R. A., Hahn, J. E., Hodgson, K. O. & Thomson, A. J. (1983) *Proc. Natl Acad. Sci. USA 80*, 393 – 396.
15. Emptage, M. H., Dreyer, J.-L., Kennedy, M. C. & Beinert, H. (1983) *J. Biol. Chem. 258*, 11106 – 11111.
16. Robbins, A. H. & Stout, C. D. (1989) *Proteins, structure, function and genetics 5*, 289 – 312.
17. Robbins, A. H. & Stout, C. D. (1989) *Procl. Natl Acad. Sci. USA 86*, 3639 – 3643.
18. Kennedy, M. C., Kent, T. A., Emptage, M. H., Merkle, H., Beinert, H. & Münck, E. (1984) *J. Biol. Chem. 259*, 14463 – 14471.
19. Piszkiewicz, D. (1982) *Fed. Proc. 41*, 890.
20. Plank, D. W., Kennedy, M. C., Beinert, H. & Howard, J. B. (1989) *J. Biol. Chem.*, in the press.
21. Kennedy, M. C., Emptage, M. H. & Beinert, H. (1984) *J. Biol. Chem. 259*, 3145 – 3151.
22. Emptage, M. H., Kent, T. A., Kennedy, M. C., Beinert, H. & Münck, E. (1983) *Proc. Natl Acad. Sci. USA 80*, 4674 – 4678.
23. Rabinowitz, J. C. (1972) *Methods Enzymol. 24*, 440 – 442.
24. Kennedy, M. C. & Beinert, H. (1988) *J. Biol. Chem. 263*, 8194 – 8198.
25. Petering, D., Fee, J. A. & Palmer, G. (1971) *J. Biol. Chem. 246*, 643 – 653.
26. Thomson, A. J. (1985) in *Top. Mol. Struct. Biol. 6*, 79 – 120.
27. Moulis, J.-M. & Meyer, J. (1982) *Biochemistry 21*, 4762 – 4771.
28. Tsibris, J. C. M., Namtvedt, M. J. & Gunsalus, I. C. (1968) *Biochem. Biophys. Res. Commun. 30*, 323 – 327.
29. Wood, J. L. (1987) *Methods Enzymol. 143*, 25 – 29.
30. Rydén, L., Öfverstedt, L.-G., Beinert, H., Emptage, H. H. & Kennedy, M. C. (1984) *J. Biol. Chem. 259*, 3141 – 3144.
31. Plank, D. W. & Howard, J. B. (1988) *J. Biol. Chem. 263*, 8184 – 8189.
32. Hagen, K. S., Watson, A. D. & Holm, R. H. (1983) *J. Am. Chem. Soc. 105*, 3905 – 3913.
33. Hausinger, R. P. & Howard, J. B. (1983) *J. Biol. Chem. 258*, 13486 – 13492.
34. Martin, A. E., Burgess, B. K., Stout, C. D., Cash, V., Dean, D. R., Jensen, G. & Stephens, P. J. (1989) *Proc. Natl Acad. Sci. USA*, in the press.
35. Girerd, J.-J., Papaefthymiou, G. C., Watson, A. D., Gamp, E., Hagen, K. S., Edelstein, N., Frankel, R. B. & Holm, R. H. (1984) *J. Am. Chem. Soc. 106*, 5941 – 5947.
36. Dreyer, J.-L., Beinert, H., Keana, J. F. W., Hankovszky, O. H., Hideg, K., Eaton, S. S. & Eaton, G. R. (1983) *Biochim. Biophys. Acta 745*, 229 – 236.
37. Kennedy, M. C., Spoto, G., Emptage, M. H. & Beinert, H. (1988) *J. Biol. Chem. 263*, 8190 – 8193.
38. Kent, T. A., Emptage, M. H., Merkle H., Kennedy, M. C., Beinert, H. & Münck, E. (1985) *J. Biol. Chem. 260*, 6871 – 6881.
39. Cammack, R., Dickson, D. P. E. & Johnson, C. E. (1977) in *Iron-sulfur proteins* (Lovenberg, W., ed.) vol. 3, pp. 283 – 330, Academic Press, New York.
40. Debrunner, P. G., Münck, E., Que, L. & Schulz, C. E. (1977) in *Iron-sulfur proteins* (Lovenberg, W., ed.) vol. 3, pp. 381 – 417, Academic Press, New York.
41. Bill, E., Haas, C., Ding, X.-Q., Maret, W., Winkler, H. & Trautwein, A. C. (1989) *Eur. J. Biochem. 180*, 111 – 121.
42. Robin, M. D. & Day, P. (1967) *Adv. Inorg. Chem. Radiochem. 10*, 247 – 405.
43. Münck, E., Debrunner, P. G., Tsibris, J. C. M. & Gunsalus, I. C. (1972) *Biochemistry 11*, 855 – 863.
44. Rius, G. & Lamotte, B. (1989) *J. Am. Chem. Soc. 111*, 2464 – 2469.
45. Kanatzidis, M. G., Ryan, M., Coucouvanis, D., Simopoulos, A. & Kostikas, A. (1983) *Inorg. Chem. 22*, 179 – 182.

46. Johnson, R. E., Papaefthymiou, G. C., Frankel, R. B. & Holm, R. H. (1983) *J. Am. Chem. Soc. 105*, 7280–7287.
47. Noodleman, L., Norman, J. G. Jr, Osborne, J. H., Aizman A. & Case, D. A. (1985) *J. Am. Chem. Soc. 107*, 3418–3426.
48. Meyer, J., Gaillard, J. & Moulis, J.-M. (1988) *Biochemistry 27*, 6150–6156.
49. Schloss, J. V., Porter, D. J. D., Bright, H. J. & Cleland, W. W. (1980) *Biochemistry 19*, 2358–2362.
50. Telser, J., Emptage, M. H., Merkle, H., Kennedy, M. C., Beinert, H. & Hoffman, B. M. (1986) *J. Biol. Chem. 261*, 4840–4846.
51. Kennedy, M. C., Werst, M., Telser, J., Emptage, M. H., Beinert, H. & Hoffman, B. M. (1987) *Proc. Natl Acad. Sci. USA 84*, 8854–8858.
52. Rose, I. A. & O'Connell, E. L. (1967) *J. Biol. Chem. 242*, 1870–1879.
53. Gawron, O., Glaid, A. J. & Fondy, T. P. (1961) *J. Am. Chem. Soc. 83*, 3634–3640.
54. Münck, E. & Kent, T. A. (1986) *Hyperfine Interactions 27*, 161–172.
55. Noodleman, L. (1988) *Inorg. Chem. 27*, 3677–3679.
56. Noodleman, L, Case, D. A. & Aizman, A. (1988) *J. Am. Chem. Soc. 110*, 1001–1005.
57. Orme-Johnson, W. H., Hansen, R. E., Beinert, H., Tsibris, J. C. M., Bartholomaus, R. C. & Gunsalus, I. C. (1968) *Proc. Natl Acad Sci. USA 60*, 368–372.
58. Schloss, J. V., Emptage, M. H. & Cleland, W. W. (1984) *Biochemistry 23*, 4572–4580.
59. Gahan, L. R., Harrowfield, J. M., Herlt, A. J., Lindoy, L. F., Whimp, P. O. & Sargeson, A. M. (1985) *J. Am. Chem. Soc. 107*, 6231–6242.
60. Glusker, J. P. (1980) *Acc. Chem. Res. 13*, 345–352.
61. Emptage, M. H. (1988) in *Metal clusters in proteins* (Que, L. Jr, ed.) ACS Symposium Series 372, pp. 343–371, Am. Chem. Soc. Washington, DC.

Subject Index

acetoin dehydrogenase, enantiomeric specificity 184:10
acetylcholine 179:259
aconitase 186:5, 13, 14
–, apoenzyme 186:6, 7, 14
–, ENDOR spectroscopy 186:10–13
–, Mössbauer spectroscopy 186:8, 9
–, purple-, linear cluster 186:7, 8, 14
actin 179:264
–, transcription factor 185:494
acyl-CoA dehydrogenase, α, β-dehydrogenation 181:4
–, mechanism 181:4
acylphosphatase, antigenic sites 183:489
–, nuclear Overhauser effect 183:484
–, twodimensional correlation spectroscopy 183:484
adrenaline 179:256
L-alanine dehydrogenase 184:2
–, application for L-amino acid synthesis 184:2, 3
–, characterization 184:2
–, production 184:2
alcohol dehydrogenase 184:8
–, liver, resolution of racemic compounds 184:8
–, Sulfolobus solfataricus 184:9
–, Thermoanaerobium brockii, stereospecificity 184:9
–, –, coenzyme regeneration 184:12
–, yeast 184:8, 9
alkylhydroperoxides, oxidants, monooxygenation 184:278, 279
N-alkylation, cytochrome P-450 haem 184:274, 275
D-amino acid oxidase, carbanion mechanism 181:3
aminoacyl-tRNA synthetase, splicing 182:481
amphetamine 184:270
apocytochrome c, NMR 183:489
arrestin, binding to rhodopsin 179:263
ataxia telangiectasia 185:12
–, association with leukaemia 185:13
ATPase 182:489
–, subunit 9, mRNA 182:479, 480
Azotobacter vinelandii, Fe-S cluster 186:8

B-cell tumors, chromosome abnormalities 185:8–12
–, follicular lymphoma 185:11
Bacillus sphaericus, L-alanine dehydrogenase 184:2
Bacillus stearothermophilus, lactate dehydrogenase 184:7
benzodioxole derivatives, oxidation by cytochrome P-450 184:273, 274
benzyl halides, reduction by cytochrome P-450 184:271
bioluminescent reaction 181:14
BM-40 see osteonectin
bombesin-like peptides 184:491
Brevibacterium sp., L-phenylalanine dehydrogenase 184:5
Burkitt's lymphoma, chromosomal translocation 185:8–10
–, endemic 185:8, 9

–, sporadic 185:8, 9
–, variant 185:10
calcium, binding to basement membrane 180:495
–, binding to osteonectin 180:496
–, calmodulin, NMR 183:493
–, guanylate cyclase 179:256
–, inositol trisphosphate 184:487
–, prothrombin, NMR 183:494
–, release from rod-outer-segment 179:259
calmodulin, NMR 183:481, 486, 493
–, –, calcium binding 183:493
capping, nascent RNA 185:496
catabolite repression 182:479, 485
cGMP phosphodiesterase, activation by transducin 179:262
–, switch off mechanism 179:265
Chaga's disease 180:491
chondroitin sulfates, basement membrane 180:495
chromagranin A, NMR 183:484
chromatin, in vitro artifacts 185:487
–, native 185:490
chromosome abnormalities, oncogene activation 185:7
–, lampbrush-, structure 185:488
–, translocation, B-cell tumors 185:8–12
–, –, T-cell tumors 185:12–14
coenzyme, regeneration systems 184:10, 11
collagen IV, composition, domain model 180:487, 488
– –, electron microscopy 180:489
– –, evolution 180:490
– –, in cellular recognition 180:498
– –, oligomers, assembly process 180:488
– –, VII, structure 180:496
colony-stimulating factors (CSF) 184:487, 492
CSF see colony-stimulating factors
cysteine, aconitase, cluster ligands 186:7, 8
–, catalytic role, mercuric reductase 181:9, 10
–, T-cell receptor 185:6
cytochrome b_5, NMR 183:481
–, mRNA 182:479, 480, 483
–, –, stability 182:482, 483
cytochrome c, antigenic sites 183:490
–, gene expression 182:485
–, NMR 183:481, 485, 486, 489, 490
–, nuclear Overhauser effect 183:489, 490
cytochrome c oxidase, gen expression 182:488
cytochrome P-450 184:267
–, active site 186:267, 268
–, catalytic cycle 184:268
–, epoxidation of alkenes 184:273, 274
–, isozymes 184:269, 270
–, model systems 184:276–282
–, –, applications 184:281, 282
–, –, regioselective 184:280, 281
–, reductase, electron transfer, multi-redox-center 181:11

dehydrogenases 184:1
–, NAD(P)H-dependent 184:2

–, –, alcohol- 184:8, 9
–, –, L-amino acid- 184:2
–, –, hydroxy acid- 184:6, 7
–, –, diketone reductases 184:10
dehydrogenation, flavin-catalyzed 181:2
–, –, carbanion transfer 181:2, 3
–, –, hydride transfer mechanism 181:5
–, –, radical mechanism 181:5
denaturation, aconitase 186:6
diacylglycerol, growth stimulation 184:487
dioxygen, activation by cytochrome P-450 184:268
disulfide bond, cross linking, NMR 183:492
–, NC1 segments of collagen IV 180:488
–, redox-active 181:6
–, –, glutathione reductase 181:7
–, T-cell receptor 185:6
DNA endonuclease 182:480
–, mitochondrial (mtDNA) 182:477, 478
–, –, gene order 182:478
–, –, regulation of transcription 182:478, 479
–, nuclear, structure 185:448, 490
–, polymerase 185:492
– rearrangement, lymphoid cells 185:1
Drosophila, collagen IV 180:490
–, position effects in genes 185:497

EGF see epidermal growth factor
Ehrlich ascites carcinoma, antineoplastic activity of leucine dehydrogenase 184:3
electron microscopy, collagen IV 180:489
electron-nuclear double resonance (ENDOR) 186:10
–, aconitase, exchangeable and non-exchangeable protons 186:11
–, –, substrate binding 186:12
–, coupling constants for Fe-S cluster 186:13
electron transfer, rate, NMR 183:489
ENDOR see electron-nuclear double resonance
enhancer, binding to nucleoskeleton 185:497
–, H and L chain genes 185:5
entactin see nidogen
enzyme, flavoprotein- 181:1
–, –, multi-redox-center 181:11
– membrane reactor 184:2, 4, 10
epidermal growth factor (EGF) 184:487
–, basement membrane 180:497
epidermolysis bullosa acquisita 180:497
epinephrine see adrenaline
EPR spectroscopy, cytochrome-P-450-iron-metabolite complexes 184:271
–, Fe-S cluster 186:9
Epstein-Barr virus, Burkitt's lymphoma 185:8, 9
erythromycin 184:270
erythropoietin 184:491
Escherichia coli, leucine dehydrogenase 184:3

FE-S cluster, ENDOR spectroscopy 186:13
– –, linear, aconitase 186:8

– –, Mössbauer spectroscopy 186:8, 9
– –, nomenclature 186:5, 6
ferrodoxin, removal of iron 186:7
FGF see fibroblast growth factor
fibroblast growth factor (FGF) 184:490, 492
flavin, chemical properties 181:1
–, hydroperoxide 181:12, 13
flavocytochrome b₂, active center 181:2
flavodoxin, electron transfer 181:11
flavoenzymes see flavoproteins
flavoproteins 181:1
–, dehydrogenation, mechanism 181:2
–, flavin-substrate, stereospecificity 181:5, 6
–, pyridin-nucleotide-linked, stereospecificity 181:6
–, redox-active disulfide 181:6
FMN reductase, NAD regeneration 184:11
formate dehydrogenase, coenzyme regeneration 184:4, 6, 11
frameshifting, ribosomal 182:484

gene, bcl-2, tumorigenesis 185:12
– expression 182:484, 490
– –, ARS sequences 182:486
– –, attachment hypothesis 185:495, 496
– –, cytochrome c 182:485, 486
– –, immunoglobulin 185:5
– –, transcriptional activators 182:485, 486
– rearrangement, immunoglobulin 185:3, 4
– –, T-cell receptor 185:6, 7
glioma cells, PDGF 184:487
glucose oxidase, mechanism 181:6, 7
glucose repression 182:479
glutathione reductase, active site, mechanism 181:6, 7
–, stereospecificity 181:6
glycosylation, laminin 180:491
growth factor 184:487, 488
–, autocrine 184:492, 493
–, bombesin-like peptides 184:491
–, colony-stimulating factors (CSF) 184:491
–, epidermal (EGF) 184:490
–, fibroblast (FGF) 184:490
–, insulin-like (IGF) 184:490, 491
–, melanoma-growth-stimulatory activity (MGSA) 184:491
–, platelet-derived (PDGF) 184:489
–, transforming (TGF-α) 184:489, 490
GTPase, inactivation by transducin 179:264
–, rate, cycle of transducin 179:264
guanylate cyclase, calcium, phototransduction 179:256

haem, binding to transcriptional activators 182:485
–, cytochrome c, NMR 183:489, 490
–, cytochrome P-450, N-alkylation 184:274
–, models 184:270–274
haemoglobin, X-ray crystallography 183:480
haemoproteins, electron transfer 181:11
Halobacterium halobium, bacteriorhodopsin 179:258
HeLa cells, globin premessenger 185:496
–, globin sequences 185:490

heme see haem
heme proteins see haemoproteins
heparan sulfate proteoglycans, basement membrane 180:493
– –, self assembly 180:495
heparin, binding to laminin 180:492
histidinyl-tRNA synthetase, cytoplasmic 182:487
–, mitochondrial 182:487
HTLV-I see human T-cell leukemia virus I
human T-cell leukemia virus I (HTLV-I), interleukin-2 184:491
hydratases/dehydratases, iron-sulfur proteins 186:5
hydrazines, oxidation by cytochrome P-450 184:272
hydrogen/deuterium exchange 183:482
–, epoxidation of alkenes 184:273, 274
hydrogen peroxide, monooxygenation 184:278, 279
hydroperoxide, flavin- 181:14
2-hydroxy acid dehydrogenases 184:6–8
–, characterization 184:8
–, production 184:7
p-hydroxybenzoate hydroxylase, mechanism 181:12, 13, 15
hydroxylases, flavin-dependent, mechanism 181:15

IGF I, II see insulin-like growth factor I, II
imidazole, cocatalyst, monooxygenation 184:278, 279
immunoglobulin, gene organization, man 185:1–3
–, – rearrangement 185:3, 4
–, – variability 185:4, 5
initiation factors, translation 182:484
inositol trisphosphate, growth stimulation 184:487
insulin-like growth factor I, II 184:490, 491
integrins, binding to laminin 180:498
interleukins 184:491, 492
intron, structural role in transcription 185:497
iodosylarenes, oxygen donors 184:277, 283
iron-sulfur proteins 186:5
–, electron transfer 181:11
isomerization, retinal 179:255, 258, 259

D-lactate dehydrogenases, synthesis of D-phenyl lactate 184:7
L-lactate dehydrogenase see flavocytochrome b₂
L-lactate oxidase, active center 181:2
laminin, binding protein 180:498
–, –, nidogen complex 180:493
–, polymerization, binding to collagen 180:492
–, regulation of translation 180:491
–, structure 180:490, 491
leishmaniasis 180:491
L leucine dehydrogenase 184:3
–, characterization 184:3, 4
–, production 184:3
–, synthesis of L-amino acids 184:4
lipoyldehydrogenase, active site, catalytic cycle 181:7
luciferase, mechanism 181:14
lysozyme, hydrogen/deuterium exchange 183:482
–, NMR 183:481, 483
–, nuclear Overhauser effect 183:487

magnesium
–, light-activated process, transducin 179:260
melanoma-growth-stimulatory activity (MGSA) 184:491
membrane
–, basement membrane 180:487, 499
–, cGMP-cascade 179:256
–, Reichert's- 180:495
–, rod-outer-segment 179:255
mercuric reductase, catalytic mechanism 181:8–10
methylation, nascent RNA 185:496
MGSA see melanoma-growth-stimulatory activity
mitochondria 182:477
–, genetic system 182:488, 489
monooxygenases 184:267
–, cytochrome-P-450-dependent 184:267
–, –, mechanism 184:268
–, flavin-dependent 181:12
–, –, mechanism 181:14, 15
Mössbauer spectroscopy 186:8
–, aconitase 186:6
–, isomer shift of Fe-S cluster 186:8, 9
mutation, CBP1 protein 182:482, 483
–, cytochrome c, NMR 183:490
–, effects on splicing 182:482
–, NAM2, leucyl-tRNA synthetase 182:481
c-myc, Burkitt's lymphoma 185:10
–, chromosomal translocation 185:8, 9, 14
–, T-cell tumors 185:14

NADP⁺, absorption spectrum, mercuric reductase 181:9
nidogen, laminin-nidogen complex 180:492, 493
–, structure 180:492, 493
Neurospora crassa 182:480
NMR see nuclear magnetic resonance
NOE see nuclear Overhauser effect
novobiocin, RNA polymerase 185:490
nuclear magnetic resonance (NMR) 183:479, 480
–, acylphosphatase 183:484
–, apocytochrome c 183:489
–, calmodulin 183:481, 486, 493
–, chromagranin A 183:484
–, correlation time 183:484
–, coupling constants 183:485
–, cytochrome b₅ 183:481
–, cytochrome c 183:481, 485, 486, 489, 490
–, enzymes, general aspects 183:496
–, frequency shifts 183:480
–, lysozyme 183:481, 495
–, phosphoglycerate kinase 183:491
–, plasminogen 183:487, 494
–, prothrombin 183:494
–, troponin C 183:493
–, zinc finger protein 183:495
nuclear Overhauser effect (NOE) 183:480, 486, 487, 496
–, acylphosphatase 183:484
–, cytochrome c 183:490
–, two-dimensional NOE difference spectroscopy 183:481
nucleoskeleton, molecular nature 185:492
–, transcription complex 185:494

oncogene activation, chromosome
abnormalities 185:7
—, growth factors 184:489, 490
osteocalcin, cross-linked by calcium
183:494
osteonectin, structure, basement membrane
180:495, 496
oxidases, flavin-dependent 181:12

Parodi-Irgens feline sarcoma virus, PDGF
transformation 184:489
PDGF see platelet-derived growth factor
L-phenylalanine dehydrogenase 184:5
—, application 184:5, 6
—, characterization 184:5
—, —, Brevibacterium sp. 184:5
—, —, Rhodococcus sp. 184:5
phosphoglycerate kinase, NMR 183:491
phosphorylase, binding of phosphate, NMR
183:494
—, X-ray crystallography 183:494
phosphorylation, photoexcited rhodopsin
179:258, 263
plasminogen, NMR 183:487, 494
platelet-derived growth factor (PDGF)
184:487, 492
—, isoforms 184:489
—, virus-induced transformation 184:489
polyadenylation, nascent RNA 185:496,
497
polyhalogenomethanes, reduction by cytochrome P-450 184:271
porphyrin, N-alkyl- 184:274, 275
—, Fe(III)-, models for cytochrome P-450
184:277–280, 283
—, iron-, models 184:267–275
—, —, chiral 184:281
—, Mn(III)-, models for cytochrome P-450
184:277–280, 283
pre-mRNA, processing 182:480
processing, mitochondrial RNA 182:480
promotor, immunoglobulin gene 185:5
—, mitochondrial DNA 182:478
propoxyphene 184:270
prothrombin, γ-carboxyglutamate 183:494
—, NMR 183:494
protein, iron-sulfur- 186:5
—, kinase C 184:487
—, seleno-, aconitase 186:7
—, —, calorimetry 183:482
—, —, cross linking 183:492, 493
—, —, domain/domain interaction
183:489
—, —, NMR 183:481–483, 486, 487
—, —, random coil 183:483
—, trigger- 183:484, 493
—, tyrosine kinase 184:487
G-protein 179:255, 259
—, transducin 179:259
—, effects of fluorides 179:262
proteoglycans, basement membrane
180:493, 495
—, structure 180:494, 495
Pseudomonas putida, cytochrome P-450
184:268

receptor, adrenergic 179:256, 259
—, epidermal growth factor (EGF)
180:498; 184:487, 490

—, Epstein-Barr virus 185:9
—, T-cell 185:5, 6
—, integrin 180:498
—, muscarinic 179:256
—, platelet-derived growth factor (PDGF)
184:487, 489
—, substance K 179:256
recombinase, B-cell follicular lymphoma
185:11
—, gene rearrangement, immunoglobulin
185:3
—, T-cell receptor 185:7
redox potential, flavin 181:1, 2
retinal, interaction with rhodopsin 179:258
—, isomerization 179:255, 258
Rhodococcus sp., L-phenylalanine dehydrogenase 184:5
rhodopsin, bacterio- 179:258
—, bovine, receptor for retinal 179:257
—, GTP dependence 179:261
—, interaction with retinal 179:258
—, photoexcited 179:255
—, —, coupling with transducin 179:260
—, —, phosphorylation 179:263
—, —, release of transducin 179:261
rhodopsin kinase, inactivation of photoexcited rhodopsin 179:263
RNA, m-, mitochondrial 182:479–483
—, —, processing 182:480
—, —, splicing 182:480, 481
—, —, stability 182:482
—, maturases 182:480
—, polymerase, mitochondrial 182:478
— —, processing 182:480
— —, soluble 185:488
— —, transcription complex 185:494, 496
RNase P, Escherichia coli 182:480
rod-outer-segment, calcium release 179:259
—, —, cGMP dependent 179:263
—, cGMP cascade 179:256

Saccharomyces cerevisiae 182:486
—, mitochondrial DNA (mtDNA)
182:477
Schwann cells, laminin 180:492
Simian sarcoma virus (SSV), PDGF transformation 184:489, 493
Soret peaks, cytochrome-P-450-ironmetabolite complexes 184:270, 271
splicing, mitochondrial RNA 182:480, 481
—, nascent RNA 185:496
—, self- 182:480
SPARC see osteonectin
stereospecificity, substrate-flavin interaction
181:5, 6
stopped-flow, mercuric reductase 181:9
sydnones, oxidation by cytochrome P-450
184:275, 276

T-cell receptor 185:5
— —, chromosomal abnormalities
185:12–14
— —, gene organization 185:5–7
TGF-α see transforming growth factor-α
topoisomerase, mutants 185:491, 492
—, transcription 185:494, 496
transcription 182:478; 185:487
—, activator proteins 182:485, 486
—, attachment model 185:494–498

—, complex 184:491, 492, 494, 496
—, mitochondrial DNA (mtDNA)
182:478, 479
—, —, promotor 182:478
—, processing 182:480
—, rate 185:488
—, splicing 182:480, 481
—, 'text-book' model 185:487, 488, 498
transducin, activation of cGMP phosphodiesterase 177:262
—, GDP-bound holoenzyme 179:259
—, replacement of GDP by GTP 179:261
transformation, activation of growth factor
expression 184:492
—, B chain of PDGF 184:489, 493
—, T-cells by HTLV-I 184:491
transforming growth factor-α (TGF-α)
184:489, 490, 492
translation 182:483
—, initiation factors 182:484
—, frameshifting 182:484
troleandomycin 184:270
troponin C, NMR 183:493
Trypanosoma cruzi 181:8
trypanothione reductase 181:8
trypsin, proteolysis of phosphodiesterase
179:262
tubulin 179:264
tumor, B-cell 185:8–12
—, —, chronic lymphocytic leukaemia
185:11, 12
—, —, follicular lymphoma 185:11
—, —, multiple myeloma 185:11
—, —, prolymphocytic leukaemia 185:11
—, Engelbreth-Holm-Swarm (EHS), mouse
180:487, 492
—, T-cell 185:12–14
—, —, acute lymphocytic lymphoma
185:14
tumour see tumor
tyrosine kinase, growth factor receptors
184:487

ubiquinol-cytochrome c reductase, assembly
of subunits 182:488
urokinase, domain 183:494

virus, Epstein-Barr-, Burkitt's lymphoma
185:8, 9
—, human T-cell leukemia- (HTLV-I)
184:491
—, Parodi-Irgens feline sarcoma-, PDGF
184:489
—, simian sarcoma- (SSV), PDGF
184:489, 493
von Willebrand factor, basement membrane
180:497

X-ray crystallography, phosphorylase
183:480
—, haemoglobin 183:480
X-ray spectroscopy, cytochrome P-450
184:268
xanthine oxidase, electron transfer 181:11
xenobiotics, oxidative metabolism 184:270,
276

zinc finger protein, NMR 183:495